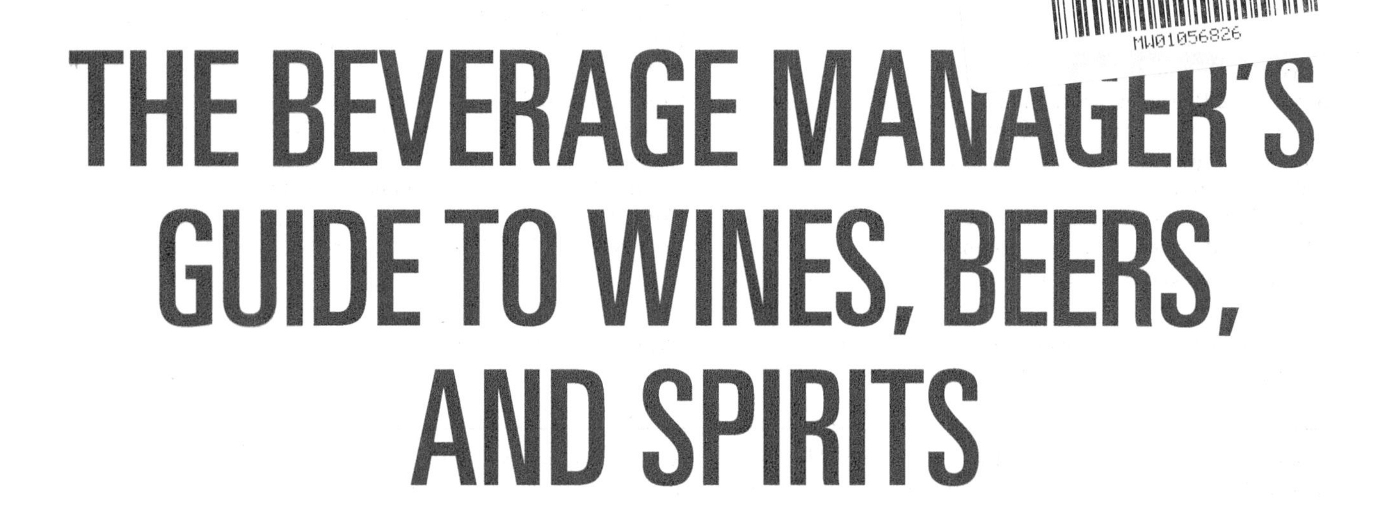

THE BEVERAGE MANAGER'S GUIDE TO WINES, BEERS, AND SPIRITS

Fourth Edition

John Peter Laloganes
The Wine and Beverage Academy of Chicago
Kendall College and the Wine Professional (WP) Program

Albert W. A. Schmid
Guildford Technical Community College

Pearson

330 Hudson Street, NY, NY 10013

Vice President, Portfolio Management: Andrew Gilfillan
Portfolio Manager: Pamela Chirls
Editorial Assistant: Lara Dimmick
Development Editor: Pamela Chirls
Senior Vice President, Marketing: David Gesell
Marketing Coordinator: Elizabeth MacKenzie-Lamb
Director, Digital Studio and Content Production: Brian Hyland
Managing Producer: Cynthia Zonneveld
Managing Producer: Jennifer Sargunar
Content Producer: Rinki Kaur
Manager, Rights Management: Johanna Burke
Manufacturing Buyer: Deidra Smith
Full-Service Project Manager: Rajiv Sharma, iEnergizer Aptara®, Ltd.
Cover Design: Studio Montage
Cover Photos: Courtesy of Carrie Alyssa Schuster
Composition: iEnergizer Aptara®, Ltd.
Printer/Binder: LSC Communications, Inc.
Cover Printer: LSC Communications, Inc.
Text Font: 10.75/12, Sabon LT Pro

Credits and acknowledgments borrowed from other sources and reproduced, with permission, in this textbook appear on appropriate page within text.

Library of Congress Cataloging-in-Publication Data

Names: Laloganes, John P., author. | Schmid, Albert W. A., author.
Title: The beverage manager's guide to wines, beers, and spirits / John Peter Laloganes, the Wine and Beverage Academy of Chicago, Kendall College and the Wine Professional (WP) Program, Albert W. A. Schmid, Guildford Technical Community College.
Other titles: Hospitality manager's guide to wine, beer, and spirits
Description: Fourth edition. | Boston : Pearson, [2017] | Revised edition of:
The beverage manager's guide to wine, beer, and spirits / Albert Schmid, John Laloganes. 2008; itself an edition of: The hospitality manager's guide to wines, beers, and spirits / Albert W. A. Schmid. 2004.
Identifiers: LCCN 2016057097 | ISBN 9780134655307 | ISBN 0134655303
Subjects: LCSH: Alcoholic beverages. | Alcoholic beverage industry. | Bartending.
Classification: LCC TP505 .S355 2017 | DDC 663/.1—dc23 LC record available at https://lccn.loc.gov/2016057097

1 17

ISBN 10: 0-13-465530-3
ISBN 13: 978-0-13-465530-7

BRIEF CONTENTS

CONTENTS

PREFACE

The Goals of This Text

The Beverage Manager's Guide to Wines, Beers, and Spirits, Fourth Edition, intends to serve as an authoritative guide for those individuals pursuing or enhancing a career in the food and beverage industry. The book is written in a lively and engaging style that is both comprehensive and yet concise, yielding the fundamental aspects of managing beverages. The book strives to be intellectually appealing with an engaging literary style and stimulating photography while providing the necessary knowledge on building and sustaining a profitable beverage program. This text is a thorough course not only on understanding and appreciating the varying categories of "drink," but it also provides useful concepts that beverage managers can incorporate into their daily tasks of educating their employees while making key business decisions as they manage their establishments.

While this work is targeted towards inspiring those individuals in education and training programs, the book will be equally fascinating to the beverage enthusiast. Most significantly, this work will make a lasting contribution toward creating a foundation for a new generation of beverage connoisseurs. It strives toward enlightening and instilling a greater appreciation in society at large about the sometimes perplexing, expansive, and eternally fascinating subject of wines, beers, and spirits . . . and other critically important beverages.

Organization of Content

The Beverage Manager's Guide to Wines, Beers, and Spirits strives to provide relevant, to-the-point content that saves the reader from personal or long-winded stories. This fourth edition of *The Beverage Manager's Guide* includes an immense quantity of updated and expanded material. The contents have been designed and tested through trial and error over a period of several years of classroom and "real life" scenarios. Keep in mind that these kinds of content heavy books are never perfect, as changes occur continuously and trends evolve daily, making it impossible to maintain a real-time reality.

The content within this book has been arranged in the following manner: **Chapter 1 – The Origins of Alcohol and Beverage Trends** is a **NEW** initial chapter that presents an overview of the history of alcohol from its beginning to the present, briefly illustrating the origin and roles these beverages have played as they have evolved over time. The second half of this first chapter focuses on beverage trends as they relate to the strategic mindset of the beverage manager. **Chapter 2 – Introduction to Wine and Service** is a **NEW** section that presents the basics of wine with an emphasis on understanding fundamental differences between white, rosé, and red wine along with distinguishing among still, sparkling, and fortified wine. It then covers glassware, closures, and approaches to labeling. **Chapter 3 – Viticulture: Outside in the Vineyard** focuses on the grape-growing practices with a discussion on the current trend of "Green Friendly," including sustainable, organic, and biodynamic growing and production. **Chapter 4 – Enology: Inside the Winery** details enology, or the process of wine making.

Chapter 5 – The Wine Styling Approach: White Wines and **Chapter 6 – The Wine Styling Approach: Red Wines** offer an expanded coverage of grape varietals as they are illustrated by the wine-styling approach, rather than by region or country, showing readers the most user-friendly method for communicating wine to others.
Chapter 7 – Other Wines: Sparkling, Fortified, and Dessert wines is an immersion into the production processes and styles of these wines that are available in the marketplace.
Chapter 8 – The Brewery: Beer Production and Service and **Chapter 9 – Beer Styles: Ales and Lagers of the World** include a discussion of the major brewing countries, production process of beer, and then present beer by its two major categories with their style derivatives.
Chapter 10 – Sake, Cider, and Mead is a NEW section and quite timely due to each of these beverages seeing immense growth unlike ever before.
Chapter 11 – The Distillery: Spirits and Liqueurs of the World breaks down spirits into clear (non-aged) and browns (aged) categories and details the major types of spirits and liqueurs.
Chapter 12 – Mixology: The Art and Science of the Cocktail is NEW with the addition of the most current practices behind the bar including 22 classic cocktail recipes tested by mixologist Kai Wilson.
Chapter 13 – Coffee and Tea is a NEW section that covers an in-depth explanation of the two most profitable beverages in the industry.
Chapter 14 – Constructing the Beverage Concept discusses how the food and beverage concept is a combination of various factors that form its foundation and uniqueness as a business. Effort is made highlighting ways in distinguishing the establishment as different and better from the competition.
Chapter 15 – Managing for Profit discusses such important concepts and techniques used in the process of controlling costs and in building and sustaining a profitable beverage program.
Chapter 16 – Marketing the Beverage Establishment examines the foundational considerations and helpful advice on developing an effective wine menu.

The Appendix Section

This section is used as a collection of key topics that either don't fit into the bulk of the text, or need further expansion without "weighing down" the main pages of the book. **Appendix A – The Science of Fermentation** begins with a more in depth (for the science minded) look at the process of fermentation, then progressing to **Appendix B – Alcohol Safety and Liability** as it relates to lessening the risk of the beverage establishment. **Appendix C – The Tasting Process** is an in-depth section on sensory analysis and how to deconstruct the tasting experience and make sense of what one is imbibing. Then **Appendix D – Drink and Food Pairing** offers a practical and straightforward 3-step approach for making intelligent, well-informed wine- and food-pairing decisions. This time-tested approach can be used as a template for those looking to successfully pair drinks with food and for managers to use in training their service staff. **Glossary** – This section is a lexicon of beverage terms that act to support all the chapters throughout the book. It allows the learner to gain maximum benefit and depth of content through providing quick access to understanding or clarifying a certain term or concept.

Whom This Book Is for

The Beverage Manager's Guide to Wines, Beers, and Spirits, Fourth Edition has been written for everyone who desires a strong, solid foundation in understanding drinks. The book is targeted not only at students who are presently enrolled in hospitality or culinary programs, but it's also intended for individuals working within the wholesale, retail, and food and beverage industry.

One may not be a fan of all drinks, but as a beverage manager, it's critical to grasp the basics of all beverages if the intention is to successfully manage in today's food and beverage industry. As an avid wine drinker, one may be more passionate about wine than any other, however, it's important to value and respect all beverages. Ultimately, it's our job is to please others, and in knowing about all beverage options, it provides us a better opportunity to meet and even exceed the customer's expectations.

The text can serve as the sole resource in a single course or as a supplement in conjunction with other hospitality, wine, culinary, or pastry classes. The book can work in classes such as "Introduction to Wine," "Wine Appreciation," "Food and Beverage Operations," or as a supplementary text in "Guest Service" or "Introduction to Hospitality."

Additional Tools and Features

Like working in many capacities within the hospitality industry, this text has been produced as a collaborative effort. Thanks to my co-author Albert W. A. Schmid, we are unified in our approach for designing a learner-centered textbook that offers numerous features that attract, engage, and navigate the visual, auditory, and kinesthetic learner. The writing style and format of the book have been designed in a way that allows simplicity to guide the student. The tools presented below offer a bridge for the student by reinforcing techniques learned. Curriculum-based activities and pedagogical concepts are imbedded within the text and are additionally available through accessing the instructor's supplement. These learning tools are devised to enhance the classroom setting and to act as a natural extension for the student outside of the traditional learning environment. Some of the additional tools include the following:

- **Full-Color Text and Photographs** depicting the beverages as artistic and unique, and creating visual appeal to engage learners.
- **Thoughtful Use and Placement of Colorful Maps** identifying key geographic locations within the context of the text.
- **Clearly Stated Chapter Learning Objectives** are placed at the beginning of each chapter and should be used as roadmaps to help readers track learning.
- **Pronunciation Guides** for "foreign" terms are placed in the text when they appear for the first time. Some beverage terms can be challenging to articulate, because they are taken from French, Dutch, Italian, German, or Spanish (or even, in some cases, from Portuguese or Hungarian). The phonetic guide is included for English speakers and provides an approximate pronunciation in English.
- **Glossary** a lexicon of terms from the beverage industry are highlighted within the text, along with phonetic spelling to assist the reader in ease of communication.
- **Check Your Knowledge Quizzes** are placed at the end of each chapter for assessing comprehension and retention of important chapter topics, including discussion questions that encourage in-class discussion between the instructor and among the students. One of the foundational principles of adult learners is their desire to bridge previous knowledge and experiences, with that of the content currently being learned. The incorporation of these discussion questions can provide an additional source of increased cognitive growth while promoting greater retention of relevant concepts.

- **Recorded podcasts** for most chapters are available to listen or downloaded to any smart phone or similar digital device accessible at *www.johnlaloganes.com*. The podcasts are topical discussions between Albert, myself, and other guests, and are based on each chapter of the book.

The Beverage Manager's Guide strives to provide an insight of the beverage world through discovering the enduring influence of wines, beers, and spirits that has been inseparable since the inception of civilization. We believe that grasping a full understanding and appreciation for these beverages can lead to living a more enriching and fulfilling life. The roles of these beverages have evolved over time—shifting from an important source of nutrition to becoming an acceptable cultural and social complement to food and conviviality, symbolic of a healthy lifestyle. If I could make one suggestion, it would be the following: Pour yourself a glass of your favorite beverage—imbibe—and contemplate the words in these pages. Allow this book to uncover all the intrigue and complexity that drinks can offer. If it hasn't already, it is my hope that you permit wine, beer, spirits (or whatever your preference) to enrich your life as it has mine.

ACKNOWLEDGMENTS

For the wonderfully talented individuals who directly assisted us with this endeavor: Pamela Chirls for your patience and support throughout each stage of the editing process; Erika Cespedes for your incredibly enchanting photography; Meghan Vandette for being incredibly photogenic; Thomas Moore for your timely map making abilities; Kai Wilson for your classic cocktail recipes Adam Herbert; Janel Syron; Tim Coonan; Suzette (Sooz) Hammond; Denise and Martin Cody; Carrie Alyssa Schuster; Leo Alaniz; Michael Howe; Clayton Chapman; David Vance; Sam Baxter; Carolyn Corley Burgess.

- The Radler – Chicago, IL
- Liquor Park – Chicago, IL
- The Grey Plume – Omaha, NE
- Cellar Angels – Chicago, IL
- Big Shoulders Coffee – Chicago, IL
- Terra Valentine – Saint Helena, CA
- Kendall College – Chicago, IL
- The Dana Agency – Harper Edgcumbe & Angelica Galan, Miami, FL
- American Social Bar & Kitchen – Miami, FL
- IL Mulino New York South Beach – Miami Beach, FL

ABOUT THE AUTHORS

JOHN PETER LALOGANES, M. Ed, Level III Sommelier
Proprietor of The Wine and Beverage Academy of Chicago
Director of Beverage Management and Assistant Professor at Kendall College

John Peter Laloganes is an award-winning sommelier, author, and educator and has been a contributor to the hospitality industry since the mid-1980s, avidly collaborating and strategizing with restaurants and related beverage establishments. He has traveled and tasted extensively throughout the major wine and beer-producing regions of Europe and the United States. In 2012, Mr. Laloganes was awarded in New York City with the "Sommelier/Wine Steward of the Year" by the IACP (International Association of Culinary Professionals).

Currently, John is the Proprietor of The Wine and Beverage Academy of Chicago (a Chicago-based training and consulting firm); *Director of Beverage Management* and *Assistant Professor* at Kendall College in Chicago, IL.

In 2016, John was recognized as the *Educator of the Year* from Kendall College and additionally in 2016 was awarded the highest organizational honor in being the recipient of the Laureate Global Products and Services (GPS) *Award for Innovation* for his work in creating the Wine Professional (WP) Program at Kendall College. In 2014, he created Kendall's Wine Professional (WP) Program, an accelerated 3-tiered sommelier educational track that is designed for industry professionals and serious beverage enthusiasts.

From 2012 to 2014, John was Wine Director at the Grey Plume restaurant in Omaha, NE, and Sommelier / Management Instructor at the Illinois Institute of Art – Chicago where he was honored with the *Excellence in Teaching Award* in 2012.

In 2007, John was awarded the distinctive Sommelier Diploma (level III) through the International Sommelier Guild (ISG). And from 2007–2009, he taught wine fundamentals level I and II for industry professionals through the ISG. Before then, he was an Associate Professor at Le Cordon Bleu, Chicago where he created their wine class and earned the *Customer Service Award* from North American Le Cordon Bleu Schools in 2005 as well as the *Educator of the Year* award in 2004.

John has authored several highly-acclaimed books: *The Essentials of Wine with Food Pairing Techniques* Amazon #1 Best Seller in Wine Pairing; *The Beverage Manager's Guide to Wines, Beers and Spirits,* Third Edition, awarded the Best Wine Book for Professionals in the United States in 2013 by the Gourmand International Book Awards; and Managing the Beer-Centric Operation in 2014.

Mr. Laloganes has taken extensive coursework in Culinary Arts, earned a Bachelor's Degree in Hospitality and Tourism Management from the University of Wisconsin–Stout, and a Master's Degree in Vocational Education in Human Resource Development from the University of Minnesota.

John is a current member of the Society of Wine Educators (SWE) and the International Association of Culinary Professionals (IACP).

ALBERT W. A. SCHMID, M.A. CEC, CCE, CCA, CCP, MCFE, CFD, CHE, CFBE, CHIA, CSS, CSW, COI

Director, Culinary Arts & Hospitality Management, Guildford Technical Community College Jamestown, North Carolina

Albert W. A. Schmid is a chef, bartender, sommelier, and award winning author. He is the author of *The Kentucky Bourbon Cookbook* published by the University Press of Kentucky, which won the 2010 Gourmand Award for Best Book for Cooking with Wines, Beers and Spirits in the USA and later won Best Book for Cooking with Drinks in the World. He is also the author of *The Beverage Manager's Guide to Wines, Beers and Spirits,* published by Prentice-Hall/Pearson Education, which won the 2012 Gourmand Award for Best Wine Book for Professionals in the USA, The Hospitality Manager's Guide to Wines, Beers and Spirits, The Old Fashioned: An Essential Guide to the Original Whiskey Cocktail, and The Manhattan Cocktail: A Modern Guide to the Whiskey Classic.

In 2016, Albert joined the faculty at Guildford Technical Community College (GTCC) in Jamestown, North Carolina. Before joining GTCC, he was a member of the faculty at Sullivan University's National Center for Hospitality Studies where he was the Director of the Hotel-Restaurant Management, Event Management, Beverage Management, and Hospitality Management departments. He was granted the title Professor by the Sullivan University Academic Rank Committee in 2010. In 2011, Albert received an Award of Excellence from the International Association of Culinary Professionals (IACP) as the Wine Steward or Sommelier of the Year and was awarded membership in the Wine Century Club, which is reserved for wine connoisseurs who have tasted wines from a minimum of 100 different grape varietals. He is also a Master Knight in the Brotherhood of the Knights of the Vine. Albert holds a Bachelor of Science degree from Charter Oak State College. He is the 2002 recipient of the International Association of Culinary Professionals' *Le Cordon Bleu/ University of Adelaide* scholarship for graduate study in Gastronomy. In 2006, Albert was awarded a Master of Arts in Gastronomy, with a dissertation titled *The History of Beer in Recipes in the United States*. In 2011, Albert earned a Professional Certificate in Distance Education form the University of Wisconsin – Madison. Albert holds continuing education from the University of South Dakota, Purdue University, and The Culinary Institute of America. In addition to his education, Albert holds 12 professional certifications including Certified Culinary Professional, Certified Executive Chef, Certified Specialist of Wine, Certified Specialist of Spirits, and Certified Food and Beverage Executive, as well as four educator certifications. Albert has been a featured speaker on bourbon and wine at the Society of Wine Educators, the American Wine Society, the International Association of Culinary Professionals, the American Hotel and Lodging Association, and the International Council on Hotel, Restaurant, and Institutional Education. Also, Albert has guest lectured at Purdue University and Western Kentucky University.

ABOUT THE CONTRIBUTOR

Erika Cespedes is a Costa Rican born photographer. Her photography career has allowed her to work as a commercial photographer and now as a fine-art artist creating compelling imagery. Her work is distinctive and at times mysterious. One of her biggest inspirations is her son, Liam Bleu, whom she takes everywhere with her while photographing. Her work ranges from traditional analogue, digital to mixed media. She currently resides in Palm Beach, FL.

THE BEVERAGE MANAGER'S GUIDE TO WINES, BEERS, AND SPIRITS

CHAPTER 1

The Origins of Alcohol and Beverage Trends

CHAPTER 1 LEARNING OBJECTIVES

After reading this chapter, the learner will be able to:

1. Recall key moments in the history of wine, beer, and spirits in the ancient world
2. Discuss the basic production process necessary for distillation and fermentation
3. Identify the significance of the religious movement and its influence on the early production and development of alcoholic-based drinks
4. Explain how Prohibition had impacted the beverage industry
5. Recognize how trends can act as part of a successful strategic approach
6. Identify the six influencers of broad-based and/or localized trends

If you would understand anything, observe its beginning and its development.

—Aristotle (ancient Greek philosopher)

The Essential History of Alcohol

Learning Objective 1
Recall key moments in the history of wine, beer, and spirits in the ancient world

Regardless of one's personal beverage preference, there is no mistaking that each drink—whether it's wine, beer, spirits, or coffee and tea—has played a pivotal role in the evolution of society from ancient times to current day. As we imbibe in our preferred drinks, many of us unfortunately fail to take a moment to appreciate that each beverage has its own sordid past complete with adventurous tales of heroes and villains, trials and tribulations. Collectively, these experiences have all contributed in shaping the course of each beverage and parallel that of the rise and advancement of civilization.

The human consumption of alcohol began unintentionally around 10,000 years ago (approximately 8000 B.C.). In all probability, alcohol originated from the storage of overripe and decaying fruits, most likely grapes, honey, or apples. Figure 1.1 depicts a cluster of red wine grapes—likely one of the first intentionally produced fermented beverages. Over time, the sweetened fruit was affected by airborne yeasts, which initiated fermentation, and ultimately the fruit was transformed into an intoxicating product. This "accident" and its lasting impact on history is immeasurable; it may have been one of the most significant factors in the switch from a hunter-gatherer subsistence base to one of cultivation.

Archeological evidence dates the intentional production of beer and wine to the first civilization that arose around 6000 B.C. in Mesopotamia and Egypt (largely corresponding to modern-day Iraq). These two parallel civilizations were founded on a surplus of cereal grains produced by organized agriculture on a massive scale. The production

FIGURE 1.1
Clusters of grapes. Courtesy of Carrie Alyssa Schuster.

of beer (which relies on a large amount of grain) and wine (which similarly requires a large amount of grapes) could not have taken place prior to the domestication of agriculture around 8000 B.C. in the Near East, and the consequential agricultural surplus and capability of storage. This allowed other members of the population to pursue alternative areas of specialty such as potters, writers, and philosophers. Beverages, like many other products, became a commodity for trade and a source of monetary influence.

The origin of distillation of spirits is far more recent, and is traced to Middle East or China at about 700 A.D. Spirits could not be produced until there was enough capable knowledge to determine exactly how to extract and purify it from the fermented mixture of grains or fruits. Ultimately, the creation of an apparatus known as a *still* was used to extract and concentrate the alcohol and create what became known as a *distilled spirit*.

While all three drinks (wine, beer, and spirits) were initially created by accident, ultimately they became deliberately reproduced and integrated as a daily necessity of life. As the production of drinks evolved, they were increasingly replicated according to the preferences of the maker and/or the people consuming them. The benefits of these drinks began to provide more than just the obvious allure of alcohol. Throughout the ages, these libations have relinquished many benefits as they often acted as a source of philosophical enlightenment, social lubrication, and symbolic of many religious or political rituals. They were a commodity to be traded, prescribed as medicinal remedies, and acted as a measure of social status. Drinks have been used to celebrate life, forge new partnerships, and even pay tribute to those who have died. As time progressed, the cultural rituals and ceremony surrounding drinks have become almost as momentous as the beverages themselves. Beer, wine, and spirits represent special meaning to some and continue to provide pleasure to many.

The Role of Fermentation

Learning Objective 2
Discuss the basic production process necessary for distillation and fermentation

Alcoholic beverages (also referred to as "drinks") are relatively distinguishable from one another, as they each look, smell, and taste quite different. Despite these obvious differences, however, all alcohol is produced using the initial method of fermentation. The unintended natural process of fermentation precedes human history. Since ancient times, however, humans have been attempting to understand and control the phenomenon of fermentation. Louis Pasteur (1822–1895), a French chemist and biologist, made significant contributions to chemistry, medicine, and indirectly the food and beverage industry and has subsequently greatly benefited civilization. In 1849, Pasteur began studying fermentation—a chemical process that breaks down organic materials and converts them into alcohol.

The understanding that yeast is a living organism means it operates on the same principle as any other living organism—food is needed (in the form of sugar) for continued survival and reproduction. The biochemical process of fermentation in wine, beer, or spirit is the result of a chemical reaction that turns sugar into an alcoholic beverage. During fermentation, yeast interacts with sugar to create ethyl alcohol (also known as ethanol) and carbon dioxide (as a by-product). In regard to beer, fermentation occurs when yeast breaks down sugar obtained from malted grain, while in wine, it breaks down sugar obtained from grape juice. The fermentation process is also essential in many aspects of the food and beverage industry—the production processes of bread, cheese, and yogurt all rely on the chemical conversion of fermentation. Other scientists in the early 20th century contributed additional knowledge to the understanding of the complex chemical processes involved in the conversion of sugar to alcohol.

Fermentation requires three basic ingredients: *water*, *yeast*, and *sugar*. As simple as this seems, however, there are endless variations in which these ingredients can alter the final product in countless ways. Additional ingredients such as fruits, nuts, and herbs may also be added to achieve different results.

Water

Water is the predominant ingredient in any alcoholic product, though its composition varies greatly from source to source. Varying mineral and pH levels that are present in water from one part of the country can be quite distinct from another. Water from a mountain spring has uniqueness apart from water out of the tap, and fresh water formed in a limestone basin possesses different characteristics than fresh spring water. During the production process, vintners, brewers, and distillers are particular about their water source as differences in levels of purity and minerals can alter the fermentation process, and alter the personality of the finished product.

Yeast

Yeast is a single-celled organism that lives and thrives on simple sugar. There are endless varieties and strains of yeast throughout the world but the most common species is *Saccharomyces cerevisiae* (sack-a-roe-MY-sees sair-ah-VIS-ee-eye). Many strains exist naturally (known as wild or ambient yeast) in the air or on the exterior of a yeast's food source; yet most often, yeast is commercially created and is known as cultured yeast. It can be purchased by brewers and vintners to impart specific characteristics to their products. Any strain of yeast used for fermentation must be alive; fortunately for alcohol producers, yeast is very hard to kill. Heating it above approximately 137°F is deadly to yeast, but its adaptability allows it to be frozen or freeze-dried, and revived for later use by thawing or reconstituting.

FIGURE 1.2
Bottle of Chopin Vodka. Courtesy of Erika Cespedes.

Sugar

The type of sugar(s) chosen as a food source for the yeast can dramatically alter the taste and alcohol content of the end product. There is surprising flexibility in this ingredient; some sugars are sweeter than others, and many have other subtle or obvious taste differences. Fructose, maltose, and glucose are the main sugars used in fermentation. Each has a different chemical structure and source that will be appropriate to the product being made, such as grape juice (fructose and glucose) for wine, germinated barley (maltose) for beer, or potatoes (glucose) for the creation of vodka. Figure 1.2 depicts a bottle of Chopin, a premium brand of potato-based vodka. Like water, sugars have varying chemical components aside from their obvious sweetness content that modify the fermentation process and create distinctions in the drinks.

The Role of Alcohol in Religion

Learning Objective 3
Identify the significance of the religious movement and its influence on the early production and development of alcoholic-based drinks

Historically, wine has also played an important role throughout many religious ceremonies. The ceremonies according to both the Biblical and Christian tradition teach that alcohol is a gift from God that makes life more joyous—yet, gluttony that leads to intoxication is a sin. Both joyous and drunkenness appear in literal and poetic passages throughout the Bible and are intended as means of storytelling. Important mythological and religious figures acknowledged the significance of drinks in their cultural rituals of everyday life. Early on, the Greek god *Dionysus* and the Roman equivalent *Bacchus* symbolized wine, and the drink also became used in Catholic Eucharist and the Jewish Kiddush rituals. Figures 1.3 and 1.4 display the Church's prominent role in the evolution of alcohol.

Between the 6th and 14th centuries, the wines of France acquired a reputation through the founding of numerous monasteries. The monks became famous for their viticultural and vinification skills. During the war-like times, religious communities were important as they were known as safe passage to some extent. The Benedictine monks founded the Abbey of Cluny in 909 and its offshoot the Cistercian monks who founded the Abbey of Cîteaux in 1098 became the first truly large vineyard owners of Burgundy, France, over the subsequent centuries. The monks of Cîteaux built their first château (completed in 1336) enclosed by a *clos*, or a walled vineyard. The château is situated in Burgundy, named Clos de Vougeot. Figures 1.5 and 1.6 depict the historic "Clos de Vougeot" built by monks from the Abbey of Cîteaux.

The early Roman Catholic clergy were not only conducting religious worship and performing other spiritual functions, but many were also winemakers, brewers, and distillers as part of their daily ritual and contribution to a communal lifestyle and service to the community. The early Roman Catholic Church contained the highly literate priests and monks. This made it more likely for the clergy to transfer their practices down throughout the generations. The Church's ability to maintain a supply and surplus of wine was very important because of the necessity that wine existed

FIGURE 1.3
Church. Courtesy of John Peter Laloganes.

FIGURE 1.4
Influence of the Church in a Burgundian Vineyard. Courtesy of John Peter Laloganes.

FIGURE 1.5
Clos de Vougeot. Courtesy of John Peter Laloganes.

FIGURE 1.6
Close-up of Clos de Vougeot. Courtesy of John Peter Laloganes.

FIGURE 1.7
Clos. Courtesy of John Peter Laloganes.

FIGURE 1.8
Champagne from 1874. Courtesy of John Peter Laloganes.

for the mass. The assurance of this "in-house" supply of wine for the ceremonies was important to the morale of the congregation who were practicing their spiritual beliefs. Because the Catholic Church required wine in the Eucharist, wherever Catholicism spread, the missionaries also brought grapevines so that they could commemorate the Mass and replicate the blood of Christ. Figure 1.7 depicts the existence of Burgundian walled vineyards (known as "clos") that were built for protection back in the Middle Ages.

The monks kept detailed notes on their improvements to wine, beer, and spirits that they discovered or accidentally stumbled upon. The most famous of these accidental discoveries is credited to the Benedictine monk *Dom Pierre Pérignon* (d. 1715). Even though most scholars on the subject agree that Dom Pérignon did not invent Champagne, they do agree that he performed great volumes of research and contributions on the subject of sparkling wine. He maintained detailed vineyard records that allowed for the technical expertise of blending that led to the significance of consistency and complexity in the finished bottle of Champagne. Dom Pérignon was also highly instructive to the pickers to harvest grapes in cool conditions and to harvest slightly underripe grapes to preserve their acidity. He promoted low-yielding vineyards to achieve better quality of grapes and the practice of pressing the grapes as close to the vineyards in order to minimize any color contact with the juice. Lastly, he is noted for ultimately being able to control the secondary fermentation that would occur in the bottle, which ultimately led to what was known as the *méthode champenoise* or more recently *méthode traditionelle*. Figure 1.8 identifies a historic bottle of Pommery Champagne from 1874. Over time, several monastic groups would specialize their alcohol production, and become known for particular alcoholic products. For example, the Benedictine Abbey of Fécamp in Normandy, France, became known for its Bénédictine liqueur, an herbal, medicinal sweetened spirit. Figure 1.9 depicts the ancient Bénédictine liqueur.

The Trappist Order, another monastic group, had taken their name from the La Grande Trappe, another Abbey in Normandy,

FIGURE 1.9
Bénédictine liqueur.
Courtesy of Erika Cespedes.

FIGURE 1.10
Westmalle Trappist. Courtesy of John Peter Laloganes.

France. These monks became quite illustrious for their crafted beers, some of which still exist today. Many of these early monastic orders distilled, brewed, and vinified products that were used by the early Church as both a medicine (for which it was not very effective) to sterilize wounds (which it does rather well) and a source of prosperity. Figure 1.10 depicts a bottle of Westmalle Trappist beer. As of 2016, Westmalle is one of eleven remaining brewing monasteries (known as Trappist Beers) found around the world that is still in existence today.

Wine is also used in Jewish ceremonies and celebrations, including Passover, weddings, Shabbat, or the circumcision ceremony. Shabbat is a day of rest, considered as a holy and festive day. It is symbolic of when the Jewish are freed from the regular labors of everyday life, in remembrance of the Israelites' liberation from slavery in ancient Egypt. Kosher wines are produced for those who follow orthodox dietary rules.

The Effects of Prohibition

Learning Objective 4
Explain how Prohibition had impacted the beverage industry

Prohibition was one of the most turbulent periods in the history of the United States. This restriction brought about unforeseen consequences and long-term implications to the American beverage industry. Although it has been mentioned previously in this chapter, Prohibition deserves a more detailed discussion. On January 16, 1920, Congress passed the 18th Amendment (Prohibition) to the Constitution that made the production, transportation, and sale of alcohol illegal in the United States. Prohibition continued for a period of fourteen years up until December 5, 1933, when the 21st Amendment (repeal of Prohibition) went into effect.

The seeds of Prohibition were planted in America long before passage of the 18th Amendment. In the 1830s, the Temperance Society advocated only moderate, if any, consumption of alcoholic beverages. Later, the group took the more radical stance of total abstinence, and they acquired the not so flattering name "teetotalers," referencing their preference to consume tea or other nonalcoholic beverages. Their work laid the groundwork for the legislation to follow. In 1851, Neal Dow of Maine wrote the country's first prohibition law that was signed by Maine's Governor John Hubbard. When it was passed by the Maine legislature, Neal Dow became known as the "Napoleon of Temperance" and the "Father of Prohibition." By 1855, similar state laws were passed in Rhode Island, Massachusetts, Vermont,

Minnesota, Michigan, Connecticut, New York, New Hampshire, Nebraska, Delaware, Indiana, Kansas, and Iowa. These laws were never enforced, however, because most people were more concerned with the impending Civil War than with Prohibition. Ironically, Maine's law was repealed in 1856 in reaction to the 1855 Portland Rum Riot during which Dow, who was the mayor of Portland, Maine, ordered the militia to open fire on the crowd. One person was killed and seven injured. Later, Dow was prosecuted (and acquitted) of violating the law he penned, for improperly acquiring alcohol and illegal liquor sales. Later, Dow would serve as a Brigadier General for the Union Army in the Civil War, and in 1880, Dow ran for President of the United States as the Prohibition Party's candidate. The Prohibition Party is still active today and remains the oldest third party in U.S. politics.

The Temperance movement, however, had sympathizers at the highest levels of government during the Civil War. In 1862, the advisors of President Abraham Lincoln asked him to dismiss the successful General Ulysses S. Grant from his command because of Grant's excessive drinking. As Grant continued to be successful, Lincoln went against his advisors, suggesting that a barrel of General Grant's preferred whiskey should be sent to all the Union generals. General Grant later became the eighteenth president of the United States, but the alcohol debate continued and intensified.

The anti-alcohol movement was unrelenting in American society until 1917, when World War I began. By September of that year, President Woodrow Wilson, acting under special powers granted by the Food Control Act, was allowed to lessen the production of beer by not allowing grain to be utilized for beer production. President Wilson also limited the alcohol content in beer to less than 2.75 percent by weight. His reasons may have reflected darker motives than just helping the war effort. Many of the large brewery owners were of German descent, making it an easy time for unfavorable ethnic groups to be the subject of additional pressures since the United States was at war with Germany.

In December 1917, Congress proposed the 18th Amendment to the Constitution. The amendment was known as the Volstead Act, and it outlawed the "manufacture, sale, or transport of intoxicating liquor," but not consumption and/or use for purposes of medicinal and sacramental reasons. Within thirteen months, two-thirds of the states had ratified this new amendment. It became law on January 16, 1919, and took effect the following year. All the states ratified the amendment with the exception of two: Connecticut and Rhode Island. Interestingly, these two states had had earlier prohibition laws that had already been abolished. However, they were still required to uphold the new constitutional amendment. The votes in the state legislature were a landslide with 85 percent of the senators and 79 percent of the members of the House of Representatives voting for the measure. Prohibition's long-term effect on the country was devastating as an entire industry had been abolished almost overnight. In the opinion of the late Max Allen (who was awarded in 1997 the International Bartender of the Year and Bartender Emeritus at the Seelbach Hilton Hotel in Louisville, Kentucky), Prohibition was at least partially responsible for the Great Depression: People who had jobs in the liquor industry were suddenly without work and there were no new jobs to replace them. Allen also noted that Kentucky was especially devastated because one in every three jobs was somehow tied to the beverage industry.

Prohibition was repealed fourteen years later by the 21st Amendment, but until then, crime and corruption related to Prohibition were widespread. Bootlegging, which originally referred to the concealment of a pint-size flask in a boot for a trip, began in earnest between the United States and Canada. Canada had also imposed prohibition in 1918, but it lasted only one year. Once Canadian prohibition was repealed and the United States' prohibition was imposed, the door opened for illegal

FIGURE 1.11

Bertillon photographs of Al Capone (17 U.S.C. § 101 and 105). Courtesy of PD-USGOV-DOJ.

smuggling from Canada to the United States. Figure 1.11 is a mug shot of Al Capone, one of the most notorious gangsters in American history.

The careers of many infamous American gangsters ascended during this turbulent period. Al Capone and other gangsters smuggled a variety of alcoholic beverages over the Canadian border with Elliot Ness and other U.S. Treasury officers in pursuit. In the end, Al Capone did go to jail, but the charges were related to tax evasion rather than any infractions related to the 18th Amendment. The alcohol that was successfully brought across the border was usually sold at a "speakeasy," or an illegal bar. To enter a speakeasy, customers had to pass a guarded door through the use of a secret password. During the first years of the speakeasy, the owners sold unregulated alcohol. The Prince of Wales, who would later become King Edward VIII (and later the Duke of Windsor), was drinking at a speakeasy when the police raided the establishment. The future king was fortunate to have a quick-thinking host who moved Prince Edward into the kitchen, put a chef's toque on his head, gave him a pan, and told him to cook eggs until the raid was over. The police never knew the prince was there. An illuminating comment supposedly said of Al Capone, "When I sell liquor, it's bootlegging, when my patrons serve it on a silver tray on Lakeshore Drive, it's hospitality."

Another common practice during Prohibition was the making of homemade gin. Makers would acquire a basic neutral alcohol, and after placing the neutral spirit in the bathtub, they would add extracts or oils of juniper berries to the spirit, giving the mixture the flavor and kick of gin. After the mixture was finished, it would be bottled and become known as "bathtub gin." Ultimately, this led to the foundation of many of our fashionable cocktails, as the addition of flavoring agents helped disguise the crude form of alcohol.

One of the more covert practices people undertook to acquire alcohol and ultimately break the law involved collaboration with American wineries. Some wineries continued to grow grapes and many switched from making wine to making unfermented grape juice concentrate. The makers of this concentrate made certain that buyers could read the following warning label on packages of grape juice concentrate:

> "WARNING: If sugar and yeast are added fermentation will occur."

The alcoholic beverage industry strived to remain operable during Prohibition. Some wineries were allowed to remain open and produce wine for medicinal or sacramental reasons. Many of the major distilleries and breweries switched to making industrial alcohol, or "near beer." Some of the breweries made other products, such as candy, malted products, soda, or cheese.

In 1928, Democrat Alfred E. Smith ran for president against Republican Herbert Hoover. Hoover used the campaign slogan "Rum, Romanism, and Rebellion," because one of Smith's issues was the repeal of Prohibition. Smith did not win, but he did generate widespread debate—by the next presidential election, the country was ready for a drink. Democrat Franklin D. Roosevelt beat Hoover, and within the first nine days of his administration, he asked Congress to amend the Volstead Act. Congress proposed the 21st Amendment on February 20, 1933, and the states went to work. Two-thirds of all states are needed to adopt a new amendment to the Constitution. On December 5, 1933, the 18th Amendment was history, and by the end of 1933, people were legally drinking alcohol again. When someone asked Elliot Ness, the U.S. Treasury agent who had worked to stop Al Capone and other bootleggers, what he would do now that Prohibition was over, he answered, "I think that I am going to have a drink."

Post-Prohibition

When Prohibition was repealed on December 5, 1933, Congress, for the very first time in history, directly created business opportunities for the citizens of the United States. While some American wineries had been able to hold on through fourteen years of Prohibition by selling grape juice or industrial alcohol, most of them failed. Daniel Okrent, a researcher and author of "Last Call: The Rise and Fall of Prohibition," stated that one year after the end of Prohibition in 1933, 90 percent of federal revenues were generated from the excise tax on the sale of alcoholic beverages.

Americans are increasingly interested in a lifestyle that embraces a robust culture of wine and food. Today, wine is produced in every one of the fifty states in the United States. California maintains well over 90 percent of U.S. wine production followed by Washington State, New York State, and then Oregon. According to the Wine Institute, American wine drinkers consume much less wine per capita (about 2.6 gallons per person) compared to Italy or France (about 14 gallons per person), yet ever since 2011, the United States had become and has since remained the largest consumer of wine in the world and remains roughly the fourth largest wine-producing country in the world. Recent gains for wine consumption have been driven by many factors, including the adoption of wine in early adulthood by the large Millennial generation, the availability of quality wine at varying price points, and the acceptance of moderate wine consumption as compatible with a healthy lifestyle.

The beer industry was just as damaged as the wine industry by Prohibition. Many of the breweries that existed before Prohibition were not able to reopen their doors, largely because of being undercapitalized. However, new breweries opened, and it has taken much less time (as compared to wine) for the beer industry to bounce back from the effects of Prohibition. According to the Brewers Association, 4,656 breweries operated for some or all of 2016, the highest total since before Prohibition.

Currently, wine, beer, and spirits are widely discussed and consumed globally. The wine and beer and premium spirit industry has undergone a massive boom in popularity over the last few decades resulting in many varieties and types of products from which to choose.

Beverage Trends: The Strategist Mindset

Learning Objective 5
Recognize how the significance of trends acts as a part of a successful strategic approach

The strategic mindset of the manager is one that should strive to identify emerging trends in the marketplace. Through gaining an understanding of the numerous external forces that impact their business, the managers are more likely to keep their business competitive and persist in the forefront for satisfying the needs of their customers. Beverage trends can be divided into what's happening in the mainstream (or broad-based) trends impacting the larger communities and what's happening in cutting-edge niche (or localized) markets within cities like Chicago, New York, Portland, and San Francisco. In their efforts to identify the relevancy of broad-based and localized trends, managers can recognize how they can proactively sculpt a competitive advantage through harnessing a company's core capabilities.

- *Broad-based trends* are those that influence a larger percentage of people in the wider marketplace.
- *Localized-based trends* are influencing and capturing smaller niche markets within a localized area.

There are many people and businesses that are "late" to trends, and mistakenly adopt a trend after it's no longer in fashion. They exist in music. They exist in mutual funds. They exist in clothes and in cars. Recognizing growing trends as it relates to the application of beverages within a given business is an imperative effort on behalf of the business. These efforts help companies to capture market opportunities, spur innovation, and allow for better problem solving and decision making.

Trends and Fads

Trends and fads are often used incorrectly as interchangeable terms. Both trends and fads could be something in vogue; however, trends have a more sustainable nature due to a foundation that grew from a consumer need or movement from some type of external influence. Fads are simply a cute or novel idea or concept—sort of a "one and done" type scenario. Once a consumer has experienced a given fad, their curiosity is often satisfied yielding a lack of long-term sustainability for the given fad. A fad or trend is neither bad nor good, it just calls into question as to whether the business should choose to address and capitalize on said fad or trend. Being able to identify the distinction allows an operation to more fully devote its resources and address the changing needs of the marketplace.

- *Fads* are when consumers are having some interest in a phenomenon (new product, concept, service, etc.) with exaggerated enthusiasm for a brief period of time. Generally, fads are not something that a food and beverage establishment may invest too many resources—they are fleeting bits of consumer interest.
- *Trends*, on the other hand, are a prevailing tendency of fondness, a style or preference, or the general movement that has been developing over time of a detectable long-lasting change in the marketplace.

Identifying and keeping current with trends are important aspects of developing and implementing any operational and marketing plan. Being able to identify fads and trends in their early stages allows an operation to more fully devote its resources and address the changing needs of the marketplace. Paying attention and/or being ahead of the curve can work to illuminate a distinction and a competitive advantage—ultimately leading to profitability.

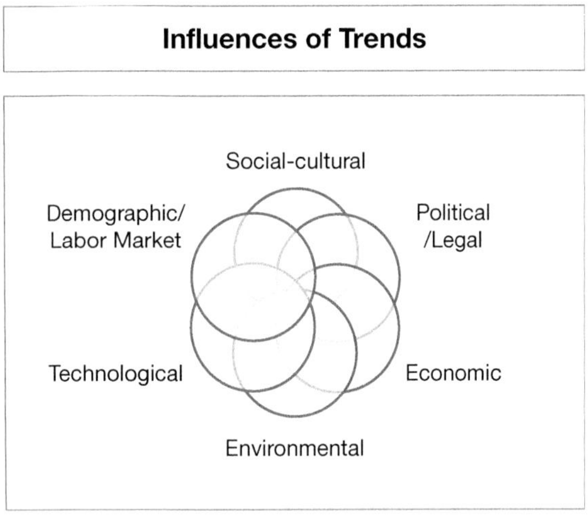

Influences of Trends

Social-cultural
Political /Legal
Economic
Environmental
Technological
Demographic/ Labor Market

FIGURE 1.12
Influence of trends. Courtesy of John Peter Laloganes.

External Variables

Trends are often founded (having evolved) from the needs in the marketplace due to one or several of the common six external variables. These aspects can include, but are not limited to, six external variables: social and cultural trends, political and legal trends, economic trends, environmental trends, technological trends, and demographic and labor market forces within the marketplace. Figure 1.12 depicts the six external variables that can influence marketplace trends.

1. **Social and Cultural Trends** Signifying the combination or interaction of social and cultural values within society or specific to a geographical area or demographic.
2. **Political and Legal Trends** Relating to politics or government control through federal or local laws and policies.
3. **Economic Trends** Pertaining to the production, distribution, and use of income, wealth, and commodities. The prevailing direction in which the broader or local economy is operating.

4. **Environmental Trends** Relating to the natural world and the impact of human activity on its condition or sustainability.
5. **Technological Trends** The application of newer tools, techniques, and processes used for better efficiency and/or effectiveness.
6. **Demographic and Labor Market Trends** The developments in population as they relate to changes in age, gender, marital status, educational levels, employment status, race, and so on.

For example, food and beverage purchases are largely influenced by demographics—attitudes and awareness about health. The aging baby boomer generation (those born in 1946–1964) is approaching points in life where dietary concerns and health become much more paramount than in years previous. Therefore, monitoring these factors as they relate to a target market over time can provide a comprehensive understanding of the current consumer trends for that particular group.

Not only do environmental variables transform throughout time, but consumer attitudes and desires also change continually; therefore, it is important for operators to remain connected on the pulse of society. Just as in the world of fashion, beverage products come in and out of vogue. In the 1970s and early 1980s, wine bars were the rage from coast to coast. The late 1980s and early 1990s saw a marked increase in the consumption of the so-called white spirits—gin, vodka, tequila, and some rum. Classic drinks such as the Martini, Manhattan, Old-Fashioned, and other such retro cocktails that were popular in the 1950s were once again in style. The mid- to late 1990s ushered in a return of the popularity of the dark spirits. No serious beverage operation could be without a variety of fine, aged Kentucky bourbons and imported single malt Scotches in the 1990s. And now in 2017, the rage for the past two years has been all about American whiskeys. Consumers today expect a wide variety of imported and domestic wines, and operators who offer beer have discovered that one or two domestic brands in bottles and on tap no longer satisfy today's sophisticated consumer. Instead, craft beers and a wide variety of imported beers are the norm today.

One should be aware, however, that while trends will often seem to be taking hold or losing ground, this is not necessarily a good reason to make dramatic changes in the business's methods of operation. Different regions of the country have different values, and what is true of one end of the United States may not be true of the other. Truly, the only way to determine what trend may or not be appropriate is to be intimately in tune with the selected target market.

Recognizing Trends as Part of Strategy

Learning Objective 6
Identify the six influencers of broad-based and/or localized trends

Given the often-fragile state of the local or world economy, many hospitality establishments have experienced challenging profit margins, particularly some dramatic volatility over the past decade. The more effective businesses continue to search for strategies that increase opportunities to stabilize and to grow the base of their business. With increased competition in the marketplace, these effective strategies help a business to differentiate themselves and to remain as a viable alternative against the competition. Managers play pivotal roles in leading the strategy-making process. Their mindset should involve staying in touch with the latest fashions and innovations that allows their analytical skills to identify and implement the appropriate strategies that incorporate trends (or fads) successfully within their given establishment.

Appropriate strategies are the ones selected that enables a particular company to achieve *superior performance* and provide their company with a *competitive advantage*. Food used to be the main source of creating distinction among the consumer, but in current times, beverages have become just as viable, as a source of differentiation. Traditionally, wine has been the driver of alcoholic beverages in full-service restaurants. However, modern restaurants and bars have "tapped" into some alternatives to wine, featuring beer from local breweries and craft cocktails. These venues are making significant progress as a differentiator with the surging demand of the craft segment. In some instances, businesses have used food as a loss leader in order to get customers to purchase drinks, which have a more favorable profit margin. For example, McCormick & Schmick's (an upscale national restaurant chain) offers a $1.99–3.99 food menu between the hours of 4–6 pm and 9 pm–close in their bar area. This is a great opportunity to fill up their bar area, and to "bridge the gap" during off-hours of the restaurant; the requirement to order off the special discounted menu is the purchase of a 2-drink minimum. According to *USA Today*, McCormick & Schmick's offers the number one happy hour in America. A slightly different example involved a fast food restaurant capitalizing on the healthy beverage margins in order to offset increasing labor and food costs. Back in 2005 Oak Brook, Illinois-based McDonald's Corporation had procured and then promoted premium roast coffee and other made to order coffee drinks as a differentiator among their competitors. Simultaneously, this allowed them to benefit from the higher margins (difference between cost of the coffee and their selling price) offsetting their need to increase food prices due to increasing costs. It brings to notion the concept of a *loss leader*, an item that is strategically sold by an operation that has a reduced to no margin between the cost of its ingredients and its selling price. The intention is to attract the consumer with the appealing selling prices; while the buyer experiences or consumes the loss leader, it encourages them to purchase or pay for additional products or services.

Heightened Opportunities in the Beverage Industry

The chef has always been the centerpiece—the one who comes out and speaks with the customers about the food at the table; now the beverage person is just as exciting. The past generation has brought about an elevated role of the beverage professional (sommelier, bartender, brewer, barista, etc.) in the hospitality industry. Now many customers are at least as interested in the beverage professional as they have always been in the chef. At one point in history, very few people would consider going to school and study beverage management, or obtain beverage certifications. However, the times have changed and changed quite dramatically. What the cooking network had done for the chef, the movies and documentaries—*Bottle Shock*, *Red Obsession*, *Somm*, *Sideways*, and *Blood into Wine*—are doing for sommeliers and other jobs in the beverage profession. The image of sommelier has changed from older white male to that of a younger, hipper, and more diverse individual—there are just as many females pursuing and succeeding as sommeliers as there are men. Simultaneously, what used to be called "bartending" has been rechristened "mixology" as the craft cocktail craze accelerates and as ingredient lists and techniques grow in sophistication. What's more, the demand for $16 cocktails shows no sign of slowing in big cities or elsewhere: As the economy has improved, so does the fortune of the restaurant and hospitality industry, whose performance is tied closely to discretionary income levels.

According to the Beverage Information & Insights Group (a research and trade group for all segments within the beverage industry), overall consumption of alcoholic beverage has increased, though with a slight decrease of restaurant/bar (on-premise) consumption. Although restaurant/bar consumption may have slightly decreased, a consistent trend is reflected of people buying more premium beverages: drinking less—but selecting higher quality products. More consumers are also making purchases and spending more from retail stores (off premise) for in-home consumption. The fundamental beverage consumer base has and will continue to change—the largest and most frequent buyers have traditionally been the Baby Boomers. However, with the coming of age, younger generation known as the Millennials has proven intense intrigue in the beverage industry—both from an employment standpoint and from a consumer one. Never before has a generation been so passionate about exploring the vast world of beverages and food. With their surprisingly sophisticated palates at such a young age and a demand for local and "natural" products, many trends of yesteryear have been renewed and/or reinvented whether for nostalgic reasons or having been adapted to appeal to their modern sense of values. The surging movement of wines being packaged in a can is possibly most telling. Millennials are not only ditching cork closures but also slipping their cans of wine into to-go bags. What used to be a perception and reality of cheap wines has all but changed in modern day. Some quality-oriented and environmentally conscious wineries are now producing both a bottle and a canned version of their wines for consumers seeking an alternative that suits their lifestyle. Millennials have prioritized premium beverages: wine, beer, spirits, sake, cider, coffee, and tea as a compulsory element of their culture. The new way people (whether it's Millennials or Baby Boomers) drink—less quantity, yet more frequently and often of better quality.

The Wine Segment

According to the California Wine Institute, in 2016, there are well over 10,000 wineries in all fifty states in the United States. Wine has grown some 2.0 percent in 2015, similar to the beer and spirit segments, and consumers have shifted toward selecting more premium brands (Beverage Information & Insights Group). As the world of wine broadens and American wine drinking culture becomes more firmly established and accepted in the socioeconomic mainstream, wine sales seem likely to see continued growth at all levels of price point. There has been a dramatic increase of wine purchases through off-premise, nontraditional online sources. One notable example is *Cellar Angels*, a virtual company that promotes and sells exclusive, small production, nondistributed wines while being socially conscious along the way. On-premise operations have seen an increase of quality wines being offered through draft systems for their by-the-glass programs. More options in the red and white blended (non-Bordeaux style) wines and rosé wines have garnered immense demand, with production coming from just about every wine-producing country. There has been awareness with natural, sustainable, and biodynamic wines as more producers throughout both the Old and New World have converted their vineyards from reliance on synthetic fertilizers and pesticides to create a safer path for demands of the modern consumer and to preserve the land and resources for future generations. Wine-producing countries have come out of obscurity, such as China quickly becoming one of the larger producers of wines in less than a half a dozen years. Figure 1.13 depicts the Chinese wine *Silver Heights* that is located on the eastern slopes of Mount Helan in the Ningxia region. Winemaker Emma Gao is a holder of the Diplôme National d'Oenologue from Bordeaux, France, and one of few female Chinese winemakers in the industry.

FIGURE 1.13
Silver Heights wine from China. Courtesy of John Peter Laloganes.

Additionally, wines coming from Central Coast of California, notably Paso Robles and Santa Maria Valley, and others, have gained worldwide notoriety as quality production alternatives to the traditional iconic powerhouses of Napa Valley and Sonoma County. Second-tier wine-producing countries of Greece and Portugal have gained greater traction in recent years, as their reliance on indigenous grape varietals is becoming more accepted in restaurants and bars. Something notable for existing and new beverage managers

FIGURE 1.14
Wine Professional (WP) Program. Courtesy of Kendall College, LLC.

is the increase of industry professionals seeking wine credentialing from recognized institutions—acquiring advanced knowledge and skill set of wine production, grape varietals, wine and food pairing, and proper service for wine-oriented establishments. Wine Professional (WP) Program at Kendall College, Wine & Spirit Education Trust (WSET), and Court of Master Sommeliers (CMS) have offered foundational and advanced training for the beverage professional to obtain a recognized sommelier certificate at varying levels. Figure 1.14 identifies the logo for the Wine Professional (WP) Program at Kendall College.

The Beer Segment

Despite growing slightly in 2015, beer's beverage alcohol market share has decreased due to increased market share in other alcoholic beverages such as the exploding cider and flavored malt beverage categories. However, craft beer continues to mobilize the beer category, resulting in a 14.1 percent increase (Beer Information Group). According to the Brewer's Association, American breweries have reached an all-time high of 4,656 breweries as of June 30, 2016. The beer industry has seen tremendous growth ever since the 1970s but at a more intensive rate over the past several years. The number of breweries has increased 25 percent with an additional 917 breweries just from June 30 of 2015.

FIGURE 1.15
Sour beers. Courtesy of John Peter Laloganes.

Over recent years, the consumer has been captivated by the local craft brewer and intrigued in the vast availability of beer styles. There have been eye-opening types of innovation in the beer industry with the creation of Dogfish Head Brewery (from Delaware) creating beer and wine hybrids along with Goose Island Beer Co. (from Chicago, Illinois) creating a beer and sake hybrid. There has been a surge in the production of session (lower alcohol) beers while many local breweries have begun collaborating with local chefs from nationally recognized and acclaimed restaurants. Some breweries have increased their production of barrel-aged beers, with their Stouts or Porters often spending many months in old bourbon barrels, while other brewers are showcasing the tartness of their sour beers made with wild yeasts and/or bacteria. Figure 1.15 identifies a trio of barrel-aged sour beers from Goose Island Beer Co. with *Juliet* (rye beer made with blackberries), *Lolita* (Belgian-style Pale Ale beer brewed with raspberries), and *Halia* (Saison beer brewed with peaches).

Similar to the wine segment, many existing or new beverage professionals are increasingly seeking beer credentialing from recognized institutions—acquiring advanced knowledge and skill set of beer production, beer styles, and proper service for beer-centric types of establishments.

The Spirit and Cocktail Segment

Distilled spirits grew by 2.0 percent in 2015, marking the nineteenth consecutive year of growth (Beer Information Group). This growth is largely led by the resurgence in American whiskey and tequila along with consistent growth in vodka. Mixologists have been championing simplicity, using three or four ingredients, so

FIGURE 1.16
Craft Cocktail. Courtesy of Erika Cespedes.

that base spirits are allowed to come through, and with some exceptions, they are moving away from the seven- to eight-ingredient cocktails. Figure 1.16 illustrates a bourbon-based cocktail prepared with fresh lemon juice and thyme.

Over the past decade, there has been a rise of microdistilleries, allowing for a dramatic departure from the classic originating distilleries from around the world. Due to some changes in laws now permitting distillery licenses, locally owned distilleries producing handcrafted spirits have blossomed across the United States.

Over the past decade, there has been some experimentation with an increase of molecular mixology, largely in first-tier cities like New York, San Francisco, Los Angeles, and Chicago. Classic cocktails have remained a constant over the past decade plus with the Martini, Old-Fashioned, and Manhattan being the renewed favorites of many of today's consumers. Bartenders have borrowed the chefs' practice of using seasonality to drive their cocktail menus, while others have incorporated exotic ingredients such as lychee fruit and elderflower. Many bartenders have adopted various in-house productions of syrups, infusion, barrel-aged spirits, and bitters.

The Sake, Cider, and Mead Segment

Sake is no longer confined to the shelves of sushi bars and Japanese restaurants, but is now earning an ever-present fixture on beverage menus around the country. Consumers, most notably Millennials (and those under 35 years), are gravitating toward sake. Overall, consumers are becoming more educated about this Japanese beverage and embracing all sorts of upscale products. As the sake category grows in the United States, many consumers are opting for sake-based cocktails.

Cider is one of the fastest growing beverage categories with a vast array of styles and production techniques showing a more diverse perception to the consumer. Cider has shown as a strong competitor for the beer segment also serving as a useful alternative to the consumers favoring gluten-free products.

Mead (honey wine) has been reawakened and is moving beyond medieval fairs and the Halloween season. Like all other beverage segments, there has been a movement to showcase specialty types of honey (deriving from different flowers) and incorporating different types of barrel-aging techniques during its production.

FIGURE 1.17
Military Latte: Matcha tea with espresso. Courtesy of John Peter Laloganes.

The Coffee and Tea Segment

The surge in quality coffee consumption can be compared to the shift in beer, spirit, and wine consumption—less quantity, yet better quality. Starbucks has helped to propel the consumer into becoming more sophisticated with their coffee preferences. Consumers have increasingly become intrigued as to how coffee is grown, roasted, and prepared. Many consumers don't think twice about spending $5 or more dollars on a cup of coffee or coffee concoction skillfully prepared from your local barista. Coffee shops have forged ahead with offering single-origin coffee beans, microroasting, and applying their creativity often regarded for mixologists. Figure 1.17 depicts a "cocktail" like latte made with Matcha green tea and espresso. Furthermore, the consumers are taking their coffee experience home when they purchase a 1 lb. bag of Guatemalan natural (or dry) processed coffee beans for $15–20. The consumers will enthusiastically transport the

FIGURE 1.18
Pouring tea. Courtesy of Suzette Hammond, for World Tea Academy.

beans home and place them into their $200 burr grinder—allowing themselves maximum control using any of several manual brewing methods, from French press to pour-overs. The presence of local specialty coffee roasters with their emphasis on microroasting has brought them recognition in many small and large cities across the United States.

Tea is one of those beverages that has not always been given the degree of respect in the United States as it deserves—unfortunately, it has often been relegated as the last drink that the beverage professional thinks about when constructing a menu. Teavana and Argo Tea (both are national tea sellers and cafes) are doing for tea, what Starbucks has done for coffee. They are both procuring high-quality loose leaf teas from around the world. Loose leaf teas are on the rise in restaurants as an alternative to the ground of tea "dust" consumers had been used to throughout much of the past century. Figure 1.18 depicts a serving of an oolong loose leaf tea that was brewed from a single tea kettle. Additionally, tea is being elevated to a contemporary status behind the bar as mixologists and spirit bars across the country are using it to enhance their gin-, rum-, and vodka-based cocktails. Increasingly, bartenders are discovering that tea offers a broad range of appealing flavor profiles along with a vibrant spectrum of visual characteristics. As they help to move tea into the 21st century, there has been a focus on promoting a more appealing presentation with modern-day applications.

Check Your Knowledge

Directions: Use these questions to test your knowledge and understanding of the concepts presented in the chapter.

I. MULTIPLE CHOICE: Select the best possible answer from the options available.

1. The Greek god who represented wine was known as
   a. Bacchus
   b. Pliny
   c. Dionysus
   d. Obama
2. Which alcohol beverage is likely to have been intentionally created first?
   a. Wine
   b. Beer
   c. Spirits
   d. All of the above
3. Where are wine and beer likely to have originated?
   a. France and Germany
   b. America
   c. Mesopotamia and Egypt
   d. Italy
4. Prohibition was an infamous period in American history that
   a. made it unlawful to consume alcohol
   b. made it unlawful to purchase alcohol

c. made it unlawful to taste wine in church
d. made it unlawful to use spirits, but not wine and beer

5. Brandy is distilled from
   a. grains
   b. wine
   c. barley
   d. sugarcane
6. Which is derived from the fermented juice of grapes?
   a. Beer
   b. Cider
   c. Wine
   d. Spirits
7. Which trends are important for the beverage professional to consider as a viable strategy?
   a. Localized trends
   b. Broad-based trends
   c. World-based trends
   d. Both a and b
8. The concept of a *loss leader* is an item that is strategically sold by an operation that has
   a. a reduced margin
   b. no margin
   c. a loss of money when it's sold
   d. both a and b
9. Which answer best describes a demographic and labor market external variable?
   a. relating to the natural world and the impact of human activity on its condition or sustainability
   b. concerned with the prevailing direction in which the broader or local economy is operating
   c. concerned with the developments in population as they relate to changes in age, gender, marital status, educational levels, employment status, race, and so on
   d. all of the above
10. When customers have the use of handheld tablets or iPads that contain the restaurant wine list is an example of which external variable?
    a. Economic trends
    b. Environmental trends
    c. Technological trends
    d. Sociocultural trends
11. When a restaurant offers a selection of wine-by-the-glass options that are dispensed from a draft system, it is NOT an example of which external variable?
    a. Economic trends
    b. Environmental trends
    c. Political trends
    d. Sociocultural trends
12. When a restaurant begins offering happy hour drink options between 3 and 5 pm in the bar area is an example of which external variable allowing them to partake in such practice?
    a. Political/legal trends
    b. Environmental trends
    c. Technological trends
    d. Economic trends

II. DISCUSSION QUESTIONS

13. What is the major distinction between beer and wine and spirits? Explain.
14. Discuss the reasons for Prohibition. How long did it last? Why was it repealed? What were the repercussions?
15. What purpose did the religious communities serve during the Middle Ages?
16. Identify some trends associated with the wine segment of the beverage industry.
17. Identify some trends associated with the beer segment of the beverage industry.
18. Identify some trends associated with the spirit and cocktail segment of the beverage industry.
19. Identify some trends associated with sake, cider, and mead segment of the beverage industry.
20. Identify some trends associated with coffee and tea segment of the beverage industry.

CHAPTER 2

Introduction to Wine and Service

CHAPTER 2 LEARNING OBJECTIVES

After reading this chapter, the learner will be able to:

1. Discuss the rise of wine and recall key moments in its evolution to modern day
2. Recognize key grape-growing and production areas in the Old World
3. Provide an overview of the Judgment of Paris
4. Understand the basic components of a grape and their contributions to a wine
5. Identify the famous Old and New World wine production areas
6. Recognize the distinctions between stemware for the different categories and types of wine
7. Distinguish the four methods of labeling table wine
8. Discover the five most significant pieces of information on wine labels
9. Perform key steps and etiquette in table wine and sparkling wine service
10. Explain the reasons why wine may be decanted

Wine makes every meal an occasion, every table more elegant, every day more civilized.

—Andre Simon, *Commonsense of Wine*

The History of Wine

Learning Objective 1
Discuss the rise of wine and recall key moments in its evolution to modern day

Wine is the fermented juice of grapes (unless otherwise specified). Figure 2.1 illustrates wine as the enduring and evocative beverage that it has been through the ages. The history of wine spans thousands of years and is closely entangled with the history of agriculture, gastronomy, civilization, and humanity itself. Throughout time, wine's influence on western culture has been transformational for civilization, a sentiment that is evidenced in the above quote. From its earliest development, wine has had a special place in the world's customs, at the dinner table, and at social gatherings. Few areas of the world remain untouched by the many virtues of the vine and its popularity closely resembles the development of the western world.

The Early Years

Wine has a rich history dating back to roughly 6000 B.C. and, similar to beer, is thought to have originated in the Middle East. Archaeological evidence suggests that the earliest wine production came from sites in Georgia (a former kingdom and province of Russia) and the *Caucasus* (kaw-keh-ses) mountain range between the Black Sea and the Caspian Sea that form part of the traditional border between Europe and Asia Minor. As societies moved around, civilizations transported and cultivated the vines wherever they could grow. Figure 2.2 depicts a map of the fertile crescent—believed to be the origins of the grapevine.

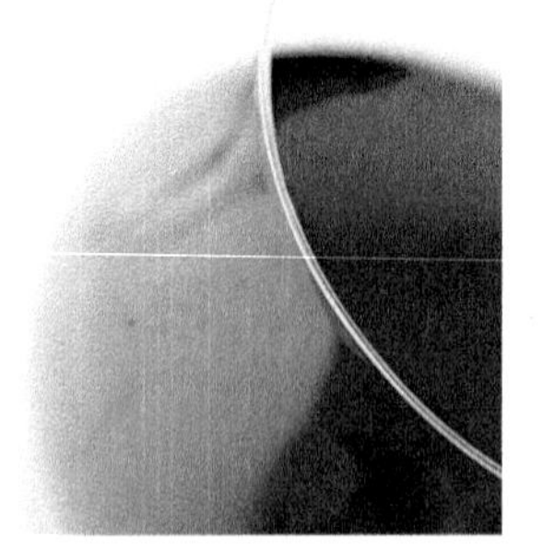

FIGURE 2.1
Glass of red wine. Courtesy of Erika Cespedes.

Ancient Egypt

Historical records illustrate Egyptian vineyards dating back to 2900 B.C. At that time, the consumption of wine was the drink of the prestigious and privileged, limited to the nobility and clergy in ancient Egypt, while the "working class" was drinking beer. It is clear from hieroglyphics and archaeological evidence that wine was used in religious ceremonies and was buried with the dead. When tombs of the pharaohs were opened, clay jars used for holding wine were discovered. The jars were marked in a manner that was similar to the way bottles are labeled today, indicating where the grapes were grown, the year the grapes were harvested, and who made the wine. Apparently, this wine was part of the deceased pharaoh's provisions for the afterlife.

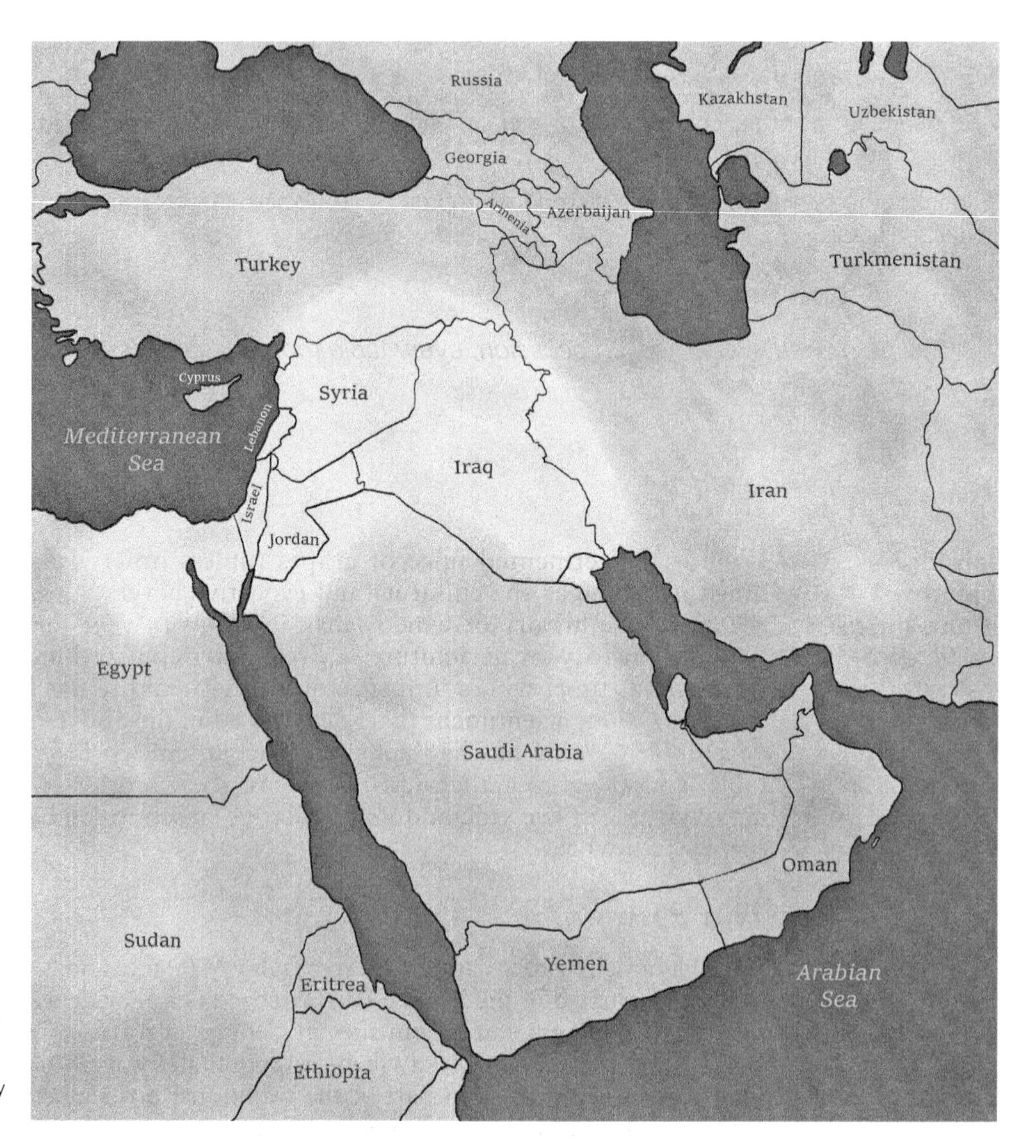

FIGURE 2.2
Fertile crescent. Courtesy of Thomas Moore.

FIGURE 2.3
Tribute to ancient wine. Courtesy of John Peter Laloganes.

FIGURE 2.4
Ancient wine press. Courtesy of John Peter Laloganes.

The Greeks and Romans

To say that winemaking and drinking played an important role in ancient Greek and Roman life is an understatement. The significance of wine in their culture can be illustrated initially by looking at the ancient meaning of the term *symposium.* Its Greek derivative means "to drink together," which was often conducted with a gathering of individuals where dialogue was used to foster contemplation and enlightenment. Much of the modern wine and social culture derives from the practices of the ancient Greeks. The prosperous culture that developed within ancient Greece in the first millennium B.C. gave way to important advances in philosophy, politics, science, and literature. Figure 2.3 depicts one of the many tributes to grape growing and wine production.

Wine was so strongly embedded in Greek culture that in Greek mythology, there was a designated youthful deity of vegetation, wine, and ecstasy—Dionysus (die-uh-ny-suhs). The importance of wine in Greek life was celebrated each year by a festival to honor Dionysus. The participants would sacrifice live animals, drink wine, watch plays, drink wine, and then drink more wine. According to Hugh Johnson's book *Vintage: The Story of Wine*, their wine also may have included pinesap, hallucinogenic mushrooms, and a natural form of the hallucinogenic drug known as LSD. Due to its intoxicating effects, the Greeks believed that drinking wine allowed them to consume "their god" as it provided a state of exhilaration—it was reasoned that any ill feeling consequences were credited to the god departing their bodies. Figure 2.4 depicts an ancient wine press, likely similar to the ones used by the Greeks and Romans.

During the reigning period of the ancient Greeks, wine maintained not only an obvious role in culture but also a prominent one in literature and science. Many famous poets and philosophers made mention of this as Homer included wine in his stories about the Battle of Troy and Odysseus. In addition, Hippocrates, one of ancient Greece's scientists, proclaimed many of wine's positive and negative effects on the human body.

Greek wine was widely known and exported throughout the Mediterranean basin, as evidence of the several amphorae (wine vessels) has been found throughout the area. The Greeks introduced the *Vitis vinifera* vine and produced wine in their numerous colonies in modern-day Italy, southern Italy and their islands, southern France, and Spain. During the 8th and 7th centuries B.C., the ancient Greeks colonized in southern Italy. They found the environment so favorable to growing grapevines that they called it Oenotria (own-eet-tree-ah), meaning *land of wine.*

The Romans were pioneers of large-scale production and largely responsible for spreading the influence of the vine through their conquests and colonization after the Greeks. Figure 2.5 shows an archway in the Italian city of Verona that still exists as evidence of its once Roman occupation. Wine was such an integral part of culture that the Romans also had a god of wine whom they called Bacchus. Roman leaders, however, did not always appreciate this god or his followers. In 186 B.C., the Roman Senate banned the worship of Bacchus, or Bacchanalia, because his worshippers were accused of many sorts of crimes and vices ranging from promiscuous sex to murder.

FIGURE 2.5
Roman arches. Courtesy of John Peter Laloganes.

FIGURE 2.6
Cathedral. Courtesy of John Peter Laloganes.

The Middle Ages

In medieval Europe, following the decline of the Roman Empire and therefore of widespread wine production, the Christian Church became a staunch supporter of the wine production necessary for celebration of the Catholic Mass. Whereas wine was forbidden in medieval Islamic and other Muslim cultures, the Catholic Church, became one of the most prominent and influential forces in French winemaking during the medieval period. Figure 2.6 depicts a European church and Figure 2.7 depicts candles lit for mass in a Catholic Church; wine was a necessary component for the Eucharist to represent the blood of Christ.

FIGURE 2.7
Candles lit in church. Courtesy of John Peter Laloganes.

During the Middle Ages, the monks were the preservers of civilization—and for all things sacred. The Catholic Church was one of France's largest vineyard owners, wielding considerable influences in regions such as Champagne and Burgundy where the *terroir* (tehr-WAH) concept (which loosely translates to the connection to the land) first took priority. Due to their vast land holdings throughout France, the Christian monks contributed many advances in viticulture and enology. Their dedicated study and observations led them to the identification and classification of quality vineyards; some of the most prestigious ones are still in existence today. The monks also determined ideal site selection for noble grape varieties and discovered new and alternative methods of wine production.

FIGURE 2.8
Church from the Renaissance. Courtesy of John Peter Laloganes.

FIGURE 2.9
Church from the Renaissance (2). Courtesy of John Peter Laloganes.

FIGURE 2.10
Aging wines in Rioja, Spain. Courtesy of John Peter Laloganes.

The European Renaissance

This period occurred roughly during the 14th through 17th centuries and is marked by a renewed cultural movement that spread throughout Europe. The Renaissance is often referred to as a bridge between the Middle Ages and modern day that involved a resurgence of artistic and intelligent perspective and contributions. It was an age of exploration that was previously, in the Middle Ages, a time of constraint. Figures 2.8 and 2.9 illustrate how the architecture during this period became more ornate and artistic as Europe became much more prosperous from its emergence of the Middle Ages.

The French Revolution (1789–1799) was a period of extreme social and political upheaval in French history. France underwent an epic transformation from a monarchy to a democratic republic operated government. The revolution brought about changes of power from the Catholic Church to the state.

The period of the 18th and 19th centuries marked an era where the "Old World" references the long-established tradition of winemaking within the European countries of France, Italy, Germany, Spain, and others. They truly solidified their existence in the historical narration of wine. These countries have nurtured and developed many of the vines and winemaking techniques that form the foundation for modern practices of wine throughout the world. Figure 2.10 depicts an underground wine cellar in "Old World" Spain.

Classic Old-World Wine-Producing Countries

Learning Objective 2
Recognize key grape-growing and production areas in the Old World

Wines of France

France is one of the oldest wine-producing countries in Europe, with its origins dating back to the 6th century B.C. Through the ages, France has always been considered the benchmark for quality wine and, along with Italy, has remained one of the largest producers and consumers of wine throughout the world. France is one of the most revered and often imitated wine-producing countries that has served as the standard of excellence of wine for centuries. Additionally, France acts as the spiritual source of many international grape varietals that are ubiquitous around the world. Throughout its history, the French wine industry has been largely shaped by the influences of three of the more prominent and pervasive authorities: from the British through both commercial interest and political forces, the Dutch who were significant contributors of technology and traders in the wine industry for much of the 16th and 17th centuries, and the *Catholic Church* that held considerable vineyard properties up until the French Revolution.

The oldest vineyards in France are believed to have originated from the southern French region of Provence. They were planted either by the Greeks, or in a slightly earlier period, by the Phoenicians around 600 B.C. It is generally agreed that the Romans introduced viticulture to Burgundy around the middle of the 1st century and that Champagne, Alsace, and Bordeaux were developed shortly afterward. The French city of Paris has for ages acted as a thoroughfare and world cultural icon for the development of drinks as well as the evolution of the bistro concept. Figure 2.11 identifies a bistro in Paris, France. Paris has a long history of well over 2,500 years and has since played a significant role in shaping the culture of food and wine.

FIGURE 2.11
Bistro in Paris, France. Courtesy of John Peter Laloganes.

The novice and intermediate wine consumer may find French wine (or wine from the Old World) intimidating, primarily because the labeling system is based largely on geography. Most French wines are labeled by the name of the place or appellation (which is registered and legally defined under French law) where the grapes are grown, rather than by varietal labeling as is done in the New World. The French term *appellation* refers to a viticultural area distinguished by geographical features that produce wines with shared characteristics. In France, the appellation term is legally applied to specific and stringent grape-growing and winemaking requirements. In simple, broad terms, an appellation is a place where the grapes are grown.

Created in 1935, France founded the *Institut National des Appellations d'Origine*, or INAO. France became the first nation to set up a countrywide system based on geography for controlling the origin and quality of its wine. The INAO is part of the French government that is officially authorized to regulate the French wine industry. This plan originated during the Great Depression as a preventative measure to protect French winemakers and consumers from fraudulent and inferior wine-blending methods practiced by some unethical French wine brokers. The *Appellation d'Origine Contrôlée* system, or AOC, is a French term meaning "controlled appellation of origin" and is applied to standards of production for various kinds and types of products such as wine, cheese, and butter. The French AOC system became the model in the wine industry and parallels the regulation systems in other major wine-producing countries throughout the world.

Wine Regions of France

For the purpose of simplicity, the major wine regions of France can be divided into three broad areas on the basis of grape varietals—which are often an indication of the overall climate. Of course, these are generalizations, and there are obvious exceptions within each area. However, it can be helpful to discuss the complexity of Old-World wines initially with more generalities. There are eight significant (for our purposes) French wine regions that specialize in certain grape varieties that are

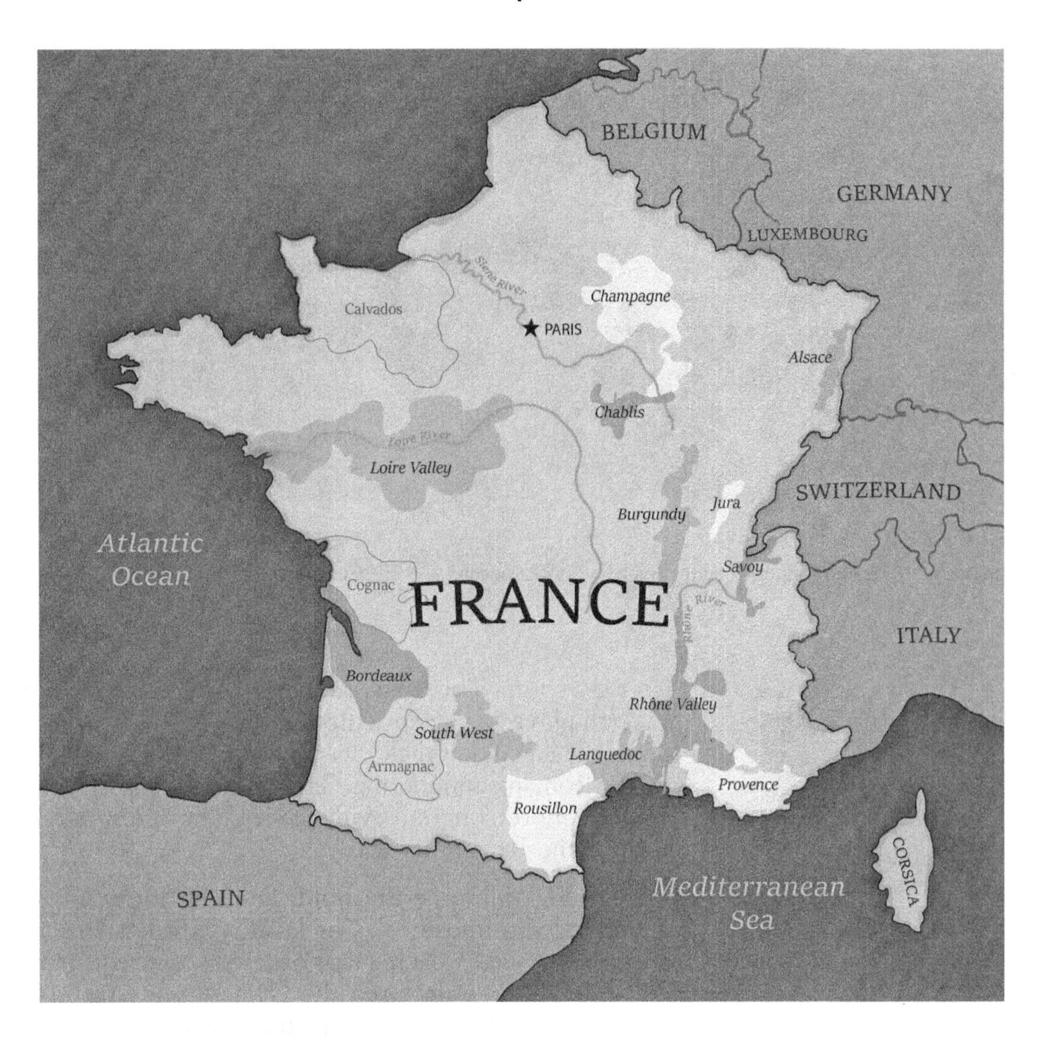

FIGURE 2.12
Wine map of France. Courtesy of Thomas Moore.

known for producing world class wine. Wine map of France in Figure 2.12 identifies the most significant French wine and brandy regions.

1. *The northeastern section of France (Alsace, Burgundy, and Champagne)* is subject to a continental climate that consists of four distinct seasons with short summers and harsh winters. This type of climate contributes to creating less ripe, moderate-to-highly acidic grapes, often illustrating some prevalent mineral qualities with moderate levels of alcohol in the finished wine.

   - **Alsace** (al-SASS) For many centuries, France and Germany fought over the region of Alsace. After World War II, this small bordering strip of land once again became controlled by France. This region produces mostly single varietal white wines from grapes that are of German origin but the wine is made in the dryer French style. The most prolific grapes include Riesling, Gewürztraminer, Pinot Blanc, and Pinot Gris grapes. Pictured in Figure 2.13 is the Alsatian city of Strasbourg, which has served as a thoroughfare for both French and German cultures throughout history.
   - **Burgundy** (BER-gun-dee) The Burgundy region specializes in single varietal white wines based on Chardonnay and single varietal red wines made primarily from Pinot Noir with smaller amount of Gamay-based wines

FIGURE 2.13
Strasbourg in Alsace, France. Courtesy of John Peter Laloganes.

FIGURE 2.14
Road signs in Burgundy. Courtesy of John Peter Laloganes.

from the south. The Celts, Romans, Cistercians, and Dukes, to some degree, have all played a role in sculpting the towns and villages of Burgundy. It has been speculated that the Celts may have been growing vines in the region prior to the Romans conquering Gaul (which we now call Burgundy) in 51 B.C. Though more concrete evidence suggests Burgundy's vineyards existed as early as the 1st century A.D.

In 1395, Philippe the Bold, Duke of Burgundy, was perhaps the first individual to impose rules or laws regarding what could be grown in his duchy. He issued a decree that ordered strict requirements of the growing of grapes. The duke declared the Gamay grape (prominent throughout Burgundy during this period) was unfit for human consumption and that it be removed from Burgundian vineyards and replaced with the more complex Pinot Noir grape varietal. The Burgundian village of Gevrey-Chambertin is noted as the largest and one of the most famous villages specializing in Pinot Noir-based wines. Figure 2.14 depicts a road sign indicating some of the famous Burgundian villages.

- **Champagne** (sham-PAYN) Champagne is both a region and a type of sparkling wine. More precisely, Champagne is often considered the most famous of all sparkling wines in the world. The Romans were the first known inhabitants who planted vineyards in Champagne. Champagne (the wine) is produced from varying blends of Pinot Noir, Pinot Meunier, and Chardonnay grapes—while applying the méthode traditional technique as the manner of incorporating the wine's well-known and alluring carbonation. Figure 2.15 shows Ruinart (HWEE-nahr) that is credited with being one of the first Champagne houses (founded in 1729) that is still in existence today. Figure 2.16 shows one of the well-preserved Champagne crayères (underground chalk cellars) originally dug from the Roman occupation. Figure 2.17 depicts the Bollinger logo and Figure 2.18 depicts the Bollinger estate—representing one of the most famous Champagne houses and has even been subject of dialogue in a James Bond movie franchise.

2. *The western section of France (Bordeaux and most of the Loire Valley)* has a maritime climate of mild winters and cool summers, created from the moderating influence of the Atlantic Ocean. The moist and cool climate ultimately

FIGURE 2.15
Ruinart Champagne. Courtesy of John Peter Laloganes.

FIGURE 2.16
Champagne cave. Courtesy of John Peter Laloganes.

FIGURE 2.17
Bollinger logo. Courtesy of John Peter Laloganes.

FIGURE 2.18
Bollinger estate. Courtesy of John Peter Laloganes.

creates inconsistency and significant vintage variation for the wine from year to year.

- **Bordeaux** (bohr-DOH) The Bordeaux region of France produces blended red wine (primarily blended in varying quantities of Cabernet Sauvignon, Merlot, Cabernet Franc, and others) and white wine and dessert wine (both from a blend of various quantities of Sauvignon Blanc and Sémillon varietals). Bordeaux (the wine, city, and region) has maintained an intimate connection to England ever since the 12th century, when Eleanor of Aquitaine (the earlier Bordeaux and southwest France region) wed Henry Plantagenet from England. When Henry became king of England, Eleanor's Aquitaine, which included Bordeaux, became an English dominion. Gradually, the wines of Bordeaux came to dominate the important English market. Figure 2.19 depicts a bottle of Chateau Latour that is one of five wines ranked as a first-growth Premier Cru.

FIGURE 2.19
Chateau Latour. Courtesy of Erika Cespedes.

FIGURE 2.20
Bottle of Sancerre. Courtesy of John Peter Laloganes.

- **Loire Valley** (LWAHR) The Loire Valley, another famous wine region of France, is known primarily for their extraordinary single varietal white wines (mainly from Chenin Blanc or Sauvignon Blanc grapes), but also produces single varietal red wines (from Cabernet Franc and Pinot Noir), dessert wines, and sparkling wines. The vineyards flourish and sit among the picturesque castles built for the aristocracy along the banks of the Loire River. Figure 2.20 shows a bottle of Vincent Delaporte from the appellation of Sancerre located in Loire Valley, France—devoted to single varietal Sauvignon Blanc or Pinot Noir-based wines.

3. *The mid-central and southern sections (Rhône Valley, Languedoc-Roussillon, and Provence)* of France maintain a Mediterranean climate. In warmer and hotter southern climates, grapes ultimately have the capability of producing wines with higher alcohol levels, riper fruit, and denser medium- to full-bodied red wines.

- **Rhône Valley** (ROHN) The Rhône Valley is located toward southern France with a probability that grapevines were first planted around 600 B.C. The region produces mostly red wines (either single varietal or blended wines) from Syrah, Grenache, Mourvèdre, and others, with white wines produced from the Viognier, Marsanne, and Roussanne grape varietals. The northern half of the Rhône Valley specializes in single varietal-based red wines from Syrah—home of the famous appellation *Hermitage* (EHR-mee-tahj). This area was named after Henri Gaspard de Sterimberg, a knight who fought in Pope Innocent III's crusade. When he returned from the crusade, he became a hermit and spent the next thirty years dedicated to viticulture. The southern half of the Rhône Valley specializes in blended red wines dominated by Grenache, Syrah, Mourvèdre, and others with the presence of an intriguing feature of large stones called galets covering the topsoil of many vineyards. This soil works to absorb the heat of the sun during the day, while keeping the vines warm at night and continuing the grape's ripening and development process. One of the most famous growing areas is the *Châteauneuf-du-Pape* (shah-toh-nuhf-doo-PAHP), or CDP. CDP is a blended red wine containing up to thirteen red and white varieties. Châteauneuf-du-Pape means "Pope's new castle" and was named after the relocation of the Italian papal court to the French Rhone city, Avignon, in the 14th century to house Pope Clement V, the first French Pope.
- **Languedoc-Roussillon** (lahng-DAWK roos-see-YAWN) **and Provence** (praw-VAHNS) Located in southern France, these wine regions border the Mediterranean Sea with Languedoc-Roussillon to the west and Provence to the east. The majority of production is blended red wines from Syrah, Mourvèdre, Grenache, and numerous other varietals in smaller quantities. These regions, also referred to as the *midi* (mid-ee), are the most extensive in France and represent 40 percent of the total vineyard area and produce most of France's "bulk" production wines. In addition, these regions produce some of France's most famous versions of fortified wine known as Vin Doux Naturel (VDN) (van doo nah-tew-REHL)—regarded as France's version of the famous fortified Port wine).

Wines of Italy

Italy is one of the oldest and largest wine-producing countries in the world—yet is only three-fourths the size of California. The people of Italy have been making wine for thousands of years as tradition and culture are entrenched in everyday life. Wine and food have clearly evolved parallel to one another and truly reflect the uniqueness of each of Italy's twenty distinct wine regions. The Etruscans were early inhabitants of Italy who had been cultivating grapes for well over two thousand years ago; though it wasn't until the Greek colonization that winemaking began to flourish. Viticulture was initially introduced in southern Italy and the Island of Sicily around 800 B.C. It was in 2nd century B.C. when the Roman Empire began spreading the grapevine throughout much of the rest of modern-day Italy.

Italy is one vast vineyard that produces a variety of grapes of both international and indigenous types. The country is well suited for grape growing from north to south, with over 80 percent of the land being mountains or hilly and having close proximity to the ocean. Italy's extensive latitude spans as far north as the Alps (bordering Austria, Switzerland, and France), which have a cool, alpine, continental climate, to the warmth of southern Sicily (near North Africa), which maintains more of a Mediterranean-type climate. Combined with varying soil types and topography, Italy can produce a variety of grapes of both international and indigenous types. It has been noted that Italy has well over 400 authorized grape varieties—contributing to a huge range of wine style options, but also to some confusion among the international markets. Figure 2.21 depicts Italy's twenty wine regions.

FIGURE 2.21 Wine map of Italy. Courtesy of Thomas Moore.

FIGURE 2.22
Transportation in Venice. Courtesy of John Peter Laloganes.

FIGURE 2.23
Passito grapes. Courtesy of John Peter Laloganes.

FIGURE 2.24
Amarone wine. Courtesy of John Peter Laloganes.

It is generally regarded that some of the most prestigious Italian wines come from the northern half. Piedmont is in the northwest part of Italy and produces large amounts of sparkling wine along with numerous indigenous red wine grapes such as Barbera, Dolcetto, and the highly tannic and ageable Nebbiolo that goes into the reputable wines of Barolo and Barbaresco. Tuscany is located in north-central Italy and is home to one of Italy's most prolific grapes, Sangiovese. The Veneto, Friuli Venezia Giulia, and Trentino-Alto-Adige are the three regions in the northeast part (called Tre Venezie) of Italy, partly famous due to the iconic city of Venice. Figure 2.22 pictures a gondola, one of the famous methods of transport around the city of Venice. These three regions of northeast Italy produce a mix of wines from the famous sparkling wine, Prosecco, to the white wines of Soave (made from Garganega), Pinot Grigio, and Pinot Bianco, to the red wines of Bardolino, Valpolicella, Amarone, and Recioto (all produced from indigenous Italian red grapes of Corvina, Molinara, and Rondinella). Pictured in Figure 2.23 is the ancient *passito* method of drying grapes that go into the famous Amarone and Recioto wines. Figure 2.24 shows a bottle of Amarone wine, regarded as the most famous red wine of northeast Italy.

Throughout Italy, each wine region is governed by the laws according to its quality level as granted by the Italian government. In 1963, Italy adopted a comprehensive, nationwide, regulatory quality-control system very similar to the French AOC. The top two levels of the system are called the *Denominazione d'Origine Controllata e Garantita (DOCG)* and the *Denominazione d'Origine Controllata (DOC)*. The system was loosely modeled after the French AOC system; however, the Italian system has been highly criticized for its overgenerous awarding of high classification levels to wine areas that, arguably, are not necessarily deserving of it. All Italian wines awarded the highest status of quality, the DOCG designation, are required to be identified with a special paper strip. Figure 2.25 showcases a Chianti Classico—arguably Italy's most famous wine ranked at the DOCG level and therefore containing the paper strip on the upper portion of the bottle.

FIGURE 2.25
Chianti DOCG label. Courtesy of Erika Cespedes.

Wines of Germany

Germany is a significant wine (and even more so beer)-producing country located in the heart of Europe. It shares a border with Denmark, Poland, the Czech Republic, Austria, Switzerland, France, Luxembourg, Belgium, and the Netherlands. Germany is one of the northernmost (and coolest) wine-producing countries in Europe—as a result, most of the thirteen wine regions, or *Anbaugebiete* (AHN-bough-geh-BEET-eh), are concentrated in the southwestern part of Germany, along the River Rhine and its tributaries to assist in tempering weather extremes. Figure 2.26 identifies the significant German wine regions.

Due to its cold northerly location, white grapes are most prized and account for roughly 64 percent of production versus 46 percent for red wine grapes. Because of the cool continental climate (except in small pockets), red wine grapes do not flourish to the degree that white wine grapes do. Therefore, the majority of wine produced derives from white wine varietals, predominantly Riesling. Other white wine grapes found throughout Germany include *Muller-Thurgau* (MOO-lehr TOOR-gow), *Silvaner* (sihl-VAH-ner), *Gewürztraminer*, *Grauburgunder* (GROUW-buhr-goon-dair) or *Rulander* (otherwise known as Pinot Gris), and *Weissburgunder* (VICE-buhr-goon-dair) (also known as Pinot Blanc). A small percentage of red wine grapes are grown in Germany, with the most notable being the up-and-coming *Spätburgunder* (SHPAYT-buhr-goon-dair) (also known as Pinot Noir).

Grapes struggle to grow in the cool German climate (though it's helpful for maintaining the grape's natural acidity), and in addition, the grapes obtain greater

FIGURE 2.26
Wine map of Germany. Courtesy of Thomas Moore.

FIGURE 2.27
German vineyard. Courtesy of Leo Alaniz.

FIGURE 2.28
German vineyard located above the river Mosel. Courtesy of Leo Alaniz.

ripeness by being harvested later in the season as compared to other wine regions. These ripeness levels (specifically for white wines) directly determine the natural sweetness of the grapes and ultimately—in combination with winemaking techniques—the sweetness level and cost of the final wine. Most of the land consists of steep hillsides with an angle of about 60 degrees or even 45 degrees in some areas. This type of terrain is not ideal for harvesting grapes, but is necessary for optimal sun exposure in such a cool climate. During the Roman occupation of present-day Germany, Romans determined which grapes did best in this hilly land, based on sites where the snow first melted at the end of winter. Because of Germany's geographic location, there are approximately one hundred sunny days in the country. Since there is so little direct sunlight (and a consistently cool climate), the grapes must rely on maximizing the sun's rays through the angle via the steep hillsides. Figure 2.27 depicts a typical steep vineyard in the Mosel region.

The highest quality wines in Germany are classified according to the prädikat wines (formerly Qualitätswein mit prädikat or QmP). These wines make up the top level of German wine classification. Similar to the French and Italian laws, the German's classification of growing of grapes and production of wine are also held to a specific set of standards based upon the particular growing region.

Many of the famous vineyards are established along the Mosel and Rhine rivers benefiting from the water's moderating influence and reflection of the sun's rays back onto the vines. The tempering influence of the rivers allows high-quality wine grapes to grow this far north and creates a long growing season that allows the flavors within the grapes to mature slowly—the sugars to develop, and yet, the acids to remain high. Figure 2.28 depicts a Mosel vineyard located just above the river.

Wines of Spain

Spain is a significant wine-producing country located on the Iberian Peninsula in southwestern Europe. Spain has a long rich history of winemaking, possibly reaching as far back as 3,000 years. This country maintains more vineyards than any other country in the world, yet is only the third-largest wine producer (after Italy and France). This disparity exists because of the overall dry, warm air that reduces vineyard yields. Figure 2.29 identifies some of the major wine regions of Spain.

The country contains an abundance of indigenous grape varietals with well over 600 varieties planted throughout Spain—though majority of the country's wine production derives from only a couple dozen grapes, namely the red varietals of Tempranillo and Garnacha (aka Grenache) and the white varietals of Albariño and Macabeo.

The Spanish government's Instituto Nacional de Denominaciones de Origen (INDO) (equivalent to France's INAO) guarantees the authenticity of its wine by designating each with a region classification. The top-quality wines are classified according to a DO or a DOCa, in which each one is overseen by a *Consejo Regulador* (cohn-SAY-ho ray-goo-lah-DOOR), or administrative body. These agents ensure that each winery acts in accordance with the individual quality requirements according to each designated location. Each individual consejo regulador within each DO/DOCa region issues *contraetiquetas* (con-trah-ett-ee-kAY-tahs), or back labels, as a stamp of approval.

FIGURE 2.29
Wine map of Spain. Courtesy of Thomas Moore.

The Infamous Mite

Phylloxera (fil-LOX-er-uh), a tiny aphid-like organism, native to the United States, caused one of the most infamous pest outbreaks in the history of the wine industry. The infestation began in the early 1860s when the pest was unknowingly introduced from North America into European vineyard sites. In 1863, the aphid initially began invading two prominent French areas: the Gard (southern France) and the Gironde (southwest France). By 1865, phylloxera had spread to vines in the Rhône Valley, then Bordeaux, and over the next three decades, it inhabited and devastated an estimated nearly two-thirds of vineyards throughout Europe. Figure 2.30 depicts the infamous pest phylloxera.

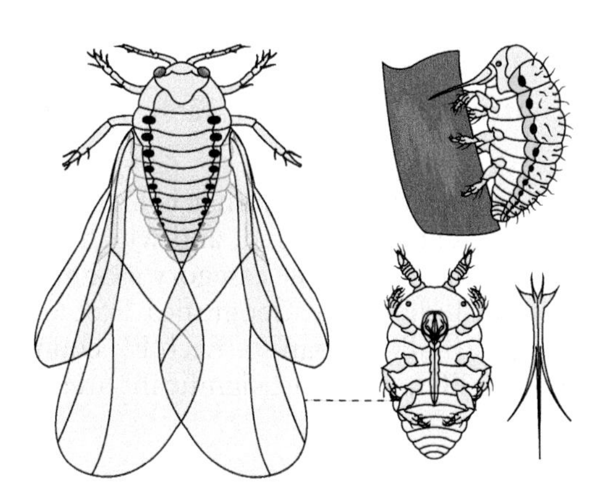

FIGURE 2.30
Phylloxerra (From *Meyers Konversations Lexikon* [1885–90], fourth edition). Meyers Konversationslexikon.

Phylloxera feeds on a vine's roots and leaves, causing them to starve and eventually the plant dies, driving the pests in search of new live hosts and spreading relentlessly through entire viticultural regions. It transports itself relatively easy through the soil, farm equipment, and the like, and has the capacity to reduce a grape crop by one-quarter in the first year and may render an entire vineyard infertile in only three years.

Many attempts were made at eliminating or at least slowing the spread of phylloxera. In 1869, Professor Gaston Barzille proposed that if French vine cuttings were grafted (or connected and allowed to grow together) onto the American rootstock, it might solve the problem of phylloxera destroying the vineyards. Barzille was correct, as most of the world's vineyards are now grafted in this manner. This laborious solution became the only one that had any lasting effect since it was found that native American rootstocks had evolved with the bug and developed thicker and tougher root bark and became relatively resistant to damage. Once the graft took place, the vine grew the European fruit with the benefits of the hardy resistance in the roots.

Modern Wine Laws: The European Union

The European Union (EU) is the world's largest wine economy with twenty-seven EU member states producing wine to some extent. Each state maintains its individuality with its own language, traditions, and wine classifications. However, the EU's best interest is to also encourage consistency across the entire economic zone with wine quality classifications and production laws. In 2011, they have created two important designations that existing individual wine classification systems can fall under:

1. PDO (Protected Designation of Origin)
2. PGI (Protected Geographical Indication)

Although the PGI production rules are not as stringent as those applied to PDO wines, there are examples of PGI wines commanding more respect (and higher prices) than their PDO counterparts.

The Protected Designation of Origin (PDO) category is named Appellation d'Origine Protégée (AOP) in French, Denominazione di Origine Protetta (DOP) in Italian, and Denominación de Origen Protegida (DOP) in Spanish. Each EU country has its own existing quality categories that correspond to PDO. The most significant are:

- *France:* AOC (Appellation d'Origine Contrôlée)
- *Italy:* DOCG (Denominazione di Origine Controllata e Garantita) and DOC (Denominazione di Origine Controllata)
- *Spain:* DOCa (Denominación de Origen Calificada) and DO (Denominación de Origen)
- *Portugal:* DOC (Denominacão de Origem Controlada) and IPR (Indicação de Proveniência Regulamentada)
- *Germany:* "Prädikatswein" (formerly known as "QmP" or Qualitätswein mit Prädikat) and QbA (Qualitätswein bestimmter Anbaugebiete)
- *Austria:* DAC (Districtus Austriae Controllatus) and Qualitätswein and Prädikatswein

The Protected Geographical Indication (PGI) designation is one that is linked to the geographical area in which it is produced, processed, or prepared, and which has specific qualities attributable to that broader geographical area. The category is named Indication Géographique Protégée (IGP) in French, Indicazione Geografica Protetta (IGP) in Italian, and Indicación Geográfica Protegida (IGP) in Spanish. Each EU country has its own quality categories that correspond to PGI. The most significant are:

- *France:* VDP (Vin de Pays)
- *Italy:* IGT (Indicazione Geografica Tipica)
- *Spain:* VT (Vino de la Tierra)
- *Portugal:* VR (Vinho Regional)
- *Germany:* Landwein
- *Austria:* Landwein

Wines of the New World

Learning Objective 3
Provide an overview of the Judgment of Paris

The foundation of the American wine industry began in California when Father Junípero Serra began spreading the Christian faith on behalf of the Spanish missionaries. His work led him to travel north from Mexico and eventually in 1776, set up the Chapel at Mission San Juan Capistrano located in Southern California. The *Criolla* or "Mission grape" was first planted at the chapel's vineyard in 1779 out of necessity for sacramental purposes. In 1783, the first wine was produced in Alta California from the Mission's winery.

Prior to 1919, New York, Missouri, Michigan, Pennsylvania, Ohio, Iowa, and North Carolina were well known for their wines as early as the early 18th century. Thomas Jefferson (d. 1826), one of the founding fathers of the United States, had attempted to grow vineyards and produce wine in Virginia before the Revolutionary War in 1775. Jefferson promoted American wine quite enthusiastically and was a noted wine connoisseur. As Secretary of State (1789–1793), he selected the wines for President George Washington's table and is distinguished for keeping the cellars of the White House stocked with wine. When Jefferson became the third president of the United States in 1801, he maintained extensive vineyards at his personal residence at Monticello, just outside Charlottesville, Virginia. Thomas Jefferson was one of the earliest and outspoken advocates for the beverage industry. He considered wine to be, as he once said, "a daily necessity."

Nicholas Longworth (d. 1863) is considered the founding father of American wine. He owned the first commercially successful winery in the United States in (of all places) Cincinnati, Ohio. Longworth's accomplishments went on to inspire a generation of grape growers. Longworth experimented with hundreds of different grape varietals and several vine species in his attempts at making wine an egalitarian beverage. He is best known for his sparkling Catawba (a hybrid grape varietal). By the mid-1850s, he was producing nearly 100,000 bottles annually according to Paul Lukacs, author of *American Vintage: The Rise of American Wine.* Longworth never conceived of wine as an elitist, aristocratic beverage, just another form of an agricultural product.

Indirectly, the California gold rush was a defining moment in forming a critical foundation for the wine industry. Gold was first discovered in modern-day El Dorado County in Sierra foothills in January 1848. This drew vast numbers of immigrants and created a large immigrant population of Germans, Italians, and Asians. As grapevines were planted during the gold rush, each of these immigrant groups became either winemakers or vineyard workers. By 1852, as the gold dried up, population in and around San Francisco had already surged. It didn't hurt that abundant sunshine coupled with low rainfall and mild winters could provide motivation to stay. The region of North Coast (home of Napa Valley and Sonoma County) ultimately became the anchor to America's most important wine production area. Figure 2.31 depicts the major California wine regions.

Grape growing and wine production spread and continued to expand throughout the United States until Prohibition in 1920 placed a long-lasting dampening effect upon the wine industry. When Prohibition was repealed just shy of 1934, few people had the financial capital or technical experience to resume production. In addition, most states continued to enforce Prohibition locally, and World War II delayed the full return of production levels that existed prior to the war. Large bulk wine producers such as Almaden (California's oldest winery, established in 1852) and E & J Gallo focused on quantity jug production meeting a demand for the consumer's desire for easy-to-drink, sweet wines. At this point, wine production of the United States had for the most part been an afterthought.

The late 20th and early 21st century brought considerable change to the world of wine with the emergence of "New World" wine producers. There was a growing interest around the world for higher quality domestic options. As the 1960s approached, American tastes and attitudes toward wine were beginning to change as new consumers started to approach wine as something sophisticated. Robert Mondavi (d. 2008) was one of the most influential winemakers as he brought worldwide recognition to California wine. From an early period, Mondavi assertively

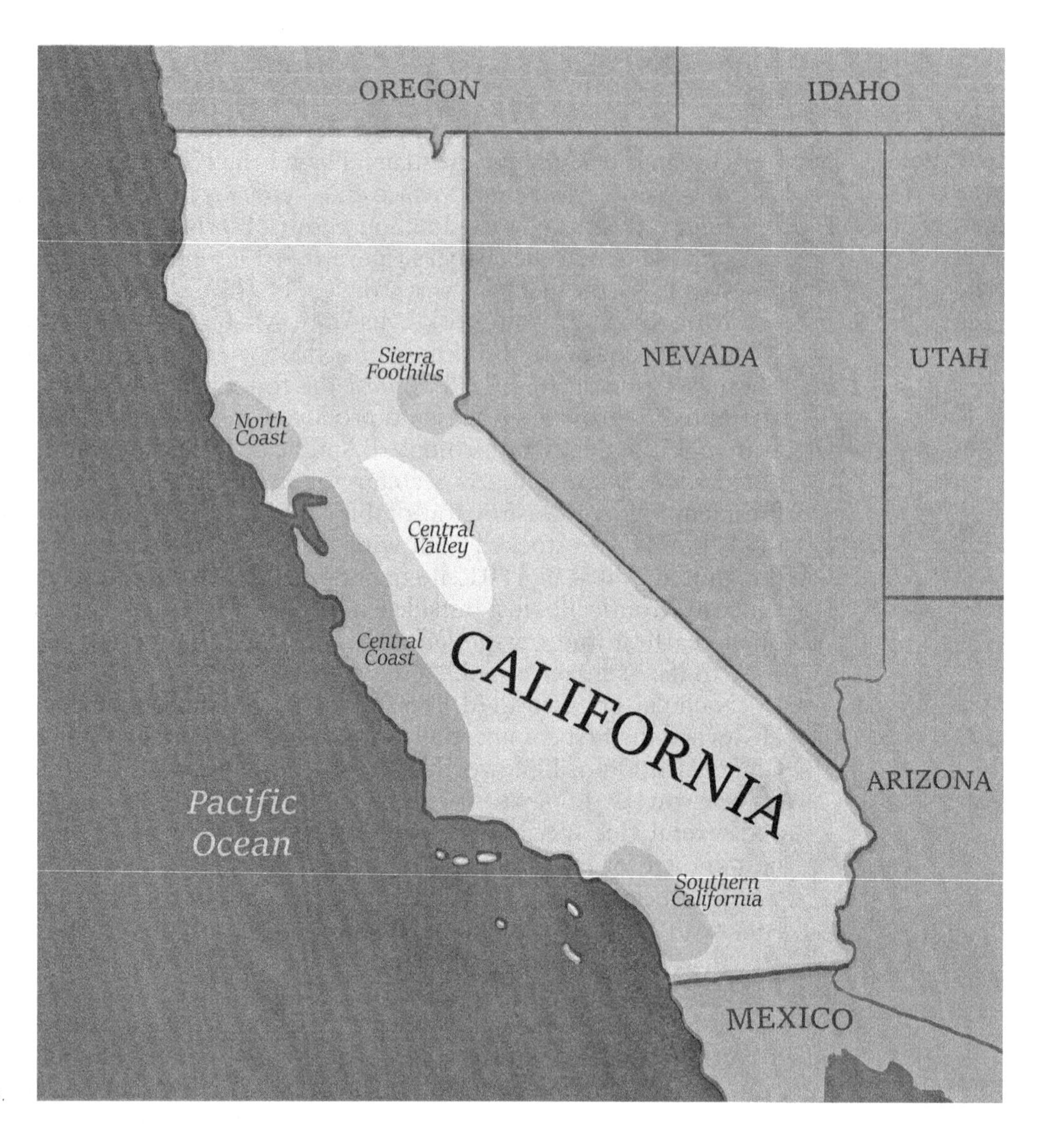

FIGURE 2.31
Wine map of California.
Courtesy of Thomas Moore.

promoted varietal-based labeled wine as opposed to generically labeled as was the norm in the 1950s. Robert Mondavi Winery in Oakville (within Napa Valley) was the first major winery built in 1966 since post-Prohibition. For decades, Mondavi went on to serve as a source of innovation and inspiration.

The Judgment of Paris: The 1976 Tasting

As time went on, a shift from jug, generic wines to varietal-based wines started to become more favorable. It is believed that wine production did not return to pre-Prohibition levels until 1975. The wine industry in the United States would not recover for over forty years until the famous "1976 Judgment of Paris," the famous wine tasting event that shocked the world and became the significant defining point for the American (and, for the most part, the entire New World) wine industry. This tasting was held in Paris and organized by a British wine merchant, *Steven Spurrier.* The competition was judged by nine French wine professionals that involved blind tasting and scoring the quality of ten French and Californian Cabernet Sauvignon wines and ten California and French Chardonnay wines. The American winners were *Warren Winiarski's* Cabernet Sauvignon from Stag's Leap Wine Cellars and *Mike Grigich's* Chardonnay from Chateau Montelena. These wines won both of their respective categories over their prestigious French counterparts. Although Spurrier had invited numerous members of the press to the tasting, the only reporter to attend was George M. Taber from *Time magazine*, who quickly revealed the outcome to the world.

French Paradox

In the 1980s, medical studies found a paradox in that people of France who had a diet higher in saturated fat also had a low incidence of heart disease. The study concluded that people who consume moderate amounts of red wine are less likely than nondrinkers to suffer from cardiovascular disease. One of the phenolic compounds found largely in grape skins is resveratrol (rez-VEHR-ah-trawl) that has beneficial effects on cholesterol levels and cancer preventative qualities.

Development of Appellation Areas

In 1978, the United States implemented the officially designated grape-growing areas—American Viticultural Areas (AVAs). These geographical designations are intended to be a means of showcasing a grape-growing area's distinctions from another. More specifically, for a potential AVA to be approved, the U.S. government entails that evidence exists that its growing conditions, such as climate, soil, and topography, are distinctive. American wine labels may identify a grape-growing location or official AVA when a minimum of 85 percent of the grapes used for the wine come from the location identified on the bottle. The first and second AVAs recognized were surprisingly Augusta, Missouri, in June 1980 and then Napa Valley, California, in January 1981. Figure 2.32 is a wine label depicting one of the most famous and reputable AVAs in the United States.

As of March 2016, the TTB had recognized 234 AVAs throughout the United States with the majority being documented in California with 138 AVAs. Lately, there have been an increasing number of sub-appellations designated to showcase even further distinction and specificity in a growing area. For example, there are approximately 16 sub-appellations within the larger Napa Valley.

The American wine industry has evolved quite a bit since its early days. As of 2014, the Wine Institute states that there are 10,417 wineries in the United States—up from 440 back in 1970. In 2014, there were 4,285 in California with each of the fifty U.S. states growing grapes and producing wine—even Texas and Florida!

FIGURE 2.32
Napa Valley Label. Courtesy of Erika Cespedes.

Wine Defined

Learning Objective 4
Understand the basic components of a grape and their contributions to a wine

Wine is, very simply, an agricultural product derived from the fermented juice of grapes (unless otherwise specified on the label). The sugar source comes from the pulp, found within the grape, which contains juice and sugar along with acids and chemical compounds that go to create the essential personality of the wine. During fermentation, yeasts interact with sugars to create ethyl alcohol (also known as ethanol) and carbon dioxide (as a by-product). *Saccharomyces cerevisiae* yeast is a single-celled organism that lives and thrives on simple sugar. There are endless varieties and strains, and many strains exist naturally (known as wild or ambient yeast); however, more often, cultured yeast is used due to its predictability and reaction during the fermentation process.

The Personality of a Wine

Wine can be made from a single grape varietal or a blend of different complementary grapes. The grape variety, or blend of grapes, will impart a specific style or personality to a wine. Table grapes are a basic agricultural product that yields about $1.50 per pound—wine grapes can deliver a complexity that can yield anywhere from $3.00 to $16,000 a bottle, and sometimes even more. There are some 10,000 different grape

FIGURE 2.33
Zoetic winery red grapes. Copyright David Vance.

varietals used for winemaking; however, most beverage establishments rely on about 30–40 significant ones. The most significant six grapes (white wine grapes include Sauvignon Blanc, Riesling, and Chardonnay, and red wine grapes include Pinot Noir, Merlot, and Cabernet Sauvignon) are what all wine stores, restaurants, and bars build the wine selections around. Out of the big six, it is Cabernet Sauvignon and Chardonnay that are the most important wine grapes throughout the world. If Cabernet Sauvignon is the king of red grapes, Chardonnay is the queen of white wine grapes. These varieties are widely planted around the world and far and away remain the most popular ones in the United States. Figure 2.33 depicts some Cabernet Sauvignon grapes hanging on the vine from the vineyards of Zoetic winery.

The personality of a wine is derived from several different stages throughout the production process, wherein thousands of different compounds influence its aroma, flavor, body, and more. This is what makes each grape unique, with its own set of performance characteristics. Some of the personality-influencing compounds are inherent in the original grape and its juice; others are created during fermentation or through processing or aging techniques.

- *Grape(s):* An influence of the core constituents of the wine
- *Location:* An influence of the geographical origin of the grapes
- *Winemaking Techniques:* An influence of the person and techniques used to transform the grapes into wine

Grape Composition

Grapes are the most important raw material used for making wine and creating its identifiable personality characteristics. The grapes (or berries) make up a cluster or bunch and are attached to a stem. The essential parts of the berry include three major components: the pulp, skin, and seeds/stems.

1. **Pulp** Pulp is located on the inside of the grape, where the juice (containing acid, sugar, and flavonoids) can be found. Once the berry has been pressed, the juice is commonly referred to as free run. It makes up approximately 75 percent of a grape by weight and plays a major role in providing not only the sugar for yeast but also the acid, which is present in the juice, and is pivotal to giving both red and white wines a lively structural sensation. A wine without acid falls flat on the palate and has a difficult time standing up to food when they are paired together. When a wine has inadequate acid, it is frequently referred to as flat or flabby.
2. **Skins** Skin is located on the exterior of the grape, where the tannin, flavor, and color can be found. It makes up approximately 20 percent of the grape by weight and plays a significant role in the style and structure of a red wine, which are achieved when the skins are allowed to ferment with the juice. Anthocyanins and other natural pigment chemicals found in the skins of red wine grapes are responsible for contributing the color to a wine.

   Grape skins also contribute tannin to a red wine. The tannin (a natural chemical compound) is pivotal to providing a red wine with good structure and aging potential. Tannin is a compound that causes the same dry feeling on the tongue and around the gum line that one feels after drinking black, heavily steeped tea. Tannin content varies with grape variety and wine style. Like a wine with insufficient acid, a red wine without adequate tannin falls flat on the palate and is often referred to as flat or flabby. Figure 2.34 depicts a cluster of red wine grapes.

FIGURE 2.34
Bunch of red wine grapes. Courtesy of John Peter Laloganes.

3. **Seeds and Stems** Seeds are found inside the grape while stems are found on the outside. They contribute approximately 5 percent of the grape by weight. If handled poorly, or if the grapes were not allowed to ripen properly, both stems and seeds may contribute to an overly bitter component if crushed or used in excess and can have a pronounced negative influence.

Colors of Wine

Wine grapes are generally described as either white or red grapes, which can go on to make a white, rosé, or red wine. Since a wine's color is derived from their skins, the fundamental difference between a white wine versus a rosé and a red wine is due to the treatment of the grape skins in connection and in contact with their juice.

White wine grapes are not actually white, but any shade between green and an amber-yellow, and in a few cases, a light pink in color. White wines are typically made from these "white" wine grapes, but can be made from any red grapes if the skins are removed from the juice to prevent any color extraction. Therefore, white wines are fermented without their skins.

Rosé wines are made either from a blend of red and white wine or from the more common process known as the French saignée (san-YAY) method, allowing some of the color from red grape skins to bleed into the fermenting juice, creating a pinkish color. Rosé wines are fermented with the skins of grapes for a short period of time from several hours to at most a couple of days.

Red wine grapes are generally not actually red, but instead can range from blue to a deep purple-black. Red wine can also be made from a blend of red and white wine grapes, but most often not. The red grape skins remain with their juice and pulp during fermentation for a period of approximately two weeks to extract maximum color, flavor, and tannin from the skins.

Categories of Wine

For the purpose of simplicity, all wine can be categorized in one of three ways. It could be categorized as a table (or still) wine, sparkling wine, or fortified wine.

1. **Table/Still Wine** This category of wine gets its name because it's made to be drunk at the table with meals, as well as a wine without the presence of bubbles. The alcoholic content of table wine generally is between 8 and 15 percent. Table wines are white, pink, or red wines that can be vinified to be dry, sweet, or somewhere in between. Figure 2.35 illustrates the three colors or types of table/still wine.
2. **Sparkling Wine** This category of wine consists of a table/still wine as the base, with the addition of large amounts of CO_2 for carbonation. Sparkling wine typically contains between 10 and 13 percent alcohol. The most prestigious of all sparkling wines is Champagne. However, many other sparkling wines of varying quality are produced throughout the world and can rival the excellence of Champagne.

FIGURE 2.35
Trio of wines. Courtesy of Erika Cespedes.

Sparkling wine can be made as a white, rosé, or red wine, and can be transformed into a dry, off-dry, or sweet wine—most sparkling wines are white and made into a dry style. It's possible to find sparkling wine options of varying levels of sweetness, particularly from Italy and Germany.

3. **Fortified Wine** This category of wine consists of table wine as the base, with additional alcohol in the form of a distilled spirit—often an unaged brandy. Therefore, fortified wine typically contains between 15.5 and 22 percent alcohol and can be transformed into dry, off-dry, or sweet wine. If a fortified wine is consumed prior to, or in the beginning of a meal, it is known as an apéritif (if dry or bitter)—the fortified wine served at the end or after the meal is a digestif (if it contains some sweetness).

Wines of the Old and the New World

Learning Objective 5
Identify the famous Old and New World wine production areas

In many of the Old-World wine regions, viticulture and vinification date back centuries with the Phoenicians, Greeks, and Romans establishing some of the earliest vineyards. Grapes and wines have been traded internationally since ancient times. Many grapes that are considered "home" in Western Europe were transported through ancient trade routes from the Eastern Mediterranean and the Black Sea and brought to their new "spiritual" homes. More recently in history, the world of wine has expanded from its European origin to new possibilities in the Far West and the southern hemisphere of Australia, New Zealand, Chile, Argentina, and South Africa. The map in Figure 2.36 identifies the significant wine-producing locations within both the Old and the New World.

There are two broad schools of thought and practice in the wine world, and they are identified by broad geographical concepts—the Old World and the New World. These terms are used to identify an obvious geographical distinction—but often a philosophical distinction as well. These differences (generally affiliated with viticulture and vinification methods) lead to perspectives that may stylistically affect the personality of the wines. What follows are broad generalizations between the two worlds that can assist the novice to intermediate wine consumer in the broad understanding of wine concepts and styles.

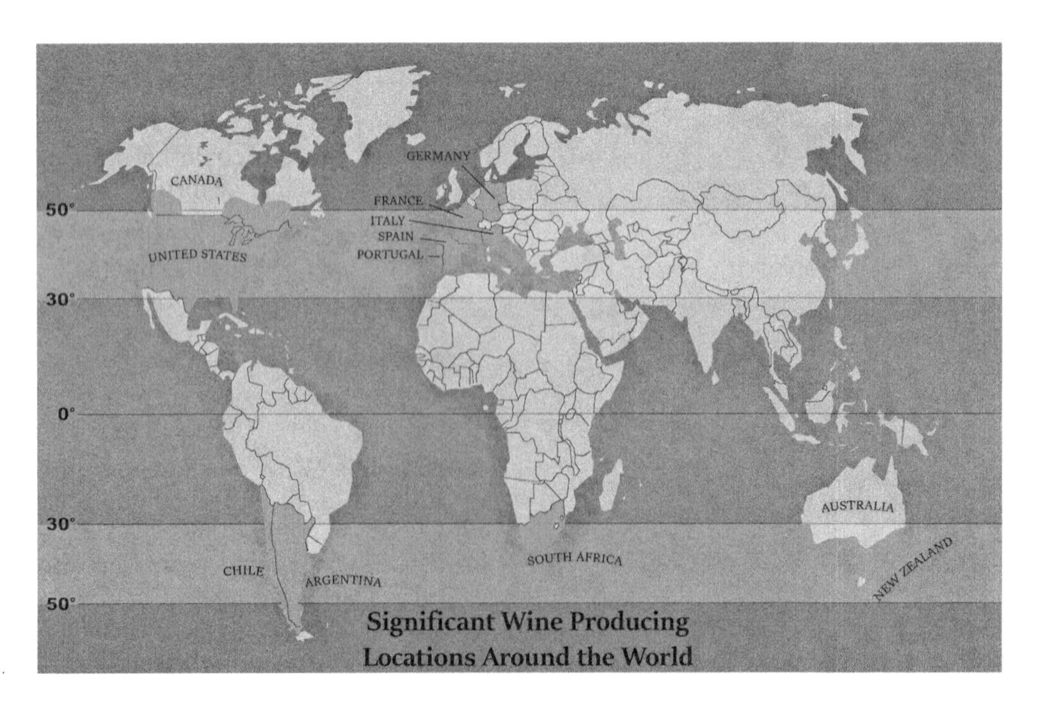

FIGURE 2.36
Wine map of the world.
Courtesy of Thomas Moore.

Old World: Primarily Inside Europe

The Old World references the long-established tradition of winemaking within the European countries of France, Italy, Germany, and Spain but can also include other countries located around the Mediterranean basin. These countries have a long history of growing grapes and making wine. They are largely responsible for the nurturing and development of the grapevine. In addition, many of the winemaking techniques practiced in these countries helped to form the foundation for the modern wine industry.

Tradition and *Terroir* (a French word for "a sense of place") are two significant and defining influences for the Old World. While tradition refers to collecting several hundreds or even thousands of years of refinement, terroir communicates the unique sense of location that cannot be duplicated elsewhere in the world. Many Old-World wine producers believe in these concepts so passionately that majority of their wines are labeled according to the origination of the grapes (geographical labeling) as opposed to the popular New-World method of varietal labeling.

New World: Primarily Outside Europe

The New World references the significant countries that have a relatively brief history and culture associated with grape growing and wine production. In the New World, grapevines arrived by way of European settlers through immigration, exploration, trade, and war. The significant New-World wine-producing countries include the United States, Australia, Argentina, Chile, South Africa, and New Zealand. These countries were, at the earliest, settled within the last 500 years or so.

Unlike the Old World, New-World wine producers aren't bound by tradition—instead they emphasize *science* in the vineyards and wineries. They offer a *freedom of legalities* and a sense of a somewhat *renegade spirit*. These different perspectives can radically alter the style of wine in comparison with what may be produced in the Old World. Instead of trying to replicate a style reminiscent of the Old World, climate variations (beyond the philosophical ones) will also play a role in distinction. New-World vineyards are generally in warmer climates coupled with New World's love affair with *hang-time*, a technique that intentionally leaves the grapes on the vine for an extended period. Hang-time leads to riper fruit, and higher sugar content that leads to a "fruit forward" wine with ample alcohol. This process creates a bolder and richer style of wine than their European counterparts. Being less dependent on geography, New-World wines have placed more emphasis on branding the *varietal* as a marketing tool. This is evidenced by the grape's often prominent identification on the wine label. As the New-World winemaking has evolved, winemakers have made a greater effort with thoughtful site selection—the practice of matching appropriate grape varietals to a given location. This practice is evident as winemakers list more precise origins of the grape-growing areas on wine labels with their typical and obvious use of varietal labeling.

Stemware

Learning Objective 6
Recognize the distinctions between stemware for the different categories and types of wine

Wine glasses are often referred to as stemware or glassware. They are constructed with a focus on three crucial parts: bowl, shape, and the rim. It's generally believed that the construction and shape of a glass can significantly impact both its aesthetics and functionality. These variables of glassware can significantly improve the personality characteristics of the wine that is poured in it. Some companies go as far as to create specific shapes suited for different grape varietals, in addition to creating several price points based on varying production methods of quality. Technically speaking, stemware can be made from glass or crystal. Glass provides a more durable and inexpensive alternative to crystal stemware, which is delicate and expensive, but viewed as better quality. In handling stemware, it's important to hold the glass by the stem to avoid smearing the glass with fingerprints and to avoid warming the wine.

Varietal Specific Stemware

Various kinds of red wine glass shapes are available that are tailored to the specific types of wine (or grape varietal) being served within them. According to Riedel, an 11th generation Austrian glass company, "The same wine displays completely different characteristics when served in a variety of glasses." Varietal specific glassware is designed in a way that relates to the essential DNA of the wine category and/or grape variety. From firsthand experience, an identical wine can display varying characteristics when served in different types of stemware.

Generally, the more luxurious high-end restaurant will be likely to invest in this additional glassware, definitely signaling a wine-focused establishment. The "Bordeaux glass" shape is characterized with an expansive bowl and large surface area. It is made specifically to catch and hold aromas as well as promote a great degree of aeration, helping the drinker maximize the enjoyment of the wine. This glass is designed for bold red wines that historically originate from the French region of Bordeaux with grape varieties such as Merlot and Cabernet Sauvignon. The "Burgundy glass" is made specifically for the Pinot Noir grape varietal that originated from the French region of Burgundy. This glass works in much the same way as the Bordeaux glass but the design of this particular glass is intended to direct wine to the tip of the tongue to de-emphasize the wine's dryness and allow for greater expression of the wine's fruit aromas and flavors.

Broadly speaking (and for simplicity), wine glasses can be divided into four types: red wine, white wine, sparkling wine, and fortified wine. Additionally, wine tumblers (without stems) are also increasing in popularity but negate the significant purpose of the stem—to minimize temperature fluctuation and limit smudging of the glass. Generally, for most wines, the glass should be considerably larger than the desired volume it's intended to hold. Except for sparkling wine, the actual amount of wine poured in a glass should be half (or less if possible) of the glass's capacity. The extra capacity is ideal to properly swirl and release a wine's desirable volatile aromas into the nasal cavity of the taster.

The wine category and type and the grape variety are the key factors in determining the best glass; however, the operation should consider the large investment necessary for the variety and types of stemware available in today's beverage industry.

Still Wine Stemware

Many restaurants utilize an all-purpose wine glass that can be used for both white and red wines. For those restaurants that choose a more extensive wine list or have the budget to allow for better glassware, they can opt for the minimum white wine and red wine glasses. Some high-end restaurants with extensive wine budgets carry several different styles of stemware for individual grapes or types of table wine. Not only is there a variety of various shaped glassware, but there are various quality levels as well. Crystal stemware is made with lead to provide a higher index of refraction than normal glass affords. It gives the wine a greater "sparkle" at a higher price.

An effective size for wine glassware should be large enough (about 10–24 ounces) to allow for the standard portion size of wine (about 5–6 ounces) to be swirled in the glass for increased introduction of oxygen without being spilled. Figure 2.37 illustrates a collection of wine glasses mostly used for table/still wines.

FIGURE 2.37
Glassware for table/still wines. Courtesy of John Peter Laloganes.

FIGURE 2.38
Sparkling wine glasses. Courtesy of Erika Cespedes.

- *White Wine Glassware:* White wine glasses are tulip shaped, with a small to moderately sized bowl, generally narrow, and with a slightly inverted tapered lip, which allows for enhanced aroma concentration of a white wine's delicate nuances after being swirled. The narrow surface area also allows the wine to retain its chilled temperature by reducing surface area.
- *Red Wine Glassware:* Red wine glasses are usually characterized by their large rounded bowl and wider surface area, which allows the wine to have greater air contact to cause the softening of tannin and integration of aromas.

Sparkling Wine Stemware

These wines exude elegance; they are often perceived to be the epitome of femininity and therefore ideal for consumption in the long and delicate, slender flute glass. Sparkling wine glassware should be designed to maximize the idea of what a sparkling wine is about—its bubbles. But just like still wine, it's generally believed that the construction and shape of a glass can significantly impact both its aesthetics and functionality. Figure 2.38 depicts the three types of sparkling stemware.

- *Flute and Tulip:* The flutes and tulips are the most suitable stemware for all types of sparkling wine, as they are tall, slender, and designed to bring the delicate aromas toward the nose. The length of the flute allows the preservation of carbonation as it slowly rises to the surface, though its narrow surface opening can challenge the sommelier in a dimly lit dining room coupled with poor depth perception.
- *Saucer/Coupe:* The saucer glass (sometimes called coupe) is considered an inferior option from a functional standpoint as it causes a faster rate of bubble dissipation. In addition, it has a large surface area that makes it difficult to drink from (yet easier for the sommelier to pour) without spilling the wine. However, its appearance can convey an appealing French Renaissance feel.

Fortified Wine Stemware

Fortified wine is a table wine to which a distilled spirit has been added to increase its alcohol content to an amped up 18–20 percent by volume. The act of fortifying with additional alcohol renders the wine more suitable for smaller portions to imbibe without the overindulgence and debilitating effects of intoxication. Therefore, their portion size (about 2 ounces) is much smaller than other types of wine. For practical and functional purposes, the fortified wine glassware is corresponding with its portion size—generally small and capable of holding about 2–4 ounces. It resembles a miniature white wine glass. Another variation is the Spanish tulip-shaped glass called a *copita* (koh-pee-tah) that is most appropriate for consumption of Sherry, Spain's most famous fortified wine. Both fortified wine glasses work toward contributing their aroma-enhancing narrow taper toward the rim.

Labeling Wine

Learning Objective 7
Distinguish the four methods of labeling table wine

Wine labels provide customers with useful information in regard to the contents within a given bottle. The labels assist the buyer in selecting a wine that meets or doesn't meet their needs—brand name or producer, grape varietal, origin of the wine, vintage date, possible vineyard designation, and so on. Different countries have different standards for label information; generally, the labels from United

States are relatively straightforward and easy to understand as compared to labels from most of France, Italy, and Spain. In the United States, wine labels are regulated by the Alcohol and Tobacco Tax and Trade Bureau (TTB) of the U.S. Treasury Department. Law mandates most of the information offered on a label, but additional information is sometimes provided to assist the consumer with making a well-informed decision.

For ease of comprehension, wine can be categorized according to three types—table wine, sparkling wine, and fortified wine. Table wine could possibly have acquired its name from Old-World Europe, suggesting the cultural concept that wine is meant to be drunk with food—at the table with a meal. Table wine is the category of wine (separate from sparkling and fortified wine) that is most commonly consumed and potentially tricky to understand. Largely, this confusion begins with the numerous approaches to labeling table wine. Table wines can be labeled per one of four methods: (1) varietal-based labeling; (2) geographically based labeling; (3) generic labeling; or (4) the proprietary-based labeling.

Varietal-Based Labeling

The varietal-based labeling approach is applied to most non-European wine labels, including those from Argentina, Australia, New Zealand, South Africa, South America, and the United States. However, parts of France, Italy, Spain, and most of Germany utilize some varietal labeling. The names of wines in this category are derived from their predominant grape variety. For example, the wine will have a prominent wording such as "Pinot Noir" or "Chardonnay." Figure 2.39 depicts a varietal-based label on a bottle of Tablas Creek Vineyard, *Roussanne*, from Paso Robles, California.

All wine-producing countries and regions regulate the minimum amount of a particular grape within each bottle. Some wines (certainly not most) are a combination or blend of two or more compatible grape varietals. Throughout the United States, any wine with a designated grape varietal must contain at least 75 percent of that grape within the bottle. Legally, the label only has to reveal the name of a single grape varietal if at least 75 percent (the U.S. federal minimum requirement, but often 85 percent in the rest of the world) of the wine is made from that particular grape variety. The other 25 percent of the wine can be made from one or more complementary grape varietals and those do not have to be listed. However, some U.S. winemakers voluntarily list all grape varietals used in a wine blend on their label. Furthermore, some producers may also identify the percentages of each varietal on the back label. Each state within the United States can choose to be stricter with this rule. For example, most varietals in Oregon must contain minimum of 90 percent (for most varietal wines) of the grape identified on the label.

FIGURE 2.39
Varietal label of Roussanne. Courtesy of John Peter Laloganes.

Don't be mistaken—varietal labeling identifies the geographical origin of the grapes as well—it's just not the most prominent feature on the label. Currently, though, winemakers among the New World are showing a greater interest in specifying very precise locations. As non-European countries establish reputations for the wines of certain regions, they often add the more specific precise section of the growing area in combination with the name of the varietal. Examples are Cabernet Sauvignon from Oakville (which is located within Napa Valley) and Pinot Noir from Russian River (which is located within Sonoma County), California.

FIGURE 2.40
Geographical label of Sancerre. Courtesy of John Peter Laloganes.

Geographically Based Labeling

The geographically based labeling method applies to most European (Old World) wine labels, where the wines are named after the place the grapes were grown. Europe's wines are intimately linked to their terroir (geographical origin) concept rather than to the name of the grape variety. They often believe that the grape used is the conduit for expressing the sense of place, therefore placing greater emphasis on geographical location. For example, the label may read "Burgundy" (a French region that specializes in Pinot Noir) rather than identifying the name of the grape. To further complicate the geographical nature of labeling, locations can be broad, such as the identification of a region, or as precise as a vineyard. A general rule that can assist with reading and understanding a label is to identify how much specificity is given on that label. A broader place on the label generally indicates less quality—the more specific label indicates better quality. Figure 2.40 depicts a geographically based label on a bottle of Sancerre Blanc (100 percent Sauvignon Blanc) from Loire Valley, France.

Generic Labeling

This labeling method has a reality of poor quality wines that became widely popular throughout much of the 20th century. It wasn't until the mid-1980s that varietal labeling began to overtake and replace generic wines based on consumer demand for more authentic and quality-oriented wines. These wines have been commonly referred to as jug wines because they were often sold in a large jug or box. Although this labeling practice has diminished, these wines still exist in the United States—largely sold at the local drug or grocery store.

Producers of these kinds of wines have freely "borrowed" the names of world class European regions to label their jug or generic wines, rather than give them grape varietal names, as is the more common and higher quality practice throughout the United States. Generic wines generally consist of a blend of different nondescript grapes that often are of lower quality. In most cases, the grape(s) is not even known, because these wines may be labeled so such as, "Burgundy," or "Chablis"; all are names based on famous wine-growing regions in the Old World, although these wines are clearly unrelated to the wines from those coveted European designated areas. Figure 2.41 depicts a collection of jug/generic wine.

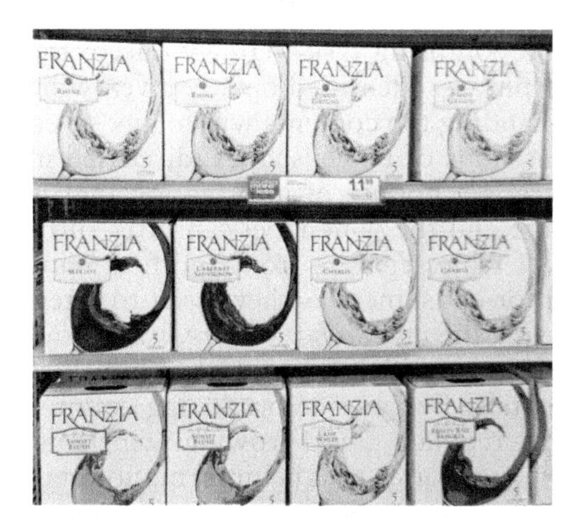

FIGURE 2.41
Generic labeling. Courtesy of John Peter Laloganes.

Proprietary-Based Labeling

Some select wine producers have been applying a proprietary-based labeling approach—one that uses a branded name under the producer's portfolio of products. Instead of the typical varietal or geographical labeling, these producers have chosen to create and market alternative names that sound prestigious or unique to the winery. These wines are labeled in a manner that allows for maximum freedom and creativity. Many of these proprietary blends have opted to create wines through blending various complementary grape varietals to distinguish their wines from the traditional varietal-based wines. Many (but not all) of the proprietary-based wines are respectable high-quality American versions of the classic French red

Bordeaux style. These wines (as they are in the French region of Bordeaux) are a blend of Cabernet Sauvignon, Merlot, and Cabernet Franc grape varietals. Some producers have also opted for other types of blends as the popularity of white and red Rhône-style wines has increased dramatically.

Wines labeled as per the proprietary manner are less subject to the somewhat limiting legal requirements of location and varietal labeling methods. Sometimes a proprietary name may refer to an entire estate such as in "Harlan Estate" or "Opus One" wineries or other producers may opt for an element of uniqueness such as in "Papillon" (meaning butterfly in French) that is produced by Orin Swift. The producers of these brands hope to gain the consumer's acceptance over time as they associate quality and prestige with their product. The types and kinds of proprietary blends are endless—while they don't offer any clues to the contents within the bottle, their "concept" adds an element of intrigue.

These wines are often made with an intention of high quality and can be truly a unique expression of the winemaker's artistry. Craig Williams, winemaker at Joseph Phelps Vineyard, was the first winemaker given credit for producing the first proprietary red—Insignia—in 1974. These wines in many cases are a winery's flagship option, and prices often start at around $80 to $200+ a bottle. In addition to possessing unique and clever names, they also often have stylish labels and are sold in large Bordeaux-type bottles to convey a sense of power and prestige.

Meritage wines' (rhymes with the word heritage) intention of high quality can be truly a unique expression of the winemaker's artistry. The name Meritage is a combination of two words, "merit" and "heritage," to symbolize the quality and history associated with the origination of these wines made in a Bordeaux style. According to the Meritage Association, the wine must contain at least two of the approved grapes (classic Bordeaux), with no single variety constituting more than 90 percent of the blend. The approved grape varietals for red Meritage include Cabernet Sauvignon, Merlot, Cabernet Franc, Petit Verdot, and Malbec. As long as the basic requirements are met, the combination and proportions of these grapes are completely determined by the individual producer. However, most often, the wines tend to be dominated by either Cabernet Sauvignon or Merlot, with smaller amounts of the other approved varietals.

Reading a Typical Wine Label

Learning Objective 8
Discover the five most significant pieces of information on wine labels

Selecting a wine can be quite a daunting and confusing task—largely due to the foreign (literally and figuratively) terms and phrases identified on any given wine label. An effective approach to begin understanding the contents within any given bottle of wine can commence with the distinctions of bottle shapes (discussed in one of the following chapters). Second, there is a bounty of information that is located on a wine label. Both the bottle shape and the wine label can offer sufficient clues for the buyer to discern a style of wine separate from another. Wine labels are important sources of information for consumers as they assist to interpret the contents within the bottle by identifying the *who*, *what*, *where*, *when*, and potentially *how* of the wine. All those clues can provide a great deal of information about the wine's personality characteristics. Every wine-producing country has its own set of government wine laws that regulate grape growing, winemaking, and labeling; however, the following five categories of label information will often be found on all labels, willing to provide basic clues to understand a given wine. Law mandates most of the information offered on a label, but additional knowledge is sometimes provided to assist the consumer with making a well-informed decision.

Typical New-World Wine Label

New-World wine labels are often more simple to decipher than Old-World labels. With many New-World wines, the grape varietal is likely to be the prominent identification on wine labels. In addition, the grape's geographic origin will be listed to provide clues about the contents within the bottle. In many Old World wine-producing countries, there are some select locations that label the grape variety in a prominent manner.

Keep in mind, when percentages are identified, these are federal minimum requirements, and therefore, each state can choose to create more strict requirements. The most prominent items on a varietal-based wine label are the following:

- The *WHO* is the name of the winery or vintner that produced the wine. The producer's name usually is the largest text on the label and the easiest element to identify.
- The *WHAT* addresses the type of grape(s) the wine is made from. In the U.S., a wine predominantly from a single grape varietal identified on the label means that at least 75 percent of that varietal is used in the wine. There can always be more than 75 percent as is often with the case of Pinot Noir and Chardonnay that tend to always be 100 percent because they like to single express themselves. But most often Cabernet Sauvignon and Merlot are commonly blended taking advantage of the play with being able to add 25 percent of whatever is complementary. When the wine is blended from several varieties and neither one reaches the required minimum, it will not be labeled as a varietal wine.
- The *WHERE* addresses the geographical location as to where the grapes were grown. Some growing areas are as broad as "California," whereas others are narrowly defined as a subsection of a mountain top. In the United States, if a legal American Viticultural Area (AVA) is listed, then at least 85 percent of the grapes (regardless what varietal or blend of grapes) must derive from the stated area. If a county AVA such as Sonoma, Lake or Mendocino County is listed, the 85 percent rule drops down to 75 percent. Furthermore, in the United States, vineyards can be named if a minimum of 95 percent of the grapes came from that vineyard.
- The *WHEN* addresses the vintage "year" in which the grapes were harvested and the wine was made. If a vintage year is displayed, it means that at least 95 percent or more of the wine was produced from grapes grown in the stated year.

Typical Old-World Wine Label

Old-World wine labels are often more challenging to decipher than New World ones. With many European wines, the geographical origin of the grapes is the only prominent identification on wine labels—not the grape variety as associated in the New World. The most prominent items on a geographical wine label are the following:

- The *WHO* identifies the name of the winery (often referred to as a vigneron) that produced the wine. The producer's name may be difficult to find on the label.
- The *WHAT* identifies the wine grape(s) but is generally not listed on the wine label in the Old World. In Old World France, Italy, and Spain, the primary source of labeling is based on the geographical approach. Therefore, the "what" is also the "where."
- The *WHERE* indicates the geographical location where the grapes were grown and could be very broad or incredibly precise as a specific vineyard. The degree of specified location may also offer clues as to the potential for quality of the given wine.

- The *WHEN* indicates the vintage date, which is the "year" in which the grapes were harvested and the wine was made. If a vintage year is displayed, it means that at least 95 percent or more of the wine was produced from grapes grown in the stated year.
- The *HOW* identifies the *level of quality* or some other notation or classification. Whereas these are not always listed or known, there may be clues that imply (but do not necessarily guarantee) quality. For example, a wine labeled with the term Grand Cru (or great growth) is held to specific geographical designation with legally required ranges of viticulture and vinification techniques. Theoretically, the presence of a "HOW" term indicates that some standards of production are regulated and could equate to a better-quality wine as opposed to a wine that doesn't have the right to label the Grand Cru term.

Other Potential Label or Marketing Terms

- *Alcohol Content (%):* This information is given a tolerance of the following that will be permitted either above or below the stated percentage on the label:
  - 1 percent, when a wine contains more than 14 percent of alcohol by volume
  - 1.5 percent, when a wine contains 14 percent or less of alcohol by volume
- *Estate Bottled:* The term estate bottled means that 100 percent of the wine came from grapes grown in a vineyard on land owned or controlled (or where they have significant control with long-term contracted growers) and that is adjacent to the winery estate. The winery must be located in the same viticultural area as the vineyard. The winery must crush and ferment the grapes and age and bottle the wine on their premises. An identical concept is coined in French as *Mis en bouteille au château* or *Mis en bouteille au domaine. Château* or *Domaine* are the French words for "castle" or "house."
- *Declaration of Sulfites:* Required on all wines that contain 10 or more parts per million of sulfur dioxide.
- *Reserve or Vintner's Reserve:* In the New World, this terminology doesn't have any legal meaning (with some exceptions in Argentina and Chile). In the United States, this terminology is purely for marketing purposes as it may or may not have meaning on behalf of the winery, although legally it has no definition. In the Old World, this terminology has special legal designation in Spain where the terms Crianza, Reserva, and Gran Reserva identify aging minimums.
- *Vieilles Vignes:* (vee ay veen-yuh) A French term for "old vines." Due to their naturally reduced yield, old vines should produce better-quality fruit with smaller berries and thicker grape.

Wine Service

Learning Objective 9
Perform key steps and etiquette in still wine and sparkling wine service

Appropriate beverage service should be provided each time a guest orders wine. It's important to not lose sight that wine service is carried out not much different than buying any other product—a business transaction that applies rituals and processes to make certain the accuracy of the customer's order and to ensure the health of any given wine. Any beverage establishment that chooses to offer wine, regardless of scope or depth of options, has numerous requisite considerations to first consider. It's imperative for the management to delineate and correlate wine service to the corresponding vision and mission of the restaurant concept. The vision and mission are foundational agents that dictate the level of formality and must be translated via the service staff to the customer experience. Careful planning and effective training should define what professional service is for any wine-focused establishment. Today's consumer has come to expect wine (or beer and spirit) service along with that of their meal. Wine endeavors to heighten the overall service experience, assuming it's conducted with appropriateness of the beverage concept in mind.

Wine Openers

The sommelier has a few special tools needed to perform the necessary etiquette and service rituals when opening and serving wine.

- *Corkscrew:* The sommelier's primary tool is a corkscrew, which is used to remove a cork from a bottle of wine. The tool consists of a metal spiral called the worm, a lever used for attaching on to the neck of the bottle, and a small, hinged knife that is housed in the handle end for removing the foil that wraps around the neck of many wine bottles. The most common type of corkscrew is known as the single- or double-hinged wine key or waiter's tool.
- *Ah-So:* The ah-so is an additional wine opener exclusively used for older wines with more fragile corks. The ah-so is a double-pronged device that is inserted in the neck of a wine bottle that extracts the cork by grasping onto its sides. This lessens the risk of the cork breaking or crumbling into pieces if it were to be pierced in the center.

Serving Temperature for Wine

Before a wine reaches the dining-room table, service staff should ensure that the wine is delivered at the optimal serving temperature. Proper temperatures allow the personality of any given wine to best illustrate itself in both the nose and the palate. Wines of the same color or style for the most part are served near an identical range of temperatures.

Sparkling Wine Sparkling wines should be served at approximately 40–45°F or below, because such a temperature promotes the wine's desirable personality traits of acidity and effervescence. As sparkling wine warms, its perception of acidity tends to lessen and bubble life begins to dissipate.

White Wine When white wine is presented to the customer, it should be slightly chilled to approximately 45–55°F. If white wine comes directly from the refrigerator (typically about 40°F), many of the flavors are subdued and work toward minimizing many of the aromatic characteristics. Allowing a white wine to warm up to its optimal temperature range encourages its aromatic compounds to be unmasked and allows for greater expression of its personality. Conversely, lesser-quality jug white wine can be served very cold to mask undesirable flavors and aromas. *Caution*: If a white wine warms up much beyond 55°F, the white wine's acidity becomes less pronounced and the wine may be perceived as lacking vibrancy upon entering the palate. Figure 2.42 identifies the optimal temperature range of serving white wines.

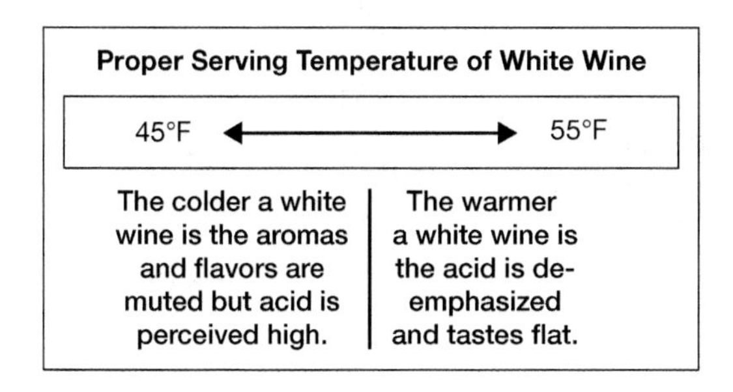

FIGURE 2.42
White wine serving temperature range.

Red Wine Much confusion surrounds the appropriate serving temperature of red wine. Red wine should NOT be served at room temperature despite what many people read. The often-misquoted recommendation of serving a wine at "room temperature" is truly intended to be

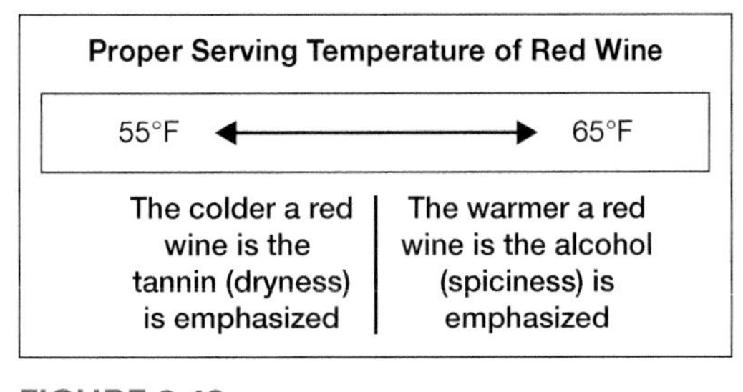

FIGURE 2.43
Red wine serving temperature range.

understood as the room temperature of the underground cellars that would naturally maintain a wine's temperature near 55–65°F. A small quantity of simple, fruity red wines (Beaujolais Nouveau from France's Burgundy region and Bardolino from Italy's Veneto region) should be served at temperatures like white wine. The slight chill assists to provide a bit more structure and ultimately a dash more character to benefit the wine. Figure 2.43 identifies the optimal temperature range of serving red wines.

Service of Still Wine

Present the bottle of wine in full view of the label to the host (the individual who ordered the wine). It's important to always confirm the accuracy with the host by stating the producer, grape varietal (or name of wine), and vintage date. Allow time for the host to respond as to the accuracy of the information. Figure 2.44 outlines the steps in a simplified format. Figure 2.45 illustrates the presentation of the bottle.

Next, cut off the capsule just below the top groove located in the neck of the bottle and remove the upper portion of the capsule. Place the capsule inside the apron pocket. Be cautious not to remove the lower part of the capsule, as it is part of the bottle's decoration. Figure 2.46 illustrates the cutting of the foil capsule.

Insert the point of the corkscrew just off-center of the cork and twist once clockwise, then continue to turn the corkscrew while straightening it upright until it's almost fully inserted into the cork. Do not turn the bottle. *Note*: Try not to pierce the opposite end or sides of the cork. Figures 2.47 and 2.48 illustrate the insertion of the corkscrew (or the worm) into the center of the cork.

Eight Simplified Steps of Still Wine Service

1. Present the bottle of wine in full view of the label to the host
2. Cut off the capsule just below the top groove in the neck of the bottle and remove the upper portion of the capsule.
3. Insert the point of the corkscrew in the cork and twist once clockwise, then continue to turn the corkscrew while straightening it upright until almost fully into the cork.
4. Attach the lever onto the rim of the bottle and apply pressure with one hand. With the other hand, lift up firmly, but slowly, until the cork emerges.
5. Remove the cork from the corkscrew and place on a small plate next to the host. Clean the neck of the bottle to remove any remaining mold or cork dust.
6. Pour approximately 1 oz into the host's glass. Twist the bottle slightly before lifting away from the glass in order to leave the last drop in the glass.
7. Once the host approves the wine, proceed with pouring wine into the glasses of all remaining guests.
8. Top up the host and place the partially empty bottle to the right hand of the host.

FIGURE 2.44
Simplified steps to wine service.

FIGURE 2.45
Presentation of the bottle. Courtesy of Erika Cespedes.

Attach the lever onto the rim of the bottle and apply pressure with one hand. With the other hand, lift firmly, but slowly, until the cork emerges, while holding the neck of the bottle and the lever together with the other hand. Figures 2.49 and 2.50 depict the cork being slowly pulled from the neck of the bottle.

Once the cork has been removed from the corkscrew, unscrew the cork from the worm and place the cork on a small plate on the right-hand side of the host. Figure 2.51 illustrates the cork being unscrewed from the worm. This ritual is intended for the host to inspect the cork for moistness—a sign that the wine was properly stored on its side. An old-school tradition consists of smelling the cork, even though most often the smell of a cork will tell little about the quality of the wine inside the bottle. Then, with a cloth napkin (or serviette), clean the neck of the bottle to remove any remaining dust, mold, or cork particles. Figure 2.52 shows the serviette being used to wipe out any cork dust or mold on the inside of the neck.

FIGURE 2.46
Cutting the foil capsule. Courtesy of Erika Cespedes.

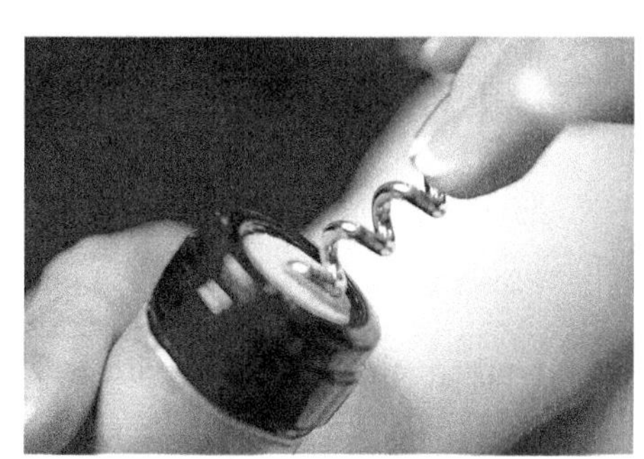

FIGURE 2.47
Insertion of the corkscrew (1). Courtesy of Erika Cespedes.

FIGURE 2.48
Insertion of the corkscrew (2). Courtesy of Erika Cespedes.

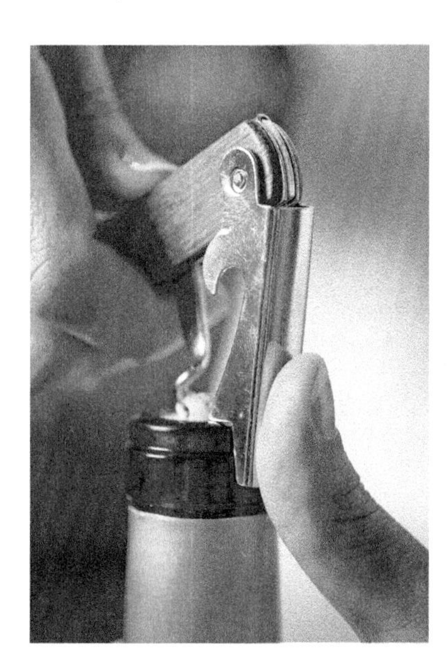

FIGURE 2.49
Cork is pulled from the neck of the bottle. Courtesy of Erika Cespedes.

FIGURE 2.50
Cork is emerging from the neck of the bottle. Courtesy of Erika Cespedes.

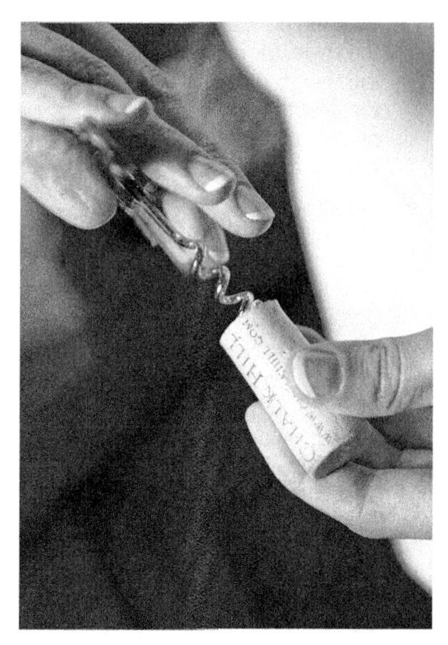

FIGURE 2.51
Cork is removed from the worm. Courtesy of Erika Cespedes.

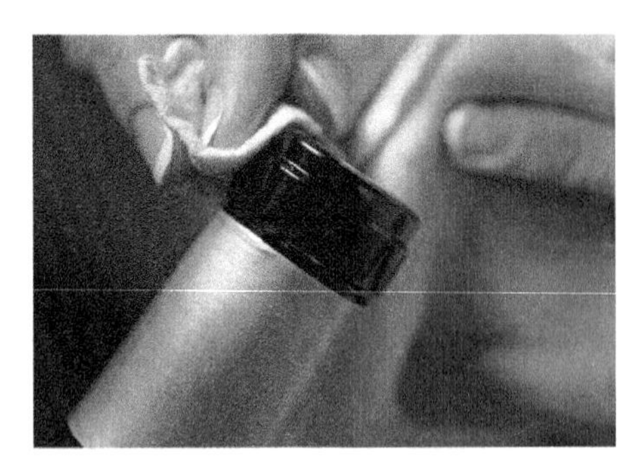

FIGURE 2.52
Serviette being used to wipe out cork dust. Courtesy of Erika Cespedes.

Pour approximately 1 oz. into the host's glass. Twist the bottle slightly before lifting away from the glass to leave the last drop in the glass, with napkin ready to wipe the lip of the bottle in order to catch any additional droplets. As the host tastes, hold the bottle with the label facing the host. If the wine is not approved, follow these three steps:

1. Listen carefully to the explanation as to why the wine is unacceptable
2. Acknowledge the explanation, and remove the tasting glass
3. Ask whether it's acceptable to bring another bottle or, instead, the wine list to make an alternative selection

Once the host approves the wine, proceed with pouring wine into the glasses of all the guests (ladies first), approximately one-half full; start with the guest to the left of the host, and continue clockwise, finishing by refilling the host's glass.

Top up the host's glass and then place the partially empty bottle to the upper right-hand side of the host (with the label facing the host). Offer the guest the option of having the wine chilled in the ice bucket (white wine) or left on the table or service station.

Service of Sparkling Wine

Present the bottle of sparkling wine in full view of the label to the host (the individual who ordered the wine). It's important to always confirm the accuracy with the host by stating the producer, grape varietal (or name of wine), and vintage date. Allow time for the host to respond as to the accuracy of the information. Figure 2.53 outlines a simplified list of sparkling wine service. Figure 2.54 illustrates the presentation of the bottle to the host.

Seven Simplified Steps of Sparkling Wine Service

1. Present the bottle of sparkling wine in full view of the label to the host and confirm for accuracy.
2. Next, slit the foil capsule with the knife of the wine key or pull the tab and remove the upper portion of the capsule that is enclosed over the wine muzzle and cork.
3. Grip the cork and wire hood through the towel (or underneath the towel with the thumb placed over the top of the towel while holding down the cork) and proceed to untwist and loosen the wire hood that covers the cork.
4. Firmly twist or wiggle the bottle and maintain a firm grip with the other hand on the cork to prevent it from flying until the cork is liberated with a soft gasp. Wipe the rim of the bottle with a clean serviette.
5. Pour a 1 oz taste into the host's glass located on the right side of the place setting. Allow the host a moment to tasting the sample in order to provide approval.
6. Proceed to pour other guests to the left of the host.
7. Top up the host and then place the bottle in an ice bucket and drape with a folded serviette.

FIGURE 2.53
Simplified steps of sparkling wine service.

FIGURE 2.54
Presentation of the bottle. Courtesy of Erika Cespedes.

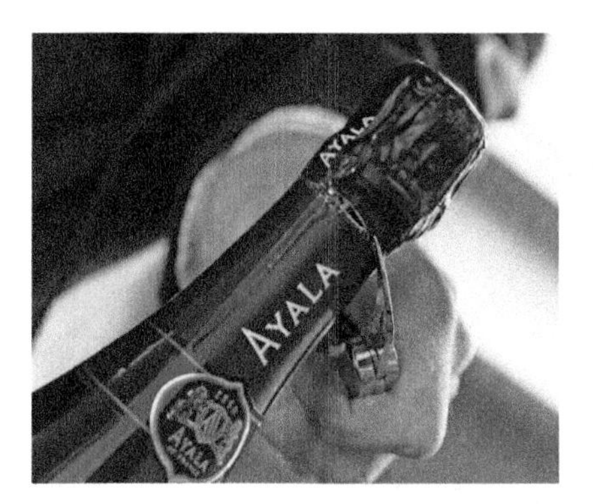

FIGURE 2.55
Cutting the capsule. Courtesy of Erika Cespedes.

FIGURE 2.56
Removing the capsule (1). Courtesy of Erika Cespedes.

FIGURE 2.57
Removing the capsule (2). Courtesy of Erika Cespedes.

Next, slit the foil capsule with the knife of the wine key or pull the tab and remove the upper portion of the capsule that is enclosed over the wine muzzle and cork. Figures 2.55, 2.56, and 2.57 illustrate the most professional manner to remove the capsule of the sparkling wine bottle. Ensure minimum damage to the bottom half of the foil located around the neck of the bottle.

Place a clean folded towel or serviette over the top of the bottle (the cork and wire hood). Grip the cork and wire hood through the towel (or underneath the towel with the thumb placed over the top of the towel while holding down the cork) and proceed to untwist and loosen the wire hood that covers the cork. Figures 2.58 and 2.59

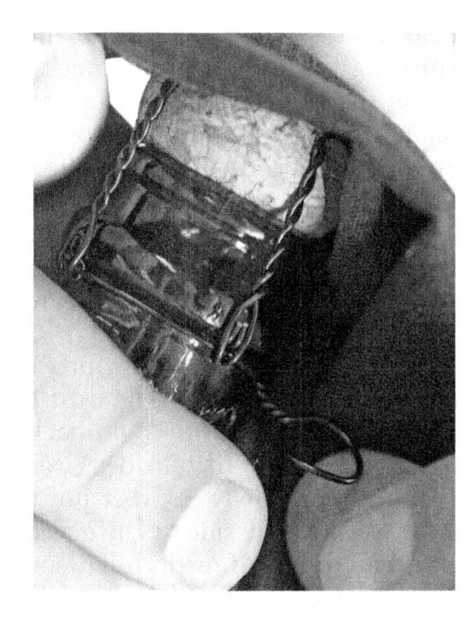

FIGURE 2.58
Untwisting of the wire hood. Courtesy of Erika Cespedes.

FIGURE 2.59
Loosening of the wire hood. Courtesy of Erika Cespedes.

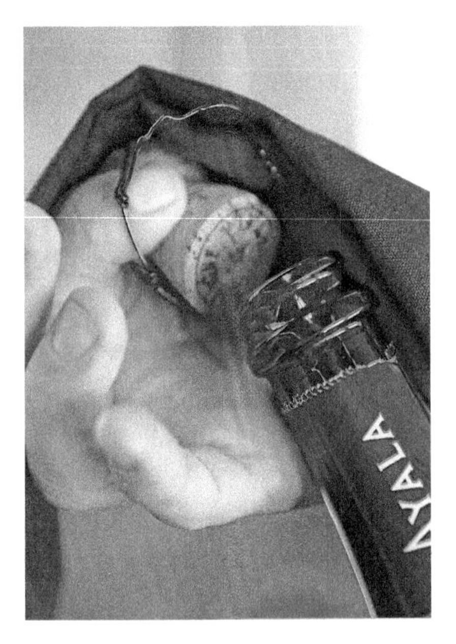

FIGURE 2.60
Removing the cork from the neck of the bottle (1). Courtesy of Erika Cespedes.

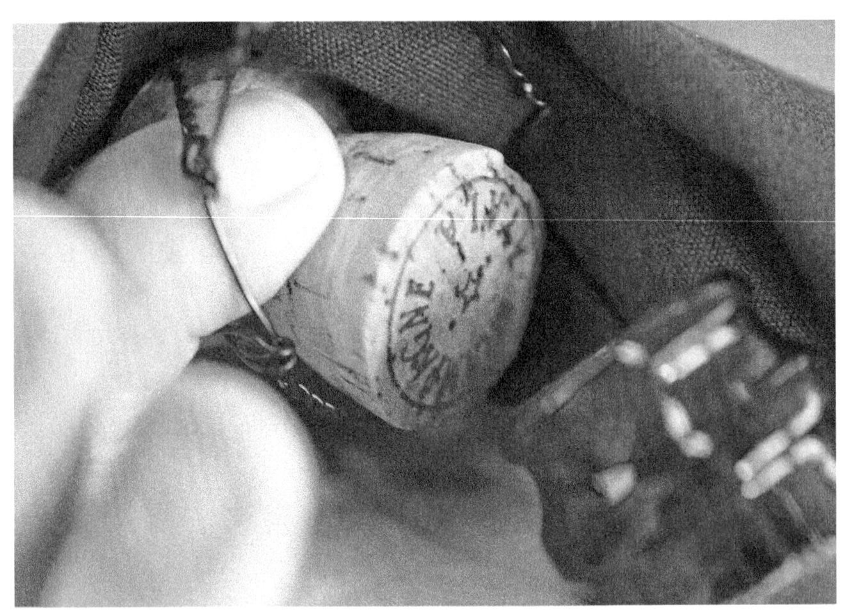

FIGURE 2.61
Removing the cork from the neck of the bottle (2). Courtesy of Erika Cespedes.

illustrate the loosening of the wire hood while simultaneously applying pressure on the cork to ensure it doesn't unexpectedly pop off. Some prefer to remove the wire hood before progressing to the next step, while others choose to leave it on for safety reasons.

Hold the bottle at an angle of 45° in order to increase the wine's surface area and decrease its pressure. Firmly twist or wiggle the bottle while simultaneously gripping the cork. Ensure to grip the bottle from underneath and maintain a firm grip with the other hand on the cork to prevent it from flying until the cork is liberated with a soft gasp. It's very critical to take one's time when removing the cork to ensure a slow release and minimal loss of carbon dioxide. Figures 2.60 and 2.61 illustrate the slow methodical manner of removing the cork from the neck of the bottle. Place the cork on a side plate, placed on the right of the host for possible inspection. Figure 2.62 illustrates the cork being placed on a liner plate for possible inspection by the host. Proceed to wipe the rim of the bottle with a clean cloth napkin.

FIGURE 2.62
Cork placed on a liner plate for inspection. Courtesy of Erika Cespedes.

Begin by pouring 1 oz. of wine for taste into the host's glass located on the right side of the place setting. Allow the host a moment to taste the sample in order to provide approval.

Once the host approves the wine, proceed to pour to other guests to the left of the host (ladies first), approximately one-third full; hold back for a moment and allow the foam to subside. Great care should be taken to pour slowly in order not to permit the wine to overflow the side of the glass. Continue to fill glass half to three-fourths full (depending upon how many glasses need to be poured). Continue clockwise finishing by refilling the host's glass. Figure 2.63 illustrates the pouring of the guest's glass with sparkling wine.

FIGURE 2.63
Sparkling wine poured into coupe. Courtesy of Erika Cespedes.

FIGURE 2.64
Sparkling wine being chilled. Courtesy of Erika Cespedes.

Once wine for all the guests has been poured, top the host's glass last. Then place the bottle in an ice bucket near the table and drape the bucket with a folded serviette. Figure 2.64 illustrates the bottle placed in an ice bucket and stand near the guest's table.

Decanting Wine

Learning Objective 10
Explain the reasons why wine may be decanted

Decanting is the process of carefully transferring wine from its original bottle into another serving container, known as a decanter. The decanter is a glass vessel used to receive a wine during the elaborate decanting process. Decanting is often associated with upscale restaurants that offer aged and more expensive wine selections that would dictate the care and expectation of this process. Truly, it is a process (traditionally, it was a ritual and necessity prior to large-scale clarification methods) that allows a wine to more effectively express itself.

The decanting process can take place for multiple reasons; the most obvious one is to remove the sediment associated with an older red wine. Historically, etiquette would dictate that a wine's liquid portion be removed from its sediment to allow for a clear pristine wine prior to consumption. Though in more modern times, some believe this process can partly strip the wine of its essential characteristics, and therefore would not desire to have their wine decanted. The second reason for decanting includes the intention of providing some additional aeration to benefit a young red wine. The added and intended oxygen allows the wine's tannins to soften and aromas and flavors to more effectively integrate. Another reason to decant is purely for hedonistic purposes in order to provide a sense of value and prestige, both of which may contribute to a sense of distinction to any wine-focused establishment.

Removing Sediment

An old red wine has color pigment and tannin particles that separate out, causing sediment and a loss of color intensity and an evolution of color hue. The sediment is harmless; however, etiquette dictates that the sediment be removed from the wine before the wine is poured into a glass and prior to consumption. The server or sommelier should place a lit candle and a *decanter* (a glass vessel designed to hold and pour wine) on the table. The bottle should be held above the candle as the server should slowly pour the wine into the decanter until the sediment appears near the neck of the bottle. While pouring, the candle is used to help illuminate the neck of bottle to identify any unwanted sediment that would communicate to the sommelier to slow the flow of pouring the wine. At this point, the server should continue to follow the general principles of serving wine, using the decanter in place of the bottle.

Allowing Aeration

Tradition holds that red wine must be allowed to breathe (be aerated), or be exposed to a small amount of oxygen prior to serving to facilitate its repressed aromas and flavors. All wines can benefit from even a little aeration because the characteristics of many wines are tight, or closed from their time in the bottles and may not emerge immediately after the cork is pulled. Aeration is generally considered desirable and intentional where the positive effects of oxygen allow the aromas and flavors to integrate and tannin compounds to soften, as opposed to oxidation where it is generally considered a fault.

Young red wine often needs some time for aging, or laying down, so the components within the wine will assimilate with one another. If a youthful red wine is opened too early, the tannin often is higher; therefore, the longer period of aging allows the tannin to naturally soften over time. Decanting and swirling wine in a glass are techniques that mimic the process of aging by exposing the wine with a large dose of oxygen. The tannin, which, until this point, has acted as a preservative, now begins to soften, and simultaneously, the flavors and aromas are heightened. The benefits of aeration begin almost immediately and continue for hours. Figure 2.65 outlines a simplified approach to decanting wine.

Three Simplified Steps of Decanting Red Wine

1. Allow the bottle to stand upright so that the sediment falls to the bottom. Position a lit candle (or other light source) next to the decanting vessel.
2. Slowly pour the wine from the bottle over the lit candle and into a decanter (or simply a separate vessel).
3. If the sediment begins to float into the neck of the bottle, stop the decanting process and let the bottle rest to allow the sediment to settle back into the bottom of the bottle once again. The end result will be a clear decanter of wine.

FIGURE 2.65
Simplified decanting service.

Three Steps of Decanting Red Wine

1. For as long as time permits, allow the bottle to stand upright so that the sediment falls to the bottom. After gently removing the foil cap and uncorking, position a lit candle (or another light source) next to the decanting vessel.
2. Slowly pour the wine from the bottle over the lit candle and into a decanter (or simply a separate vessel). Allow the candle to illuminate the neck and shoulder of the wine bottle so that the sediment can be located as the wine is being poured. Continue to be cautious to ensure that any potential sediment is left behind in the original bottle. Figures 2.66 and 2.67 illustrate the process of decanting wine over a candle to illuminate the neck of the bottle.
3. If the sediment begins to float into the neck of the bottle, stop the decanting process and let the bottle rest upright for about ten minutes to allow the sediment to settle back into the bottom of the bottle once again. The result will be a clear decanter of wine. For most bottles requiring decanting for sediment removal purposes, about a half-inch to an inch of wine should likely remain in the bottle. Figure 2.68 illustrates a decanted bottle of wine.

FIGURE 2.66
Decanting wine. Courtesy of Erika Cespedes.

FIGURE 2.67
Decanting wine. Courtesy of Erika Cespedes.

FIGURE 2.68
Decanting wine. Courtesy of Erika Cespedes.

Check Your Knowledge

Directions: Use these questions to test your knowledge and understanding of the concepts presented in the chapter.

I. MULTIPLE CHOICE: Select the best possible answer from the options available.

1. The Judgment of Paris was a wine tasting event that shocked the wine world and became the significant defining point for
   a. French wine
   b. Italian wine
   c. American wine
   d. none of the above
2. The world's largest consumer of wine is
   a. France
   b. Italy
   c. United States
   d. Germany
3. Which is derived from the fermented juice of grapes?
   a. Beer
   b. Cider
   c. Wine
   d. Spirits
4. Phylloxera is an aphid that
   a. attacks barley
   b. attacks a grapevine's root system
   c. is a form of grapevine species
   d. none of the above
5. On an American wine label, if a varietal is listed, what is the minimum percentage of grape that must be found within the bottle?
   a. 100 percent
   b. 75 percent
   c. 85 percent
   d. 95 percent
6. On an American wine label, if a vintage date is listed, what is the minimum percentage of wine that must have come from the stated vintage that can be found within the bottle?
   a. 100 percent
   b. 75 percent

   c. 85 percent
   d. 95 percent
7. On an American wine label, if a specific AVA location is listed, what is the minimum percentage of grapes that must have come from the stated location within the bottle?
   a. 100 percent
   b. 75 percent
   c. 85 percent
   d. 95 percent
8. On an American wine label, if a vineyard is listed, what is the minimum percentage of grapes that must have come from the stated vineyard that can be found within the bottle?
   a. 100 percent
   b. 75 percent
   c. 85 percent
   d. 95 percent
9. The terms reserve and vintner's reserve hold
   a. special meaning for an American wine
   b. special meaning for a California wine
   c. no legal meaning in America
   d. no legal meaning anywhere in the world
10. Generically labeled wines
    a. are of exceptional quality
    b. are often of poor quality
    c. often steal the names from famous Old-World wine locations
    d. both b and c
11. Proprietary labeled wines
    a. are often of exceptional quality
    b. are often of poor quality
    c. are often high-quality respectable versions of an Old-World wine
    d. both a and c
12. Varietal-based labeling means
    a. there will be no identification of a place of origin on the label
    b. the most significant grape in the bottle will be identified
    c. a location of where the grapes originated from will be identified
    d. both a and c
13. A geographically-based label
    a. is only associated with the New World
    b. is associated with all wines from France
    c. is most often associated with Old-World wines
    d. is mostly used in the New World
14. Generally speaking, the more specific the origin of the grapes
    a. the wine can be of better quality
    b. the wine can be of lesser quality
    c. it depends
    d. none of the above
15. The *WHERE* on a wine label will legally indicate
    a. where the wine was made
    b. where the grapes were fermented
    c. where the grapes were grown and harvested
    d. all of the above

16. The reason that wine is opened in a methodical process is based on
    a. ritual
    b. ensuring the health of the wine
    c. ensuring accuracy of what the customer ordered
    d. all of the above
17. Which is not a reason that decanting a wine is done
    a. to aerate the wine
    b. for display purposes
    c. to remove sediment
    d. to allow the color to become brighter and more intense
18. The proper temperature range for serving red wine is
    a. 40–45°F
    b. 45–55°F
    c. 40°F or below
    d. 55–65°F

II. DISCUSSION QUESTIONS

19. Identify the four methods of labeling table wine.
20. Identify and explain the five significant labeling clues that may be found on a typical wine label.

CHAPTER 3

Viticulture: Outside in the Vineyard

CHAPTER 3 LEARNING OBJECTIVES

After reading this chapter, the learner will be able to:

1. Identify the five aspects of viticulture
2. Explain how site selection and grape varietals are important considerations in the process of viticulture
3. Explain the four most significant influencing factors associated with a location and defining to the concept of terroir
4. Identify at least two common vineyard hazards
5. Discuss the important considerations for determining grape harvest
6. Explain how the concept of hang-time changes the style of wine
7. Identify three ways that grape growers can be "green" friendly

Great wines taste like they come from somewhere. Lesser wines taste interchangeable; they could come from anywhere. You can't fake somewhereness. You can't manufacture it . . . but when you taste a wine that has it, you know.

—Matt Kramer, *Making Sense of Wine*

Introduction to Grape Growing

Learning Objective 1
Identify the five aspects of viticulture

Although there are a multitude of factors that impact a wine's taste and character, the personality of any given wine is fundamentally determined by three significant factors. The first fundamental factor is the grape(s) used to make the wine. For example, Chardonnay grapes create a very different wine than one made from Riesling grapes. The second fundamental factor is the growing location by region, vineyard, and even specific section in a vineyard. For example, grapes grown in a cool region, less sun exposure or in chalk-based soil create very different wines than grapes grown under other conditions. Last are the winemaking processes, called enology, where several vinification techniques can be implemented within the winery. For example, the winemaker can use French oak versus the more aggressive American oak. Collectively, these variables perform to create distinctions from wine to wine. Figure 3.1 shows a full cluster of Zinfandel grapes from 117-year-old vines.

FIGURE 3.1
Zinfandel grapes. Courtesy of Cellar Angels, LLC.

Viticulture, deriving from the Latin word for vine, refers to the study and practice of cultivating grapes—specifically when the grapes are used for winemaking. This practice involves the science and study regarding the production of grapes that endure a series of events that occur within the vineyard.

FIGURE 3.2
Wairarapa vineyards in New Zealand. Courtesy of Carrie Schuster.

Figure 3.2 shows a textured-looking vineyard in Wairarapa, New Zealand. The vineyard is considered one of the most influential variables in the characteristics of a finished wine. Grape growing, literally and figuratively, merges science and art through the expertise of the farmer. The farmer or vineyard manager is commonly responsible for monitoring and controlling pests and diseases, for fertilization and irrigation, for monitoring fruit development, for conducting pruning throughout the season, and, finally, for determining when to harvest the grapes.

From a beverage manager's perspective, it is important to have a basic understanding of the intricacies involved with growing wine grapes in a vineyard. The influences in the vineyard will most certainly impact the finished wine. The viticulture aspects that are discussed within this chapter include:

1. grapevines and wine grapes (site selection and grape varietals)
2. location (climate, soil, water, and topography)
3. grapevine maintenance and training (canopy management and pruning)
4. common vineyard hazards (including microorganism, animals/pests, and weather-related issues)
5. harvesting of the grapes (hang-time and methods of harvesting)

While the history of grape growing goes back at least 10,000 years, it is largely the Romans who are recognized with being largely responsible for spreading the grapevine to some of our most famous present-day Old-World wine regions. The Romans practiced a crude form of viticulture (though more modern than any of their predecessors) as they attempted to train and domesticate the wild vine. It was then during the Middle Ages, when many of the Cistercian monks became prominent viticulturists, as they made significant contributions to growing the grapevine. They experimented with finding suitable grapes for appropriate locations that would best allow the grapes to fully express themselves. Many of the vineyard practices developed during this period would become standards of excellence up until the 18th and 19th centuries.

FIGURE 3.3
Old vines from Terra Valentine. Courtesy of Terra Valentine and Carolyn Corley Burgess.

Modern-day grapevines are not created from seed, but rather vine cuttings attached to rootstocks that are obtained from specialized nurseries. Since a grapevine is genetically unstable (as are apple trees), planting a grapevine from its seeds would only create a mutated variety; therefore, all grapevines around the world are created through a cloning process. Clones are a replication of an original or mother vine with the intention of duplicating its desirable traits and recreating those in another vineyard. Grafting is a method of plant propagation widely used to produce these cloned varieties throughout the wine industry. It involves fusing the tissues of one plant with those of another. In most cases, one plant is selected for its roots (called the rootstock) while the other plant is selected for its stems (called the scion). The scion contains the desired traits to be duplicated in future production by the newly propagated grapevine. The methods of grafting and cloning were also discovered to be the most significant solutions to the *Phylloxera* epidemic that plagued much of the wine world in the 19th and 20th centuries. Figure 3.3 shows old vines from Terra Valentine with the thick and gnarled trunk associated with old rootstock. Older vines are considered prized if the intention is to produce high-quality grapes.

Grapes can be found growing on a vine within designated growing areas, known as vineyards. A vineyard is the place where a wine is conceived, and may consist of several plots or parcels of land that are characterized by their geographical, climatic and geological elements. Similarly, a collective group of nearby vineyards generally creates an appellation, or simply a larger grape-growing area that is often defined by specific legal delimited boundaries. Figure 3.4 illustrates the Terra Valentine Wurtele Vineyard in California's Napa Valley. Napa Valley is largely considered one of the most coveted grape growing areas in the world for quality Cabernet Sauvignon. Appellations refer to a viticulture area that in most cases consists of numerous vineyards that all share similar distinctive geographical features that produce wines with shared characteristics. The same grape can produce very different characteristics depending on where it is grown and how the vintner handles the grapes. In France, the term appellation holds a very specific meaning that is unlike its use throughout much of the rest of the wine world. France has requirements for all its appellations, identifying specifics on how grapes can be grown, limits on yield, type of grape varieties, and even the potential of vinification techniques. Each appellation is unique. Therefore, learning about the appellations are critically important, yet can be a little intimidating when trying to navigate the requirements for each one.

FIGURE 3.4
Terra Valentine Wurtele Vineyard. Courtesy of Terra Valentine and Carolyn Corley Burgess.

Grapevines and Wine Grapes

The predominant grape varietal is the most influential factor in determining the personality of any given wine. The universal belief that "great wine is made in the vineyard" is absolutely accurate in that better-quality grapes, theoretically, make better wine. Wine grapes are shaped and influenced by the copious factors brought about by unforeseen forces (largely Mother Nature) throughout the entirety of the growing season.

There are about twenty different species of grapevines, but only one of them, *Vitis vinifera* (vin-if-EHR-ah), the European species, produces all the grapes used in high-quality wine as displayed on restaurant wine lists and retail shelves throughout the world. *Vitis vinifera* wines include all of the major grape varieties, such as Sauvignon Blanc, Riesling, Chardonnay, Pinot Noir, Merlot, and Cabernet Sauvignon. Other species of grapevines native to the Americas are also used to make wine; these include *Vitis labrusca* (lah-BROO-skah), *Vitis riparia*, *Vitis aestivalis*, and *Vitis rotundifolia*. These other grapevine species are not known for their ability to produce quality wine grapes, but their base or rootstock is extremely valuable because of their resistance to the infamous phylloxera pest. While there are many different species of grapevines, it is the *Vitis vinifera* that yields wine grapes with potential for complexity and expressions of distinction. Within the *vinifera* species, there are estimated to be as many as 10,000 strains, clones, and hybrids of different variations or subspecies of grapes. Most of the wine world depends on perhaps thirty to forty so-called international grapes, of which an even smaller number are considered classics or noble grape varietals that are most popular to the consumer. Some of the more important varietals (listed alphabetically) that are represented on most American wine lists and retail boutiques include:

- *White Wine Grapes:* Albariño, Chardonnay, Chenin Blanc, Gewürztraminer, Grüner Veltliner, Marsanne, Muscat/Moscato, Pinot Blanc, Pinot Gris/Grigio, Riesling, Roussanne, Sauvignon Blanc, Semillon, Torrontés, and Viognier
- *Red Wine Grapes:* Barbera, Cabernet Franc, Cabernet Sauvignon, Carménère, Dolcetto, Gamay, Grenache, Malbec, Merlot, Mourvèdre, Nebbiolo, Pinotage, Pinot Noir, Sangiovese, Syrah/Shiraz, Tempranillo, and Zinfandel

Figure 3.5 depicts some red grapes hanging on the vine.

Grapevines are fairly adaptable, growing in a wide range of soils and temperature ranges that can alter a basic grape's DNA to offer slight variations in its aroma/flavor and structural components. The most successful wine grapes are grown in temperate climate bands in the range from 30° to 50° north and south of the equator. The areas located in these temperate climate bands generally provide the right combination of sun, rain, and temperature. Therefore, grapevines have an advantage of variability that expresses itself based on their location. Vine cuttings are taken from a mother vine and may be transplanted in several locations, each having different growing conditions—the young vines will adapt themselves to the new environment. This cloning of vines accounts for a great deal of the spread of wine varieties from one place to another. This technique has been conducted for centuries and explains why there is so much confusion over grape varieties having many synonyms for the identical varietal. For example, the indigenous Italian grape varietal Sangiovese goes by different names, Brunello and Prugnolo Gentile, largely based on the vine's adaptation to its varying locations.

FIGURE 3.5

Cluster of red grapes left on the vine. Courtesy of John Peter Laloganes.

The work of the ampelographer (amp-pehl-ah-gruh-fer)—an individual who studies the identification of grapevine botany—has come into significant importance over recent decades to assist in learning more about grapes around the world.

The Annual Life Cycle of Wine Grapes

The grapevine is a deciduous plant that loses its leaves in the fall, becomes dormant (below a temperature of 50°F) in the winter, and follows the basic process of budbreak, flowering, fruit set, summer pruning, and véraison throughout the spring and summer. Grapevines follow a growing season based on the combination of sunlight and weather, and with each stage in the cycle come a series of actions that growers must take to promote healthy vines and ultimately tasty wines. In the northern hemisphere, grapes usually are harvested in the late summer or early fall (September or October). In the southern hemisphere, harvest time (six months ahead that of the northern hemisphere) occurs in the late winter to early spring (February or March). Keep in mind, harvest varies based on grape type and the given climate of that particular vintage. Identified below is the life cycle of a typical vine from the northern hemisphere.

- *Dormancy:* At this stage of the life cycle, the dried vines are cut back during winter pruning to assist in conserving their energy throughout the season. Winter pruning will also train the vine for the approaching growing season.

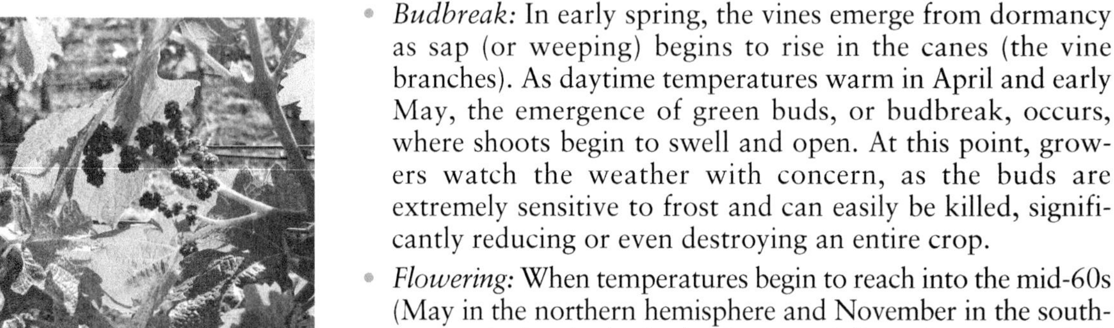

FIGURE 3.6
Green berries forming. Courtesy of Cellar Angels, LLC.

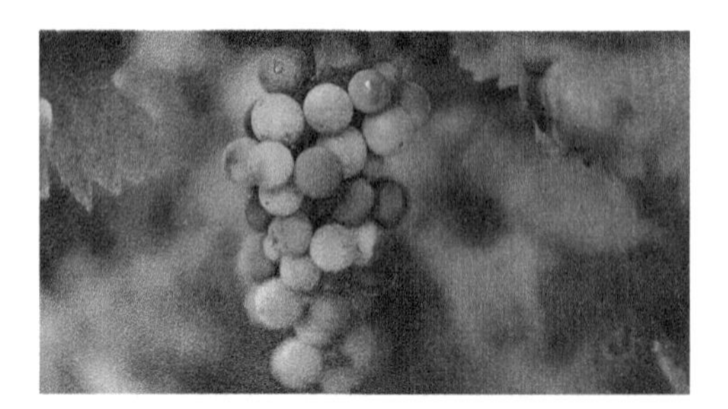

FIGURE 3.7
Grapes going through véraison. Courtesy of Cellar Angels, LLC.

- *Budbreak:* In early spring, the vines emerge from dormancy as sap (or weeping) begins to rise in the canes (the vine branches). As daytime temperatures warm in April and early May, the emergence of green buds, or budbreak, occurs, where shoots begin to swell and open. At this point, growers watch the weather with concern, as the buds are extremely sensitive to frost and can easily be killed, significantly reducing or even destroying an entire crop.
- *Flowering:* When temperatures begin to reach into the mid-60s (May in the northern hemisphere and November in the southern hemisphere), the buds bloom and flowering occurs. It is during this phase that self-pollination and fertilization of the grapevine take place. At this point, excessive rain or hail can prevent proper flowering and the future potential for fruit set.
- *Fruit Set:* Fruit set will occur during summer as the grape's flesh and skin begin to develop as the flowers convert into green, hard berries. The berries continue to gain sugar and ripen throughout the summer. Figure 3.6 shows some green, hard berries forming.
- *Véraison:* Near the middle to end of summer, ***véraison*** (vehr-ray-ZOHN) occurs, where the green berries begin to change color and become recognizable as grapes. Toward the end of the summer to early fall, depending on grape varietal and climate, the grapes are at the optimal level of ripeness (level of sugar and acid) and flavor (phenolic) ripeness to begin the harvest. Figure 3.7 shows a cluster of grapes going through the conversion of véraison.

Site Selection and Grape Varietals

Site selection is the process of choosing the appropriate grape varietal for a given location, per the desired business model of the winery—producing wine of quantity or wine of quality. Grape varietals are significantly influenced according to their surroundings. Therefore, the location of a vineyard will undoubtedly be one of the most determining

Learning Objective 2
Explain how site selection and grape varietals are important considerations in the process of viticulture

factors in the personality characteristics of the grapes. The grape varieties that are chosen to be planted depend on what the given vineyard location will allow the ultimate intent of the winemaker. While certain grape varieties thrive in many climates and soil conditions, each variety will have its own specific parameters and therefore will not suit every location. Typically, a producer selects grape varietals specifically for their style and suitability to a particular site—varieties that will perform best under the climatic conditions within a set of specific vineyards. Winemakers often must question themselves: What kind of grape can potentially be planted successfully given this location—or what kind of location is necessary to plant a given grape varietal? This question might be quite necessary in many New-World wine-producing vineyards—but the Old World often has different philosophical and legal processes. Through hundreds of years of trial and error, the French invented and perfected this system of matching varietal to location. In many Old-World wine-producing countries, local legislation often dictates which grape varieties are selected, how they are grown, whether vineyards can be irrigated, and exactly when and how grapes can be harvested, all of which serves to reinforce tradition. Figure 3.8 showcases the steep vineyards in the Mosel region of Germany where the grapes obtain their much-needed sun exposure from being on steep-angled hillsides to maximize the sun's rays.

FIGURE 3.8
Steep vineyards in the Mosel region of Germany.
Courtesy of Leo Alaniz.

The length of the growing season is an essential consideration that can determine what grape variety will perform best. Some grape varieties require longer growing seasons to fully ripen while other grape varieties ripen sooner and therefore can handle being planted in a different climate. It is not only a matter of having to decide whether white wine or red wine grapes are planted or a matter of personal taste, instead choosing a variety that not only will make good wine but will ripen and mature properly given the location. Once the grower has decided which grape varieties to plant, a process begins that will take several years (typically at least three years) for the vine to turn out quality fruit that is ready for wine production.

Location

Learning Objective 3
Explain the four most significant influencing factors associated with a location and defining to the concept of terroir

The geographical location and related climate can significantly dictate the type of grapes a vineyard will be able to successfully grow. Therefore, wine is a product that can truly reflect terroir, or the growing conditions from its place of origin. Terroir is a French concept that represents a sense of place—an element of distinction that reflects the way local influence is expressed in the wine. This concept is a driving force that separates artisanal wine versus factory-made mass-produced ones. While both kinds of wines have their appropriateness, it's the terroir that ultimately explains the individual appeal, subtle nuances, and their price. Terroir (or simply location) contains four basic elements that intersect in intricate complexity: climate, water, soil, and topography. According to Jonathan Nossiter's work on Liquid *Memory: Why Wine Matters*, "... without terroir—in wine, cinema, or life—there is no individuality, no dignity, no tolerance and no shared civilization." The concept of a location having a dramatic influence on wine grapes goes back to connect with the French and their controlled appellation of origin that was created in the 1930s. The significance of this system was so critical that they created laws tied to the origins of grape, whether broad or specific, to guarantee that their products (not only wine but also cheeses, lentils, and honey) have been held to a set of rigorous standards. As a general guide, when grapes come from a broader location, such as California, the grapes do not carry a very defined sense of place. However, if grapes derive from a very defined growing area within California, such as the Napa Valley, the grapes have a more precise point of origin, and are often going to carry both a perceived and a real value of quality.

The influence of location on grapes can be compared to the idea of the housing market and the value of location. The value of real estate in some neighborhoods within the same city can take a dramatic shift in perceived and real value of quality.

Climate

Climate refers to the general weather conditions prevailing in an area over a long period. Thankfully, grapevines are adaptable, growing in a wide range of climate and soil variations. The vast majority of the grapevines are found in the parallels of 30° to 50° north and south of the equator. Climate remains one of the most important variables that determines what type of grapes a vineyard will be able to grow successfully. It's within these temperate bands that annual mean temperatures range between 50 and 68°F. This temperature range can potentially provide sufficient warmth and sunshine to produce quality wine grapes. In addition to temperature, other mitigating factors such as the presence of large bodies of water (which assist to moderate temperature extremes and lengthen growing seasons), elevation, and mountain ranges (which assist to minimize cool temperatures, rain, and wind) can have additional effects on the climate and vines within the location.

Climates are broadly classified to better understand the likelihood of particular grape varietals and styles of wines produced. In the world of wine, there are four major types of climates (often referred to as macroclimates) that are found throughout the major producing areas:

- *Maritime climates* have large bodies of water to moderate the temperatures throughout the year by keeping cool summers and mild winters and overall a moist environment. For example: Bordeaux, France, and New Zealand.
- *Continental climates* have four distinct seasons with short, hot summers and cold winters. For example: Burgundy, France, and the Willamette Valley, Oregon.
- *Mediterranean climates* have long, warm to hot growing seasons with mild winters, low moisture, and low rainfall. For example: California's Napa Valley and south of France (Languedoc-Roussillon and Provence).
- *Alpine climates* have influences of altitude from vineyards being perched upon mountainous areas. For example: Savoie and the Jura in France.

The grapevine needs approximately 1,300–1,500 hours of sunshine throughout the growing season to produce grapes suitable for winemaking. In ideal circumstances, the vine will experience maximum sunshine in the day (to develop a grape's sugar and flavor ripeness) and cool nights (to help preserve a grape's acidity). This temperature differential between day and nighttime temperatures is referred to as diurnal range. In warmer climates, diurnal range becomes very important to maintain the structure and desired complexity in wine grapes. In cool climates or fog-prone areas, diurnal range may become less important as the daylight hours may be too limited and the grapes desire for continuing warmth and ripening hours throughout the nighttime period. Figure 3.9 illustrates a fog-prone vineyard that lessens the amount of sunlight hours for the grapes to ripen.

FIGURE 3.9
Morning fog in the vineyard. Courtesy of Cellar Angels, LLC.

Extremes in temperatures—whether cold or hot—can alter the grapevine's ability to perform its basic function of growing and ripening grapes. If the climate is too cold (or inadequate sunshine), the grapes will never develop ripe flavors and adequate sugar requirements necessary to create a wine with optimal expression of personality and alcohol content. If the climate is too warm, the grape may develop overripe flavors with an overabundance of sugar content and potentially a wine that is out of balance and is overtly alcoholic in character. Furthermore, the grapevines may shut down if temperatures consistently exceed 95°F. These high temperatures interfere with ripening and can adversely affect quality through its underripe aromas and flavors and harsh structural tannic compounds. Below is a simplified example of how the overall climate of a growing environment might alter the aroma/flavor components of white and red wine grapes.

- *Cool Climate Whites:* White wines from cool climates provide aromas and flavors associated with tree fruits like apples and pears, citrus fruits like lemon and grapefruit, and minerals/chemicals such as wet stone nuances.
- *Warm Climate Whites:* White wines from warm climates yield aromas and flavors associated with tropical fruits like mango, banana, and pineapple.
- *Cool Climate Reds:* Red wines from cool climates promote fresh red and black fruits such as cranberries and red cherries or black berry and black cherries.
- *Warm Climate Reds:* Red wines from warm climate encourage dried and stewed fruits such as figs, plums, and dried black cherries and black berries.

These examples only highlight aroma and flavor distinctions between grapes grown in different climates, but other variables such as acidity and sugar content of grapes will also be influenced and will ultimately affect the structural components in the finished wine.

Dr. Albert J. Winkler was a professor at the University of California–Davis, who is often credited as being the father of modern-day viticulture throughout California. One of his contributions was the development of the Winkler heat index system. The system divides the wine world into five regions based on overall heat received during the growing season, using a minimum temperature of 50°F (grapes don't grow under this temperature), which is typical throughout the growing season from April 1 to October 1. Throughout these months, the mean temperature is taken each day to get a daily summation—the mean minus 50 equals the summation for the day. The daily summations are added together for the span of the growing season to come up with the heat index for the regions. Regions are numbered somewhere on a scale of 1 to 5, with higher numbers representing hotter regions and lower numbers equating to cooler regions.

Region 1 includes the lowest temperature-rated areas that have a heat index of less than 2,500 degrees. These regions include Germany, areas of northern France (Burgundy and Champagne), and areas in California (Carneros, Santa Barbara, Santa Cruz, and Monterey). The suggested grapes for this type of area are Riesling and Gewürztraminer, which grow well in Germany, and Chardonnay and Pinot Noir, which grow well in Burgundy, France.

Region 2 includes areas with heat indexes of 2,501 to 3,000 degrees. These areas include northern Italy; Bordeaux, France; and areas in California, including northern Sonoma, the southern Napa Valley, and some areas spanning between Los Angeles and San Diego Counties. Grape varieties that grow well in this region are Merlot, Cabernet Sauvignon, and Sauvignon Blanc.

Region 3 includes areas with heat index ranges from 3,001 to 3,500 degrees. It includes places such as the Rhône Valley of France, central Italy, and areas in California such as northeast Sonoma County, middle Napa Valley and southern Monterey County. Grapes suitable for growing in these areas include Zinfandel, Gamay, and Syrah (Shiraz).

Region 4 includes areas with a heat index of 3,501 to 4,000 degrees. This designation refers to places such as Southern growing areas of the Rhône Valley and Italy, Central and Southern Spain or areas such as parts of Paso Robles, Lodi, Los Angeles, Orange and San Diego Counties. The recommended grapes for these areas are limited to varieties such as the Carignan, Cinsault, Grenache, Mourvèdre and Tempranillo. Many of these areas can successfully produce grapes that are listed in the first three regions, but the growing season is shorter.

Region 5 is the hottest of the five regions, having a heat index of 4,001 degrees or more. The areas that fall within this region are northern Africa, southern Spain, and a very large portion of the middle of California, spanning from Shasta County in the north to Kern County in the south. These areas are recommended for the Thompson seedless grapes and other sweet grapes.

The Winkler system offers a nice set of broad guidelines; however, it's important to recognize that many exceptions exist. For example, certain grape varietals can exist and thrive in different regions, offering unique interpretations of the same grape. Additionally, there may be climate variations within a precise growing area, therefore altering the overall growing conditions.

Water

Having sufficient water is a necessary requirement for the grapevine to survive. Water can affect a vineyard dramatically because the vines need just the right amount of water to grow—yet not too much for them to thrive. If a vineyard receives its water from rainfall, too little rain can mean the grapes will not grow properly and they can lack vibrant flavor and expected yield. Too much rain and the flavor will literally be washed away, and the bloated grape will produce a diluted wine. A vintage can be, in all respects, perfect, only to be ruined by a last-minute rainfall; in reverse, severe drought conditions can kill the vines without adequate water for survival.

Grapevines receive their necessary water supply in one of three ways: (1) rainfall, (2) irrigation, or (3) an underground aquifer. In many locations, the vine will receive this naturally from Mother Nature—preferably, it will receive most rainfall during the winter and spring months to minimize fungal diseases. In areas where rainfall is limited, the vine becomes incredibly stressed (which can be desired for a quality-driven business model), which encourages their roots to travel deep (possibly as deep as 50 feet) in search of water.

In other locations, the vine may receive such an inadequate amount of yearly rainfall that a vineyard must rely on irrigation. Irrigation is the artificial application of water delivered to the land in order to assist in the production of its associated crops. Figure 3.10 illustrates some vineyard rows where the grapevines are set up for drip irrigation. While the use of irrigation has allowed the expansion of vineyards into locations that were thought to be previously unplantable, the concept is not a new one. The Romans had knowledge of this because in certain areas they had built crude irrigation systems to ensure agricultural crop success (including grapes). Irrigation allows the grape farmer more control over the product and has helped to lessen the significant vintage variations in many wine regions. However, the crop can still vary from a huge success to a miserable failure. In drip irrigation, water is delivered at or near the vine's root system. If managed properly, this method can be the most water-efficient method of irrigation, as evaporation and runoff are minimized. The plants are only irrigated between late June and early September, with the volume of water used being decreased throughout the growing season to divert growth of foliage and instead encourage the formation and concentration of quality grapes.

FIGURE 3.10
Vineyard rows in Paso Robles. Courtesy of John Peter Laloganes.

Sometimes the farmer is fortunate because the vineyard is located next to a river or an underground aquifer that can feed the vines and the grapes regardless of the annual rainfall. Most of the time rainfall is not a problem in these regions, but every so often the extra water from the river or the underground aquifer provides the vines a real boon to the grape farmer, increasing the yield and the quality of the grapes.

Soil

FIGURE 3.11
Slate soil in the Mosel. Courtesy of Leo Alaniz.

FIGURE 3.12
Claude Riffault Sancerre and chalky soil. Courtesy of John Peter Laloganes.

Soil is another important location consideration when selecting and planting grapevines. The soil content is a mixture of minerals, organic matter, and particles that are of different sizes and textures that acts to support the root structure of the grapevine. The soil influences the drainage levels and quantity of mineral and nutrient absorption. The type of soil can also influence a vine's exposure to light and warmth or coolness. Certain soils can retain heat and/or reflect it back up to the vine, which is an important consideration that affects the ripening of the grape. Figure 3.11 identifies slate soil from the Mosel region of Germany. Slate is noted for absorbing the warmth and heat throughout the day, and reflecting that warmth at night to allow the grapes to continue their ripening. There are soils found in some famous wine regions around the world where certain grape varieties have consistently performed at their legendary best. This is not to say that the same grapes can't be grown in alternative soil types and still produce high quality. Figure 3.12 shows the chalky clay soil (referred to as terres blanches) that turns white during dry periods in the Sancerre appellation of Loire Valley, France. This soil is also of significance in the Champagne and northern Burgundy regions of France.

Planting grapes in dry, nutrient-poor soil will stress the vines, keep vine vigor down, and produce small grape berries with thicker skins, ideal for high-quality grapes. The growing of wine grapes is contrary to mainstream thinking that grapevines need an abundance of water and fertile soil like other agricultural crops. Instead, if the vines are given poor, low-nutrient soil with healthy drainage, the roots are forced to dip deep and result in better grapes, per the struggling vine philosophy. This philosophy theorizes that for the vine and the grapes to retrieve their nutritional requirements, the vine struggles to survive, forcing its roots to dig while the vine grows slowly and thus produces fewer grapes, but with thicker skins and greater flavor development. One of the aspects of struggling vine philosophy is that despite having fewer grapes that are smaller (thus providing less yield), the product is considered superior. Thus, a winemaker would have less overall yield, but what they are making typically would command a higher selling price.

Topography

Topography is referencing the land's surface and shape—simply, the lay of the land. Particularly of importance for grape growing are a vineyard's slope, aspect, and altitude. The grapevines should be located where they receive the best possible access to sunlight to encourage photosynthesis—the essential requirement for growth of any agricultural product. Photosynthesis uses energy from sunlight by converting carbon dioxide into sugar and encourages grapes' necessary ripeness upon harvest.

Slope

The slope refers to the degree of steepness or incline of a hillside. A higher slope indicates a steeper incline. For quality wine grapes, hillsides and slopes are preferred over flatter terrain, as vines growing on a slope receive a greater strength of the sunrays falling on an angle to the hillside, and steeper slopes can also aid in better drainage of water creating a soil that is low in nutrients. This action causes stress upon the vine's root system forcing it to divert its energy in producing foliage, and instead dig deeper looking for nutrients needed for survival. Grapevines have extensive root systems reaching more than 30 feet. In the process, the vine produces less yield, yet greater concentration of

FIGURE 3.13
Sloped vineyard. Courtesy of John Peter Laloganes.

aromas, flavors, and structural components. On flatter land, the strength of the sunlight is diminished as it is spread out across a wider surface area. Figure 3.13 shows a vineyard with varying slopes and aspects that contribute to better sun exposure and good drainage from rainwater.

Aspect

The aspect is a term used to describe the direction in which a slope faces. In the northern hemisphere, southern and southwest-facing orientations are ideal for wine grape cultivation as they receive the most direct sunlight and extend the growing season well into the fall time. This lengthy growing period allows for grapes to fully mature and ripen prior to harvest. Wine grapes, especially reds, demand plenty of heat and sun for maximum fruit quality.

Altitude

The altitude or elevation refers to the vertical height of vineyard location, generally referencing above sea level. The higher altitude causes a decrease in pressure and therefore the air to expand as it rises. For every 400–500 feet above sea level, the temperature drops about 1–3°F. The outcome creates cooler air that affords a grape-growing climate that might otherwise be too warm to grow better-quality grapes.

Grapevine Maintenance and Training

The grapevine is maintained at its different stages throughout the year. The use of vine training systems is aimed primarily to assist in canopy management (discussed below), in finding a balance between having enough foliage to facilitate photosynthesis, and yet without excess that would cause too much shade and prohibit air circulation. Inadequate sunlight could impede the ability for grapes to become ripe and fend off disease. Initially as grapevines are planted, the decision is made as to how they will be trained. There are two broad types of grapevine training. The first broad technique allows the grapevine to grow similar to the shape of a bush—hence Bush Vine or *Gobelet* (goh-boh-leh) training. An alternative method trains the grapevine on a wire trellis—known as *Guyot* (GEE-oh) or *Cordon* (KOHR-dahn) training. A trellis consists of firmly set, well-braced posts at intervals with wire attached along to ultimately support the vine and its increasing weight throughout the growing season. This method allows for better positioning of grape clusters as per desirable preferences of sun exposure. Each training method has its advantages and disadvantages, but it is largely decided upon by local tradition and customs.

Canopy Management

Once a grapevine has developed a stable root system (of approximately three years), the vine's canopy (which includes all the stems, leaves, and fruit clusters) will need to be managed and adjusted accordingly. Canopy management is the practice of thinning and positioning the vine's leaves, shoots, and fruit as the vine grows, to gain such beneficial advantages as increased sunlight exposure and air movement. If those same vines had not been tended, the row of vines would have grown much thicker, almost like the canopy of a tree, most likely shading too many of the grapes, resulting in underripe grapes and a greater likelihood of fungal problems. Canopy management improves varietal character and decreases problems with fungal rot and insects. Figure 3.14 illustrates a full canopy that will act to shelter the grapes from wind and excess sun.

FIGURE 3.14
Full canopy in Italy. Courtesy of John Peter Laloganes.

Pruning

In addition to adjusting and managing a vine's canopy, another form of crop control involves the adjustment of grapevine's yield through pruning—an important factor in determining the quantity and quality of the grapes and, ultimately, of the wine. Pruning is the process of removing excessive grapes and foliage from the vine for affecting yield, which influences character development in the grapes. It forces a vine to exert more energy into its fruit rather than its foliage. It is performed to optimize the production potential of the grapevine. The objective is to maintain a balance between vegetative growth and fruiting, therefore making both an economical and a quality-oriented grape crop. Pruning diverts the energy from the roots and vines into the fruit. If the amount of pruning is increased, more foliage and grape clusters will be removed, thereby configuring a vine's energies into existing clusters of grapes. If the amount of pruning is decreased, a greater quantity of less concentrated fruit (because the vine's energies are diverted) will be produced. The pruning process can also aid in adjusting the size of the grape berries. Less pruning can create larger-sized grape berries with thinner skin and more juice. Greater pruning results in smaller-sized grape berries with thicker skin and less, but more concentrated juice, which translates into more pronounced personality of the finished wine.

Typically, high-quality vineyards produce around 3 tons of grapes per acre or less—whereas vineyards of lower quality levels may produce more than 12 to 14 tons of grapes per acre. Therefore, yield of the acre has a great deal to do with the quality of the grape. As the yield of a vineyard increases, the quality tends to fall off. Depending upon the winemaker's vision of the finished wine, they will adjust the yield accordingly. Crop yield is based upon the vigor of each individual plant (very vigorous growth can result in more but lower-quality fruit) but numerous variables can influence this concept.

- *Summer Pruning:* This type of pruning is the process of removing excessive grape clusters and foliage from the grapevine for influencing yield, which ultimately affects aroma/flavor and structural development in the grapes. This type of pruning forces a vine to exert more energy into its fruit rather than its foliage.
- *Winter Pruning:* This type of pruning trims off excessive canes or old growth to avoid diverting a vine's energy in producing new growth when springtime arrives. This type of pruning forces a vine to exert its energy into new growth buds and flowers for future grape clusters.

Common Vineyard Hazards

Learning Objective 4
Identify at least two common vineyard hazards

The grape farmer knows there are many challenges in the vineyard that can destroy or at least hamper the success of a grapevine's output. Some challenges include the numerous microorganisms, pests, and diseases that can attack and kill grapes and vines. Regions vary greatly in disease and pest issues. However, both temporary and permanent solutions have been developed to combat these viticulture challenges.

Microorganism Issues

Fungal Disease

Managing fungal diseases such as odium, mildew, and grey rot is a constant concern in the vineyard. Fungal disease is often associated with vineyard locations with excessive rain or consistently moist climates without adequate sunshine. In some cases, wind can assist in drying the vines and helping to prevent some fungal disease. The two most effective means of fungal control are good canopy management and preventative fungicide treatments. Widespread control methods to lessen fungal disease have included chemical sprays such as *bouillie* (BOO-yee) bordelaise (a solution of copper sulfate, lime, and water) and better knowledge of canopy management. Several types of fungal diseases are listed below:

Powdery Mildew Grapevines infected with powdery mildew display white powder-like patches on leaves, stems, and grapes. Powdery mildew can grow well in both wet and dry regions. It can kill leaves and defoliate the vine. Grape quality suffers when leaves are unable to perform proper photosynthesis.

Downy Mildew Symptoms include light green to yellow spots scattered across the leaf. These spots appear greasy and are commonly referred to as oil spots. The biggest concern for downy mildew is leaf infection.

Black Rot Symptoms include brown circular lesions on infected leaves. Left untreated it can destroy an entire grape crop. The biggest concern for black rot is the infection of young grape clusters. Infected berries first appear light brown and then turn to near black as masses develop on the surface.

Bunch Rot Infected berries appear soft and watery. In regions with high humidity, berries become covered in a grayish growth of fungus. Tight-clustered grape varieties are most vulnerable to bunch rot.

Out of the possible solutions to the challenges associated with microorganisms identified above, canopy management is considered one of the most effective solutions. Promoting good air circulation, sunlight penetration, and uniform leaf development are all benefits of practicing proper canopy management. The removal of excessive foliage around the grape clusters will allow the sunlight and air flow to assist in drying them.

Glassy Winged Sharpshooter These pests are named after the glassy or transparent appearance of their wings. The sharpshooters have caused widespread disease by passing on a bacterial infection known as Pierce's disease. The insects spread the bacteria that can cause death of the vine. At minimum, the vine's leaves will turn yellow and the fruit will wilt. Insecticides have been used to deter the ailment, but have not worked as a complete solution. Currently, experimentation with biological control by natural enemies is underway.

Animals/Pest Issues

Phylloxera (fil-LOX-er-uh)

The most feared enemy of grapevines is a small plant louse *Phylloxera vastatrix*. This aphid feeds on the roots of grapevines (especially on the highly vulnerable *vinifera* rootstock species), causing the vine to starve and thus preventing fruit development. In the 1860s, phylloxera was unknowingly transported from hearty native American vine species (which are resistant to phylloxera) to the *vinifera* species in Europe where it effectively destroyed the grape-farming industry over the next forty years. By the mid-1870s, the vineyards of France, Spain, and other countries were nearly devastated. It took Bordeaux over three generations to recover.

The solution was to graft *vinifera* vines to the American rootstock (which was much more resistant to phylloxera). Grafting in the vineyard is the technique of securing a vine to a rootstock. In most *vinifera* vineyards (except for those in Chile and vineyards in some parts of Australia and Washington State), cuttings of the desired varieties are grafted onto rootstocks of native American varieties that are resistant to phylloxera.

Birds and Other Pests

Birds eat grapes as a source of nourishment. Large nets are often placed across the vineyards to deter birds. Scarecrows can also be used with moderate success. Other animals such as deer and raccoons have been known to consume fruit and cause vineyard damage. Deer, the largest animal problem for most grape growers, can be fenced out. Repellents can also be used, such as hot sauce and pepper, or even dogs are effective in areas where problems are moderate.

Other molds, bugs, and bacteria that affect the growth of healthy grapes exist, but these vectors—organisms that transmit pathogens—occur on a more limited basis.

Weather Issues

Wind

Wind may prevent pollination of the flowers during the early part of the season. Later in the season, the winds can knock fruit off the vine and heavier winds can knock vines over. Some of the heaviest winds occur in southern France (the Mistral) where they have been known to rip vines right out of the ground.

Frost

Frost is a serious danger in many vineyards, especially those located on the valley floors where the coldest air settles on frosty nights. Frost will kill green tissues on vines. The good news is dormant buds, canes, and trunks will usually go unharmed. Late-spring frost can cause bud damage and may affect the yield by producing fewer grape clusters. A freeze, on the other hand, can kill dormant life. An early fall freeze can be devastating. In some colder regions, grape growers actually remove the vines from the trellis and bury them just below the surface so that they will survive the winter.

Sprayers, burners, and wind machines can all be used collectively or separately where frost is a constant danger to the buds, flowers, or berries. Wind machines are used to distribute heat from a central heat source, such as a fire or chaufferettes (gas heaters), that warms the grapes (or vines) to keep them free of frost. Many grape growers located in cold climates may also use aspersion, which involves sprayers that release water into the air. The water that lands on the grapes (or buds) forms an outer ice shell, but a warm, protected state is maintained on the inside.

Rain

Heavy rains are a concern both in early spring and at harvest time. Too much rain can prevent pollination of flowers in the spring. If it rains near harvest time, the fruit may be oversaturated and the flavors, sugars, and acid that have been developing throughout the growing season may be diluted.

Select vineyard areas may be allowed to use irrigation systems created to feed off nearby mountains, rivers, or lakes. This is critical for the survival of the vine in specific locations such as Argentina, Chile, and Australia, where lack of rain is a consistent problem.

Harvesting the Grapes

Learning Objective 5
Discuss the important considerations for determining grape harvest

Throughout the growing season, farmers inspect the grapes to ensure they are developing properly. The grower mainly checks for sugar levels, or brix. Historically, the brix level is perhaps the single most important quality in grapes being grown for wine production, because a certain amount of sugar is needed for the yeast to produce alcohol during the fermentation process. Figures 3.15 and 3.16 illustrate Gewürztraminer grapes being harvested in the vineyard.

Harvest begins in September and may continue into October, more specifically when the flavor, sugar, and acid levels of the grape reach the right levels. The individual grape variety, the ripeness factor, and the weather factor have the greatest influence on when to harvest a cluster of grapes. Now, the ability to evaluate both sugar and

FIGURE 3.15
Harvesting Gewürztraminer grapes. Courtesy of John Peter Laloganes.

FIGURE 3.16
Gewürztraminer grapes in a basket. Courtesy of John Peter Laloganes.

flavor ripeness of the grape is vital for determining the suitable time of harvest. On one hand, there must be sufficient sugar and flavor in the juice of wine grapes for yeast to feed on and convert into alcohol, but there has also to be a balance of alcohol and other structural components in the finished wine. The evaluation should involve both an objective approach, by measuring grape sugars (using a refractometer), and a subjective one, by measuring flavor using taste. Flavor ripeness, otherwise known as phenolic ripeness, is represented by a group of compounds that contribute color, aroma, flavor, and tannin to a grape. This kind of ripeness allows the tannins to become softer as the growing season progresses. Phenolic ripeness often trails sugar ripeness, but is important for allowing the maximum flavor of the grape to be obtained. Grape ripeness can be compared with teenagers in this regard: Often, their bodies (a grape's sugar content) mature faster than their minds (a grape's flavor development), leading the teens to believe that they are older and more mature than they may be. Once the decision to harvest has been made, the process must happen quickly. The grapes will also continue to ripen, destroying the preferred level of sugar and acid that has been intended for all season. A freeze could also destroy the crop, so when the word is given to harvest, the grapes need to be picked quickly.

Figure 3.17 shows the collection of grapes in the vineyard. Figure 3.18 showcase grapes that have just been harvested and are awaiting transport to the winery.

FIGURE 3.17
Zoetic wines being harvested. Copyright David Vance.

FIGURE 3.18
Zoetic grapes awaiting transport. Copyright David Vance.

Hang-time

Learning Objective 6
Explain how the concept of hang-time changes the style of wine

A recent ten- to fifteen-year trend has been to extend the "hang-time" (the delay of harvest) of the grapes, with the expectation of increasing aroma and flavor development. This practice produces very ripe fruit that yields a "fruit-forward" quality in the finished wine and, consequentially, an ample supply of alcohol. In some cases, certain producers have been criticized for too much hang-time and allowing the grapes to become overripe with a surplus of sugars, yielding a wine that is higher in alcohol and can be excessively out of balance. Though arguable, it has been this style of wine that has created much of the popularity and increased consumption over the last decade or so.

Methods of Harvesting

Harvesting of grapes can be conducted in two ways—mechanically or by hand. The selected method of picking largely depends upon the grower's philosophy of the finished wine, the size of the harvest, and the vineyard's terrain.

Mechanical (or machine) harvesting has made the process of harvesting grapes more efficient, often cost-effective, and a process that is well suited for large vineyards that lie on relatively flat terrain. This type of harvesting is conducted by a large tractor that straddles a vineyard row and strikes the vines with a large paddle to dislodge the fruit from the vine. The implementation of mechanical harvesting is often stimulated by shortages and/or the expense and complications of a labor force. It can be expensive to hire labor for short periods of time, coupled with the ability to work quickly and in the nighttime or early morning hours.

Hand harvesting affords more precise selection and tends to do a better job of protecting the grape's juice content from oxidation due to damaged skins. In some vineyards, there are small or incompatible widths between vineyard rows and/or steep terrains hinder the employment of machine harvesting. In addition, many viticulturists prefer hand-harvested grapes because of the greater care and judgment associated with the ability to determine the ripeness of a cluster of grapes. When growers choose to harvest their grapes in this manner, they generally work at night when temperatures are lower and the fruit is firmer and less susceptible to bruising. After the completion of the harvest, the vines become dormant as late fall and winter creeps in. Throughout the winter, there is relatively little work to be done in the vineyards as the grapes have moved indoors to the winery. The focus of growers at this time is vine protection to ensure the remaining vines have been cut back so that they don't die once exposed to temperatures of 10°F or less.

Being "Green" Friendly

Learning Objective 7
Identify three ways that grape growers can be "green" friendly

In the present-day wine industry, there is an increasing interest in incorporating "green" practices that are considered ecologically responsible behaviors. This has become not only an important buzz word but also a defining and critical concept that wineries are using to distinguish their wines from their competitors. Being "green friendly" simply means natural farming practices—ones that were typical throughout the world in the pre-industrialization era. The popularity of this green movement is due in large part from media, health reports, and general interest and awareness by the public. The benefits and positive long-term effects of naturally farmed products have become well documented and publicized on the overall environmental health of both the earth and the body. The categories of natural farming in the wine industry can be further classified into being sustainable, organic, and biodynamic. Each classification has a governing organization and certification process giving credibility and further recognition to these practices. However, some wineries feel that the expense to seek certification is too cost prohibitive and time intensive, but they may still follow many if not all the "green" practices. To the environmental extreme, some wineries believe in trying to attempt a 100 percent carbon-neutral winery where their practices help to offset their carbon footprints.

Sustainability

Many different definitions of sustainable practices exist in the wine industry and are largely open to interpretation. Many experts believe the current sustainable viticulture movement in the United States began with sustainable agriculture, which grew out of organic farming practices and the "green revolution" of the 1950s and the earth movement of the 1970s. The term "sustainable agriculture" first came into common use in the 1980s. As defined by the U.S. Congress, it is an integrated system of plant and animal production practices having site-specific applications. Simply, it is economically viable, socially supportive, and ecologically sound. Increasingly, more wine producers are recognizing "natural farming" as a path to follow that creates wines in a holistic manner through application of sustainable practices. The trend toward sustainable practices not only benefits the earth's natural resources, but many believe it results in grapes that reflect the distinctiveness of the land. Simply, the quality of the wine will only be as good as the quality of the grapes. Figure 3.19 illustrates cover crop growing between the rows of vines to not only increase diversification but also reduce soil erosion and increase healthy microflora in the soil.

FIGURE 3.19
Cover crop between vineyard rows. Courtesy of Cellar Angels, LLC.

Practicing sustainability is definable by each organization. However, many organizations may identify with and practice some of the following philosophies: minimal use of pesticides, herbicides, and chemical inputs, composting, practicing water conservation, using solar power, use of recycled materials, incorporation of ethical business practices, utilization of wild yeasts, and so on.

Biodynamics

This concept has become increasingly more popular in viticulture throughout the world. Biodynamics is based on the same principles of organic farming, elimination of chemicals, and genetically modified organisms (GMOs) but extending it to use homeopathic mixtures that are applied as per the lunar phases and cosmic rhythms with the purpose of increasing the spiritual connection to the land. As a practical method of farming, biodynamics not only embodies the ideal of ever-increasing ecological self-sufficiency but also includes mystical–spiritual considerations. Biodynamics is a view of the land as a living system and of the vineyard as an ecological self-sustaining whole. Biodynamics takes organic farming to a new level. The concept was introduced in the teachings by Austrian philosopher Rudolph Steiner in 1924 as a way to express the authenticity of the vineyard in which the crop (grapes, in this case) has been grown. In a series of lectures, he introduced an idea for a farming system based upon on-farm biological cycling through mixing crops and livestock. While the mixed-farming approach predates Steiner's ideas, it was his idea of the farm as an organism that helped to create a new paradigm of agriculture. It bans pesticides and artificial additives and strives for a self-contained sustainable farming system in which water and organic materials are recycled to regenerate the land.

Some grape growers believe that these practices have achieved improvements in the health of their vineyards, specifically in the areas of biodiversity, soil fertility, crop nutrition, and pest, weed, and disease management. Biodynamic producers also note that their methods tend to result in a better balance in growth, where the sugar production in the grapes coincides with physiological ripeness, resulting in a wine with the correct balance of flavor and alcohol content, even with changing climate conditions. Biodynamic wines are believed by some to have a better expression of its specific place of origin through its aroma, flavor, and structural components.

This form of viticulture has been adopted by increasing numbers of high-quality wine producers in France, including Domaine Huet of Vouvray, Nicolas Joly in Savennières, Domaine Leflaive of Puligny-Montrachet, Domaine Leroy in Vosne-Romanée, Comtes Lafon in Meursault, and Chapoutier in Hermitage as well as other vineyards around the world. French vigneron Nicolas Joly describes biodynamics as "a way of helping vines catch the climate and soil in the wine."

Organic Viticulture

This type of viticulture, as compared to conventional methods, is defined in the 1990 U.S. Farm Bill as "a system of grape growing which does not use industrially-synthesized compounds on the soil or the vines to increase fertility or to combat pest problems." Organic viticulture procedures carried out in the cellar include limited handling and processing, and avoidance of chemical additives (except for the occasional use of very low levels of sulfur dioxide as a preservative). Organic grapes utilize agricultural practices that exclude the use of synthetic fertilizers, pesticides, herbicides, and fungicides along with any GMOs.

Certification from Third Parties

Producers who "embrace" the green approach may seek to achieve a certification of such practices by one of the many accrediting organizations. It is common for the credentialing process to break down the approval criteria into several categories. There are four possible stages: (1) grape growing and winemaking, (2) production and packaging, (3) transport and sales, and (4) consumption and disposal. These stages add up to 100 percent of total carbon dioxide emissions per bottle of wine. With the four stages, a producer will use them to promote green-friendly practices as they see appropriate for their organization. Below is a list of some common third-party certification organizations:

- *Leadership in Energy and Environmental Design* (*LEED*) is an internationally recognized green building certification system, providing third-party verification that a building or community was designed and built using strategies aimed at improving performance across all the elements that matter most: energy savings, water efficiency, reduction of CO_2 emissions, improved indoor environmental quality, and stewardship of resources and sensitivity to their impacts.
- *Certified California Sustainable Winegrowing* (*CCSW*) is a third-party certification program related to the California Sustainable Winegrowing Program (SWP) to increase the sustainability of the California wine industry by promoting the adoption of sustainable practices and ensuring continual improvement.
- *California Certified Organic Farmers* (*CCOF*) offers premier organic certification programs throughout North and South America for all types of processors, farms, livestock operations, retailers, private labelers, brokers, and more.
- *Demeter Biodynamic Trade Association* (*DBTA*) is a membership organization for Demeter Certified Biodynamic® farms, vineyards, wineries, dairies, food processors, traders, and distributors.

Check Your Knowledge

Directions: Use these questions to test your knowledge and understanding of the concepts presented in the chapter.

I. MULTIPLE CHOICE: Select the best possible answer from the options available.

1. Viticulture is the science and study of
   a. winemaking
   b. grape growing
   c. wine education
   d. sommelier
2. Grapes used for winemaking grow best between
   a. 20 and 50° north and south of the equator
   b. 30 and 50° north and south of the equator
   c. 30 and 50° north of the equator
   d. near the equator
3. Grafting is the process of
   a. trimming off excess growth from a grapevine
   b. fusing one plant (with desirable traits) onto another
   c. utilizing natural fertilizers throughout the vineyard
   d. none of the above
4. Cloning is the process of
   a. replicating desirable traits from a vine and recreating those in some other vineyard
   b. fusing one plant (with desirable traits) onto another
   c. spraying vines with pesticides
   d. trimming off excess growth from a grapevine
5. The classic and most significant grape-growing vine used for winemaking is
   a. *Vitis labrusca*
   b. *Vitis riparia*
   c. *Vitis rotundifolia*
   d. *Vitis vinifera*
6. The slope of a given vineyard can assist the grapes with
   a. receiving more sunlight
   b. less nutrient-rich soil
   c. having stressed vines leading to the struggling vine theory
   d. all of the above
7. The topography of a given vineyard relates to
   a. the aspect or orientation of the vineyard
   b. the slope (degree of flat land or hillside locations)
   c. the level of elevation
   d. all of the above
8. Pruning is a process used in grapevine maintenance for the purpose of
   a. adjusting a grapevine's yield
   b. adjusting the quality and/or quantity of a particular vine
   c. none of the above
   d. both a and b
9. An example of a common vineyard hazard associated with insects is
   a. phylloxera
   b. birds
   c. wind
   d. frost

10. Factor(s) used to determine when to harvest grapes include
    a. sugar levels and phenolic ripeness
    b. mechanical or hand harvesting
    c. the month of September
    d. none of the above
11. Through the concept of hang-time, grapes tend to produce
    a. more sugar ripeness
    b. greater flavor (phenolic) ripeness
    c. less sugar and more acid
    d. both a and b
12. In cooler climates, grapes tend to produce
    a. less acid and more sugar
    b. less alcohol
    c. less sugar and flavor ripeness while retaining a grape's natural acidity
    d. both b and c
13. An appellation is best described as
    a. a viticulture or grape-growing area
    b. being distinguishable by geographical and geological features
    c. having very specific legal meaning in France
    d. both a and b
    e. all of the above
14. Which statement would most likely NOT be associated with a cool climate?
    a. Higher acid levels
    b. Riper and more robust fruit aromas and flavors
    c. Lighter more subtle citrus, tree fruit, vegetables, herbs, and mineral aromas and flavors for white wine
    d. Generally lower in alcohol from less ripe grapes
15. The broader macroclimate of Napa Valley can best be described as
    a. continental
    b. alpine
    c. maritime
    d. mediterranean
16. The classic soil type associated with Cabernet Sauvignon in Bordeaux is
    a. chalk
    b. gravel
    c. slate
    d. limestone
17. The broader the geographical label (California versus Napa Valley) on a wine label usually indicates that the grapes are:
    a. less quality oriented
    b. better quality oriented
    c. undetermined
    d. not enough information

II. DISCUSSION QUESTIONS

18. Why are machines used in the cultivation and harvesting of grapes?
19. Explain the two methods grape growers utilize to determine when the grapes are ready for harvest.
20. Explain the concept of "hang-time." How does this concept shape the finished wine? How is this concept significant for the wine-consuming public?

CHAPTER 4

Enology: Inside the Winery

CHAPTER 4 LEARNING OBJECTIVES

After reading this chapter, the learner will be able to:

1. Identify the seven steps of the winemaking process
2. Distinguish the pros and cons of hand versus machine harvesting of grapes
3. Explain the major distinction between the production of white versus red wine
4. Describe how the fermentation process can be manipulated
5. Explain the effects on a wine that has undergone malolactic fermentation
6. Identify at least two reasons that would explain the purpose of blending wines
7. Describe the significant considerations when using a wood barrel for aging wine
8. Discuss some different methods of clarification
9. Identify common variations in the wine bottle shapes

Wine is one of the most civilized things in the world and one of the most natural things of the world that has been brought to the greatest perfection, and it offers a greater range for enjoyment and appreciation than, possibly, any other purely sensory thing.

—Ernest Hemingway, *Death in the Afternoon*

Introduction to Winemaking

Learning Objective 1
Identify the seven steps of the winemaking process

Winemaking, or vinification, begins with the selection of grapes, which are then transformed through a series of events that concludes with the stimulating and evocative liquid known as wine. Traditionally known as a vintner or enologist, a winemaker is the specialist who converts a raw agricultural product the grapes—into wine. In France, there is no direct translation for winemaker; instead, the French use the term *vigneron* (vihn-yehr-RAWN), which is someone who grows grapes and cultivates a vineyard for winemaking. The word connotes or emphasizes the critical role that vineyard location and maintenance has in the production of wine.

Winemaking methods can vary greatly from country to country, region to region, and even grower to grower. Among the influential factors that shape the quality and style of a wine is the grower's philosophy and whether it is based on one of tradition or innovation. The winemaker strives to make the best-tasting wine possible from the raw product available—ultimately, however, as the philosophy goes, great wine is made in the vineyard. Giuseppe Quintarelli was largely regarded as one who produced the most coveted wines in the Veneto region of Italy. Prior to his death in 2012, he stated, "The fundamental problem in wine today is that too many producers 'hurry' to make their wines: they hurry the fruit in the vineyard and they hurry the vinification and rush to bottle. They rush to sell their product without allowing it the proper time to age. Patience—this is the most important attribute in winemaking."

Figure 4.1 illustrates one of the most historic and classic producers in northeast Italy's Veneto region.

There are numerous parallels between the two vocations of winemaker and chef. A chef can only craft a quality plate of food based on the excellence of ingredients on hand; similarly, winemakers search for the most exceptional grapes to make the greatest wine possible while considering certain revenue and cost parameters. Growing and harvesting quality grapes are just the first step in the winemaking process. Vinification entails a myriad amount of choices that will ultimately influence the personality of the finished wine. Ultimately, the two professions of chef and winemaker synergize at the table when a chef's food and a winemaker's wine are paired together. Figure 4.2 illustrates the connection of chef and winemaker with a roasted

FIGURE 4.1
Giuseppe Quintarelli. Courtesy of John Peter Laloganes.

FIGURE 4.2
Lamb shank paired with red wine. Courtesy of Erika Cespedes.

WINEMAKING PROCESS

Step #	White Wine	Red Wine
1.	Harvest/Pressing	Harvest/Crushing
	⇓	⇓
2.	Fermentation (without skins)	Fermentation (with skins)
	⇓	⇓
3.	Malolactic fermentation (possibly)	Malolactic fermentation (usually)
	⇓	⇓
4.	Blending	Blending
	⇓	⇓
5.	Aging	Aging
	⇓	⇓
6.	Clarification	Clarification
	⇓	⇓
7.	Bottling	Bottling

It's important to note that some of the steps outlined in the winemaking process may occur in a slightly different order and/or some steps may occur multiple times throughout the winemaking process.

lamb shank paired with red Bordeaux wine. Winemaking is a creative process that may involve a set of laws, traditions, and experiences—but ultimately, it's the intent and transformative powers of the winemaker that determine the success of the finished product. Winemaking consists of seven broad steps in the production process. These steps are not always exact, as there are minor adjustments made due to stylistic differences between winemakers and cultural differences from region to region.

Harvest and Pressing/Crushing

Learning Objective 2
Distinguish the pros and cons of hand versus machine harvesting of grapes

The harvest represents the culmination of the growing season, the result of nature, and the work of the farmer. As identified in the previous chapter, great care is taken to ensure that grapes are harvested at the appropriate time and in the correct manner. At harvest time, the qualities of the grapes represent the greatest potential of any wine that subsequently can be created. Since different grape varieties ripen earlier or later than others, the harvest typically will take two to three weeks prior to all desirable grapes being collected. Two methods of harvesting grapes include hand harvesting and machine harvesting.

- *Hand Harvest:* This method is generally viewed as a far superior process for harvesting grapes. It allows for the vineyard worker to be selective and to only pick ripe grapes. Labor cost can be high, and in some cases, it is difficult to find adequate labor supply.
- *Machine Harvest:* This method is viewed as a highly efficient process for harvesting grapes. While it is not always viewed as quality oriented as hand harvesting, it can address many of the labor issues associated with the other harvesting method. While there is an initial upfront investment to pay for the harvesting machinery, the long-term savings will be rewarded.

After the harvest, the grapes are transported to a winery and prepared for primary fermentation through an initial sorting process. This process is conducted by trained workers who sort through each of grape clusters to ensure the integrity of the grapes and to limit undesirable unripe or moldy berries. Figures 4.3 and 4.4 illustrate freshly harvested grapes being sorted prior to being pressed for fermentation.

Next, freshly picked and sorted bunches of grapes are placed in the crushing and de-stemming machines. Regardless of the type of wine, there are several methods used to expel must, or unfermented grape juice: (1) a mechanical crusher-stemmer, (2) a crank-operated press, and (3) a balloon press. While these contraptions vary,

FIGURE 4.3
Zoetic harvest being sorted (1). Copyright David Vance.

FIGURE 4.4
Zoetic harvest being sorted (2). Copyright David Vance.

the results are roughly the same. It's at this critical stage when white wine diverges from rosé and even further from red wine production. It may seem trivial; however, the term pressing that is associated with white wine should suggest a gentle, more delicate method of expressing juice from the skins. In contrast, grapes used for rosé or red wine are often crushed implying a more aggressive process in order to extract greater flavor and color pigment as well as tannin compounds.

Fermentation

Learning Objectives 3 and 4
Explain the major distinction between the production of white versus red wine

Describe how the fermentation process can be manipulated

The process of fermentation in regard to winemaking converts grape juice into wine—simply, the fermented juice of grapes. Fermentation involves the metabolic process of yeast consuming a sugar source and producing a by-product of alcohol, carbon dioxide, and heat. The alcohol that is created will have the personality characteristics of the initial sugar source. The fundamental distinction between white wines versus rosé and red wines is the use (or lack thereof) of their respective grape skins throughout the fermentation process. White wines are fermented without their skins as compared to rosé and red wines that are left to macerate and ferment with their skins for the length of the process. This difference could draw a loose parallel between the inherent differences of white wines being more "feminine" or delicate versus red wines being more "masculine" or robust. This is not to say that feminine white wines can't be rich and voluptuous as it's also not accurate that red wines can't be light and delicate. But it makes sense to assert that a white wine fermented without skin contact accrues a more delicate, feminine style. In comparison, rosé and red wines amass a more robust, masculine style due to the effects of skin contact throughout their production process.

Yeast

Traditionally, the fermentation process occurs through the use of natural or wild yeast that exists on the exterior of the grapes and within the winery. The bloom or white powdery yeast on the grape skin contains millions of wild yeast cells and gives the grapes their dusty look. The common application of native or wild yeasts is still evident in many Old World wine-producing regions where it is believed to connect with the concept of the terroir. It appears to be a natural extension of the "unique sense of place." However, one of the main concerns of using native yeasts is their unpredictability. If a producer has not worked with the native yeast from their vineyard, it becomes challenging to know how it will impact both the fermentation process and the outcome of the finished wine. All yeasts react and behave in different ways, almost temperamental at times when they go through the fermentation process. There are some winemakers who choose to begin with native yeasts, and in conjunction add cultured yeasts as a backup.

Now, it is common practice (particularly in the New World wine-producing countries) for winemakers to utilize the more predictable, cultured, or cultivated yeast strain. Once the grapes have been harvested and brought back to the winery, a dose of sulfur dioxide will be sprayed on the grapes in order to destroy the wild yeasts (and in the process also kill bacteria and mold) present on the exterior skins. The winemaker can make a selection of any hundreds of strains to emphasize certain characteristics that can work to emphasize or de-emphasize an aromatic compound or a structural component. These types of yeasts have been tested and are overall more predictable and understood.

The alcohol levels produced during fermentation depend primarily on the quantity of sugar present in the grapes—though there are several ways to manipulate this equation. In most wines, the fermentation process is carried out in several weeks—yeast can be either intentionally halted or naturally exhausted when the must reaches approximately 14 or 15 percent alcohol by volume. The majority of wines are made to be tasted dry—with no perceptible residual sugar (RS). However, the winemaker can also choose to intentionally halt the yeast at some point during the fermentation process leaving a certain level of residual sugar allowing some perceptible levels of sweetness. Small amounts of residual sugar may not be detectable, but can simply be used as a means of offsetting high levels of acid, tannin, or alcohol in order to bring about a better perception of balance in the wine.

The alcohol created through the fermentation process will contain the personality characteristics of whichever grape is used for the sugar source. As the grape must is fermenting and being converted into wine, the carbon dioxide is allowed to escape and the release of heat is controlled through refrigeration.

- *White Wine:* In the case of white wine, the juice is separated from the skin and seeds immediately after being pressed. The skins if left in further contact with the juice may only lead to a greyish hue, undesirable bitter, and overpowering elements in the finished wine.
- *Rosé Wine:* These wines can be made through the saignée (sahn-yeah), press method, or through blending red and white wine together. The saignée method is constructed from the use of red grape skins being induced to bleed some slight red color through several hours of contact time between the juice and skin. Another method that is known to produce delicate looking rosé wines is the press method. This method has an even briefer contact time of juice and skins compared to the saignée method. Simply, the grapes are pressed and as the juice is released and separated from the skins, it picks up a very light, pale pink color. The blending method is the mixing of white with just enough red wine to create the desirable level of pink color in the finished product.
- *Red Wine:* These wines are allowed to ferment with the presence of grape skins—the release of carbon dioxide produced during fermentation pushes a thick layer of the purple skins, pulp, stems, and seeds to the surface of the fermenting vessel forming a cap (or chapeau in French). Figure 4.5 illustrates a vat of grape must and skins being converted into wine.

FIGURE 4.5
Red wine grapes and skins fermenting at Terra Valentine. Courtesy of Terra Valentine.

Several components are extracted throughout the fermentation process, in particular during the production of rosé and red wines. The lengthier and warmer fermentation process will extract greater aromas/flavors through the release of flavonoids from the skins and seeds while the color pigment from the skins will bleed into the juice. Additionally, tannin compounds are another substance extracted through fermentation—this particular component is extremely significant to the structure and characterization of a red wine. Tannin is technically classified as a phenolic compound found in a grape's skins, seeds, and stems that acts to aid red wine in providing structure, texture, and ageability.

Malolactic Fermentation

Learning Objective 5
Explain the effects on a wine that has undergone malolactic fermentation

Malolactic fermentation (MLF), or malolactic (mahl-low-lak-tic) conversion, is a process used in winemaking that converts the naturally present tart "malic acid" into softer "lactic acid." This technique is based on a biochemical reaction that historically would occur unknowingly; now, this process is intentionally induced through the use of modern vinification techniques. MLF conversion is typically initiated by an inoculation of desirable bacteria shortly after the completion of the initial fermentation process.

MLF is used to alter the personality of a given wine by transforming the tart and slightly abrasive malic acid—commonly found in fruits such as apples—into softer and creamy lactic acid—commonly found in dairy products. MLF is generally thought to enhance the body and soften the texture by lowering the levels of acidity as well as altering its sensation. The result is an overall reduction of tartness with an increased weight or mouthfeel. In addition, this winemaking technique releases an obviously detectable "butter-like" and "banana-like" aromas and flavors that derives from a natural chemical known as *diacetyl* (die-ASS-it-ahl). The winemaker may perform malolactic fermentation on an entire batch of wine or only on a portion. Ultimately, it depends on the stylistic vision of the winemaker. Malolactic fermentation is applied universally to most red wines (except for Gamay based wines) and a select few white wines—typically useful for Chardonnay.

Blending

Learning Objective 6
Identify at least two reasons that would explain the purpose of blending wines

Blending is another vinification technique that can occur throughout many of the various stages of the winemaking process. In some cases, it can take place immediately after fermentation or delayed until clarification. Blending (or lack thereof) should not be thought of as good or bad; instead, it can be best understood as just another technique available to the winemaker. Some of the most famous wines in the world can consist of either stand-alone grape varietals (such as a red Burgundy that is 100 percent Pinot Noir) or a mixture of blended complementary grape varietals (such as a red Bordeaux that is a blend of varying amounts of Cabernet Sauvignon, Merlot, Cabernet Franc, and others). Another example of a wine from the same country is the classic Champagne blend that typically contains a combination of Pinot Noir, Chardonnay, and Pinot Meunier grape varietals. Yet another type of Champagne, "Blanc de Noir," may contain only Pinot Noir. In essence, the decision "to blend or not to blend" often rests on the culture, history, and/or regulations of the geographical location and/or the intent of the winemaker. Ultimately, blending is a matter of the style of wine being produced. Figure 4.6 illustrates a bottle of Spanish Rioja, a wine that is classically blended from the grape varietals of Tempranillo, Garnacha, Mazuelo, and Graciano.

FIGURE 4.6
Bottle of Marques de Riscal from Rioja Spain. Courtesy of Erika Cespedes.

Blending can take many different forms—it is possible to blend several complementary grapes (which is what most people think of when they hear a wine is blended) or to blend the same grape from different locations.

In the United States, a winemaker may blend with a single grape as long as a minimum of at least 75 percent is within the bottle in order to still be labeled with the stated varietal. For example, 75 percent of a Cabernet Sauvignon grape with 25 percent of a Merlot grape can still legally claim Cabernet Sauvignon on the label. The moment a wine doesn't consist of the legal minimum, no grape varietal may be listed on the label. The blending approach is utilized to adjust or fine-tune components and add complexity or greater dimension to the wine. Wines can be blended for the purpose of maintaining a certain cost parameter, by incorporating in the blend a small percentage of either an inexpensive grape or the same grape that has been harvested from a less prestigious location. This approach allows the winemaker to incorporate grapes from locations where the cost of real estate (ultimately, the cost of the grape) is cheaper.

Aging

Learning Objective 7
Describe the significant considerations when using a wood barrel for aging wine

Once the wine is fermented, it begins the journey of maturation, or aging. The winemaker selects the aging method very carefully because different characteristics can be imparted to the wine as a result of the vessel used; the aging process can significantly influence the style and personality of the finished wine. Many winemakers believe that the aging process allows the wine an opportunity to soften its rough edges while simultaneously imparting subtle aroma/flavor characteristics. The winemaker's vision is the determining factor as to whether and, if so, how a wine will be aged. The aging process can be conducted in one of two broad ways—oxidative or reductive. If the wine is aged by an oxidative technique, then it is wood aged or wood influenced. If the wine is aged by a reductive approach, then the wine is aged (or, more accurately, preserved) in stainless steel. Figure 4.7 displays a large stainless steel fermenting tank that is used for preserving the primary aromas and flavors of the wine.

FIGURE 4.7
Stainless steel vats. Courtesy of John Peter Laloganes.

Stainless steel tanks are used mainly for aromatic white (and some lighter red) wines whose primary aromas/flavors and youthful crisp acidity are intended to be preserved. Stainless steel aging doesn't truly age the wine; instead, it more accurately preserves the wine and prevents the passage of oxygen that would otherwise dramatically alter the wine's personality.

Wood barrels are a centuries-old tradition used to store and age most red wines and many full-bodied white wines. Wood-barrel aging is the process of holding the wine in wood for a maturation period of months to years, whereby various components present in the wine slowly combine to create complexity and finesse. The industry standard is to use French or American oak as the preferred wood. Oak from other places, such as Slovenian oak, is sometimes still used. In the past, wine regions have used differing kinds of wood such as mahogany, chestnut, and pine. For the most part, use of these alternative wood barrels has been disregarded over time due to their overpowering characteristics masking the personality of the wine. Figure 4.8 shows a collection of wood barrels aging wine at Justin Vineyards.

FIGURE 4.8
Aging in barrels at Justin vineyards. Courtesy of John Peter Laloganes.

The Cooper and the Barrel

The cooper or barrel maker is someone who constructs the barrels—an age-old profession that is a prestigious craft often passed down within generations of families. Coopers are skilled technicians, fashioning barrels from raw wood through many processes based on a particular set of specifications.

The cooper cuts the staves (or slats) of the barrel and allows them to dry before adhering them together. Machines are almost always used to cut the staves now, but a little over 100 years ago, the whole process was done by hand. Once the staves are dried, they are held in place on the barrels by large iron hoops. Finally, each end of the barrel has a head that is made from several pieces of flat wood that are hinged together. Figure 4.9 shows the final stages of barrel construction.

Prior to a winemaker procuring barrels, there are numerous decisions that need to be considered. Factors that affect the barrels ability to influence a wine's personality throughout the aging process include (1) level of toast; (2) type of wood; (3) size of the barrel; (4) age of the barrel; and (5) length of aging.

FIGURE 4.9
Barrel construction. Courtesy of John Peter Laloganes.

Level of Toast

After the barrel is assembled, it is "toasted" or charred on the inside (the effect of exposure of the wood to varying degrees of fire for varying lengths of time) according to the winemaker's specifications. The amount or degree of toast, or seasoning, as it is sometimes called, in the barrel has an effect on the flavor profile of the aging wine. Barrels can be ordered with varying levels of toast: Light toast contributes subtle aromas and flavors to the wine, and medium and heavy toast both contribute greater intensity of aromas and flavors to the wine. The toast decision will be made on the basis of the variety of grape to be used and the style of wine that is intended to be produced. Figure 4.10 illustrates a wood barrel being toasted in preparation for it to be filled with wine. Figure 4.11 shows some barrel chips depicting different levels of barrel toast—on the left is a medium barrel toast and on the right is a heavy barrel toast.

FIGURE 4.10
Toasting a barrel. Courtesy of John Peter Laloganes.

Type of Wood

The barrels are made from various types of wood depending on their intended use, preference of the winemaker, and tradition of a particular production area. The most common wood used for barrel making is white oak—both French and American. French oak comes from the numerous forests throughout France, and the trees usually range in age between 150 and 250 years old. (French oak grows slowly because of low levels of rainfall and cooler temperatures in its environment.) Only the upper part of the trunk is used for barrels; therefore, each tree can yield about two barrels. The remainder of the tree is used for firewood, planks, beams, and veneer wood. The type of wood can have a dramatic influence on the wine through the aging process. American oak has bigger grains that allow greater passage of oxygen and so contributes stronger, more significant aromas and flavors to a wine. French oak has smaller, tighter grains to permit less flow of oxygen and thus maintains more subtle aromas and flavors than American oak. One of the most sought-after French oak is *Limousin*

FIGURE 4.11
Toast levels. Courtesy of Erika Cespedes.

FIGURE 4.12
Red wine aging in barrique. Courtesy of Carrie Schuster.

FIGURE 4.13
Aging in foudres at Tablas Creek winery in Paso Robles California. Courtesy of John Peter Laloganes.

FIGURE 4.14
Foudre barrels at Tablas Creek winery in Paso Robles California. Courtesy of John Peter Laloganes.

(lee-moo-ZAHN) oak from the Limousin Forest near Limoges, France. Limousin oak is also used for Cognac, white Burgundy wines, and California Chardonnay. French oak is said to give a subtle tobacco shop and coffee shop aroma and flavor qualities to the wine. American oak, on the other hand, is much younger and more affordable than its French counterpart. American oak gives much more assertive vanilla, coconut, and clove characteristics to the wine, often used in Zinfandel and Rioja wines of Spain.

Size of the Barrel

Barrels come in different sizes; the smallest size is 5 gallons while larger vats contain thousands of gallons. The size of the barrel can dramatically influence the aroma, flavor, and color of the wine being held in the barrel. The smaller the barrel, the greater the flavor imparted, because the wine has more direct contact or surface area with the wood of the barrel—while a larger barrel has less surface area and therefore contributes less aroma and flavor. The typical wine barrel holds 225 liters (almost 60 U.S. gallons) and is commonly called a *barrique* (ba-REEK). Figure 4.12 illustrates a barrique, the standardized wine barrel. However, between wine regions barrel names and capacities will slightly vary. Figures 4.13 and 4.14 show larger size barrels known as foudre.

Age of the Barrel

Typical aroma and flavor characteristics from wood aging fall in the category of bakeshop/coffee shop (coffee, chocolate, caramel, vanilla, almond, and toasted nut), cigar shop (tea and tobacco), and bakeshop (nutmeg, cinnamon, and all spice). These aromas and flavor nuances can be more or less assertive dependent upon the number of times a barrel has been used. If the barrel has never been used before, the wood will yield greater aromas and flavors—in contrast, any previous use will impart less influence from the wood, and instead, more impact through the slow passage of oxygen. The distinction and prominence of wood tannin is most pronounced in new, unused barrels and becomes less significant with older barrels that have been previously used. By the time a barrel is about five years old, it is virtually neutral in its influence on the taste of the wine. Every time that a barrel is reused (for each yearly vintage), it contributes less flavor and fewer components and becomes more of a holding vessel rather than a contributor to the wine.

FIGURE 4.15
Aging at R. Lopez de Heredia in Rioja Spain. Courtesy of John Peter Laloganes.

Length of Aging

The length of aging wines can range from a few months to several years. During this time, small amounts of evaporation will occur throughout this aging process. Lengthy aging assists to soften harsh tannins and allow desirable flavors to develop. Additionally, as red wine ages, its tannins and color compounds polymerize (PUH-lym-err-ize), or attach to each other forming large molecules. Eventually, these particles fall out of the suspended wine solution, forming sediment at the bottom of the barrel. Figure 4.15 shows wine being aged for a lengthy period of time in a wine cellar located in Rioja, Spain.

In lower-quality, large-production wineries, shortcuts have been created to gain the benefits of oak flavor without actually going through the time or expense of traditional oak barrel aging process. Such methods include using oak chips or oak shavings in a large "tea bag" placed inside stainless steel tanks or neutral wood vats.

Clarification

Learning Objective 8
Discuss some different methods of clarification

Clarification is the process of both removing undesirable particles in the wine and making it more stable by eliminating the chance for refermentation. Clarification is a major consideration because, on one hand, many believe that a wine should be free of particles and, on the other hand, clarification can run the potential risk of stripping the wine of desirable aromas, flavors, and structural components. Therefore, many quality-oriented winemakers opt for the softest, gentlest method and the least amount of clarification. In fact, it has become common practice at the higher-end of the market that red wines leave some existing sediment from the production process. The clarification process can be carried out by several different methods, possibly, in combination with one another, depending on the grapes or the traditions associated with a winemaking region. The four common methods are racking, cold stabilization, fining, and filtering. In many cases, methods such as fining and racking are used together.

Racking

The racking method is considered one of the gentlest and most common methods for limiting the loss of desirable aspects in the wine. Racking involves periodically draining the sediment, or decomposing yeast cells (called lees) by transferring the wine from one container to another, leaving the sediment behind in the original container. Racking is a natural method because it relies on gravity to pull the unwanted particles to the bottom of the original container. Racking can be conducted once or several times before bottling, for greater clarification.

Cold Stabilization

This clarification process is used largely in white wines to remove excess tartaric acid that would otherwise later form potassium bitartrate crystals, or tartrates. Tartrates have the appearance of shards of glass, but are completely safe and edible. Although a common practice is to remove this type of sediment, not all producers do so, and it seems less common in producers who believe in a "hands-off" type winemaking philosophy. Cold stabilization is accomplished through chilling a wine down to 40°F, causing the tartaric acid to crystallize, which allows the wine to then be racked, leaving the crystals behind.

Fining

Often used in conjunction with the racking method, this is another form of clarification method that incorporates a fining agent that forms a chemical bond with the undesirable particles, causing them to precipitate out at a faster rate to the bottom of the storage vessel. Then the wine is racked, leaving the particles behind in the original container. Some fining agents can include egg whites, bentonite clay, bull's blood, gelatin, and isinglass (an extract from fish bladder) that is a gelatin-like substance. In addition to clarification, this process can soften harsh astringent tannins and allow desirable flavors to develop.

Filtering

This clarification method passes wine through tubes and filters containing a fine mesh filter with small holes. These holes are smaller than the particles to be removed. Thus, the particles are collected and disposed of. The wine flows through a series of filters, which take most, if not all, of the sediment out. This method is common in large-scale production in order to produce a wine that is free of any particles and appears as pristine and consistent as possible.

Bottling

Learning Objective 9
Identify common variations in the wine bottle shapes

During the Greek and Roman periods, wine was originally transported in a two-handled vessel called an *amphora* (ahm-FOR-uh). Possible evidence suggests that this was not only a crude form of a wine bottle, but also part of an early appellation system. The shape of the container (and any etching identified on the container) could indicate the city or region, the winemaker, and the vintage of the wine. During the late 1600s and early 1700s, the glass bottle evolved. Once it became more durable, it was evident that wine could be aged in glass bottles to evolve much more effectively than in the past.

Eventually, various bottle shapes and colors were developed to hold different types of wine—which made it easy to identify the type or style of wine that was within each bottle. The shape and color of the bottle can communicate a great deal of information about the region or country of origin and the grape varietal within the bottle. This can guide a wine drinker in the general direction of understanding the grapes or particular style of wine that might be found within the vessel.

Wine bottles come in a variety of sizes, shapes, and colors—though many wine regions have their traditional bottle shapes. Winemakers throughout the world typically respect the traditional wine bottles associated with their regions and the wines placed within them. Even though winemakers can choose any bottle shapes, for the purpose of recognition, wine is typically sold in one that is consistent with the wine region. Figure 4.16 depicts some different wine bottle sizes being showcased in a restaurant.

FIGURE 4.16
Varying bottle sizes on display. Courtesy of John Peter Laloganes.

Wine Bottle Sizes and Names

Wine bottles have been standardized to generally contain 25.4 oz. (750 ml) of liquid. The half-bottle (contains 12.7 oz./375 ml) has become increasingly popular over the last decade as an alternative to provide a source of high-quality wine for the solo diner. Generally, a standard bottle of wine contains between four and six glasses of wine (depending upon portion size). Bottles of other sizes range from 187 milliliters

Name of the Bottle	Size of the Bottle	Typical Glasses of Wine per Bottle	# of Standard Bottles
Split	187 milliliters	1	¼
Half-bottle	375 milliliters	2–3	½
Standard bottle	750 milliliters	4–6	1
Magnum	1.5 liters	8–12	2
Jeroboam (jehr-OH-boam)	3 liters	16–24	4
Methuselah (imperial) (Mehth-OOHS-ehl-ah)	6 liters	32–48	8
Salmanazar (sahl-MAHN-ah-zahr)	9 liters	48–72	12
Balthazar (BAHL-tah-zahr)	12 liters	64–96	16
Nebuchadnezzar (neb-ah-kahd-NEHZ-her)	15 liters	80–120	20

(6.3 oz.) to 15 liters (507 oz.). Each size of bottle is identified by a particular name and several are named after biblical kings and other significant figures.

Wine Bottle Shapes

Bordeaux Bottle

The Bordeaux wine bottle has straight sides with steep, tall shoulders. It's an excellent shape for wines that tend to exude sediment (typically, old red wines and Bordeaux reds are known to age well) because the steep shoulders can serve to hold back the sediment as the wine is poured. The Bordeaux bottle shape is the most common shape used for red wines around the world. It is often found in dark red or black colored glass. Figure 4.17 shows a William Cole Cabernet Sauvignon found in a Bordeaux bottle shape.

Red wines commonly found in this style of bottle include *Cabernet Sauvignon* (KAB-er-nay SOH-vin-NYOHN), *Merlot* (mare-LOW), and *Zinfandel* (ZIN-fun-del). The Bordeaux bottle shape is also used for white wine, though it will be found in light green or clear uncolored glass. White wines commonly found in this style of bottle include *Pinot Grigio* (PEE-know GREE-joe), *Sauvignon Blanc* (SOH-vihn-yohn BLAHN), particularly if the wine is from Bordeaux or California, and *Semillon* (SEM-ee-YAHN).

FIGURE 4.17
Bordeaux shaped bottle: Maroon Cabernet Sauvignon from Napa Valley California. Courtesy of Cellar Angels, LLC.

Burgundy Bottle

The Burgundy wine bottle typically is sturdy and heavy, with shallow, gently sloping shoulders. The Burgundy bottle shape used for red wine can often be found in a light green or black colored glass. Figure 4.18 shows a collection of Chardonnay from famous locations around the world, found within the Burgundy bottle shape.

Red wines found in this bottle include *Pinot Noir* (PEE-know-NWAHR), *Gamay* (gam-may), and *Syrah* (SEAR-ah) and Syrah blends. The Burgundy bottle shape can also be used for white wine and is often found in a light green, yellow, or clear uncolored glass. White wines commonly found in this bottle include *Chardonnay* (SHAR-duh-nay) and Sauvignon Blanc when the wines derive from New Zealand or Loire Valley, France.

German Bottle

The German (or Hock) wine bottle is narrow, thin, and tall and maintains a very gently sloping shoulder. The color of the bottle is typically light green or brown. The bottles with brown glass often (at least in Germany) identify wines that have been produced from the Rhine (RINE) region, and green glass for wine from the *Mosel-Saar-Ruwer* (MOH-suhl sahr ROO-vayr) region in Germany. Figure 4.19 shows German bottle shapes.

The German bottle is most frequently used for white wines. They use the most popular German and Alsace grape varietals that include *Riesling* (REEZ-ling), *Gewürztraminer* (guh-VERTZ-trah-mean-er), and *Pinot Gris* (PEE-noh GREE) when it comes from Alsace, France.

Sparkling Wine Bottle

The sparkling wine bottle is made from a very thick glass, with gently sloping shoulders and a long neck. The sparkling wine bottle also contains a rather large punt, or indentation in the bottom of the bottle, to assist in durability. The punt is needed to help reduce the pressure felt along the bottom of the bottle—the bottle's weakest point. Classic Champagne (and other famous sparkling wines around the world) commonly uses a blend of three grapes, including Chardonnay, Pinot Noir, and *Pinot Meunier* (muh-NYAY). Other sparkling wines from around the world may use the same or similar grapes but will often still be found in the sparkling wine bottle shape. Figure 4.20 shows the sparkling wine bottle shape.

Sparkling wine bottles are designed to withstand the high pressure exerted by the carbonation development after bottling. Pressure can often exceed 90 pounds per square inch (or PSI), approximately 2–3 times the pressure of a car tire.

FIGURE 4.18
Burgundy bottle shapes. Courtesy of Erika Cespedes.

FIGURE 4.19
German bottle shapes. Courtesy of Erika Cespedes.

FIGURE 4.20
Sparkling wine bottle shape. Courtesy of Erika Cespedes.

FIGURE 4.21
Fortified bottle shape. Courtesy of Erika Cespedes.

Fortified Wine Bottle

The fortified wine bottle shape contains the four most significant fortified wine types: Port, Madeira, Marsala, and Sherry. These wines are contained within this typically sturdy, bulky bottle that often has tall shoulders and a larger bulge in the neck to help capture or hold back the potential of sediment. Often, these are wines that need many years to properly age and tend to contain some sediment. With the exception of vintage or late-bottled vintage port, these bottles usually have a cork stopper rather than the traditional larger corks typically used for other wines. The cork stoppers allow easy opening and closing of a bottle after each serving. Figure 4.21 shows a fortified wine bottle shape.

Preservation Options for Wine

Wine is a product that is quite perishable. Therefore, there are some considerations the winemakers can contemplate. One of the most universal practices is the usage of sulfur dioxide. But another consideration is to alter the type of packaging through using box wine, aseptic packaging, or wine on draft.

Sulfur dioxide (SO_2) or sulfites are chemical compounds found within most bottles of wine. It is created naturally at low levels near 10–20 ppm (parts per million) during the process of fermentation. Additional sulfur is commonly added throughout the winemaking process to act as an antibacterial and antioxidation agent. Sulfur's antimicrobial and antioxidant properties assist in preventing a wine from refermenting within a bottle and prohibit oxygen exposure throughout the winemaking and bottling processes.

Producers are required to state "Contains Sulfites" (meaning sulfur dioxide) on the label of every bottle of wine whether it's domestic or imported in the United States if it contains 10 ppm or more. Almost every bottle will contain this amount, whether the winemaker has added sulfur or not, because 10–20 ppm is quite common to occur naturally through the fermentation process. The maximum limit of sulfites that a wine can contain is 350 ppm with most averaging around 125 ppm. Even if a wine states "No sulfites," it still may contain sulfites, but less than 1 ppm of wine.

Box wine (cask wine or boxed wine) is wine packaged in a bag-in-box that was created in the early 1900s. This packaging has wine contained in a plastic bladder typically with an airtight valve emerging from a protective box. Another type of box wine is aseptically packaged in an enclosed, nonreactive container. These methods serve as alternatives to traditional wine bottled in glass with a cork closure. The benefit of buying wines in these types of packaging is that they allow for greater shelf life and less concern over damage from light and the need to store wines on their sides. The downside is that typical box wine has been of low quality, relegated to kitchen use or for sale at grocery stores. However, some producers have upgraded with a few quality options, but the vast majority is mass produced.

Wine on draft consists of the use of wine being packaged in a metal container or keg—similar to the concept of draft beer. This packaging not only protects the wine from all of its environmental concerns of light, humidity, temperature, and oxidation, but also has environmental advantages. Kegs are reusable (therefore reducing waste and cost) and reduce the weight of shipments. This type of packaging has had a resurgence in recent years due to the commitment by high-quality wine producers recognizing the viability for a wines-by-the-glass program in restaurants and bars.

Wine Closures

Once the wine is bottled, it will be sealed with some form of closure for protection. The most common closure is the cork or screw cap as they have been proven most effective in preventing oxygen from entering and destroying a wine or at the very least minimizing a wine from oxidizing prematurely. If desired, a wine will be bottle-aged in order to integrate the wine's aroma/flavor components. The fruit characteristics in wine tend to develop slowly into more complex characteristics with time in the bottle. These changes may take between six months and five years to become noticeable. In addition to altering of aroma/flavor, bottle aging has a softening and mellowing effect on the wine's structural components. Most wines are given a period of months (and in some cases years) to allow for integration of the wine's components.

There are various types of closures or stoppers available for a winemaker to seal a bottle of wine. There are advantages and disadvantages for all, but each one must perform the essential function of preserving the wine and, if necessary, promote conditions conducive to appropriate development. Most wines are produced for early consumption shortly after purchase, and only a small percentage of wines are created to benefit from and to be enhanced by bottle development through long-term aging. Bottle development occurs when a wine closure allows the optimal amount of oxygen (too much air can lead to oxidation) to positively affect the wine. The type of wine closure can affect the outcome of the finished wine by determining the personality and overall quality of the finished wine. Ultimately, a winemaker's vision of the finished product will determine the appropriate wine closure.

Cork Closure

Cork is a natural material—specifically, the bark of a tree that has been used as the primary closure for wine and beer bottles since the late 17th and early 18th centuries. Authentic wine corks are made from the bark of an evergreen oak tree, predominately found in Portugal, Spain, North Africa, and other Mediterranean countries. Cork production is extremely slow as the bark grows at the rate of 1.5 mm per year and eventually slowing down to 1 mm. After the tree reaches maturity (sixteen to twenty-five years), it is harvested by hand every nine years in a labor-intensive process that strips the bark, only for it to regenerate throughout the coming centuries. Typically, extractions of bark can be made until the tree reaches 150 years of age, allowing for about twelve or fifteen extractions before its productive cycle comes to an end. Figure 4.22 shows part of a bark stripped from a Spanish oak tree used for cork production.

FIGURE 4.22
Tree bark used for cork closures. Courtesy of John Peter Laloganes.

Corks are flexible, elastic, lightweight, and natural, and when the cork is wet, it swells to form a tight hermetic seal within the neck of the bottle. Therefore, wine bottles closed with corks must be stored either upside down or on their sides in order to keep the cork wet and the bottle tightly sealed. If a wine bottle closed with a cork is stored upright, over time the cork can dry out and contract, allowing air to enter the bottle, causing a darkening of the wine and a loss of aroma and flavor.

Most corks average about 1¾ inches in length, but the size can vary from 1½ inches to 2½ inches. Longer corks are reserved for wines that will age well. For example, the wines of Bordeaux (which are most known for longevity) use a 2¼-inch cork, allowing them to age longer. Longer corks can arguably provide more of a barrier between the air and the wine. Red wines tend to be bottled with longer corks, while white wines

FIGURE 4.23
Display of corks. Courtesy of Erika Cespedes.

have shorter corks as they are generally intended for earlier consumption. Figure 4.23 shows a collection of wine corks.

Besides proper storage, another concern is the development of an off-flavor from tainted corks. During the corks' preparation for use as wine closures, they are sanitized with chlorine, and if a certain mold is present in the cork, a highly aromatic compound called *2,4,6-trichloroanisole* (try-clore-AN-iss-all), or TCA, is formed. This TCA has a disagreeable smell that is detectable in very low concentrations and will destroy a bottle of wine by imparting a dank "wet cardboard" character to wine. Winemakers refer to a wine having detectable levels of TCA as being corked. It has been estimated that between 3 and 5 percent of the corks are tainted with TCA, and unfortunately, there is no efficient way to determine whether a cork is tainted until a bottle of wine is opened.

Screw Cap

Many New-World (and some select Old-World) winemakers are leading a campaign to replace the traditional wine cork with a high-tech aluminum screw cap, named the Stelvin after the company that created it. Screw caps first appeared in the 1970s; however, the connotation that they were "cheap" didn't promote their success. Over the last twenty years, there has been a renewed excitement for screw caps because they are inexpensive, easy to open (not requiring a special tool—the corkscrew), and easily resealable and because they limit the passage of oxygen.

Philosophically, it seems suitable to use the screw cap, particularly if the wine has been stored only in stainless steel prior to bottling and is destined to be consumed early. This carries on the intended style of the winemaker of pure essence of primary fruit aromas and flavors and preservation of structural component of acid while maintaining the youthfulness of the wine that oak aging and a cork might otherwise alter.

The Debate Between Cork Versus the Screw Cap

Wine traditionalists find it difficult to accept the screw cap because of the lost romance surrounding the opening of a bottle of wine sealed with a cork closure. Many wine purists will still promote the cork because the tradition and symbolism that are pervasive in the wine industry tend to associate a screw cap with inexpensive wines of low quality. Nonetheless, the more adventurous winemakers in Australia and California, and even some Old-World producers in Germany and Chablis, France, have begun to bottle some of their prestigious wines with a high-quality screw top. Ultimately, the biggest test will be the acceptance by the consumer and their perceptions associated with each type of closure.

Other Closures

The popularity of artificial (synthetics) or plastic corks for early-drinking wines has been on the upswing in response to the problem of cork taint found in natural corks. But synthetic corks are not without problems of their own and have not been widely embraced by the industry. For long-term storage, the biggest problem has been the quick passage of oxygen, which, after a period of time, can result in oxidized wines that exhibit symptoms of aging sooner than if sealed with other closures. Others are hesitant to put their wines in contact with the elastic polymers that make up a synthetic cork for fear that some undesirable compounds may be extracted from the corks and alter the wine. Technicals, or composites, are formed with pieces of natural cork and bonding materials and usually incorporate disks of natural cork at each end. The glass stopper is a recent creation that can be made out of either glass or Plexiglas.

Check Your Knowledge

Directions: Use these questions to test your knowledge and understanding of the concepts presented in the chapter.

I. MULTIPLE CHOICE: Select the best possible answer from the options available.

1. Enology is the science and practice of
   a. winemaking
   b. grape growing
   c. wine education
   d. sommelier
2. Fermentation is the process of
   a. yeast consuming sugar to produce alcohol
   b. yeast consuming alcohol to create sugar
   c. yeast consuming sugar while producing a by-product of alcohol and CO_2
   d. none of the above
3. A significant distinction between the production of red wine versus white wine is that
   a. red wines are macerated and fermented with their skins
   b. red wines are not macerated and fermented with their skins
   c. white wines are fermented with their skins and red wines are not
   d. white wines are macerated with their skins while red wines are fermented with their skins
4. Malolactic fermentation is the process of
   a. converting lactic acids to malic acids
   b. converting malic acids to lactic acids
   c. increasing the mouthfeel of a wine and softening its acids
   d. both b and c
5. The reason for aging a wine in stainless steel is to
   a. preserve the natural acidity in the wine
   b. eliminate the passage of oxygen
   c. preserve a wine's youthfulness
   d. all of the above
6. The reason for aging a wine in wood barrels is to
   a. allow the slow passage of oxygen to evolve the wine's character
   b. introduce additional aromas and flavor components to the wine
   c. allow time for a wine's components to integrate
   d. all of the above
7. The purpose of clarification of a wine is to
   a. assist in stabilization of the wine until consumption
   b. integrate additional aromas and flavors in the wine
   c. preserve the wine's youthfulness
   d. alter the wine's acidity
8. The standard sized bottle of wine
   a. is called a magnum bottle
   b. is called an imperial bottle
   c. contains 24.5 oz.
   d. contains 25.4 oz.
9. A Bordeaux shape bottle will likely contain which of the following grapes?
   a. Chardonnay
   b. Pinot Noir
   c. Riesling
   d. Cabernet Sauvignon

10. A Burgundy shape bottle will likely contain which of the following grapes?
    a. Merlot
    b. Chardonnay
    c. Pinot Noir
    d. Both b and c
11. A major distinction between French and American oak is that French oak
    a. expresses more vanillin components
    b. breathes slightly more oxygen through the barrel
    c. is a harder wood
    d. is only used in Europe
12. Aromas and flavors of "bakeshop spice" and "tobacco shop" are often associated with
    a. stainless steel aging
    b. malolactic fermentation
    c. oak aging
    d. grape type
13. Which is NOT an influence of malolactic fermentation (MLF)?
    a. Imparts the aroma/flavors of vanilla
    b. Imparts diacetyl
    c. Rounds out the acids
    d. Heightens the body
14. Which grape/wine is MLF usually NOT applied?
    a. Gamay
    b. Riesling
    c. Chardonnay
    d. Malbec
    e. Both a and b
15. Which wine is NOT a famous blend of various grape varietals?
    a. Red Rioja
    b. White Bordeaux
    c. Châteauneuf-du-Pape
    d. Red Burgundy
16. When red wine is aged in a barrel for an extended length of time, what effects will take place?
    a. Tannins and color compounds polymerize
    b. Small amounts of evaporation occur
    c. Harsh tannins soften
    d. Red wines lose color intensity
    e. All of the above
17. The process of aging the wine on the lees can
    a. contribute greater mouthfeel and texture
    b. contribute a yeast and bread aroma and flavor
    c. act to lessen oxygen exposure
    d. all of the above

II. DISCUSSION QUESTIONS

18. Differentiate between a white wine aged in stainless steel versus the same wine aged in a wood barrel.
19. Explain the decision-making factors that need to be considered when a winemaker uses wood aging.
20. Discuss the five different bottle shapes and common grapes (or wines) found within each one.

CHAPTER 5

The Wine Styling Approach: White Wines

CHAPTER 5 LEARNING OBJECTIVES

After reading this chapter, the learner will be able to:

1. Explain the wine styling approach
2. Explain the concept of typicity
3. Explain characteristics of the crisp and youthful white wine style category
4. Explain characteristics of the silky and smooth white wine style category
5. Explain characteristics of the rich and voluptuous white wine style category

". . . from so simple a beginning endless forms most beautiful and most wonderful have been, and are being, evolved."

—Charles Darwin

Grape Varietals and the Wine Styling Approach

Learning Objective 1
Explain the wine styling approach

To more effectively make sense of wine, the consumer has to first understand some of the most popular grape varietals that are found throughout the world. While there may be an estimated 10,000 different *vinifera* grapes, the astonishing diversity of wines in the marketplace can leave the consumer bewildered when making a purchasing decision in the wine aisle or from the restaurant wine list. Even more difficult is when the wine buyer has to make hundreds and thousands of purchasing decisions when building a wine menu for a given restaurant or bar. This book is concerned with the most prevalent international varieties (with a smaller selection of local indigenous varietals) and ones that are likely to be experienced at most food and beverage establishments. An international grape variety is often referred to as a "classic variety" or "noble variety" which has both a long established reputation and adaptability for producing high-quality wine throughout the world. As the wine industry expands across the globe, select grape varieties, ones that have always been thought to be connected with specific locations or homelands, referred to as indigenous grape varietals, have now spread out and begun to gain international recognition.

One approach to making sense of the world around us—whether it's understanding music, art, or wine—is to find some means of broad categorization for the purpose of comprehension and communication. Many can relate to the idea of organizing taste in music; someone who may enjoy the sounds of Pearl Jam may also appreciate the band Nirvana, both of which can be broadly lumped into alternative rock known in the day as grunge. Not unlike music, the beverage enthusiast can determine meaningful ways to communicate their favorite wines by referencing a "style" recognizable by others. The wine styling approach is an attempt to neatly arrange the numerous

FIGURE 5.1
Glass of white Burgundy from Chassagne-Montrachet. Courtesy of Erika Cespedes.

FIGURE 5.2
Aerating a red wine. Courtesy of Erika Cespedes.

grape varietals (and the wines they produce) according to their predominant sense of personality, based primarily on specific weight of the wine or other structural components. Arranging wines according to "style categories" is a unique and user-friendly approach to the framing and making sense of the numerous wines that surround us. This approach can assist in painting broad brush strokes of a wine's style by creating categories useful for understanding the range of structural components associated with the numerous wines available. Figure 5.1 shows a glass of Chardonnay from the village of Chassagne-Montrachet in Burgundy, France.

Application of this useful approach can reduce intimidation for the novice and intermediate wine consumer, allowing them to explore other wines with similar but not identical characteristics. The wine styling approach is an unpretentious method that allows one to understand, describe, and convey their likes and dislikes to others. This approach can also arm service staff and consumers alike with a sensible template that allows them to easily peruse a wine list or wine store and identify a selection that best suits their intended preferences. Application of the styling approach also becomes a helpful training, communication, and selling tool for the service staff of any given food and beverage establishment.

The process of assessing a wine's structural components is the main source of determining a wine's style and its corresponding category. These components consist of a wine's mouthfeel or tactile sensations that are detectable on the palate through application of the tasting process at varying levels of (1) dryness/sweetness, (2) acidity, (3) tannin (only perceptible in rosé and red wines), (4) body (or weight of the wine), and (5) alcohol. Each one of these potential five aspects cumulatively creates the distinctive structural components of a given wine and thus its style category. Figure 5.2 identifies a glass of wine being aerated through swirling it in order to soften tannin components and integrate its aromas and flavors—all of which allow the taster to better assess any given wine.

This chapter begins with an examination of grape varietals and applying "typicity" as a means to understand them. The chapter then proceeds with the application of the wine styling approach to white wines—then in the following chapter, progressing to rosé and red wines. The subsequent chapter will then cover the two alternative categories of wine: sparkling and fortified wine—and in addition, dessert wines. The style approach divides the white wines into three broad structural categories beginning with "crisp and youthful whites," then "silky and smooth whites," and finally "rich and voluptuous whites." Continuing with the style approach, the rosé and red wines will be categorized into three broad structural categories as well, beginning with "fruity rosé and vibrant reds," then "mellow and complex reds," and finally "bold and intense reds."

White Wine Style Categories	**Red Wine Style Categories**
Crisp and Youthful Whites	Fruity Rosé and Vibrant Reds
zesty ... clean ... vibrant	youthful ... lively ... charming
Silky and Smooth Whites	Mellow and Complex Reds
refreshing ... bright ... velvety	rich ... smooth ... velvety
Rich and Voluptuous Whites	Bold and Intense Reds
lavish ... elegant ... voluptuous	complex ... concentrated ... evolved

Understanding Typicity

Learning Objective 2
Explain the concept of typicity

Typicity is the term used to describe a wine that reflects or expresses its "typical" or classic personality traits and thus exhibiting its signature characteristics of the grape (or blend of grapes) it derives from. It is the process of profiling and using stereotypes to help make sense of wine. Doing this helps with understanding the typicity of any given grape varietal and becomes quite useful for navigating through the vast world of wine. This approach can showcase the stylistic distinction between grape varietals and "like" varietals from different locations and varying winemaking techniques. Typicity helps narrow the confusion and makes the process of understanding the world of wine less intimidating. In essence, how much a Cabernet Sauvignon "smells and tastes like a Cabernet Sauvignon." If a particular Cabernet Sauvignon, for example, does not express its typical characteristics, question, "Why does this wine not express its typicity?" If a wine lacks typicity, it can be a potential "yellow flag" that should question the quality of the particular drink on hand. In the wine world, understanding the concept of typicity is an approach that can be used to demystify wine through an established and understood range of acceptable variations among a grape varietal as they are shaped by the thousands of wine producers around the world.

The practice of classifying a wine and assessing some stereotypes can assist the wine consumer in understanding and more effectively being able to communicate what they see, smell, and taste within a glass. The concept of typicity allows for any wine enthusiast or beverage manager, no matter their level of expertise, to maintain a starting point of comprehension and to develop the complex skill of analyzing and communicating how specific grape varietals or wines should generally perform most of the time.

The process of beginning to assess a wine starts with identifying or stereotyping its two broad components or elements: the aroma/flavor components and the structural components. A wine's aroma/flavor components (fruit, coffee shop, and earth, etc) are just one segment of defining a wine's profile. Second, a wine's structural components (dryness/sweetness, acid, alcohol, bubbles, tannin, and body levels) are an additional source of detecting the personality through textural sensory abilities in the mouth. Collectively, these two components define a wine's typical personality otherwise known as typicity. Once a consumer can discover the type of wine they enjoy, it's possible to now explore different producers, regions, and similar grape varietals, just as it's possible to seek out a new genre in art, film, or music. Figure 5.3 illustrates the locations where a taster can experience a wine's structural components versus its aroma and flavor components.

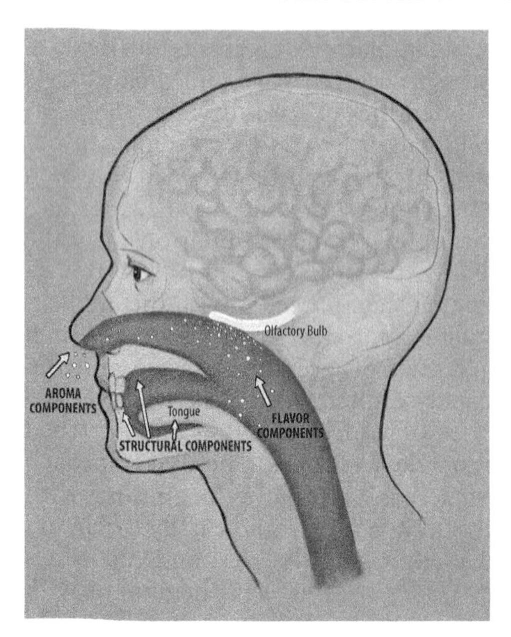

FIGURE 5.3
Tasting diagram. Courtesy of Thomas Moore.

The significance of typicity is quite valid to use as a measuring post to know that one may have been somewhere before—to know what to look, smell, and taste for in the wine glass. Stereotyping provides a guide to a wine's personality and offers possibilities as to what should be expected when a wine is examined. Typicity will never yield a definitive personality of a particular grape type or wine (nor is it necessarily desired), but it will provide some useful range of personality attributes.

White Wine Style Categories

FIGURE 5.4
Hugel Riesling from Alsace, France. Courtesy of Erika Cespedes.

There are at least twenty major white-wine grapes commonly offered in beverage establishments, but the three most significant international grapes used for producing high-quality wine are Sauvignon Blanc, Riesling, and Chardonnay. Figure 5.4 displays iconic Hugel Riesling from Alsace. Other distinguishable white-wine grapes renowned for producing great wine in select locations include Albariño, Chenin Blanc, Gewürztraminer, Grüner Veltliner, Pinot Blanc, Pinot Gris/Grigio, Sémillon, Torrontés, and Viognier. It is very important to note that the wines within each category (or style) have been stereotyped for illustrative and simplification purposes. They have been arranged into categories based on some common structural characteristics; however, these wines can vary dramatically from producer to producer, varying growing locations, and the different applications of winemaking techniques.

Crisp and Youthful Whites

Learning Objective 3
Explain characteristics of the crisp and youthful white wine style category

zesty ... clean ... vibrant

The "crisp and youthful whites" visually appear watery to pale in color intensity with a greenish to straw yellow color hue—illustrating its youth and use of reduction techniques throughout its creation. This style category produces wines that often contain pronounced fruit, mineral, and floral aromas and flavors. These wines have aromas/flavors that are fresh, clean, crisp, and lively, often referred to as "fresh smelling." The crisp and youthful white wines tend to accentuate their medium to highly acidic nature. They offer ample malic acid that provides tartness and sourness—similar sensations that would be found in a granny smith apple. This wine style category utilizes stainless steel and/or neutral oak aging so as not to overpower the desirable, more simplistic primary characteristics of the grape varietal and the connection to its place of origin. Crisp and youthful white wines are most often light to medium in body with a moderate alcohol percentage, usually between 11.5 and 13.5 percent. This style category intends to produce wines that are best consumed in their youth, within one to three years from their vintage date.

Albariño (ahl-bah-REE-nyoh)

Albariño is a small, thick-skinned grape considered to be Spain and Portugal's premier white wine varietal. It's grown primarily in northwest Spain's *Galicia* (gah-LEE-thee-yah) region but is also prevalent in the *Minho* (MEE-nyoh) region in northwest Portugal where the grape goes by the alternative name—*Alvarino* (ahl-vah-REE-nyoh). Albariño has flourished around this part of the world for quite possibly 900 years—but has only in the more recent era, been introduced to the American wine-drinking public. Figure 5.5 shows an Albariño from the Rías Baixas area of Spain.

- *Aroma/Flavor Components:* Albariño is often described as being fairly to highly aromatic in floral, citrus fruit (lemon), tree fruits (apricot, peach, and apple), and bakeshop (almond).
- *Structural Components:* This grape is vinified dry with medium to high acidity and quite possibly a slight spritz of bubbles in some Portuguese versions. Albariño is typically light plus to medium-bodied. This grape's inherent tartness is typically intended to be embraced in its youth, anywhere from one to three years from harvest. Some producers have begun to experiment—applying subtle use of malolactic fermentation, neutral oak aging, and/or aging the wine on the lees. The use of these techniques attempt to add a touch of additional complexity and rounding out a touch of its tartness.
- *Significant Locations:* Albariño is Spain and Portugal's most prolific white wine grape. In the northwest Galicia region of Spain, the cool and damp area of *Rías Baixas* (REE-ahs by-SHAHS) produces roughly 90 percent of their plantings as the varietally labeled wine—Albariño. While in the Portugal's Minho region, Alvarino may be a stand-alone varietal but is more often blended with other indigenous white-wine varietals to produce the white version (as opposed to the alternative red wine version) of Vinho Verde wines (literally "Green Wine" referring to its youthfulness). These wines offer a more affordable and lighter version of the typical Albariño found in Spain.

FIGURE 5.5
Albariño from Rías Baixas, Spain. Courtesy of Erika Cespedes.

Grüner Veltliner (GREW ner VELT lee ner)

Grüner Veltliner is the most important indigenous grape variety in Austria. As with most grapes, its sites and production yield are crucial to the quality. These wines can range from a more austere—spicy, peppery version to ones yielding more stone fruit characteristics. Grüner Veltliner is most often fermented and aged in stainless steel tanks or at most in very old, neutral casks—allowing the wine to showcase its transparency of both the grape's personality and the nuances of location. Figure 5.6 displays a Grüner Veltliner from Weinviertel region in northeastern Austria.

FIGURE 5.6
Grüner Veltliner from Weinviertel region in northeastern Austria. Courtesy of Erika Cespedes.

- *Aroma/Flavor Components:* Citrus and tree fruits (lemon, lime, grapefruit, and nectarine) and vegetal (asparagus and green pepper with some white pepper).
- *Structural Components:* Dry wine with medium plus to high acid. Depending on the quality level and style, Grüner Veltliner delivers a range of light to medium body.
- *Significant Locations:* Austria (most widely planted in Niederösterreich and in areas rising north or west of Vienna and running along the Danube in the Wachau, Kamptal, and Kremstal growing areas).

Pinot Grigio/Gris (PEE-noh GREE-joh/GREE)

This varietal offers clusters of grayish-blue colored grapes that have mutated from the Pinot Noir varietal. Pinot Gris means "gray" in which the skins yield a brassy colored white grape juice. There are two basic styles of Pinot Gris/Grigio. In Italy, the grape is known as Pinot Grigio, where its largely grown in the north central and northeast regions. Traditionally, these grapes were harvested early in order to preserve ample acidity even though flavor ripeness wasn't always fully developed. When picked in this manner, the wines tend to be lightly aromatic with subtle aromas and flavors. The structural components consist of light body and higher acidity.

FIGURE 5.7
Pinot Grigio from Italy's Friuli region. Courtesy of Erika Cespedes.

This approach has started to slowly shift with higher quality versions experiencing extended hang-time along with application of some brief maceration with its skin for additional viscosity, and perhaps some contact with lees for a couple of weeks to months.

France and Oregon—truly the rest of the world—identifies this grape as Pinot Gris. Typically, Pinot Gris expresses quite a different style of wine as compared to Pinot Grigio. Pinot Gris is often left on the vine for a slightly extended time period in order to obtain greater flavor development. This yields a greater "fruit-forward" quality to the wine. In this process, the grapes sacrifice a small amount of acid, but in return, the grapes gain greater aromatics and a bit more body. Structural components offer light to medium body and medium acidity with moderate alcohol content. Figure 5.7 displays a Pinot Grigio from northeastern Italy.

- *Aroma/Flavor Components:* Pinot Grigio/Gris often has subtle, somewhat light to fairly aromatic nuances that include citrus fruit (lemon), tree fruits (pear and apricot), bakeshop (almond and honey), and mineral (wet stone).
- *Structural Components:* Pinot Grigio (the Italian version) tends to be light-bodied as the grapes are often harvested early to ensure a successful crop prior to any inclement weather. These grapes may be medium to high in acidity and slightly underripe—hence their inability to develop as much ripeness, maintaining a lightly aromatic quality. These wines are commonly aged in either stainless steel or neutral oak barrels. Pinot Gris (the French, Oregon, and California versions) offers greater aromatic intensity. Due to the extended hang-time, the grapes have greater intensity and contain a medium body with medium acidity.
- *Significant Locations:* Pinot Gris/Grigio thrives in cooler regions that ideally allow for a lengthened fall time to ripen. Some of the most significant locations include France (Alsace), Italy (Trentino-Alto-Adige, Veneto, and Friuli), New Zealand (Marlborough and Martinborough), and Oregon (Willamette Valley).

Sauvignon Blanc (SOH-vihn-YOHN BLAHN)

Sauvignon Blanc is a green-skinned grape variety that originated from southwestern France in the Loire Valley and Bordeaux regions. The grape gets its name from the French word sauvage (Sah-VAHJG) that means "wild," referring to its ability to produce excessive foliage. At some point in the 18th century, through natural cross pollination, the Sauvignon Blanc grape crossed with Cabernet Franc to parent the Cabernet Sauvignon varietal in Bordeaux, France.

Most wines made from Sauvignon Blanc can be broadly classified into two distinct styles: the Loire Valley style and the Bordeaux style. This categorization is largely based on stylistic considerations of tradition, culture, or simply the preference of the winemaker. The "Loire" styles are less manipulated, showcasing a more transparent sense of the grape and location. On the other hand, the "Bordeaux" style offers some greater influence of vinification technique whether it's a blend of a secondary grape varietal, or a use of barrel fermentation and/or barrel aging, incorporation of lees aging, or converting a portion of the wine's malic acid to lactic acid.

- *Aroma/Flavor Components:* Sauvignon Blanc is a fairly to highly aromatic grape varietal that provides nuances of garden (grassy, herbs, and asparagus), citrus fruits (lemon, grapefruit, gooseberry, and lime), and tropical fruits (guava, cantaloupe, and honeydew melon). The aroma/flavor profile can vary depending upon the climate and vineyard practices. Less common in modern-day, Sauvignon Blanc can contain an aggressive "cat-pee like" odor when the grapes lack sun exposure or are harvested early. More appropriate clone selections and viticultural practices that expose the grapes to greater sunlight now produce wines that are more ripe and citrus to melon-like in aroma and flavors.

- *Structural Components:* Sauvignon Blanc is now planted in many wine regions of the world, producing a dry, youthful and refreshing white wine with a light to medium body and ample acidity. The alcohol content is moderate and generally hovers near 12–13.5 percent. Most Sauvignon Blancs are intended to be consumed in approximately one to three years from harvest date.

  The rarer, sweetened versions of Sauvignon Blanc are famously produced in Sauternes (sow-TEHRNS), an appellation in the Bordeaux region of France. These wines are always blended in varying amounts with the Semillon varietal. The wine is named after its appellation—arguably the world's most famous "Rot" wine where this particular location is very susceptible to "noble rot" (or the Latin term Botrytis cinerea). This desirable mold works to dramatically concentrate a grape's characteristics while rendering a highly aromatic dessert wine with a viscous, rich mouthfeel.
- *Significant Locations:* Sauvignon Blanc is grown throughout the world. The varietal is most well known in particular areas in France (Loire Valley and Bordeaux), Italy (Friuli and Venezia), New Zealand (Marlborough and Martinborough), California (Sonoma County), and Chile (Casablanca).

The Loire Valley Style

The Loire Valley style is characterized by the fragrant, zingy freshness reminiscent of cut-grass, fresh herbs, grapefruit, gooseberry, and wet stone and smoke. These wines are 100 percent single-varietal Sauvignon Blanc and range from light to medium body. Loire style wines are produced in either stainless steel or neutral oak in order to maintain the wine's natural crisp malic acid and preserve the wine's charming youthfulness. Historically, the two most famous production areas are located in Loire Valley, France: the appellations of Pouilly Fumé (poo-yee foo-MAY), and directly across the Loire River in Sancerre (sahn-SEHR). It is these two appellations where the grape expresses its most understated complexity, with the best wines grown on limestone hills containing a high proportion of limestone and flint, which is believed to impart a subtle "gun smoke" element. Figure 5.8 shows Claude Riffault Sancerre from Loire Valley, France.

For the purpose of categorizing, the Sauvignon Blanc wines from New Zealand (most famously from the Marlborough region) can fall under the Loire Valley style. This is due to the similar approach to showcasing the transparency of grape and location. New Zealand is notably credited with the varietal's modern-day success when, back in 1985, Cloudy Bay Winery produced Sauvignon Blanc grapes with greater sun exposure and hang-time that yielded an increased ripeness and an overall easy-to-like aroma/flavor profile containing bursting nuances of a ripe, red-ruby grapefruit. While New Zealand styles are commonly recognized as a riper, more aromatic version of the wine from Loire Valley, they certainly share the same sense of purity with minimal intervention from the winemaker, rendering a wine that speaks of the grape's characteristics and its unique sense of place. Pictured in Figure 5.9 is Mohua Sauvignon Blanc from Marlborough, New Zealand.

FIGURE 5.8
Claude Riffault Sancerre from Loire Valley, France. Courtesy of John Peter Laloganes.

The Bordeaux Style

The Bordeaux style of Sauvignon Blanc is characterized as a bit more rounded and luscious—almost medium plus in body as compared to the lighter Loire style. These wines are often blended with Sémillon, another grape that is indigenous to Bordeaux. This combination is often referred to as the ultimate

FIGURE 5.9

Mohua Sauvignon Blanc from Marlborough, New Zealand. Courtesy of Erika Cespedes.

"odd-couple" as Sémillon contributes richness (fatty and less acidic mouthfeel) and softens the sometimes abrasive acidity found in Sauvignon Blanc. The Bordeaux style may also utilize barrel-fermentation and/or aging the wine on the lees (in contact with its yeast cells) to soften its acid, build complexity and body.

Fumé Blanc is a wine (that can fall under the Bordeaux style) that was given its nickname from Napa Valley's Robert Mondavi who wanted to rebrand his Sauvignon Blanc back in the 1960s. At the time, Sauvignon Blanc was a rather unpopular variety in America—much in contrast to the French Sauvignon Blancs that were very popular but were labeled by their appellation names, Sancerre and Pouilly-Fumé from the Loire Valley. Mondavi decided to tame the wine's herbal and vegetal characteristics through imparting a hint of smoke, toast, and vanilla from barrel aging, as smoke was reminiscent of the aromas derived from the flinty soils in the wines of Pouilly Fumé. In the process, the wine increases in dimension and body. He released the wine under the name Fumé Blanc (Fumé literally translates to "smoke") as an allusion to the French Pouilly-Fumé. Although there are no legal constraints, the usage of the name Fumé Blanc is a marketing distinction that usually indicates a Sauvignon Blanc that has been oak-aged. This alternate name can allow a winery to offer both unoaked and oaked versions of the Sauvignon Blanc while creating a distinction of the two styles for the consumer.

Torrontés (toh-RON-tehs)

The emergence of Torrontés is today's signature Argentine white wine varietal. Torrontés is best expressed in Argentina's higher, elevated sites, where the grapes slowly develop their intense phenolic characteristics and retain their signature crisp levels of acid. The wine is fermented and aged in stainless steel tanks where its primary aromas and flavors are preserved. It is a varietal that offers intense aromatics that are reminiscent of Gewürztraminer or Muscat (believed to be one of its parents); on the other hand, the structural components are suggestive of Sauvignon Blanc.

- *Aroma/Flavor Components:* Torrontés offers intense aromatics of tree and tropical fruits (peach and lychee) and floral (rose, jasmine, and geraniums).
- *Structural Components:* Torrontés is vinified dry, and it produces a light-bodied, medium to highly acidic wine with moderate levels of alcohol.
- *Significant Locations:* The grape is cultivated in locations that assist in preserving the grapes' natural high acidity, most notably the Argentine growing areas of Catamarca, La Rioja, Mendoza, Salta, San Juan, and Rio Negro.

Silky and Smooth Whites

Learning Objective 4
Explain characteristics of the silky and smooth white wine style category

refreshing ... bright ... velvety

The "silky and smooth whites" visually appear pale to medium in color intensity with straw yellow to golden yellow in color hue. This style category produces wines that are often concentrated with a pronounced smell of highly aromatic dried fruits, floral, and honeyed aromas and flavors. The grapes that create this style category have likely been allowed to remain on the vine for extended hang-time to achieve greater ripeness.

Silky and smooth whites are often a bit more complex and rounder than wines of the crisp and youthful whites wine category. These wines will ultimately yield greater mouthfeel due to the slight potential for residual sugar remaining after the fermentation process. The residual sugar can yield a slightly off-dry wine style that will assist to soften and round-out some of the sharp malic acid. Silky and smooth white wines are most often medium in body and can range in alcohol percentage from as low as 9.5 percent to upward of 14 percent. This style category intends to produce wines that can be consumed in their youth, but additionally gain complexity as they age for several years from their vintage date. Figure 5.10 displays Riesling from Alsace that works to reflect this style category.

FIGURE 5.10
Riesling from Alsace. Courtesy of Erika Cespedes.

These wines offer the greatest spread of diversity; they can be found in varying spectrums of dryness/sweetness, body, and alcohol levels. Silky and smooth white wines may have experienced some modest neutral oak aging and/or undergone slight aging on its yeast cells in order to provide some additional richness and to soften some of the wine's tart malic acid.

Chenin Blanc (SHEN-ihn BLAHN)

Chenin Blanc is arguably the most versatile of all white wine varieties, at least in terms of its various styles and types, as it can successfully produce dry to off-dry table wines, sparkling wines, and sweetened dessert wines.

- *Aroma/Flavor Components:* Chenin Blanc can offer fairly aromatic characteristics of tropical fruit (melon), tree fruits (apple, pear, and peach), floral, bakeshop (honey and tea), and mineral (wet stone).
- *Structural Components:* These wines can offer a varying range of dryness and sweetness. Depending upon the quality of grapes and the quantity of residual sugar, Chenin Blanc's body will range from a light to full body. Regardless of style, Chenin Blanc always contains medium to high acidity. This grape can stand up to modest oak-aging but is more often left to age in either stainless steel or neutral oak barrels in order to accentuate the grape's primary characteristics and the emphasis of soil type.
- *Significant Locations:* Chenin Blanc expresses its most distinctive and reputable personality in France's Loire Valley region. It reaches its pinnacle in the appellations of Savennières (mostly dry versions) and Vouvray (mostly off-dry to sweetened versions). In the New World, Chenin Blanc is mildly popular in California (Central Coast) and South Africa (Coastal Region) where the grape is locally referred to as Steen. Pictured in Figure 5.11 is a Vouvray, arguably one of the world's best representations of Chenin Blanc.

FIGURE 5.11
Vouvray. Courtesy of Erika Cespedes.

Gewürztraminer (guh-VERTZ-trah-mean-er)

Gewürztraminer is an unusual white wine varietal that contains pink to red skin but yields white juice with a slight brassy color. This grape is believed to have originated in the Italian village "Tramin" (located in northeastern region of Alto-Adige) in the

16th century. Possibly sometime in the 19th century, the grape mutated to resemble more its current-day expression. Gewürztraminer is a German word for "spicy grape" and occasionally referred to as Gewürz or Traminer. Figure 5.12 illustrates some recently harvested clusters of Gewürztraminer grapes.

FIGURE 5.12
Clusters of Gewürztraminer. Courtesy of John Peter Laloganes.

- *Aroma/Flavor Components:* Many quality-oriented Gewürztraminer growers leave the grapes on the vine well into October in order to build maximum aromatics in the grapes. The personality of Gewürztraminer is characteristically highly aromatic with intense spicy aromas and flavors of citrus fruit (grapefruit), tropical fruit (lychee), bakeshop (cinnamon), and floral (rose).
- *Structural Components:* Gewürztraminer is typically medium to full in body, with the ample alcohol content giving a slight spice sensation at the back of the throat. While leaving Gewürztraminer on the vine in order to gain aromatics, the late harvest can unfortunately reduce the grape's acidity and creates an obvious flat or flabby wine. The variety's high natural sugar content means that the wine can range from dry to off-dry to sweet depending upon how the yeast and fermentation process is handled. If the wine is intended to be dry, it will take on a considerable high level of alcohol that can reach levels of 14 percent. Otherwise, the wines can be made into off-dry or sweet style, to lower the potentially out-of-balance alcohol content and provide silkiness in the mouth. Figure 5.13 shows a Domaine Weinbach Gewürztraminer from the Alsace region of France.
- *Significant Locations:* Gewürztraminer performs best in cool moderate climates with a slow, long growing season for gradual development of aromatics and preserving of its somewhat deficient acidity levels. All of Gewürztraminer's styles, dry to sweet, are prevalent in Alsace, France, including full-bodied dessert wine versions. Though Gewürztraminer reaches its pinnacle in Alsace, the grape can also be found in Germany and Italy (Trentino-Alto-Adige and Friuli). In the New World, Gewürztraminer is found in New York state, Canada, California, and Washington state.

FIGURE 5.13
Domaine Weinbach Gewürztraminer from the Alsace region of France. Courtesy of Erika Cespedes.

Pinot Blanc (PEE-noh-BLAHN)

Pinot Blanc is a white wine grape that was created from a mutation of the dark-skinned Pinot Noir. It's easy to understand how this grape remains relatively unknown largely due to its many synonyms throughout the world. In Germany, Pinot Blanc is called *Weissburgunder* (VICE-buhr-gun-dehr), in France it's called Klevner, and alternatively—Pinot Bianco in Italy.

- *Aroma/Flavor Components:* The fragrance of a Pinot Blanc is typically fairly neutral that ranges from muted to lightly aromatic. The delicate aromas and flavors most often include tree fruits (apple and pear), citrus fruit (lemon), tropical fruit (melon), bakeshop (almond and yeast), and earth (wet stone).
- *Structural Components:* Pinot Blanc commonly produces a dry white table wine—though it also acts as the base for Crémant d'Alsace—the sparkling wines of Alsace, France. The varietal tends to produce a light-to-medium body

wine with medium acidity. For increased richness and complexity, some producers apply extended contact with lees during the production process.

- *Significant Locations:* This grape is grown primarily in the northernmost wine reaches of the world. Some of the most prominent locations for Pinot Blanc are in the Old World: France (Alsace), Germany, Austria, and Italy (Friuli, Veneto, and Trentino-Alto-Adige); and in the New World, Oregon remains one of the few places to truly adopt this grape.

FIGURE 5.14
German Riesling. Courtesy of Erika Cespedes.

Riesling (REEZ-ling)

Riesling is a white-wine grape variety native to the Rheingau region of Germany where it has been cultivated since the early 16th century. Though often consumed young, Riesling's substantial acidity, aromas, and concentration of flavors are suitable for extended aging, particularly of wines that contain higher residual sugar content. Riesling is a variety that is highly expressive of its place of origin—it prefers to be a "stand-alone" varietal as it establishes its distinctively seductive personality without the need to be blended with other grapes.

In recent history, Riesling had become unfashionable due to some less than quality-oriented winemakers making overly sweet versions with inadequate levels of acidity. Currently, Riesling may be experiencing a bit of a renaissance, as producers have sought better site selection (Riesling prefers cooler areas with temperate climates to allow the grapes to ripen slowly), and are providing drier or sweeter options with ample acidity to provide a more balanced expression of the grape's personality. Figure 5.14 illustrates a Joh. Jos. Prüm Riesling from the Wehlener Sonnenuhr vineyard in Mosel region of Germany.

Riesling can be categorized in two broad styles: the French style and the German style. Winemakers either will ferment the wine dryer, achieving higher alcohol levels, as in Alsace, France, or will often leave noticeable residual sugar (RS) through partial fermentation, yielding a wine with varying levels of sweetness, as in many German styles. The density and body increase with greater levels of sweetness, providing an effective pairing with more robust, fatty, spicy, or sweet food items. Figure 5.15 shows two bottles of Riesling from the same producer, yet illustrating different levels of ripeness—the one on the left is a Spätlese and the one on the right is an Auslese.

FIGURE 5.15
Riesling from Joh. Jos. Prüm. Courtesy of Erika Cespedes.

Riesling is also known for producing some of the world's most celebrated dessert wines. These wines can be made by a combination of methods. Three of the most well-known Riesling dessert wines are from

1. **Late Harvest Wine** When the grapes are left on the vine for extended hang-time
2. **Rot Wine** When the grapes are attacked by a friendly fungus (Botrytis Cinerea) that concentrates the aromas/flavors and structural components
3. **Ice Wine** When the grapes are frozen on the vine in order to extract water content and therefore concentrate existing juice

- *Aroma/Flavor Components:* This grape is highly aromatic with concentrated aromas and flavors of tree fruits (peach and apricot), tropical fruits (pineapple and golden raisin), citrus fruits (lemon and lime), bakeshop (honey), and minerals (petroleum, flint, metal, and wet stone). The petroleum (or rubber band) aroma/flavor is associated less often with youthful wines and becomes more predominant with quality-oriented and aged ones.
- *Structural Components:* Rieslings can range from dry to sweet and light to full body—largely depending on the level of residual sugar remaining in the wine after the fermentation process. Well-made Rieslings are high in tartaric and malic acids, which are necessary (although sometimes going unnoticed) to balance the wine's varying levels of sugar content and intense fruit aromatics. The acid also acts as a preservative for long-aging capabilities. Rieslings often remain unoaked (or at minimum, stored in neutral oak barrels) in order to preserve the pure—aromatic fruit and high acidity levels.

  Determining whether a particular Riesling is dry, off-dry, or sweet can largely be based on the wine's alcohol content. The lower-alcohol versions (roughly 11 percent or lower) maintain higher levels of residual sugar, providing a richer, more viscous, medium-to-full body wine. The higher-alcohol versions (roughly 12 percent or higher) typically maintain minimal to no perceptible sugar, yielding a dry wine with a light to medium body.
- *Significant Locations:* Some prominent locations for Riesling generally offer long, steady growing seasons to allow the grapes to ripen well into the fall time. Arguably the two most significant growing locations include Germany (Mosel and Rheingau) and France (Alsace). Alternative growing areas in the Old World include Austria (Wachau) and Italy (Trentino-Alto-Adige and Friuli). In the New World, Riesling is prominent from Washington state (Columbia Valley) and California (Central Coast); Australia (Clare and Eden Valley), New Zealand (Marlborough, Martinborough, Nelson, and Wairarapa), New York (Finger Lakes), and Canada (Niagara Peninsula).

Mosel-Saar-Ruwer (MOH-zel sahr ROO-vayr)

The Mosel-Saar-Ruwer is arguably Germany's most famous wine-growing region. The Mosel River is the spine of the Mosel-Saar-Ruwer wine region, and the vineyards extend along the two small tributaries, the Saar and the Ruwer. Figure 5.16 displays a panoramic view of the Mosel region of Germany. The grapes are grown on steep hillsides (sometimes on 70-degree inclines) along the steep river banks. The Mosel region is widely known for its unique slate soil type that imparts a distinctive taste ranging from fruity to earthy, or flinty.

FIGURE 5.16
Panoramic view of the Mosel region of Germany. Courtesy of Leo Alaniz.

Rich and Voluptuous Whites

Learning Objective 5
Explain characteristics of the rich and voluptuous white wine style category

lavish ... elegant ... voluptuous

The "rich and voluptuous whites" is a style category that produces wines with a commonality of luxuriousness and depth—both in terms of aroma/flavor components and structural components. The rich and voluptuous whites visually appear pale to medium in color intensity and straw yellow to golden yellow in color hue, often affiliated with its use of oxidative aging techniques. This style category produces wines that are often concentrated with a pronounced smell of tropical fruit and bakeshop qualities.

The grapes that create these wines have often been left on the vine for extended hang-time to gain further sugar and flavor ripeness. Once in the winery, these wines have often been given extensive aging in new American or French oak for a period of months to years to provide a rich, complex personality. In some cases, these wines can be aged with their lees (decomposing yeast cells) for a period of time that offers additional aroma and flavor and textural characteristics. It is likely that these wines may have experienced an enhanced practice of lees aging called sur lie aging (in French). Another technique used is *battonage* (a French term) where the wine is stirred with its lees in order to accentuate the process of lees aging. Due to the grape's extended hang-time, these wines will likely contain higher alcohol content—often, well above 13.5 percent and in some cases, as high as 14.9 percent. Quite often the wines have gone through a large conversion of malolactic fermentation (MLF) in which a significant percentage of the wine's tart, apple-like malic acid has been converted to the creamy, dairy-like lactic acid. All of these techniques contribute to wines that are commonly medium plus, to full in body.

Chardonnay (Shar-doh-NAY)

Chardonnay is one of the most popular white wine grapes in the world and has led to incredible notoriety for many new and developing wine regions. This grape is believed to have originated in the subregion Mâconnais (mack-kohn-NAY) of Burgundy, France. The grape is extremely adaptable to different climates and winemaking techniques—virtually anywhere there are vineyards, Chardonnay is ever present. Figure 5.17 shows a Chalk Hill Chardonnay from Sonoma County.

It was a California Chardonnay wine that was responsible for bringing great fame to California (and overall the New World) back in 1976. The famous "Judgment of Paris" wine tasting placed the best white and red wines of France against California. Chateau Montelena winery took first place for the Chardonnay category beating the French and several other California wines.

- *Aroma/Flavor Components:* The Chardonnay grape itself is fairly neutral and is unusually adaptable both to its surroundings and winemaking techniques. It is sometimes thought of as a painter's "blank canvas," as the grape is quite moldable and has the ability to be influenced greatly by the winemaker. The primary flavors in cool climates include tree fruits (apple and pear), citrus fruit (lemon), and earth (mineral and wet stone) while flavors in warm climates include tropical fruits (pineapple, fig, banana, and mango). Secondary flavors derived from winemaking techniques are commonly associated with bakeshop (vanilla, butter, honey, toast, butterscotch, cinnamon, and clove).
- *Structural Components:* Chardonnay can range from a medium to full body. It showcases a medium body when aged in stainless steel or neutral oak and full-bodied and

FIGURE 5.17
Chalk Hill Chardonnay from Sonoma County, California. Courtesy of Erika Cespedes.

FIGURE 5.18
Collection of Chardonnay from around the world. Courtesy of Erika Cespedes.

rich, when aged in new oak. California versions can yield high alcohol (from the warmer climate and the extensive hang-time concept), which arguably can produce a wine to be considered out of balance.

- *Significant Locations:* Chardonnay has grown prevalent throughout the wine world and has become ubiquitous in both the Old and the New World wine-producing countries. In France, Chardonnay is most reputable in the regions of Burgundy and is vitally important in Champagne. In the New World, Chardonnay is significant in California (Carneros, Russian River, Sonoma Coast, Santa Maria, Santa Barbara, and Monterrey), Australia (Margaret River), New Zealand, and Chile (Casablanca Valley and Maipó). Figure 5.18 displays a collection of Chardonnay from around the world.

FIGURE 5.19
Blanc de Blanc Champagne. Courtesy of Erika Cespedes.

Champagne (shahm-payne)

Champagne is one of the most northerly regions of France, where it is completely committed to producing high quality and prestigious bubbly wines. Being Far North, Chardonnay is exposed to a cooler continental climate where the grapes retain ample levels of acidity necessary to create quality sparkling wine. In these wines, Chardonnay is most often blended in a trio of grapes along with Pinot Noir and Pinot Meunier. An alternative Champagne style is labeled "Blanc de Blanc" (translated to: white from white), where Chardonnay is made as a stand-alone sparkling wine where Chardonnay can be expressed in one of its most epic forms. Figure 5.19 displays a label of a Blanc de Blanc Champagne.

Côte de Beaune (koht duh Bohn)

The Côte de Beaune area is located in the south of Côte d'Or and is considered the heart of white Burgundy country. This area is diverse, offering great reds, but even greater legendary whites with a broad range of character and quality. Some of the most recognized Chardonnay from Burgundy's Côte de Beaune villages include: *Meursault* (mehr-SO), *Puligny-Montrachet* (poo-lee-NYEE mohn-rah-SHAY), and *Chassagne-Montrachet* (shah-SAHN-nyah moan-rah-SHAY). Figure 5.20 shows a white Burgundy (100 percent Chardonnay-based wine) produced from one of the most prestigious villages of Burgundy—Chassagne-Montrachet.

FIGURE 5.20
White Burgundy from Chassagne-Montrachet. Courtesy of Erika Cespedes.

Chablis (shah-BLEE)

Chablis is one of the five subregions and the northernmost part of Burgundy. This subregion is known for producing some of the most famous expressions of Chardonnay in the entire World. Chablis has its unique interpretation of Chardonnay as a completely divergent style, where the wine is left largely unoaked with minimal winemaking techniques. The wine often yields a style of citrus and tree fruit aroma and flavor, dry and fairly acidic, with a flinty, mineral quality (because of its clay and limestone soils). In Chablis, the wine is delicate in aroma and flavor; therefore, most producers apply reductive techniques in which wines are fermented and aged in stainless

FIGURE 5.21
Chablis Premiere Cru paired with oysters. Courtesy of Erika Cespedes.

steel vats or in large, older wooden tanks or barrels that impart a subtle oak flavor. Application of these techniques allows the wine to truly express the grape and its sense of place. Figure 5.21 shows a Chablis Premiere Cru paired with oysters in the half shell.

Sémillon (seh-mee-YOHN)

Sémillon is a greenish (almost yellow) skinned grape used to produce both dry and sweet wines. This well-known grape variety is used largely in the white wines of Bordeaux, France, where it's blended in varying quantities with Sauvignon Blanc. It is also used to produce some of the world's most famous dessert wines from the Bordeaux appellation of Sauternes.

- *Aroma/Flavor Components:* This grape offers fairly to highly aromatic nuances of citrus fruit (lemon), tropical fruits (fig and golden raisin), tree fruit (peach), garden (grass), and bakeshop (honey).
- *Structural Components:* Sémillon is often a brilliant gold-colored wine with a soft, full, and sometimes even oily texture with low to moderate acidity. It is most frequently vinified into a dry wine but can be left with some remaining residual sugar to produce an off-dry to sweet wine style. Sémillon typically produces a medium-to-full body wine—largely dependent upon the quantity of residual sugar and the length of extended hang-time of the grapes. Sometimes this grape is added (as is the case in most white Bordeaux wines) to complement the leaner-bodied and highly acidic Sauvignon Blanc. The alcohol content is often moderate to high ranging between 12 and 14 percent.
- *Significant Locations:* While the production of Sémillon is not widespread—the few places produce fairly extraordinary examples. Some of the most prevalent locations include France (Bordeaux), Australia (Hunter Valley and Margaret River), Washington state (Columbia Valley), and South Africa.

FIGURE 5.22
Condrieu bottle shot. Courtesy of Erika Cespedes.

Viognier (VEE-oh-NYAY)

Viognier is a white wine grape originating from southern France. In less than twenty-five years, this grape has gone from obscurity to international recognition. Viognier often produces a distinctive straw yellow to golden-yellow wine with a rich, full body. Viognier has become one of the more "fashionable" white-wine grape varietals as an alternative to Chardonnay, providing a rich, luscious, full-bodied mouthfeel. Although often a stand-alone varietal throughout much of the world, Viognier is unsuspectingly blended in small amounts to lighten the red wines of France's northern Rhône Valley and Australia. Figure 5.22 shows a bottle of Condrieu (100 percent Viognier) which is located in the northern Rhône Valley.

- *Aroma/Flavor Components:* Viognier is a highly intense and aromatic grape varietal. If the wine is aged in neutral oak barrels or stainless steel tanks as common practice in France, Viognier can produce a fragrant wine that shows off the floral, tree fruit (peach), tropical fruits (tangerine, pineapple, mango, and apricot), and bakeshop (honey). If Viognier is oak-aged, as is common in the New World, the wood barrels add further complexity by contributing elements of bakeshop (anise, vanilla, and toast).

- *Structural Components:* Viognier wines are predominantly dry, although sweet late-harvest dessert wines are made in select locations. The vinification techniques range widely depending upon the whim of the winemaker. Occasionally, winemakers will allow skin-and-juice contact for a brief period of time prior to fermentation—while others may allow the wine to undergo malolactic fermentation, oak aging, and aging on the lees. Each of these techniques contributes more richness, weight, and complexity to the wine while softening the acidity. These wines maintain a low- to medium level of acidity with a rather high alcohol content hovering around 14 percent.

  The grape prefers moderate environments with a long growing season but can grow in cooler areas as well. Viognier alcohol can easily get out of hand, so some vintners leave a touch of residual sugar (though unnoticeable) to mask the spice and heat. It's a grape with low acidity; it's sometimes used to soften wines made predominantly with the red Syrah grape to add a floral perfume quality and tame tannins.

- *Significant Locations:* France (Rhône Valley and Languedoc-Roussillon). In the New World, Viognier has increased dramatically in the Central Coast region of California as well as in Chile.

FIGURE 5.23
Condrieu label shot. Courtesy of Erika Cespedes.

Condrieu

In France's Rhône Valley, Condrieu is considered the most prestigious appellation where Viognier is the only permitted grape variety. Figure 5.23 shows a label of Condrieu. Within this appellation, there is an even smaller and more coveted site called Château Grillet (one of France's smallest appellations with less than ten acres and only one owner). Not only is this a small appellation, it is additionally prestigious because it is named after its sole winery. In addition, the northern Rhône appellation of Côte-Rôtie is known for its red wines—though it allows up to 20 percent of Viognier to add fragrance and soften the Syrah-dominated wine. Viognier is also used to create the regional, white Côtes du Rhône.

Check Your Knowledge

Directions: Use these questions to test your knowledge and understanding of the concepts presented in the chapter.

I. MULTIPLE CHOICE: Select the best possible answer from the options available.

1. The wine styling approach is a technique that
   a. simplifies the understanding of wine
   b. arranges a grape varietal's structural components into "like" categories
   c. can be useful for communicating wine to others
   d. all of the above
2. Typicity is a term used to describe
   a. a wine's typical personality characteristics
   b. the typical grape-growing climate
   c. a wine's aromas
   d. a wine's body
3. Which is NOT a significant production location for Chardonnay?
   a. Burgundy, France
   b. Margaret River, Australia

   c. Russian River, California
   d. Mosel, Germany
4. Which is NOT a significant location for Riesling?
   a. Eden Valley, Australia
   b. Mosel, Germany
   c. Alsace, France
   d. Napa Valley, California
5. Which is NOT a significant location for Sauvignon Blanc?
   a. Marlborough, New Zealand
   b. Loire Valley, France
   c. Bordeaux, France
   d. Mosel, Germany
6. Which grape varietal is largely popular in northwestern Spain and northern Portugal?
   a. Chardonnay
   b. Pinot Grigio
   c. Albariño
   d. Chenin Blanc
7. Most Chardonnay around the world experience the following vinification techniques:
   a. oak barrel aging
   b. aging on the lees
   c. malolactic fermentation (MLF)
   d. extended hang-time
   e. answers a, b and c
   f. all of the above
8. Silky and smooth whites have the potential for the following techniques during viticulture and vinification:
   a. extended hang-time on the vine
   b. wines have been vinified leaving some perceptible residual sugar
   c. new oak aging
   d. both a and b
9. Which white wine grape would typically maintain a medium plus to full body?
   a. Chardonnay
   b. Sauvignon Blanc
   c. Torrontés
   d. Pinot Grigio
10. Which white wine grape would typically maintain a light to medium body?
    a. Chardonnay
    b. Viognier
    c. Sémillon
    d. Sauvignon Blanc
11. Which white wine grape commonly produces a range of dry and sweet styles of wine?
    a. Chardonnay
    b. Riesling
    c. Chenin Blanc
    d. both b and c
12. Which white wine grapes are commonly noted for their rich, luscious mouthfeel?
    a. Viognier
    b. Sémillon

c. Chardonnay
d. all of the above
e. only b and c

13. Which white-wine grape would typically have lower levels of acid due to its necessity of longer hang-time?
    a. Gewürztraminer
    b. Sauvignon Blanc
    c. Pinot Grigio
    d. Albariño
14. Identify the significant growing area for Grüner Veltliner.
    a. Germany
    b. Washington
    c. Alsace
    d. Austria
15. Which grape varietal contains typical aromas and flavors associated with tree fruits (apricot and peach), bakeshop (honey), dried fruit (golden raisin), and mineral/chemical (petroleum)?
    a. Chardonnay
    b. Sauvignon Blanc
    c. Chenin Blanc
    d. Riesling
16. Which wine style is best represented? Most often light to medium in body with a moderate alcohol percentage between 11.5 and 13.5, these wines contain ample malic acid—providing tartness and sourness working to accentuate their youthful nature and best consumed in their youth within one to three years from their vintage date.
    a. Rich and voluptuous white wines
    b. Crisp and youthful white wines
    c. Silky and smooth white wines
    d. Mellow and complex red wine
17. What is the Italian name for Pinot Gris?
    a. Pinot Nero
    b. Pinot Bianco
    c. Pinot Greco
    d. Pinot Grigio
18. Which grape variety is Vouvray made from?
    a. Riesling
    b. Chardonnay
    c. Chenin Blanc
    d. Sauvignon Blanc

II. DISCUSSION QUESTIONS

19. Identify and explain the three styles categories of white wine. Describe at least two wines that typically fall under each category.
20. Explain the concept of typicity. How does it assist in learning and communicating about wine?

CHAPTER 6

The Wine Styling Approach: Red Wines

CHAPTER 6 LEARNING OBJECTIVES

After reading this chapter, the learner will be able to:

1. Explain characteristics of the fruity rosé and vibrant red wine style category
2. Explain characteristics of the mellow and complex red wine style category
3. Explain characteristics of the bold and intense red wine style category

Music is the wine which inspires one to new generative processes, and I am Bacchus who presses out this glorious wine for mankind and makes them spiritually drunken.

—Ludwig van Beethoven

Red Wine Style Categories

There are at least a dozen or two dozen major red-wine grapes commonly experienced in food and beverage establishments. Given the numerous array of grape varietals, the three most significant international grapes used for producing high-quality red wine are Pinot Noir, Merlot, and Cabernet Sauvignon. Figure 6.1 illustrates a small selection of red wines from multiple styling categories. Other distinguishable indigenous red-wine grapes renowned for producing great wine in select locations include Barbera, Dolcetto, Cabernet Franc, Carménère, Gamay, Grenache, Malbec, Mourvèdre, Nebbiolo, Pinotage, Sangiovese, Syrah/Shiraz, Tempranillo, Touriga Nacional, and Zinfandel. Each of these grapes offers an expected set of characteristics or expression of typicity. Given the typicity of each varietal, it is possible to categorize each one into a style category for ease of understanding. It's very important to note, the wines within each category (or style) have been stereotyped for illustrative and simplification purposes. They have been arranged into categories based on some common structural characteristics, although these wines can vary dramatically from producer to producer, varying growing locations, and the different applications of winemaking techniques. The three wine styling categories for red wines include "fruity rosé and vibrant reds," "mellow and complex reds," and "bold and intense reds."

FIGURE 6.1

Small selection of red wines.
Courtesy of Erika Cespedes.

Fruity Rosé and Vibrant Reds

Learning Objective 1
Explain characteristics of the fruity rosé and vibrant reds style category

youthful ... lively ... charming

The "fruity rosé and vibrant reds" visually are pinkish to red in color hue with medium color intensity. The wines within this style category are often produced from grapes that have been grown in a cooler to warm growing climates. The grapes may have a higher proportion of juice-to-skin ratio that may yield a lighter-bodied, higher-acidic wine with minimal tannin components. Figure 6.2 illustrates a Davis Family Vineyards Pinot Noir from the fruity rosé and vibrant reds wine styling category paired successfully with a grilled wild salmon and asparagus. Certain grapes in this category may have less color pigment in the skins, so they may appear and taste light, encouraging their early consumption within months to two years from vintage date. Oak is always light (if used at all) and serves as more of an undertone and enhancement of the wine's primary characteristics. Many times, these wines benefit from slightly chilled temperatures prior to serving and are generally an appropriate warm-hot weather drinking alternative. Served at a temperature range of 45–55°F, the wines accentuate their freshness, fruit-forward qualities, and bright levels of acidity.

FIGURE 6.2
Pairing Pinot Noir with Grilled Salmon.
Courtesy of Erika Cespedes.

Rosé (roh-ZAY) Wine

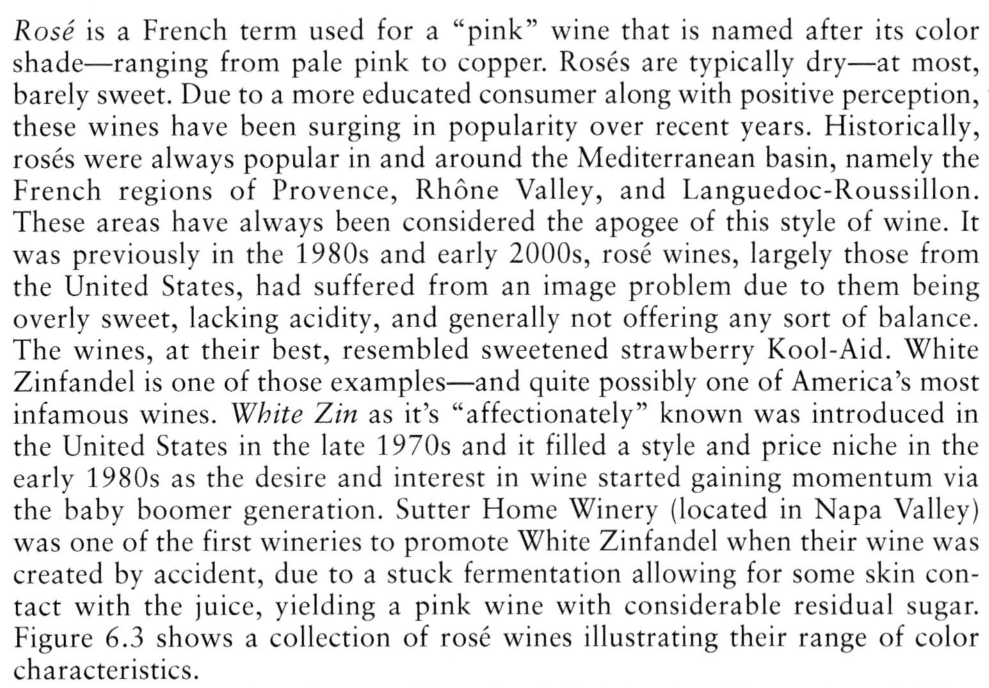

Rosé is a French term used for a "pink" wine that is named after its color shade—ranging from pale pink to copper. Rosés are typically dry—at most, barely sweet. Due to a more educated consumer along with positive perception, these wines have been surging in popularity over recent years. Historically, rosés were always popular in and around the Mediterranean basin, namely the French regions of Provence, Rhône Valley, and Languedoc-Roussillon. These areas have always been considered the apogee of this style of wine. It was previously in the 1980s and early 2000s, rosé wines, largely those from the United States, had suffered from an image problem due to them being overly sweet, lacking acidity, and generally not offering any sort of balance. The wines, at their best, resembled sweetened strawberry Kool-Aid. White Zinfandel is one of those examples—and quite possibly one of America's most infamous wines. *White Zin* as it's "affectionately" known was introduced in the United States in the late 1970s and it filled a style and price niche in the early 1980s as the desire and interest in wine started gaining momentum via the baby boomer generation. Sutter Home Winery (located in Napa Valley) was one of the first wineries to promote White Zinfandel when their wine was created by accident, due to a stuck fermentation allowing for some skin contact with the juice, yielding a pink wine with considerable residual sugar. Figure 6.3 shows a collection of rosé wines illustrating their range of color characteristics.

Quality oriented rosé wines offer adaptability to pair with a variety of dishes and applications. They offer the freshness and vibrancy found in a white wine, yet the subtle tannin present in a lighter styled red wine. These attributes allow for a suitable connection to most dishes that have a subtle amount of fat present whether it's in a sauce, or as part of the main protein in a dish. The pleasing advantage of rosé wine is that they are just as enjoyable without needing to pair them with food; they offer respite from the summer heat, or as an accompaniment to alfresco dinner parties or daytime picnics. They are most often intended to be consumed in their youth, where they best showcase their zesty freshness and fruit-forward qualities.

FIGURE 6.3
Collection of rosé wines. Courtesy of Erika Cespedes.

Production Techniques and Varietals for Rosé Wines

The wine's pinkish color identifies that rosé wines derive from some form of black-to-purple grape varieties that bleed a small amount of red color pigment into the juice when the grapes are being pressed. For some wine producers, rosé is a byproduct of their red wine production with the excess colored juice removed (the French term *Saignée*—to bleed) in order to concentrate the original red wine. A second production technique is to utilize the direct press method, where the weight of the grapes collecting in a vat expresses the juice naturally (possibly with some slight assistance) from the grapes below. While other producers may simply blend white and red wines together in order to achieve a pink-colored wine.

Rosés can be created from a single grape varietal (Pinot Noir, Zinfandel, Merlot, or Cabernet Sauvignon), while others are made from blend of grapes (Example: Syrah, Grenache, and others). This mix of varietals (or simply a 100 percent varietal rosé wine) adds to the intrigue and diversity found in this category of wine. The blend of grapes and percentages of each vary based on the creative tendency of the winemaker. Generally, grape varietals that predominate in a given growing area tend to be the ones used. Oregon relies on Pinot Noir, Bandol (located in Provence) uses Mourvèdre, Spain uses Garnacha (synonym for Grenache) and Provence prefers Cinsault, Syrah, and Grenache.

- *Aroma/Flavor Components:* Since most rosé wines experience a short period of skin-and-juice contact, they tend to combine the fruitiness of a red wine with the crispness and lightness of a white wine. Rosé wines are concentrated with fresh red fruits (cherry, strawberry, and watermelon), wet stone and floral aromas and flavors. Most rosés are minimally oak-aged, if at all, but there is potential for subtle bakeshop and tobacco notes.
- *Structural Components:* Rosé wines are almost always acidic with a light to medium body (but they can be fuller in body if there is significant residual sugar) and low to medium alcohol content. Some rosé wines are given brief lees contact, which heightens the body. Most rosés are released shortly after production as they are intended to be consumed in their youth to showcase their fresh and fruity personality.
- *Significant Locations:* France (Provence, Rhône Valley, Loire Valley, and Languedoc-Roussillon), Italy (Veneto), and Spain (Navarra and Rioja). The New World has managed to produce rosés wines in every major wine-producing country from the United States (California, Oregon, Washington, and New York), Argentina (Mendoza), Australia (McLaren Vale), and Chile (Casablanca Valley) to South Africa (Western Cape).

Barbera (bahr-BEHR-ah)

Barbera is one of the grapes most commonly planted throughout northwest Italy. It is generally regarded as producing fairly straightforward and inexpensive wine though there are some excellent high-quality examples. Better examples are found when the wine has come from lower yielding vineyards and has been allowed to see some brief aging in a French barrique. Figure 6.4 is a bottle shot of Barbera d' Alba.

- *Aroma/Flavor Components:* Fruits (black cherry, sour cherry, and dried red berry) and light tobacco shop.
- *Structural Components:* Barbera is a deep-ruby-colored, light- to medium-bodied wine with low plus levels of tannin and high levels of acidity.

FIGURE 6.4
Bottle of Barbera d' Alba. Courtesy of John Peter Laloganes.

- *Significant Locations:* This grape originates from the Piedmont region in northwest Italy—most highly regarded in Monferrato and around the towns of Asti and Alba. The wines are often varietally labeled along with the name of the town from where they derive, such as Barbera d'Asti or Barbera d' Alba.

Dolcetto (dohl-CHEH-toe)

Dolcetto means "little sweet one," in reference to its early ripening, which allows the wine to be drunk sooner than other varietals. Often considered an "everyday" drinking wine, it is a fairly straight-forward and uncomplicated wine. The wine is typically aged in stainless steel or briefly in large, neutral oak barrels; the wine is intended to be best consumed in its youth.

- *Aroma/Flavor Components:* Bakeshop (licorice), nuts (almond), fruits (plum and blackberry).
- *Structural Components:* Dolcetto produces a softer style of wine, characterized by low to medium acidity levels with soft tannin and a light to medium body.
- *Significant Locations:* This grape originates from the Piedmont region in northwest Italy—grown throughout many locations such as Dogliani and Monferrato and around the towns of Asti and Alba. The wines are often varietally labeled with the name of the town from where they derive, such as Dolcetto d' Asti and Dolcetto d'Alba.

Gamay (gah-MAY)

Gamay, or more precisely, its full name *Gamay Noir à Jus Blanc* is a thin-skinned red wine grape that is best known for its role in the wines of *Beaujolais* (BOE-zjoh-lay), a subregion located in the southern end of Burgundy, France. The grape is also planted throughout central Loire Valley, France where it's discretely grown in and around the city of Tours, specifically in the Touraine area.

The most popular rendition of Gamay is known as *Beaujolais Nouveau* where the grapes are fermented for just a few weeks through carbonic maceration—also called whole berry fermentation. In this vinification technique, entire bunches of selected grapes (with the skins, seeds, and stems intact) are placed with yeast into a stainless steel vat partially filled with carbon dioxide in order to avoid oxidation. The weight of gravity causes the lower one-third of the grapes at the bottom of the vat to be pressed from the grapes above. As the crushed grapes begin fermentation, they begin to release additional carbon dioxide causing the remaining uncrushed berries to undergo intracellular (or whole berry) fermentation while still within their grape skins. This technique produces a fresh, highly aromatic, fruit-forward, light-bodied wine, without extracting any notable quantity of tannin from the grape skins. The fermentation process takes about three to four days and the wine is only about nine weeks old when it is released for consumption. Beaujolais Nouveau is a purple-pink wine that is particularly lightweight, even by the standards of Beaujolais. Most Beaujolais and Nouveau style levels are light-bodied, low in tannin, and medium to high in acid. Beaujolais Nouveau shot to popularity in the 1970s and 1980s through the clever marketing approach of promoting the urge to "get the first wine release of the season"—via the creator and advocate, Georges Duboeuf, who still produces the most popular of all Beaujolais Nouveau wines. Figure 6.5 shows a display of Beaujolais Nouveau for the 2014 vintage.

FIGURE 6.5
Display of Beaujolais Nouveau. Courtesy of John Peter Laloganes.

FIGURE 6.6
Beaujolais Villages. Courtesy of Erika Cespedes.

- *Aroma/Flavor Components:* The Gamay grape produces extremely fruit-forward and fairly to highly aromatic wines. They are dominated by aromas and flavors of red fruits (cherry, raspberry, and watermelon) and bakeshop (chocolate and tobacco).
- *Structural Components:* Gamay produces light- to medium-bodied dry red wines. They contain ample acidity that accentuates a freshness and vibrancy with very little tannin. Beyond basic Beaujolais and Beaujolais Nouveau, the Gamay varietal has the ability to produce the less well known yet greater structured and complex *Beaujolais Villages* and *Beaujolais Crus*. Figure 6.6 displays a Beaujolais Villages. These wines are often given traditional fermentation as well as experiencing some modest oak-aging. They still produce the characteristic fruit-forward qualities with more emphasis on dried red fruits (raspberry and cherry) but with more intense aromas of earth, tobacco, and chocolate tend to balance the wines more effectively. Many of these wines are heartier (as compared to other Beaujolais) with greater color intensity, medium body, and medium tannin levels.
- *Significant Locations:* Gamay is somewhat limited in growing locations. Its spiritual home is the Beaujolais subregion of Burgundy, France where the "crus" of Beaujolais illustrate the best expression of Gamay. It also grows in Loire Valley's Central Vineyards. In the New World, Gamay is considerably limited with Oregon and California growing the grape mostly as a novelty.

The crus of Beaujolais are the highest-quality wines produced in Beaujolais. The wines come from one of the ten major villages called crus that individually each have their own special character. The crus, historically, are named after the village the grapes came from—each village offers a unique interpretation of Beaujolais while clearly showcasing its location on the labeling of the bottle. The ten Cru areas of Beaujolais are as follows:

- *Brouilly* (BREW-yee)
- *Chenas* (shay-NAH)
- *Chiroubles* (shee-ROOB-luh)
- *Cote De Brouilly* (coat duh BREW-yee)
- *Fleurie* (FLUR-ee)
- *Julienas* (ZJOO-lee-ay-nah)
- *Morgan* (mor-GAHN)
- *Moulin à Vent* (MOO-lan ah vahn)
- *Régnié* (reh-N'YAY)
- *Saint Amour* (sant ah-MOOR).

Pinot Noir (PEE-noh-NWAHR)

The "Pinot Noir" name is derived from the French words for "pine" and "black" alluding to the grape's tightly clustered dark purple pine-cone shaped clusters of fruit. Due to Pinot Noir's susceptibility to mutation, it maintains somewhat of an extended

FIGURE 6.7
Flowers Pinot Noir, Sonoma Coast, CA. Courtesy of John Peter Laloganes.

FIGURE 6.8
Drouhin red Burgundy from Beaune. Courtesy of John Peter Laloganes.

family—it's widely used relatives such as Pinot Blanc, Pinot Gris, and Pinot Meunier have become well-known varieties on their own accord. The cultivation of Pinot Noir (or *Pinot*, as it is often coined) dates back over 2000 years and today it is grown around the world. Figure 6.7 shows a glass of Flowers Pinot Noir from Sonoma Coast, California.

Pinot Noir is regarded as producing some of the most alluring and seductive red wines—and sparkling wines—in the world. André Tchelistcheff (d. 1994) is largely recognized as one of the modern fathers of California wine. He is famously quoted for declaring that "God made Cabernet Sauvignon whereas the devil made Pinot Noir." His quote references the difficulties to grow good quality Pinot without a price—it takes painstaking efforts to produce it well. Pinot Noir's thin skin makes it highly susceptible to just about every possible disease infliction known to grapes. In addition, Pinot produces fairly low yields (particularly for making quality wines), which ultimately affects the selling price. With such challenging issues and overall limited production, good quality Pinots, when found, tend to be fairly expensive. Figure 6.8 displays a glass of red Burgundy from Chorey-Les-Beaune from the well-respected producer, Joseph Drouhin.

- *Aroma/Flavor Components:* Since Pinot maintains relatively thin skins and larger berries, they tend to contribute lighter color intensity than other red-wine grapes. The aromas and flavors can alternate between garden (earth, dust, peat moss, and mushroom) and dried red fruits (cherry, raspberry, cranberry, and black cherry), coffee shop (espresso, butterscotch, vanilla, clove, nutmeg, and anise), and subtle tobacco.
- *Structural Components:* Pinot Noir tends to be of light to medium body (depending upon hang-time and yield of the grapes), with low to medium levels of tannin, and medium to high acidity.
- *Significant Locations:* Pinot Noir thrives mostly in the cooler growing areas and is chiefly associated with France's Champagne region—as well as Burgundy's subregion of the *Côte d'Or* (coat-d-OR). New World locations include Oregon (Willamette Valley), California (Sonoma, Carneros, Russian River, Sonoma Coast, Sta. Rita Hills, and Santa Lucia Highlands), and New Zealand (Marlborough, Martinborough, and Central Otago).

Burgundy

Burgundy, or Bourgogne, as it is known in France, has always produced the classic Pinot Noir style that has been so widely imitated around the world. For most of wine history, this two-mile-wide, thirty-mile-long stretch of hills, called the *Côte d'Or* "Slope of Gold," is the only French region to achieve incredible notoriety for their Pinot Noir. The Burgundian style offers a wine that is harmonious, complex, and elegant. The region is further divided into the *Côte de Nuits* (koht duh NWEE) in the north and *Côte de Beaune* (koht duh BOHN) in the south. Important villages that specialize in red Burgundy include

FIGURE 6.9
Bouchard red Burgundy from Volnay. Courtesy of John Peter Laloganes.

the famous vineyards of: *Gevrey-Chambertin* (j hevray shahm-behr-TAN), *Morey-Saint-Denis* (maw-ree san duh-NEE), *Chambolle-Musigny* (shawm-bohl moo-sih-NYEE), *Vougeot* (voo-ZHOH), *Vosne-Romanée* (vohn raw-mah-NAY), *Nuits-Saint-George* (nwee-san-ZHORZH), *Volnay* (vohl-NAY) and *Pommard* (poh-MARH). Figure 6.9 displays a red Burgundy label from Volnay, a village found in the Côte d' Beaune of Burgundy, France.

Oregon

Oregon is another significant Pinot Noir producer that typically produces wine that is stylistically similar to the Burgundian Style. Oregon's large appellation of Willamette Valley maintains a climate similar to Burgundy in which the finicky Pinot Noir grapes thrive. Over the past few decades, Oregon's climate has shown itself to be especially well suited for Pinot Noir and seemed to set the standard for North American Pinot. Oregon Pinot Noir pioneer David Lett of *Eyrie* (I-ree) Vineyards first planted Pinot Noir in 1965. With great success, several other growers followed suit throughout the 1970s. Traditionally, the Oregon versions have been compared to Burgundian Pinots although, some producers have diverged and follow a more typical California philosophy of extended hang-time to allow for greater ripeness and ultimately higher alcohol content. These riper versions lead to a more robust style of Pinot Noir. Figures 6.10 and 6.11 display Bergstrom Family Vineyards Pinot Noir from Oregon's Willamette Valley.

FIGURE 6.10
Bergstrom Pinot Noir from Willamette Valley, Oregon. Courtesy of Erika Cespedes.

FIGURE 6.11
Bergstrom bottle top. Courtesy of Erika Cespedes.

Mellow and Complex Reds

Learning Objective 2
Explain characteristics of the mellow and complex reds style category

rich ... smooth ... velvety

The "mellow and complex reds" are a style category that produces wines with a more balanced mellow mouthfeel—yet they yield complex layers of aromas and flavors. This style category often produces wines that are oak-aged for a period of months, to add enough dimensions that enhance the overall wine. The mellow and complex reds offer wines with medium to deep color intensity—moderate acidity, medium tannin, and a medium body. Figure 6.12 displays a collection of mellow and complex red wines.

Cabernet Franc (ka-behr-nay FRAHNK)

Cabernet Franc (or Cab Franc) is red wine grape varietal that is widely grown around the world. It serves as a blending grape in the famous red wines of Bordeaux, yet is also known for its effort as a stand-alone varietal in Loire Valley, France.

Through DNA analysis in the early 1990s, it was discovered that sometime in the 18th century, Cabernet Franc had crossed with Sauvignon Blanc to become parents of Cabernet Sauvignon. Therefore, it's logical to believe that Cabernet Franc would share many of the same characteristics as Cabernet Sauvignon, though in a slightly subdued version. Cabernet Franc tends to be more lightly pigmented (thinner skins) and lesser tannin with a smoother mouthfeel as compared to its notable off-spring. Figure 6.13 shows a bottle of Chinon (100 percent Cabernet Franc) from the Loire Valley.

- *Aroma/Flavor Components:* Cabernet Franc is a fairly aromatic varietal that offers nuances of garden (herbs and bell pepper), dried red fruits (strawberry, raspberry, black cherry, and cassis), tobacco shop (slightly cedar and tobacco), and floral (violets).
- *Structural Components:* This grape produces dry red wines with a medium body, medium to high acidity, and medium tannin levels.
- *Significant Locations:* Cabernet Franc is a significant varietal in the blended French red wines of Bordeaux along with Cabernet Sauvignon and Merlot. Cabernet Franc also produces single varietal wines in Loire Valley's *Chinon*

FIGURE 6.12
Collection of mellow and complex red wines. Courtesy of Erika Cespedes.

FIGURE 6.13
Chinon (100 percent Cabernet Franc). Courtesy of Erika Cespedes.

(shee-NYOHN) and *Bourgueil* (bohr-GEEL) appellation (where the grape is locally referred to as Breton). This varietal is also produced with some popularity in New World locations such as Oregon, Washington state, California, Canada (Niagara Peninsula and Okanagan Valley), and New York (Finger Lakes).

FIGURE 6.14
Chono, single vineyard Carménère. Courtesy of Erika Cespedes.

Carménère (car-men-YHER)

Originally planted in Bordeaux as a blending grape, it had eventually fell out of favor in the late 1800s and found its way to the more conduce growing climate in Chile. Carménère has since been rediscovered in Chile as a leading red grape capable of stand-alone single varietal wines or blended with international varietals such as Cabernet Sauvignon. Figure 6.14 shows a bottle of Carménère from Chile.

- *Aroma/Flavor Components:* Fruits (fresh to baked black cherry and blackberry), bakeshop (cardamom, clove, nutmeg, and dark chocolate), vegetal (green pepper) and tobacco shop (smoke).
- *Structural Components:* Medium-bodied with low to medium tannin and medium acid.
- *Significant Locations:* Chile (Rapel Valley, Maipó, and Colchagua Valley).

Grenache (Gren-AHSH)

Grenache or identified as *Garnacha* (gahrr-NAH-chah) in Spain or *Cannonau* (can-na-NOW) in Italy, thrives in dry, warm to hot growing locations. This grape is commonly blended with *Syrah* and *Mourvèdre* throughout southern France and in the United States—versus being liberally blended with *Tempranillo* in Spain. Because of the grape's easy ability to obtain sugar levels, ample alcohol levels are usually a typical characteristic in this wine. Grenache has the ability to produce simple fruity rosés—powerful age-worthy reds—and fortified wines throughout its numerous growing locations. Figure 6.15 shows a bottle of Châteauneuf-du-Pape from the highly-regarded Château de Beaucastel in the southern Rhône Valley.

FIGURE 6.15
Château de Beaucastel. Courtesy of John Peter Laloganes.

- *Aroma/Flavor Components:* This grape varietal produces aromas and flavors of concentrated baked to dried red and black fruits (strawberry and cherry), coffee shop (chocolate), bakeshop spice (cinnamon and clove), garden (earth and wet leaves), and tobacco shop (smoke).
- *Structural Components:* Grenache is known for its dry, acidic rosé wines and bold, intense red wines. Grenache typically maintains a medium to full body with medium acid and tannin levels and high alcohol content.
- *Significant Locations:* Grenache is widely planted throughout Spain and southern France. Grenache is the dominant variety in most blended red wines from southern Rhône, especially in its most famous appellations of Châteauneuf-du-Pape, Gigondas, and Côtes du Rhône. Figure 6.16 illustrates a pairing of a Gigondas (predominantly Grenache with Syrah and other red varietals) with a braised lamb shank.

FIGURE 6.16
Gigondas paired with lamb shank. Courtesy of Erika Cespedes.

Grenache also produces the epitome of classic dry rosé wines in Tavel and Provence. This varietal produces the famous French red-based fortified wines called *Vin Doux Naturel* (VDN) in southern France's Roussillon region, specifically in the appellations of *Banyuls* (bahn-YULES) and *Maury* (moe-REE). Grenache, or more aptly named in Spain, Garnacha, is a significant grape that produces the red wines of Priorat and acts as an important blending partner with Rioja. In the New World, Grenache thrives in Australia (McLaren Vale and Barossa Valley), Washington state (Columbia Valley), and California (Central Coast).

Malbec (mahl-BEHK)

Malbec is a red wine grape varietal that has traditionally been a blending grape in the red wines of Bordeaux and southwest France. Over the last decade, Malbec has emerged as the leading red-wine grape varietal in Argentina. Pictured in Figure 6.17 is La Posta Malbec from Mendoza, Argentina.

- *Aroma/Flavor Components:* The aromas and flavors of Malbec can frequently offer layers of dried black fruits (blackberry, black cherry, black currant, and plum), floral (violet and rose), bakeshop (toffee, cinnamon, chocolate, coffee, and anise) and tobacco shop.
- *Structural Components:* Malbec is a deeply intense colored grape (particularly in the New World) and is often medium-bodied with medium tannin, medium acidity, and high levels of alcohol. The Malbec from France tends to provide a less colored and slightly "green" tannic sensation—often thought of as a somewhat rustic style of wine. However, the grape clusters of Argentine Malbec show differently than their French counterpart. Argentine Malbec shows smaller berries (therefore greater color pigment) in smaller grape clusters—coupled with the New World's love affair with hang-time and greater extraction during fermentation—produces more concentrated wine with richer mouthfeel and softer tannins.
- *Significant Locations:* Malbec is the dominant red varietal in the *Cahors* (cah-OHR) appellation where it's locally known as *Côt Noir*—located in southwest France. This varietal also plays a minor role in the blended red wines of Bordeaux, France. Malbec was introduced to Argentina in 1868, producing a softer, less tannic-driven variety than the wines of Cahors or Bordeaux. Mendoza—the large growing region in Argentina has emerged as the popular and very fashionable world producer of Malbec. Smaller amounts of this grape are found in the Loire Valley and California's North Coast region, where it is blended in the New World versions of the classic red Bordeaux wines.

FIGURE 6.17
La Posta Malbec, Mendoza, Argentina. Courtesy of John Peter Laloganes.

Merlot (mehr-LOH)

Merlot is a red wine grape varietal that is thought to have derived its name from "merle"—French for young blackbird—most likely

FIGURE 6.18
"Merle" is French for Blackbird. Courtesy of John Peter Laloganes.

FIGURE 6.19
Blackbird Merlot paired with smoked BBQ pork baby back ribs. Courtesy of Erika Cespedes.

in reference to the color of the grape. Figure 6.18 depicts the "birds" of Blackbird Merlot from California's Napa Valley. Figure 6.19 depicts the Blackbird Merlot paired with smoked pork baby back ribs. Merlot is used as both a single varietal wine and a varietal for blending—along with its natural companion, Cabernet Sauvignon, and smaller percentages of other possible varietals such as Cabernet Franc, Petit Verdot, and Malbec. Merlot is considered a great complement to Cabernet Sauvignon as it works to aid in softening some of Cab's tannins while contributing greater fruit qualities (sometimes referred to as "fleshiness"), therefore making the wine a bit more approachable and adding a dimension of complexity at the same time.

- *Aroma/Flavor Components:* This grape produces intense dried and baked red fruits (cherry, berry, and plum), bakeshop (vanilla, chocolate, and spice), tobacco (nutmeg and clove), and garden (green olive).
- *Structural Components:* Merlot generally contains medium acidity and tannin. Merlot-based wines usually maintain a medium to medium plus body, though can approach full body with a higher percentage of Cabernet Sauvignon in the blend.
- *Significant Locations:* Merlot is one of the primary grapes used in the blended red wines from Bordeaux, France. It maintains the reputation of being their most widely planted varietal grown throughout the region. Merlot dominated wines can also be found largely throughout Washington state (Columbia Valley), California's Sonoma County (Dry Creek Valley) and Napa Valley, Chile, and northeast Italy's Friuli-Venezia region.

Bordeaux

The Bordeaux wine region of France is distinguished by its defining *Gironde* (zhee-RAHWN) estuary, the Dordogne and Garonne rivers which act to naturally separate the region into a left bank and a right bank. On the right bank, cool air and soil temperatures from higher concentrations of wet, compact clay, and limestone soil coordinate with faster-maturing Merlot and Cabernet Franc that are more suited for these conditions. With some notable exceptions, the wines are approachable and young compared with those of the left bank. The prominent right bank appellations are *Pomerol* (pome-EHR-all) and *Saint-Émilion* (sahn-eh-meel-YOHN). These districts produce only red wines that are focused primarily on Merlot and Cabernet Franc, with lesser amounts of Cabernet Sauvignon and Malbec. Figure 6.20 shows a bottle of Chateau Coutet (60 percent or more Merlot with the remaining Cabernet Franc) Grand Cru from Saint Emilion.

Washington State

In the 1980s, Washington state first established its reputation through Merlot, and later on Cabernet Sauvignon and then Syrah. Merlot has always been Washington's draw; it arguably produces some of the most unique Merlot in the whole world. The Columbia Valley, home to over 99 percent of all of Washington's wine acreage, is considered an "umbrella" region as it contains numerous smaller sub-appellations where several different types of sub-climates encourage various grapes to prosper. The enormous region spans from south-central Washington

FIGURE 6.20
Chateau Coutet (Merlot and Cabernet Franc Blend) from Saint Emilion, Bordeaux. Courtesy of John Peter Laloganes.

to northern Oregon encompassing more than a third of the state. The Columbia Valley lies in the rain shadow of the Cascade Mountain range, allowing for an incredible amount of sunlight, an arid climate, with an extreme difference between daytime highs and nighttime lows. These conditions go to create the uniqueness of Merlot (and other red varietals) as the grapes produce maximum flavors, incredible concentration, and rich mouthfeel with a vibrant backbone of acidity that allows the wines to maintain an air of freshness and vitality.

Pinotage (pee-noh-TAHJ)

Pinotage is a grape unique to South Africa, where it was created in 1925 as a cross between the Pinot Noir and *Cinsault* (SAHN-so) varietals. This grape doesn't get the most respect throughout the wine world, but its mellow and complex structural components along with its unique aroma/flavor profile makes an incredibly food friendly partner to hamburgers, tex-mex, and Mexican dishes.

- *Aroma/Flavor Components:* Dried to baked fruits (red and black), earth, animal (leather), tobacco shop (smoke), and chemical (acetone).
- *Structural Components:* Medium-bodied with low to medium tannin and acid.
- *Significant Locations:* South Africa is this wine's major producer.

Sangiovese (san-joh-VAY-zeh)

Sangiovese is the most famous grape found throughout central Italy, primarily throughout Tuscany where the historic Chianti region is located. This grape varietal goes by various names, depending on the type of clone, such as *Sangioveto* in Chianti, *Brunello* in Montalcino, and *Prugnolo* in Montepulciano.

Sangiovese is often a stand-alone varietal, as in Brunello di Montalcino, but is more often blended with small amounts of indigenous Italian grapes in the wines of Chianti and Vino Nobile di Montepulciano. It has also been blended with greater amounts of international varietals such as Cabernet Sauvignon and Merlot to create the Super Tuscan wines of Tuscany. Figure 6.21 shows a bottle of Chianti Classico from Borgo Scopeto that is a 100 percent Sangiovese wine.

FIGURE 6.21
Borgo Scopeto Chianti Classico from Italy's Tuscany Region. Courtesy of Erika Cespedes.

- *Aroma/Flavor Components:* Sangiovese offers fairly aromatic elements of fruits (cherry and black cherry), bakeshop (spice and nuts), earth (dirt and peat moss), tobacco shop (cigar, tea leaves, and leather), and floral (violet and rose).
- *Structural Components:* Sangiovese-based wines can range from medium- to full-bodied, with medium-to-high acid and tannin. Lighter versions may be labeled simply as Chianti or Rosso di Montalcino; medium versions may be labeled Chianti Classico; and full-bodied versions Chianti Classico Riserva, Chianti Classico Selezione, Brunello di Montalcino, and Vino Nobile di Montepulciano.
- *Significant Locations:* Italy (Tuscany), and more specifically the area of and surrounding *Chianti* (kee-AHN-tee).

Chianti

Chianti is a large wine zone located between the medieval cities of Florence and Siena—recognizable since the Middle Ages. The wine is named after its location and made primarily or solely from the indigenous Sangiovese grape, but potentially blended with smaller amounts of other local or international varietals. Some of the best Chianti derives from the Classico area where it consists of between 80 and 100 percent Sangiovese and is more recently allowed to be blended with up to 20 percent of Cabernet Sauvignon and/or Merlot.

The original designated zone of Chianti was identified in 1716 and then expanded in 1932. With the expansion, Chianti wanted to demarcate the original Chianti zone by adding the Classico name to the original area. Chianti is produced in one of the eight distinct, adjacent zones surrounded by the original core area *Chianti Classico* (KLAHS-see-koh). The Chianti Classico zone is the most prestigious, quality examples of Chianti. They have identified three designations of increasing quality standards to showcase the quality of the wines:

- *Chianti Classico d'Annata:* These wines must be at least 12 percent alcohol and aged at least twelve months.
- *Chianti Classico Riserva:* This designation must produce wines that are at least 12.5 percent alcohol and aged in barrel for a minimum of twenty-four months with three of the months spent resting in the bottle.
- *Chianti Classico Selezione:* This designation was created in early 2014, and is positioned to be the "best" quality of Chianti—only accounting for 10 percent of Chianti Classico production. The wines must be made from 100 percent estate fruit, at least 13 percent alcohol, and aged in barrel for a minimum of thirty months with three of the months spent resting in the bottle.

FIGURE 6.22
Antinori's "Tignanello"—One of the Pioneers in the Super Tuscan Movement. Courtesy of John Peter Laloganes.

Super Tuscan Wines

The rise of the popular Super Tuscan wines (an unofficial title) created one of the most revolutionary movements in Italian wine history. It began in the late 1960s and early 1970s, with some renegade winemakers experimenting in earnest by adding other grape varieties, notably Cabernet Sauvignon, Merlot, and Cabernet Franc to their Chianti wines. At the time, these "foreign" grapes were not allowed to be added due to the strict requirements of the Italian classification system. Ironically, regulations that were intended to improve quality spawned wines that were made from blends of grapes that did not conform to the rules forcing producers to declassify their wines. However, wines like Tignanello (a declassified Chianti), as well as Sassicaia and Ornellaia (coming from the Tuscan coast) were vastly superior to most Chianti wines at the time. These wines had garnered an immense amount of fanfare for their bold and intense style. Eventually, they earned the unofficial moniker as wines that were of super status. By the 1990s, these and other so-called Super Tuscans had become among the most acclaimed and coveted wines on the market. The laws of Chianti and Chianti Classico have more recently been changed to accommodate and allow for this more unconventional type of blending that had brought notoriety to the super Tuscan wines. Figure 6.22 shows a bottle of the iconic Tignanello produced from Marchese Antinori.

FIGURE 6.23
Tempranillo from Spain. Courtesy of Erika Cespedes.

Tempranillo (tem-prah-NEE-yoh)

Tempranillo is a red wine grape that is widely grown throughout its native Spain. It is often described as Spain's most noble grape due to its ever noteworthy presence in many of its regions. The name Tempranillo derives from *temprano* or "early"—referencing the grape's trait of ripening early as compared to other local varietals. Figure 6.23 shows a bottle of varietally labeled Tempranillo. In Portugal, the Tempranillo grape goes by many synonyms. It is known as *Aragonez* (air-uh-goan-ehz) and used in red table wine blends of variable quality. Specifically in the Douro region of Portugal, this grape is known as *Tinta Roriz* (ROR-eesh) and is prominently blended in the fortified wines of Port.

This grape can be produced as a stand-alone varietal but shows more complexity when blended with grapes such as Grenache, Cabernet Sauvignon, and Merlot. Tempranillo is capable of producing deeply colored and fruity wines for early consumption, or richly flavored, complex, age-worthy wine for later enjoyment.

- *Aroma/Flavor Components:* Tempranillo offers fairly aromatic nuances of dried red fruits (dark cherry, strawberry, and plum), garden (earth, soil, and wet leaves), bakeshop (vanilla, spice, and cocoa), tobacco (leather), and floral.
- *Structural Components:* This grape varietal produces dry red wines that are typically medium in body with medium acidity and tannin levels.
- *Significant Locations:* Tempranillo is native to northern Spain and widely cultivated as far south as La Mancha. Some of Spain's major areas that grow Tempranillo include *Rioja* and *Ribera del Duero.* And as mentioned previously, Tempranillo is also grown throughout Portugal.

FIGURE 6.24
R. Lopez De Heredia Rioja. Courtesy of Erika Cespedes.

Rioja (ree-OH-hah)

Rioja maintains the highest-quality ranking (DOCa) in the Spanish wine classification system. This region takes its name from the river "Rio Oja" and is located in north-central Spain between mountain ranges and along the path of the Ebro River. Rioja is one of the leading and most recognizable Spanish wine regions, most famous for its production of red wines. A typical red Rioja wine is made primarily from Tempranillo and blended with varying amounts of Garnacha and smaller amounts of *Graciano* (grah-thee-AH-no) and *Mazuelo* (mah-THWAY-low) (known as Carignan outside of Spain). Figure 6.24 displays R. Lopez de Heredia Rioja, one of Rioja's most traditional producers.

The Rioja wine region is divided into three distinct subregions: *Rioja Alta* (AHL-tah), *Rioja Alavesa* (ahl-lah-VACE-ah), and *Rioja Baja* (BAH-hah). These subregions are quite different in that they consist of varying levels of altitude, climate, and soil types. Classically, Rioja wines have been created from a blend of grapes from all three subregions, although single-vineyard wines have been gaining popularity as a way to illustrate their unique terroir differences. Pictured in Figure 6.25 is a sign from La Rioja Alta Winery.

FIGURE 6.25
La Rioja Alta Winery in Rioja, Spain. Courtesy of John Peter Laloganes.

FIGURE 6.26
Ribera del Duero. Courtesy of Erika Cespedes.

Ribera del Duero (ree-BEHRR-ah del DWAY-rroh)

This region is arguably Spain's second or third most popular region. Ribera del Duero is located north of Madrid and west of Rioja and situated along the Duero river. The region is best known for producing wines based on Tempranillo (also known as Tinta del País or Tinto Fino)—red wines that must contain a minimum of 75 percent of the varietal. Opposite of Rioja, this region allows the liberal use of French oak for aging and the potential for blending with the addition of any of the several international varietals such as Cabernet Sauvignon, Malbec, and Merlot. Figure 6.26 depicts a bottle of Ribera del Duero.

Bold and Intense Reds

Learning Objective 3
Explain characteristics of the bold and intense reds style category

complex ... concentrated ... evolved

This style category often produces robust, spicy, oak-aged wines—matured for months to years in both barrel and bottle. These reds may be associated with warm to hot climates where sugar levels allow for alcohol levels of 13.8 percent and well beyond. These red wines have aromas and mouth-filling flavors of tobacco shop, jam like, and dried fruit with structural components of moderate acidity, medium to high tannins and body. Figure 6.27 displays a typical bold and intense red wine paired successfully with a grilled porterhouse steak (prepared medium-rare) with garlic mashed potatoes, shredded fried onion rings and a red wine demi-sauce.

Cabernet Sauvignon (ka-behr-NAY soh-vihn-YOHN)

Cabernet Sauvignon is a red wine grape that remains one of the most widely recognized and popular varietals throughout the world. Despite its prominence in the wine world, the grape has risen relatively quickly given its shorter history as compared to other varietals. Sometime during the 17th century, Sauvignon Blanc and Cabernet Franc had crossed and became the parents of a newly developed varietal—Cabernet Sauvignon or "Cab" for short. Cabernet Sauvignon is frequently referred to as the "king of red wines" and is often viewed as a winery's benchmark wine—one that often gains the most prestige and notoriety. Figure 6.28 depicts clusters of old vine Cabernet Sauvignon.

Cabernet Sauvignon is often blended with "fleshy" yielding grapes such as Merlot or Shiraz in order to lower tannin and yield more balanced flavors (by contributing

FIGURE 6.27
Bold and intense red wine paired with porterhouse steak. Courtesy of Erika Cespedes.

FIGURE 6.28
Old vine Cabernet Sauvignon. Courtesy of Terra Valentine and Carolyn Corley Burgess.

a bit more fruit qualities) and softening structure. Cabernets are almost always aged in new oak for at least one year from harvest and are more likely aged several more years. Cab is also bottle-aged for years to decades in order to soften its tannin otherwise it tastes too raw and astringent. World class examples of Cabernet Sauvignon can often evolve and be aged for decades. Figure 6.29 shows Spring Mountain Vineyards Cabernet Sauvignon from Napa Valley, California.

In the famous 1976 Paris wine tasting, it was a Napa Valley Cab that was responsible for bringing great fame to California (and overall the New World) when "Stag's Leap Wine Cellars" won the top place over their Bordeaux counterparts.

- *Aroma/Flavor Components:* Cab offers intense and complex aromas/flavors of baked/dried red and black fruits (cherry, blackberry, plum, and cassis), tobacco (cedar, clove, and cigar), bakeshop (dark chocolate, coffee, and black tea), and garden (eucalyptus, mint, bell pepper, and black olives). Pictured in Figure 6.30 is a glass of Obsidian Ridge Vineyard Cabernet Sauvignon from Red Hills Lake County.
- *Structural Components:* The popularity of Cabernet Sauvignon is often attributed to its ease of cultivation—the grapes have thick skins and the vines are hardy and fairly resistant to vineyard hazards. Cabernet Sauvignon is intense in color and high in tannin (due to its thick skin), medium to full body (most often full, particularly when yields are low), warm to spicy in alcohol (often 13.5 percent for Old World

FIGURE 6.29
Spring Mountain Vineyards Cabernet Sauvignon from Napa Valley, CA. Courtesy of John Peter Laloganes.

FIGURE 6.30
Obsidian Ridge Cabernet Sauvignon. Courtesy of Erika Cespedes.

FIGURE 6.31
Bordeaux and Napa Valley. Courtesy of Erika Cespedes.

FIGURE 6.32
Bordeaux. Courtesy of Erika Cespedes.

versions versus commonly 14 percent + for New World ones), and medium in acidity.

- *Significant Locations:* Cabernet Sauvignon is grown in nearly every significant wine-producing country through a diverse spectrum of climates and locations. Cab became internationally recognized through its prominence in the red wines from Bordeaux where it's blended with smaller amounts of Merlot, Cabernet Franc, and other varietals. Arguably, "California Cabernet" is equal to or better than "Bordeaux Cabernet," depending upon your preference of style differences. Over the last hundred years or so, Cab has served as the backbone of some of the world's most renowned wines from Bordeaux, France, and Tuscany, Italy to Napa Valley, California, Washington state, Australia, and Chile. Figure 6.31 displays Cabernet Sauvignons from two of its most famous locations—Chateau Angludet from Margaux in Bordeaux, France and Hall Vineyards from Napa Valley, California.

Bordeaux

The Bordeaux wine region of France is distinguished by its defining *Gironde* and the Dordogne and Garonne rivers which act to naturally separate the region into a left bank and a right bank. Cab tends to thrive on the left bank, or west side of the river due to its desire for well draining gravelly soil that is mixed with pebbles and sand due to its proximity to the Atlantic Ocean. Pictured in Figure 6.32 is Chateau de Parenchere Bordeaux Supérieur. The left bank area enjoys slightly warmer air temperatures, and the soil remains warm as well. These factors allow for more slowly developing Cabernet Sauvignon grapes to reach optimal ripeness. The vines are old, strong, and hearty, and they produce wines with enormous power and aging potential. The most notable and prestigious growing area on Bordeaux's left bank is the *Haut-Médoc and Médoc* (may-DAWK). The red wines from the Médoc derive primarily from Cabernet Sauvignon with varying amounts of other grapes, namely Merlot, Malbec, Cabernet Franc, and Petite Verdot. The wines are full-bodied and tannic when young and more balanced and elegant when matured. Significant village appellations from Médoc and, more specifically, surrounded by the Haut-Médoc include

- Saint-Estèphe (san teh-STEHF)
- Pauillac (poh-YAK)
- Saint Julien (san zhoo-LYAN)
- Listrac (lees-TRAHK) and Moulis (moo-LEE)
- Margaux (mahr-GOH)

Napa Valley

Cabernet Sauvignon is the acknowledged king of red grapes in Napa Valley, accounting for 40 percent of total production, although Merlot has also been a fixture in the Napa Valley since the early 1970s. As in Bordeaux, Merlot is often used as a blending partner to add body and soft fruit to the more structured and tannic Cabernet Sauvignon.

Napa Valley is located in Northern California, just north of San Francisco and 36 miles east of the Pacific Ocean. This prestigious area is a small winemaking region that is just 35 miles long by 5 miles wide. It is nestled between the Mayacamas Mountains to the west and the Vacas Mountains to the east offering a sun bathed, dry, warm climate. The long growing season is ideal for wine grapes to ripen slowly and evenly, with great balance between sugar and acid development. The warm days and cool nights with lack of summer rainfall helps to contribute to consistency of vintages and reduces the risk of vineyard diseases.

There are even smaller, more defined growing areas within the Napa Valley AVA, where there are currently 16 recognized sub- or "nested" AVAs. Throughout the valley, vineyards are broadly found in three locations, they are planted on the fertile *valley floor*, the *benchlands*, with a gently sloped transition between the valley floor and steeper mountainsides, and in high elevated *mountainside* vineyards that contain poor nutrient, rocky soils.

Mourvèdre (moor-VEH-druh)

Mourvèdre is said to have originated in Spain, where it still remains very popular and is known as *Monastrel* (mohn-ah-STRELL). In southern France, this grape is extremely popular as a stand-alone varietal, in the red and rosé wines of Bandol in the Provence region. Mourvèdre is traditionally used as a blending grape in the famous Côtes du Rhône and Châteauneuf-du-Pape of France's southern Rhône Valley.

- *Aroma/Flavor Components:* Fruits (dried cherry and blackberry), tobacco shop (animal, leather, and smoke).
- *Structural Components:* Medium to medium plus body, with medium to high acid and high levels of tannin. The wines are characterized as spicy due to the grape's ability to gain considerable sugar during ripening, therefore converting it to higher levels of alcohol.
- *Significant Locations:* France (Rhône Valley, Provence, and Languedoc-Roussillon), Spain (Jumilla), and California (Central Coast and Sonoma County) are the major producers.

Nebbiolo (neh-b'YOH-loh)

The Nebbiolo name derives from the word *nebbia*, Italian for "fog," which is known to encase the Nebbiolo vineyards in Piedmont during harvest time. Nebbiolo produces some of the most ageable and long-lived wines available.

- *Aroma/Flavor Components:* Nebbiolo offers fairly to highly aromatic nuances of dried red fruits (raspberry, black cherry, plum, and prune), tobacco (earth and leather), garden (soil, mushroom, and tar), floral (rose and violet), and bakeshop (cocoa, anise, and licorice).
- *Structural Components:* Nebbiolo is dry and generally medium-bodied, with high tannins and acids. These wines can often trick consumers into thinking they are lighter-bodied wines due to the appearance of medium color intensity. However, this grape requires several years of bottle aging to tame its fierce tannins prior to drinking.
- *Significant Locations:* Nebbiolo is largely produced in northwest Italy (Piedmont). The grapes most famous appellations are Barolo, Barbaresco, Ghemme, and Gattinara. Pictured in Figure 6.33 is a Barolo label.

FIGURE 6.33
Barolo from the Piedmont region of Italy. Courtesy of Erika Cespedes.

FIGURE 6.34
Barbaresco from the Piedmont region of Italy. Courtesy of John Peter Laloganes.

Barbaresco and Barolo

These wines are both 100 percent expressions of Nebbiolo from two distinct growing areas—located a mere ten miles apart. Both offer power, intensity, and overall quality, yet Barbaresco is often considered the "queen," and Barolo as the "king." The grapes are grown in slightly varying climates where the temperatures are a few degrees warmer for Barbaresco compared to Barolo, therefore, the grapes tend to ripen earlier and produce a slightly less tannic edge. The aging minimums also recognize the difference between the two, where Barolos need more extensive aging in order to soften and become more balanced.

- Barbaresco wines must be aged a minimum twenty-six months (two years and two months) while spending nine of those months in wood. Barbaresco Riserva must be aged a minimum fifty months (four years and two months) while spending nine of those months in wood as well. Figure 6.34 depicts a label of Barbaresco.
- Barolo wines often referred to as the "king of wines and wine of kings" due to its reputation for long aging, must be aged for minimum thirty-eight months (three years and two months) with at least eighteen of those months in wood. For Barolo Riserva, the wines must be aged a minimum sixty-two months (five years and two months) with eighteen of those months spent in wood.

Syrah/Shiraz (See-RAH)/(shih-RAHZZ)

FIGURE 6.35
Doyenne Syrah from Columbia Valley, Washington State. Courtesy of John Peter Laloganes.

Syrah, as it is called in most wine-producing countries, and Shiraz in Australia, is a dark-skinned red wine grape that is widely grown throughout the world. This grape is one of the most significant emerging varietals and plays an important role in many wine areas, and in some cases, being showcased as a particular country's source of distinction.

This grape has a long documented history in the Rhône Valley region of mid-central and southeastern France. In the late 1990s, DNA profiling found Syrah to be the offspring of two obscure grapes from southeastern France, *Dureza* and *Mondeuse Blanche*. Since both of Syrah's parents derive from such a limited and confined area very close to northern Rhône, researchers have concluded that Syrah originated from northern Rhône and not from Persia in the Middle East, as many had speculated. Figure 6.35 identifies Doyenne Syrah from Columbia Valley, Washington.

Syrah is often confused with a similar-sounding grape—Petite Sirah (sometimes referred to as *Durif*) that is a completely different varietal, yet shares some connection. Petite Sirah is actually a cross of Syrah with an obscure varietal named Peloursin.

Syrah or Shiraz is often created as a stand-alone varietal but is also incredibly adaptable to being blended with other compatible grapes such as Grenache, Mourvèdre, Cabernet Sauvignon, and Viognier. As a solo or dominated varietal, Syrah is the principal style of Hermitage in northern Rhône,

California's Central Coast, or Australia. When Syrah is blended with a small amount of the white wine varietal Viognier, this traditional style is associated with the appellation of Côte-Rôtie in northern Rhône. In Australia, Shiraz can also be found as a blending component with the compatible Cabernet Sauvignon. This blended concept originated in Australia where it's often identified with both varietals on the front label as Shiraz-Cabernet. Syrah may also act as a minor blending partner for Grenache and Mourvèdre as the traditional style of Châteauneuf-du-Pape of southern Rhône. Elsewhere in the New World, this blend is occasionally identified by the grapes' initials as "GSM," referencing the trio of Grenache, Syrah, and Mourvèdre.

For the purpose of simplification, there appear to be two distinct styles of this grape: Syrah (and Syrah blends) and Shiraz. Syrah tends more toward a spicy, rustic, or earthy style that can be medium to full body with medium acids and medium to high tannins. Shiraz, on the other hand, is created in a bolder style that is intensely fruit-forward and high in alcohol due to extended hang-time. Therefore, this wine is geared for lovers of robust, full-bodied wines with intense concentration and mouthfeel.

- *Aroma/Flavor Components*: Syrah/Shiraz ranges from fairly to highly aromatic. Aroma and flavor aspects can include a range of floral (violets) to baked and dried black fruit (blackberry and blueberry), bakeshop (chocolate, espresso, clove, and black pepper), tobacco shop (leather), and earth (leaves and animal). As with most wines, over time, many of the "primary fruit" components are moderated and then supplemented with earthy or savory "tertiary" notes with a greater emphasis of tobacco and earth.
- *Structural Components:* Syrah/Shiraz tends to be densely colored, with a rich, medium to full body, medium to high tannin and acid, with high alcohol content. Syrah blends tend to be a bit softer than the solo Syrah, and with less alcohol.
- *Significant Locations:* Syrah continues to be the dominant red-wine grape of France's northern Rhone. It's associated with the famous appellations of Hermitage, Côte Rôtie, St. Joseph, and Crozes-Hermitage. In the southern Rhône, Grenache plays the dominant role while Syrah is used as the compatible blending partner in such wines as Châteauneuf-du-Pape, Gigondas, and Côtes-du-Rhône. In the New World, Syrah plays significant role in the wines of California (Central Coast), Washington State (Columbia Valley), Australia (Barossa Valley and Coonawarra), Chile (Colchagua and Maipó), and South Africa (Cape).

Touriga Nacional (too-REE-gah Nah-syon-AHL)

Traditionally, Touriga Nacional has served as one of the most significant grapes used to produce Port, the famous Portuguese fortified wine. Currently, this grape has gained renewed interest and popularity as a stand-alone varietal or blended with the traditional Port wine grapes but made in an unfortified wine style.

- *Aroma/Flavor Components*: This grape maintains the deep, concentrated flavor and aroma of baked/dried fruits (blackberry and black cherry), floral (violets), bakeshop spice (black pepper, allspice, and cinnamon), and cigar shop (smoke).
- *Structural Components*: Touriga Nacional creates a wine with substantial concentrated color, with medium plus to high tannin (from thick-skinned grapes), medium acid levels, and medium plus to full body.
- *Significant Locations:* Portugal (Douro and Dão Valleys) is the main producer.

Zinfandel (ZIHN-fuhn-dehl)

Zinfandel is a red wine grape varietal that had its origins shrouded in mystery. For years, it was believed that Zinfandel was the "authentic" American grape due to its long presence (roughly since the mid-1800s) in California. However, back in 2001, DNA fingerprinting revealed that Zinfandel is genetically equivalent to the *Primitivo* (pree-mah-TEE-voh) varietal grown in the Puligia region of Italy and the Croatian grape *Crljenak Kaštelanski* (surl-yen-ack kah-stehl-AHN-skee). Zinfandel has been traced to its homeland originating on Croatia's Dalmatian Coast.

There are several possible styles of Zinfandel—from the infamous, overly sweet rosé-style wine called White Zinfandel, to the late harvest port-like dessert wine. But the greatest quality tends to be associated with the hearty, robust, dry red wine with concentrated fruit-forward elements. Figure 6.36 shows Ridge Vineyards Zinfandel from Sonoma County's Lytton Springs.

- *Aroma/Flavor Components:* Zinfandel is fairly to highly aromatic that yields aromas/flavors of concentrated baked/dried red and black fruits (raspberry and blackberry), bakeshop (anise, vanilla, and spice), and tobacco (smoke, cigar).
- *Structural Components:* Zinfandel offers depth and complexity with medium acidity, medium tannins, and full body. The grape's high sugar content allows the wine to be fermented upward to 15 percent alcohol or more. Offering power, yet because of its restrained tannins—not too unapproachable in its youth.
- *Significant Locations:* This grape thrives throughout California, yet has been most famous from Sonoma County, Central Coast (Paso Robles), and San Joaquin Valley. Old World locations include Italy (Puglia) and Croatia (Dalmatian Coast). Figure 6.37 shows two bottles of Ridge Vineyards Zinfandel from Paso Robles in the Central Coast.

FIGURE 6.36

Ridge Vineyards Zinfandel from Sonoma County's historic vineyards of Lytton Springs in the Dry Creek Valley AVA. Courtesy of John Peter Laloganes.

FIGURE 6.37

Ridge Vineyards Zinfandel from Paso Robles in the Central Coast of California. Courtesy of John Peter Laloganes.

Check Your Knowledge

Directions: Use these questions to test your knowledge and understanding of the concepts presented in the chapter.

I. MULTIPLE CHOICE: Select the best possible answer from the options available.

1. Which red wine grape would typically be considered to have medium plus to high tannin?
   a. Pinot Noir
   b. Gamay
   c. Malbec
   d. Cabernet Sauvignon
2. Which red wine grape would typically have low to medium tannin?
   a. Pinot Noir
   b. Zinfandel
   c. Malbec
   d. Cabernet Sauvignon
3. Which red wine grape commonly produces a full body, yet commonly maintains medium tannin levels?
   a. Zinfandel
   b. Shiraz
   c. Cabernet Sauvignon
   d. Both a and b
   e. All of the above
4. When vinified dry, Zinfandel grown in California will often have .
   a. high acid and low alcohol
   b. high residual sugar content
   c. bubbles
   d. high alcohol content
5. This is a dark-skinned grape with aromatics of dark fruits (blackberry and black plum) with notes of spice (black pepper) and meat (game, smoke, bacon fat, and leather). It has medium plus to full body with medium acids and tannins.
   a. Cabernet Sauvignon
   b. Merlot
   c. Zinfandel
   d. Syrah
6. What is the Australian name for Syrah?
   a. Semillon
   b. Cabernet Sauvignon
   c. Cinsault
   d. Shiraz
7. This is a thin-skinned and lightly pigmented grape with intense aromatics of red fruits (cherry and plum) and hints of baking spice. It thrives in the heat, so wines often develop very high alcohol, and is often blended in Spain (Rioja and Priorat) and in southern Rhône Valley.
   a. Pinot Noir
   b. Cabernet Sauvignon
   c. Grenache
   d. Syrah
8. This is a thin-skinned and lightly pigmented grape with aromatics of red fruits (raspberry, cranberry and cherry), baking spices, and earth (mushroom and dust). Very susceptible to mold/fungal issues.
   a. Pinot Noir
   b. Merlot

   c. Grenache
   d. Syrah
9. Merlot is the dominant grape in which of the following wine locations?
   a. Burgundy
   b. Bordeaux
   c. Loire Valley
   d. Oregon
10. This is a small, thick-skinned, and heavily pigmented grape with aromatics of dark fruit, tobacco shop, garden, and earth. Enjoys well-drained (stone and rock-based) soils. Very hardy and one of the most planted red wine varietals in the world.
   a. Cabernet Sauvignon
   b. Merlot
   c. Zinfandel
   d. Syrah
11. Which red wine grape would typically be considered a "Mellow and Complex" style category?
   a. Pinot Noir
   b. Zinfandel
   c. Tempranillo
   d. Cabernet Sauvignon
12. This is a medium, moderate-skinned grape with aromatics of red fruit (cherry) and bakeshop (chocolate and vanilla). Maintains a medium body, acid, and tannin. Tolerates wetter (clay-based) soils.
   a. Cabernet Sauvignon
   b. Merlot
   c. Zinfandel
   d. Pinot Noir
13. Which of the following best describes the general wine style of Nebbiolo?
   a. Mellow and complex reds
   b. Fruity rosé and vibrant reds
   c. Bold and intense reds
   d. Crisp and youthful reds
14. Which Italian grape varietal is translated to "little sweet one?"
   a. Dolcetto
   b. Sangiovese
   c. Barbera
   d. Nebbiolo
15. The typical southern Rhône Valley wine consists of a blend of: Syrah, Grenache, and
   a. Pinot Noir
   b. Mourvèdre
   c. Merlot
   d. Zinfandel
16. What features of Cabernet Sauvignon contribute to its longevity in the bottle?
   a. Sugar and acid
   b. Color and alcohol
   c. Alcohol and tannin
   d. Acid and color
17. This wine style has grapes that may have a higher proportion of juice-to-skin ratio that may yield a lighter-bodied, higher-acidic wine with minimal tannin components.
   a. Fruity rosé and vibrant reds
   b. Mellow and complex reds
   c. Bold and intense reds
   d. Rich and voluptuous whites

18. This style category often produces wines that are big, often spicy, usually oak-aged—matured for months to years in both barrel and bottle. Often contains tobacco shop, jam-like, and dried fruit and structural components of moderate acidity, medium to high tannins, and full body.
    a. Fruity rosé and vibrant reds
    b. Mellow and complex reds
    c. Bold and intense reds
    d. Rich and voluptuous whites
19. A typical Bordeaux red will consist of a blend of mainly Cabernet Sauvignon, Merlot, and
    a. Pinot Noir
    b. Malbec
    c. Petite Verdot
    d. Cabernet Franc
20. In which of the following French locations is the Gamay grape most famous?
    a. Bordeaux
    b. Beaujolais
    c. Languedoc
    d. Champagne

CHAPTER 7

Other Wines: Sparkling, Fortified, and Dessert Wines

CHAPTER 7 LEARNING OBJECTIVES

After reading this chapter, the learner will be able to:

1. Explain the difference between sparkling wine and Champagne
2. Identify the categories and styles of Champagne
3. Explain the production process of sparkling wine
4. Distinguish between the different origins of bubbles in sparkling wine
5. Identify at least three other sparkling wines
6. Explain the categories and styles of Port wine
7. Explain the categories and styles of Madeira wine
8. Explain the categories and styles of Sherry wine
9. Identify some famous dessert wines found throughout the wine-producing world

A single glass of Champagne imparts a feeling of exhilaration. The nerves are braced; the imagination is stirred; the wits become more nimble.

—Winston Churchill, on Champagne

Sparkling Wine

Learning Objective 1
Explain the difference between sparkling wine and Champagne

revitalizing ... lively ... festive

Sparkling wine has the incredible ability to create a mood of revelry that sets a moment apart or makes an event special. The sound of corks popping at ritualistic events such as weddings, victory celebrations, rites of passage and, of course, New Year's Eve marks a time of both distinction and ceremony. Something about this drink has the uncanny ability to intoxicate the soul, the mind, and the body. Figure 7.1 identifies an ornate doorway in the Champagne village of Aÿ. Figure 7.2 shows a glass of Champagne being poured and is symbolic because upon entering most people's home, one is greeted with warmth and hospitality—essentially the same as being offered a glass of Champagne.

The term sparkling wine is a generic one used to identify any table wine with the addition of its obvious effervescence, or carbon dioxide (CO_2). The tiny bubbles are the discernible element that distinguishes sparkling wine from other wines. The wine should be handled with care, not only due to the contents being under pressure but also because of the respect it deserves. Sparkling wines can vary from delicate to powerful, simple to complex, white to red, and dry to sweet. They can be found at various quality levels, and price points—but the most historic, prestigious, and reputable type of sparkling wine is Champagne.

Champagne is given a special honor of being the most "important" type of sparkling wine because it has a tradition, a certain distinction, and elegance unlike all

FIGURE 7.1
Doorway in Aÿ. Courtesy of John Peter Laloganes.

FIGURE 7.2
Glass of Champagne. Courtesy of Erika Cespedes.

other sparkling wines. While sparkling wines are created in nearly every major wine-producing country and can be found in a wide range of styles, Champagne comes from a region in northern France. Along with the highly regulated growing region of Champagne (comprising its unique climate and soil type), type of grapes and method of incorporating the bubbles are three of the most defining quality factors in what separates a poor-to-average-quality sparkling wine. By understanding Champagne and how it is produced, it then becomes easier to understand almost all other sparkling wines in the world. Figure 7.3 shows a map of the Champagne region and Figure 7.4 illustrates the predominant chalky soil type that acts to preserve acidity in the grapes.

The Misuse of the "Champagne" Name

The terms sparkling wine and Champagne are often used interchangeably; however, although they can be similar, they are actually quite different. It's unfortunate that many consumers mistakenly associate any bubbly wine as Champagne. It's true that all Champagne is sparkling wine, though not all sparkling wine is Champagne. In the past, the term Champagne in the United States has been used as a generic term to capitalize on the fame of the official and authentic Champagne sparkling wine. When the term is used in America or anywhere else that is not the Champagne region, it does not relate to place of origin, as it does in France. When the Treaty of Versailles (to end World War I) was signed in 1919, France asked for limits on the use of the word "Champagne." But the United States never actually ratified the treaty, so alcohol label laws were dismissed. Out of respect, many producers in the U.S. honored this request, but others slapped on the label Champagne (and others like Burgundy, Chablis, etc) in order to promote their beverages. In 2006, the United States and the European Union signed a wine trade agreement that would respect the use of these words and what their

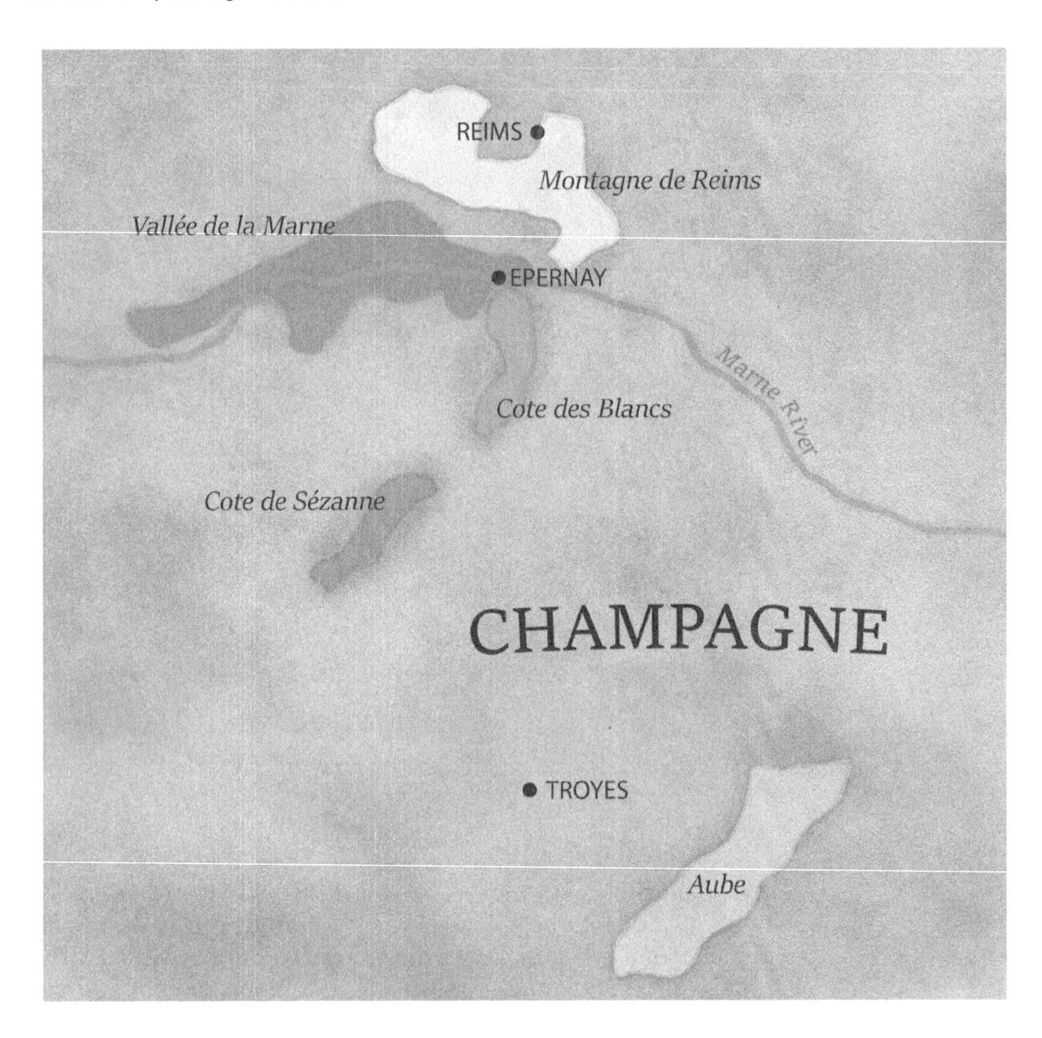

FIGURE 7.3
Map of Champagne wine region. Courtesy of Thomas Moore.

FIGURE 7.4
Chalk soil in Champagne. Courtesy of John Peter Laloganes.

intention is throughout the EU. However, anyone who already had an approved label such as Korbel and Miller High Life (the "Champagne" of beers) was grandfathered into the new rules. Beware of producers using this terminology; the products often are low-quality versions that resemble the Champagne style very little.

Categories of Champagne

Learning Objective 2
Identify the categories and styles of Champagne

Champagne can broadly be classified according to whether it's a non-vintage, vintage, or prestige. Within each of these categories, there exist various styles and varietal makeup.

Non-Vintage Champagne

Or sometimes simply referred to as NV or Multi-Vintage, it accounts for the majority of production within the Champagne market. This highly blended style of Champagne is replicated by most sparkling-wine producers all around the world. Non-vintage sparkling wine will not indicate a year on the label since it is made from a blend of several wines (or cuvées) from different years. These wines are made to achieve a "house style" that is intended to remain consistent in quality and taste from year to year. Non-Vintage Champagne is released at least fifteen months after harvest and often consumed within five years from harvest. Figure 7.5 identifies a bottle of non-vintage Ayala Champagne.

FIGURE 7.5
Ayala, Non-Vintage (NV) Champagne. Courtesy of Erika Cespedes.

Vintage Champagne

Or referred to by the French term *Millésime* (mee-lay-ZEEM), "Vintage Champagne" is only produced in years when the grapes are of exceptional quality. Vintage years are rare (maybe two to three times per decade) in Champagne because the weather is sporadic and the growing season is short in this northern region. When the grapes have a particularly good growing season, a rare vintage year may be declared by a Champagne house—though the declaration is occasionally considered quite subjective and possibly controversial. Vintage Champagne indicates that a minimum of 95 percent of the grapes from the current year's harvest are within the bottle, though not every year is declared a vintage. The wine is released at least three years after harvest, after it has gained depth and complexity through the aging process. Vintage Champagne has the capacity to be cellared for several years after purchasing, and even up to a decade or more before consumption. This type of Champagne is expensive and considered to be prestigious—priced almost three times more (and often even higher) than non-vintage. Pictured in Figure 7.6 is a bottle of Bollinger La Grande Année vintage 1999.

FIGURE 7.6
Bollinger LaGrande Année vintage Champagne. Courtesy of John Peter Laloganes.

Prestige Cuvée or Tête de Cuvée (tet duh koo-VAY)

An unofficial term, this often refers to a superior quality selection to identify a producer's best or most prestigious wine. Dom Pérignon, one of the most famous brands of Champagne, is named after an 18th-century Benedictine monk. The monk is known to have perfected the blending of different grapes and created a durable bottle to withstand the carbonation present in sparkling wine. As a brand, Dom Pérignon was launched in 1921 by *Moët et Chandon* (moh-eht ay shahng-DAWNG) as their premium level Champagne. It's a single-vineyard wine made only from grapes in a single, exceptional year and it's typically aged for six to eight years prior to release. Other similar prestige Champagnes exist in the marketplace such as *Laurent-Perrier's* (loh-RAHNG pehree-ay) Grand Siècle, *Roederer's* (ROH-duh-rer) Cristal, and *Taittinger's* (tate-teen-ZHEHR) Comtes de Champagne.

FIGURE 7.7
Prestige Cuvée. Courtesy of John Peter Laloganes.

Pictured in Figure 7.7 are bottles of prestige cuvée given extended aging to further increase their complexity.

Grower Champagnes

These Champagnes are unique in that the wine is produced by the same estate that owns the land—coming from individual, family-owned vineyards. This has been one of Champagne's biggest changes over the past decade and has risen in popularity among sommeliers in high-end restaurants. The traditional practice has mostly been for Champagne houses (normally the large estates, and ones most often represented on wine lists) to purchase all, or at least the clear majority of their grapes from growers. But now some of those growers have become winemakers. Grower Champagnes aren't necessarily better or worse than other ones, but there is a perception of having a greater sense of connection to the land. Grower Champagne can be identified by the letters RM on the label, which stand for *récoltant-manipulant* (ray-kohl-tahn mah-nip-u-lahn), meaning a winemaker who grows grapes and makes wine only from his or her own grapevines.

Styles of Champagne

Champagne distinguishes itself from the rest of the sparkling wines in many ways. Of course, some would argue that all other sparkling-wine producers around the world need to distinguish their wines from Champagne. The quality oriented non-Champagne producers will typically attempt to replicate Champagne in a respected manner, without losing their own local uniqueness. Regardless of origin, all sparkling wines share the similar characteristics in common: substantial acidity and the obvious effervescence.

FIGURE 7.8
Blanc de Blanc Champagne. Courtesy of Erika Cespedes.

Traditional Blend

In Champagne, the wine is made from a blend or cuvée of three grape varieties—Chardonnay, Pinot Noir, and Pinot Meunier are permitted for application. Since the majority of Champagne is white-colored wine, Chardonnay would be the obvious choice since it is known around the world for producing crisp white wines. The less obvious choices are the Pinot Noir and the highly obscure Pinot Meunier (muh-NYAY)—that are ironically purple-black grapes traditionally used for making red wine. Just about every grape within the *vinifera* species can produce white wine—regardless of skin color. Since all of a wine's color derives from its grape skins, the juice needs to be swiftly removed from the skins to limit any such color extraction.

Blanc de Blanc

Blanc de Blanc (translates to *white from white*) is made from 100 percent Chardonnay (non-Champagne sparkling wines can often use any other white grapes). Chardonnay is derived primarily from the smaller growing areas within Champagne such as *Côte Des Blancs* (coat day BLAHN) and *Côte de Sézanne* (coat du say-ZAHN). This grape lends considerable acidity to the wine. When blended in higher amounts or as a solo varietal in Blanc de Blanc, it produces a lean and crisp, light-bodied wine. The color is pale and light, and the wine has a certain delicate, yet crisp style. Figure 7.8 identifies a bottle of Blanc de Blanc Champagne.

Blanc de Noir

Blanc de Noir roughly translates to "white from black." This wine is made from 100 percent Pinot Noir and/or Pinot Meunier grapes. Of the two dark-skinned grapes, it's the Pinot Noir grape that reigns supreme. Pinot Noir is found in the *Montagne de Reims* (mawn-TAH-nyuh) growing area. This grape provides considerable body and some fruit qualities to the wine. When blended in higher amounts (often with Pinot Meunier) or as a solo varietal in a Blanc de Noir, it produces a full-bodied, fruitier wine. Pinot Meunier is primarily from the *Vallee de La Marne* (vah-LAY duh lah MARN) area. This grape provides considerable fruitiness and some structure of tannin to the wine. The Blanc de Noir style is full-bodied compared with Blanc de Blanc and/or the traditional Champagne blend. This wine is likely to also contain some perceptible levels of tannin and possibly a tinge of pink color because of the ever so brief contact time of the juice with the black skins.

FIGURE 7.9
Bollinger Rosé Champagne. Courtesy of John Peter Laloganes.

Rosé

This is another style of Champagne that in recent years has gained unbelievable popularity. Rosé combines the fruitiness of a red wine and the crispness and the freshness of a white wine. Its associated pink color signifies some use of black-purple grape varieties that were incorporated in the production process. There are two methods used to make rosé sparkling wine—either from the blending of red and white wines or through the saignée method. The first method is straightforward; simply use white wine as a base and carefully add red wine until the desirable color, aroma/flavor, and structural profile are obtained. The saignée method of producing rosé wine is a process very similar to making a red wine. Since all of a wine's color pigment is derived from its skins, the skins are allowed to remain in contact or macerate for a brief period of time with the juice, allowing the skins to "bleed" (or the French word, saignée) some color. Figure 7.9 identifies a Bollinger Rosé Champagne.

Production Process

Learning Objective 3
Explain the production process of sparkling wine

The production process for Champagne is typically identical when producing other high-quality complex style sparkling wine. Winemakers can use the classic Champagne method of production called méthode champenoise, to respectfully replicate the style of Champagne. Since the European Union has played a large role in improving and working on consistent use of language across the wine-producing countries throughout Europe, they have now required a modified, inclusive term identified as *méthode traditionelle.* This new term has identical meaning, as it identifies the traditional method of making Champagne or sparkling wine, involving a second fermentation in its bottle, necessary to creating its CO_2. This production method also begins the lengthy process of aging the wine in contact with its yeast cells and then attempting to expel the dead yeasts without removing the wine from the bottle. Any producer of sparkling wine around the world may choose to use the classic Champagne method, but Champagne (and a few others) is one of the few sparkling wines that must use this technique. The ten steps outlined next are necessary to produce all Champagne, while producers around the world who choose to create high-quality sparkling wines in the style of Champagne generally will mimic these steps.

THE CHAMPAGNE PRODUCTION PROCESS

1. Harvest
2. Pressing
3. First fermentation
4. Cuvée assemblage
5. Second fermentation (incorporating the carbonation)
6. Aging
7. Remuage
8. Dégorgement
9. Dosage
10. Bottling/corking

Harvest

In Champagne, the harvesting of grapes is typically conducted earlier than many other grape-growing areas. When grapes are picked earlier in their development process (typically in late September), they maintain a greater level of acidity—one of the trademark components of a well-made sparkling wine. In addition, since the wine will ultimately undergo two separate fermentations, the earlier harvest allows the grapes to contain lower amounts of natural sugar levels and therefore achieve lower amounts of initial alcohol. The first fermentation will create the initial base wine (basically, a table wine at this point) and the second fermentation produces and traps the desirable carbon dioxide.

Pressing

The process of pressing the grapes quickly after harvest is vital to maintaining the integrity of the finished wine. Once the grapes have been hand-harvested, which ensures utmost care of the delicate fruit, and sorted, they are pressed gently with a wide device to prevent the juice taking excess time traveling through skins and to limit juice and skin contact. The juice is then placed in stainless steel tanks for the initial fermentation.

FIGURE 7.10
Bottles waiting to be filled. Courtesy of John Peter Laloganes.

First Fermentation

During the first fermentation, maintaining a cool temperature throughout the process is essential to preserve the youthful, crisp, and acidic nature of the wine. This first fermentation creates a base wine (known as vin clair) that is characteristically dry and acidic with nearly 10–11 percent alcohol content. All Champagne and high-quality sparkling-wine producers reserve some of this base wine for future vintages of their house style in the non-vintage wine that is reproduced each year. Pictured in Figure 7.10 is a row of Champagne bottles awaiting being filled with the initial base wine.

Cuvée Assemblage

The blending or *cuvée assemblage* (coo-VAY ah-sahm-BLAHZH) is one of the most complicated phases of the production process. It requires the winemaker to blend dozens of wines from conceivably various vineyards and from several years to ultimately create an integrated wine. The blending strives to achieve some level of consistent (known as a house style) aroma/flavor and structural components from year to year. Once the wines are blended, they may rest or age for a period in stainless steel tanks and/or wood barrels, prior to undergoing secondary fermentation.

Secondary Fermentation

The secondary fermentation is also known as the prestigious méthode champenoise (MC) (or more recently méthode traditionelle) required to produce all Champagne and high-quality sparkling wine from around the world. Méthode Champenoise is the classic and original manner in which bubbles were imparted into a sparkling wine. The method dates back to a time well before the famous monk Dom Pierre Pérignon arrived at the Abbey of Hautvillers in 1668. However, most experts agree that upon Dom Perignon's arrival, he did perfect and contribute greatly to the process of blending and bottling sparkling wine before his death in 1715. The process begins by combining a dose of sugar and yeast known as the *liqueur de tirage* (lick-KYOOR duh tee-RAHZH) into the base wine created from the preceding initial fermentation. This addition of the liqueur will induce a secondary fermentation that creates an increased degree of alcohol (totaling around 12 percent) along with carbon dioxide. The carbon dioxide will be trapped in the bottle and create the characteristic bubble formation associated with sparkling wine.

Aging

During this lengthy process of the secondary fermentation, the wine is laid down in the cellar for a long time—minimum of fifteen months for non-vintage and three years for vintage Champagne. During this period of aging, yeast cells (or lees) break down and undergo the process of *autolysis* (aw-TAHL-uh-sihss). This decomposition of yeast cells contributes significantly to the character of any sparkling wine using this process. Autolysis is one of the fundamental aspects in the making of Champagne (and other high-quality sparkling wine), as it imparts particular types of flavor, toasty complexity, and creamy textural finesse that can be achieved in no other way. Once the wines have met their minimum aging requirements, the wine is now ready for the removal of the dead yeast cells. Figure 7.11 showcases bottles of Champagne aging.

FIGURE 7.11
Champagne aging. Courtesy of John Peter Laloganes.

Remuage

At this stage, the bottles are cellared and inverted into racks (called pupîtres) at an angle of 45° with the intention of encouraging the yeast to travel toward the neck of the bottle. Pictured in Figures 7.12 and 7.13 are Champagne bottles inverted into the pupîtres in order to encourage the lees to travel toward the neck of the bottle. Now the wine will undergo the lengthy, tedious, hand-crafted technique

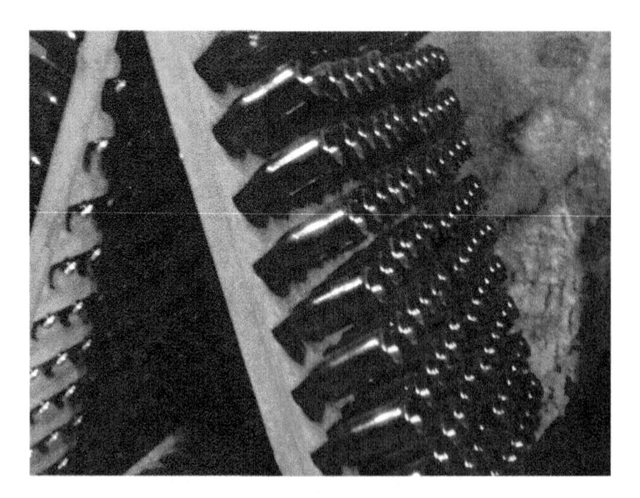

FIGURE 7.12
Pupîtres. Courtesy of John Peter Laloganes.

FIGURE 7.13
Display of bottles in Pupîtres. Courtesy of John Peter Laloganes.

FIGURE 7.14
Mechanized gyropalettes. Courtesy of John Peter Laloganes.

FIGURE 7.15
Lees in the bottle. Courtesy of John Peter Laloganes.

of remuage (RAY-moo-ahj)—the gradual process of rotating and tilting the bottle. Each day, over a period of six to eight weeks, the bottles are given a gentle shake, or riddled (referred to as the process of riddling) a quarter turn with the aim of encouraging gravity to pull the lees toward the neck of the bottle. This step allows for the eventual removal of sediment without the wine being emptied from its bottle. Traditionally, the remuage was done by hand, but increasingly it is now being carried out in large, mechanized racks known as *gyropalettes* (ZHEE-row PAH-lets) in order to increase efficiency and decrease the cost of labor. Pictured in Figure 7.14 are the mechanized gyropalettes that have become prevalent throughout Champagne.

Dégorgement

After a period of several weeks-to-months, the dead yeast cells are ready for removal through the dégorgement (day-gorge-MAWN) process. This technique removes the sediment (or lees) from the neck of the bottle without having to remove the wine. It begins with the neck of the bottle being dipped into an icy brine or glycol solution, creating a small ice plug that contains the sediment. The bottle is then placed upright, and the cap (or temporary cork) is taken off. Due to the internal pressure of the wine, the ice plug containing the sediment immediately shoots out of the bottle. At this point, the wine is clarified, but also completely dry, with no residual sugar remaining. The illuminated picture in Figure 7.15 shows the lees within the bottle of Champagne.

Dosage

After the dégorgement process, the wine is commonly, though not always, given a dose or dosage (doh-ZAHJ) of liqueur d'expédition, a mixture of sugar and wine. This sweetening agent is used to adjust the desired degree of dryness/sweetness and replenish the small amount of wine lost during dégorgement just prior to sealing the bottle. There are seven levels of dosage for Champagne. They are represented by grams of residual sugar per liter of wine or simply, g/l of RS. The first level of dosage is known as Brut Nature (a.k.a. Brut Zero). At this level, there is no added dosage, the wine is bone dry at 0–3 gram per liter (g/l) of residual sugar (RS). If any sugar is present, it is naturally leftover from the fermentation process. The second level of dosage is identified as Extra Brut—an extra brut contains anywhere from 0–6 g/l of RS. The next level of dosage is designated as Brut—this level is most recognized in the marketplace. Brut contains between 0–12 g/l of RS. This level of dosage allows a lot of wiggle room for the winemaker to use terminology and style that are appropriate to their intended market. Therefore, the winemaker can have 6 g/l of RS and have the option of using the Extra Brut or Brut designations. The next level of dosage is Extra Dry, meaning "off-dry" at 12–17 g/l of RS. The dry designation is represented at 17–32 g/l of RS. Demi-Sec is the last style that is commercially available in the United States with a range of 32–50 g/l of RS. The final dosage level is Doux, with 50 or more grams per liter of residual sugar.

LEVELS OF DOSAGE

Brut Nature (a.k.a. Brut Zero)*	0–3 g/l of RS
Extra Brut	0–6 g/l of RS
Brut**	0–12 g/l of RS
Extra Dry	12–17 g/l of RS
Dry	17–32 g/l of RS
Demi-Sec	32–50 g/l of RS
Doux	50+ g/l of RS

*Indicates there can be no added sugar in the dosage.

**Indicates a very popular style of dosage.

Bottling and Corking

The final step in the Champagne process is the bottling and corking of the wine. These wines are always distinguished by their effervescence, or CO_2, which creates intense pressure within the bottle. As the carbon dioxide builds up in the bottle, it is measured in atmosphere (or atm), normal air pressure at sea level, which is 14.7 pounds per square inch. The typical pressure in a sparkling wine is most often equivalent to 5–6 atm or 80–120 lbs. per square inch (psi), approximately two to three times the pressure of a car tire. Therefore, the bottle used for sparkling wine must be made with thicker glass than that of typical wine bottles. In addition, each bottle also contains a punt end, or indentation, at the bottom of the bottle to guarantee its stability during transportation and storage. The bottle is also sealed with a cork secured with a wire muzzle. Then the bottle is returned to the cellars for several months before being labeled for shipping.

Cork

Champagne is legally required to be closed with a cork—though very different than the ones used for still wines. The corks used for Champagne are constructed with

FIGURE 7.16
Display of Champagne cork. Courtesy of Erika Cespedes.

separate sections: the main body made of agglomerated cork, along with an addition of between one and three discs of natural cork, affixed to the portion that comes into contact with the wine. Champagne corks are larger than regular corks—up to 31 mm in diameter—and are highly compressed when they are inserted into the bottle, with a reduced diameter of only 17 mm. This provides for a snug seal, allowing the wine to safely retain its carbon dioxide.

Muselet and Plaque de Muselet

The muselet is the wire cage or muzzle that holds the Champagne cork in place, while the plaque de muselet is the metal disc affixed to the top of the cork that usually has some sort of design that is unique to each producer or to the particular cuvée. Figure 7.16 is a display of the Champagne cork, plaque de muselet, and cage.

Other Methods of Incorporating Carbonation

Learning Objective 4
Distinguish between the different origins of bubbles in sparkling wine

There are alternative methods of incorporating carbonation into a sparkling wine, though Champagne is not allowed to use any of them. These alternative methods will dramatically lower the cost and reliance on labor, as well as alter the style of a sparkling wine.

Transfer Method

This technique starts out similar to the traditional method, having contact with lees in order to obtain the complex aromas and flavors associated with Champagne. However, instead of going through the remuage and degorgèment process, the entire contents of the bottle are emptied into a large pressurized tank for bulk clarification and then finally transferred back into another bottle. This method increases efficiency and reduces production costs. The transfer method is a great alternative to the méthode champenoise process but its disadvantage is a finished wine that is slightly less intense, with larger bubbles. If a sparkling wine is created in this manner, the label will state "Bottle Fermented" or "Fermented in Bottle." Note the distinction between the champenoise method's terminology "Fermented in this Bottle" versus the transfer method's terminology "Fermented in the Bottle." This latter terminology indicates that the wine has left its original bottle to be clarified of its sediment. The final product tends to be cheaper than its méthode champenoise counterpart.

Charmat or Tank Method

The Charmat (SHAHR-maht) or tank method is a mass-producing technique in which the base wine undergoes secondary fermentation in a pressurized tank, or *autoclave* (AW-toh-klayv). Frenchman Eugene Charmat developed this method in 1910—this method is mistakenly thought to be associated with cheap sparkling wine. Instead, it's just an alternative style of production that fundamentally alters the personality characteristics of a sparkling wine. After fermentation, the wine is filtered, sweetened, and bottled, all under pressure. This method is inexpensive and is intended for wines that are not meant for long aging; therefore, it creates a light, easy-to-drink, fruit-forward style of wine. Some of these types of wine may have varying amounts of residual sugar yielding an off-dry to sweet style. Many sparkling wines are produced in this manner, including Sekt (Germany) and Asti, Moscato d'Ast, Prosecco, and Brachetto (Italy). The French call this process cuvée close, the

Italians refer to it as metodo charmat or autoclave, the Spanish call it granvas, whereas in Portugal it's called metodo continuo.

Pump Method

The pump method incorporates CO_2 into the base wine as it is being bottled. This method is like the creation of soda pop and is an inexpensive method often associated with producing a low-quality fruit-style sparkling wine. The bubbles are large, and the wine loses them quickly. The method is the least effective in adding quality carbonation to wine and is always associated with the least expensive wines. The wines are simply labeled as "carbonated" in the United States and called gazeifié in France.

Méthode Rural

By the méthode rural, the wine is bottled prior to completion of the first fermentation, allowing for a slight degree of carbonation. Petillant-naturel (meaning *natural sparkling*) shortened to "pet-net," is a phrase used for any sparkling wine made without the addition of secondary yeasts or sugars. The finished wines are light in carbonation and often contain low to moderate levels of alcohol. Asti, Moscato d'Asti, and Prosecco may be made in this manner. The Italian terms *frizzante* (freezz-AHN-tay) and *frizzantino* (freezz-AHN-teen-oh) are used to distinguish a sparkling wine with a less pronounced, softer pressure.

Other Sparkling Wines

Learning Objective 5
Identify at least three other sparkling wines

While Champagne is the most popular wine with which to toast significant milestones like the start of a new year, or birth of a child, other sparkling wines are gaining enthusiastic followings in the United States. Some of these other types of sparkling wines are considered more user friendly for everyday consumption. They are apt to be from South Africa, Australia, New Mexico, Chile, and Spain or somewhere else, including California. These other sparkling wines are yielding some notable competition to Champagne, as they offer some of the best bargains in bubbly to be found in today's market. Each production area has something to offer that is distinct and nearly all production areas have the freedom to incorporate varietals that are indigenous to their place of origin. However, many of the producers will choose to replicate the style of Champagne by using the that are classic to Champagne.

FIGURE 7.17
Crémant d' Bourgogne. Courtesy of John Peter Laloganes.

France

There are numerous sparkling wines made in France, but outside the coveted region of Champagne. They are regarded as offering an excellent value alternative to Champagne, if the notion of French bubbly is desired. These sparkling wines are referred to as *Crémant* (kray-MAHN), such as Crémant d'Alsace, meaning a sparkling wine from the Alsace region of France or Crémant d' Bourgogne, meaning a sparkling wine from the Burgundy region of France. Many of the French (non-Champagne) sparkling wines utilize local grapes (instead of/or in addition to the classic grapes) associated with that region. Figure 7.17 displays a bottle of Crémant d' Bourgogne.

FIGURE 7.18
Trio of Cava. Courtesy of Erika Cespedes.

FIGURE 7.19
Freixenet. Courtesy of John Peter Laloganes.

Spain

Spain is the largest producer of sparkling wine in the world. The highest-quality Spanish sparkling wine Cava can be made in several authorized locations throughout Spain. However, the clear majority (roughly 95 percent) is made in northeastern Spain's Catalonia region. Cava typically uses a trio of local white grapes indigenous to Spain called *Macabeo* (mah-kah-BEH-oh), *Xarelo* (sah-RE-HL-yoh), and *Parellada* (par-eh-LYAH-duh); clearly not the same as those incorporated into Champagne. Figure 7.18 displays a trio of Cava. These grapes tend to be preferred by most producers, but currently there is experimentation with the addition of some classic Champagne-type grapes (namely Chardonnay) in the blend. Cava offers a good transition between Champagne to American sparkling wine, with many reasonably priced options. All Cava must be produced using the méthode traditionelle with minimum aging requirement of nine months. If the designation of reserve or gran reserva is indicated on the label, then a minimum of fifteen months and thirty months aging is required. The Spanish Cava producer Freixenet is the largest producer of sparkling wine in Spain. Figure 7.19 displays Freixenet in Spain.

As of Fall 2016, there is a new Cava designation, Cava de Paraje Calificado that will start to appear on Cava labels. It specifies that Cava made under this new designation must be produced from estate vineyards and subject to minimum thirty-six-month ageing period on the lees. This new classification was created to heighten and promote high-end Cava. Figure 7.20 displays a bubbly glass of Cava.

FIGURE 7.20
Cava in Flute. Courtesy of Erika Cespedes.

Italy

Italy uses many indigenous grapes in the production of their sparkling wines. Most of Italy's sparkling wine uses the charmat

FIGURE 7.21
Moscato'd Asti paired with bread pudding. Courtesy of Erika Cespedes.

FIGURE 7.22
Franciacorta paired with a charcuterie board. Courtesy of John Peter Laloganes.

or tank method of production. Asti and Moscato d'Asti are produced from the Muscat grape, *Brachetto d'Aqui* (brah-KET-toe DWAH-kwee) is made from the Brachetto grape, Lambrusco is the name of both the red grape and the wine originating from the Emilia-Romagna and Lombardy region, while *Prosecco* (praw-ZEHK-koh) is produced in both the Veneto and Friuli region from the Glera grape. Figure 7.21 shows a bottle of Moscato d'Asti paired with a bread pudding.

Franciacorta (FRAHN-shah-KORT-ah) is possibly Italy's most complex and revered sparkling wine deriving from the Lombardy region of Italy. Due to its replication of Champagne, this wine has earned Italy's highest and strictest quality classification—DOCG status. Most of the wines are Chardonnay-based, with smaller amounts of Pinot Bianco and/or Pinot Noir. But what distinguishes this wine is that it's an anomaly to most Italian sparkling wines as this one is made using the classic méthode traditionelle. Figure 7.22 diplays a bottle of Franciacorta paired with a charcuterie board.

Fortified Wines

Seductive ... Rich ... Satisfying

Fortified wine derives from a table wine that has been manipulated with the addition of alcohol to achieve a bolder taste, higher alcohol content, and a longer shelf life. These wines typically range between 15 and 22 percent alcohol content and can vary in style from dry to sweet, depending on when the additional alcohol is added throughout the production process. Pictured in Figure 7.23 is a fortified wine label identifying its high alcohol content.

Apéritif (Ahp-pehr-ih-TEEF)

Apéritifs, if dry, are often consumed prior to or near the beginning of the meal, to assist with cleansing the palate and stimulating the appetite. This beverage is served in a smaller 2–3 oz. portion size because of its higher alcohol content. This category

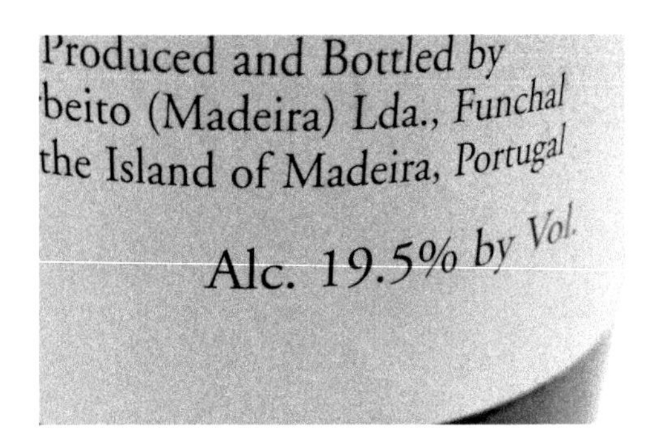

FIGURE 7.23
Ample Alcohol. Courtesy of Erika Cespedes.

can be grouped into two subcategories, depending on their base ingredient: (1) spirit based (such as vodka and gin) and (2) wine based (such as dry vermouth, sparkling wine, Fino Sherry, or Sercial Madeira).

Digestif (dee-zheh-STEEF)

Digestifs, if sweet, are often consumed toward the end or after the meal, to assist in satiety and aiding in digestion. Some examples include: Port, Cream Sherry, Malmsey Madeira. Like Apéritifs, these beverages are served in a smaller portion size because of their higher alcohol content.

Port Wine

Learning Objective 6
Explain the categories and styles of Port wine

Port wine (also known simply as Port or Porto) is a fortified wine named after the city of *Oporto* (oh-PORT-oh)—a historic port city from the Douro Valley in northeast Portugal. Port was one of the first wine regions to be officially demarcated and recognized, in 1756. These wines are typically a sweet, red wine, with a small production of a dry to off-dry white-wine version. There are over forty different varieties of grapes that can be blended and used when making Porto, with the best-quality grape varieties grown mostly on the steep hillsides in schist soil, unique to the Douro Valley. Although they may be planted separately, the varieties are normally harvested and fermented together. Over recent years, most of quality Port wine is being made from five main red grapes blended together to make the two categories of red port. These include:

FIGURE 7.24
Aged Tawny Port. Courtesy of Erika Cespedes.

- Touriga Nacional (tow-REE-gah nah syo-NAHL)
- Tinta Roriz (TEEN-tah ROR-eesh) otherwise known as Tempranillo in Spain
- Tinta Barroca (TEEN-tah bar-ROH-kuh)
- Tinta Cão (TEEN-tah cowng)
- Touriga Francesa (tow-REE-gah fran-SAY-zuh)

When the producers make the lesser known and insignificant white Port, they use the obscure, indigenous white grapes such as Esgana, Cao, Folgasao, Malvasia, Rabigato, Verdelho, and Viosinho. Pictured in Figure 7.24 is a bottle of aged Tawny Port.

During fermentation, Port is fortified with an unaged brandy at 77 percent alcohol, which halts the yeast production—leaving varying amounts of residual sugar. Port is a bold and full-bodied, sweet, and slightly tannic drink with high alcohol—at about 18–20 percent alcohol. The wine was originally fortified with high alcohol level to help preserve the wine during long sea voyages from Portugal to England and the English colonies. There are two categories of Port with numerous styles, distinguished largely by the quality of the grapes and the method in which they are aged.

Barrel-Aged Ports

Tawny (barrel or wood-aged) Port offers colors like copper, amber, and brown. Wood-aged Ports age in cask or vat, normally made of oak, and spend most of their lives, as the name suggests, maturing in a barrel. During this time, the passage of oxygen (oxidative aging) alters the color and aromatics—yielding nuts, caramel, and dried-fruit flavor due to years of slow, controlled oxidation in large wooden barrels or vats. The three common types of barrel-aged ports are identified as the following:

- *Tawny Port:* Tawny Port takes its name from its (brownish-red) color. This Port is a blend of several vintages that have been aged in wood for two to three years before being bottled.
- *Aged Tawny Port:* Age indicated Ports are some of the highest quality expressions of Port. This type of Port is mellow, silky, and rich in aromas/flavors and structural sensations. Aged tawny Ports are made from high-quality grapes and are usually bottled after decades of aging—10, 20, 30, and 40 years. These decade designations are clearly identified on the label and signify an average age of the Ports in the blend. The younger wines add freshness and vigor while the older wines add complexity. Figure 7.25 shows a bottle of Dow's ten-year-aged Tawny Port.
- *Colheita* (cuhl-YAY-tah)*:* The word colheita means "vintage" in Portuguese, therefore this wine is simply a vintage tawny Port made of wines from a single year. The date of harvest appears on the label and is aged for a minimum of seven years before it is bottled.

FIGURE 7.25
Dow's Ten-Year-Aged Tawny Port. Courtesy of Erika Cespedes.

Bottle-Aged Ports

Ruby (bottle-aged) Port offers red tonality that extends from light red, to very dark red, almost black. Bottle-aged or ruby style Ports are aged for small amounts of time in barrel, but as their name indicates, they spend most of their life maturing in a bottle with limited exposure to oxygen. Through this time of resting, they experience reductive (without oxygen) aging. This method preserves more of the youthful color and aromas/flavors of dried cherries, dark chocolate, and tobacco.

- *Ruby Port:* Ruby is a young, bright red, sweet, and peppery wine meant to be drunk young. It is a blend of several vintages that are normally aged in tanks (often not made of wood) for two to a maximum of three years and then made into a house style. Ruby is simple and offers a straightforward style of Port (as well as being most affordable). It is bottled in such a way as to limit as much oxidation as possible. Figure 7.26 displays a Ruby Port paired with a chocolate cake.
- *Single Quinta Ports:* Single Quinta (KEEN-tah) Ports are not only the product of a single harvest, but also of a single estate or quinta. These wines have gained notoriety as they showcase an element of quality control due to having single source of grapes.
- *Late Bottled Vintage (LBV):* Late bottled vintage Ports are unique in that the grapes used come from a single vintage—giving the consumer a "taste" of vintage, but without waiting for decades to age them. LBV's spend between four and six years in casks, encouraging the wine's development before being filtered and bottled. They sometimes are referred to as "the poor man's vintage Port," not because they are lower in quality, but because of their accessibility to be drunk much sooner. These wines are softer, less tannic, and not as full-bodied as vintage Port.

FIGURE 7.26
Graham's Fine Ruby Port. Courtesy of Erika Cespedes.

- *Vintage Port:* Vintage Port is one of the rarest and most sought after of all Ports because it accounts for only about 2 percent of production. The wine is made with high-quality grapes and is declared a vintage only when the crop is exceptional within a single year. For the wine to be called a vintage, the winery must seek approval of the Port Wine Institute—historically, there are only about three vintages declared in a decade. Vintage Port spends two years in casks and then matures in bottles for decades. These wines spend most their life (sometimes up to fifty years for the wine to reach its peak) evolving and maturing in the bottle to achieve great depth and complexity.

It is suggested that Vintage Ports often need to be decanted because they are unfiltered and develop considerable sediment through the aging process. In Portugal, the legends say this type of Port is often purchased when it first appears on the market to celebrate a child's birth. It is matured throughout the life of the child and is finally opened upon the maturity of both the wine and the child evolving into an adult.

Madeira

Learning Objective 7
Explain the categories and styles of Madeira wine

Authentic *Madeira* (muh-DEER-uh) comes from Portugal's Madeira Island in the Atlantic Ocean about 400 miles off the coast of Morocco. Joào Goncalvez, a Portuguese explorer, discovered the island in 1418. The island was dubbed Madeira, the Portuguese word for "Island of Wood" because it contained thick trees and brush. It was apparently so uninhabitable—the island was set on fire to remove some of the vegetation. Originally, the island served as a port for ships sailing to the East Indies and Africa and later the West Indies and the Americas.

One of the three great fortified wines of the world, Madeira and its success is attributed to the primitive shipping conditions of the 17th century. Beginning in the late 1600s, wines from Madeira's vineyards were a frequent cargo on ships sailing to the Americas, as well as to mainland Portugal, England, and India. Not all the wines were sold on arrival so that some were returned to the producers in Funchal, having completed a long sea voyage. On return to Madeira, it quickly became obvious that these wines had changed character during their time at sea. The maturing process of Madeira wine was created due to its long sea voyage where they were stored below the deck and the wines were exposed to the heat of the tropics. Madeira wine soon after became fortified when British merchants on the island began to add grape spirit to preserve it on its long voyage. This long voyage under the sun for months and even years ended up in cooking the wine. This wine ultimately has played a role in connection with major historical figures and events—Madeira was a favorite of George Washington and Thomas Jefferson and was held in high enough esteem to be used to toast the signing of the Declaration of Independence in 1776. Pictured in Figure 7.27 is a label tease identifying the history of Madeira wine.

FIGURE 7.27
History of Madeira. Courtesy of John Peter Laloganes.

Unlike any other wine, Madeira is intentionally exposed to air and heat. To replicate this process that was historically conducted through primitive shipping conditions, Madeira producers began to heat the wines in stainless steel tanks called *estufas* (es-TOO-fah), Portuguese for "stove." In the estufas, the temperature can reach heights of 120°F—conditions that normally would devastate most typical table wines. The canteiro method is considered a slower, natural, and higher-quality method that involves placing the wine in barrels in the hottest areas of warehouses for several years. This process of aging takes place over a minimum period of three years, the wine will continue to develop and enhance due to the gradual evaporation of water and process of oxidation. Through this unusual production process, the wine becomes indestructible and assumes a fairly intense nutty and caramel quality. An open bottle can last unharmed for months, while unopened ones can last for decades to even centuries.

Unlike port, Madeira is not always sweet. Fortification occurs during or after fermentation depending upon style. In the drier styles of Verdelho and Sercial, yeast is allowed to consume most of the grape sugar, while in the sweeter styles of Boal and Malmsey, yeast is halted earlier leaving varying amounts of residual sugar. After fortification, wines have approximately between 18 and 22 percent alcohol by volume.

Non-Vintage Madeira

Unlike most still wine, fortified wines such as Madeira are largely blended from multiple years. The intention is to produce more of a consistent product from year to year. There are largely four styles of Madeira that are mostly seen in the marketplace, each of which identifies its name from the grape varietal (minimum 85 percent) used in its creation. The wines are listed in graduating color (ranging from sandy through chestnut brown) in its various stages of development due to increased caramelization of sugar and oxidation and going from the driest to sweetest style.

- *Sercial* (ser-see-AHL): This is a dry version of Madeira that is named after the predominate grape used to produce this wine. This style of Madeira is dry and light golden in color with a delicate aroma of honey, floral, and light caramel with lively acidity. Sercial makes an ideal apéritif beverage to engage in foreplay with the palate.
- *Verdelho* (vehr-DEH-lyoo): This is a semi-dry version of Madeira that is named after the predominate grape used to produce this wine. This style of Madeira is dry and golden in color with aromas and flavors of lightly toasted nuts, cooked honey, and fairly acidic. The wine has a gentle, smooth, smoky flavor and is great for use in cooking. Rainwater is a variation of this semi-dry style of Madeira. This style is thought to have come about first by accident when a shipper's Madeira barrel was left outside waiting for shipment became "waterlogged" in heavy rain prior to its voyage from the island to the Americas.
- *Bual/Boal* (boo-AHL): This is a semi-sweet style of Madeira that is named after the predominate grape used to produce this wine. This style of Madeira is dark gold to brown in color and offers aromas and flavors of coffee shop (toasted nuts, coffee, chocolate, and cinnamon).
- *Malmsey* (MAH'm-zee): This is the richest and sweetest version of Madeira that is named after the predominate grape used to produce this wine. This style of Madeira is sweet with a chestnut-brown color and a greenish rim color with aromas and flavors of intense coffee shop (toasted nuts, brown sugar, coffee, chocolate, fig, and cinnamon) that maintains healthy acidity through its ample sweetness. Figure 7.28 shows a Malmsey Madeira displaying its rich brown chestnut-like color. Previously, the names "Malvasia" and "Malmsey" have been used interchangeably (as they are theidentical grape) though more appropriately

FIGURE 7.28
Malmsey Madeira. Courtesy of Erika Cespedes.

FIGURE 7.29
Vintage Dated Madeira Wines. Courtesy of John Peter Laloganes.

"Malvasia" generally refers to unfortified white table or dessert wines, while "Malmsey" refers to the sweet style of Madeira.

Vintage Madeira

This type of Madeira is a rare and very prestigious type of wine. Vintages for Madeira are not declared until the wine has aged for twenty years in the barrel and two years in the bottle. The 20th century produced only about two or three vintages a decade, making the total number well under forty. Madeira has a reputation for growing old gracefully. It is not uncommon for vintage Madeira to have the staying power of well over a full century and ironically, unlike other aged wines, these tend to be moderately inexpensive. Figure 7.29 displays some vintage dated Madeira from the 1900s.

Sherry

Learning Objective 8
Explain the categories and styles of Sherry wine

Sherry is a one of Spain's most famous fortified wines. The word Sherry is an English word for *Jerez* (Heh-REHTH) where the wine is produced and located within Southern Spain's *Andalucía* (ahn-dah-loo-THEE-yah) region from the province of Cádiz. The official Sherry-producing towns Jerez de la Frontera, Puerto de Santa Maria, and Sanlúcar de Barrameda are collectively known as the Sherry triangle because of their proximity to one another. Sherry is a blended fortified wine that usually has between 15 and 22 percent alcohol by volume.

Sherry is made exclusively from one white grape, *Palomino* (pah-loh-MEE-noh) that is used in roughly 90 percent of Sherry. Another white grape occasionally used is *Pedro Ximénez* (PEH-droh hee-MEH-neth), often referred by its nickname "PX," which is associated with many of the sweeter styles of Sherry. Both grapes are fairly neutral varietals that grow in a special chalky soil, rich in limestone, known as *Albariza* (ahl-bah-REE-thah). This unique soil contributes to these grapes growing successfully in an otherwise overly hot and dry climate. The soil retains moisture and preserves high acid in the wine grapes, both of which would otherwise be overly lost in such a dry and sweltering growing location.

When the Sherry grapes are pressed, only the first 85–90 percent of the liquid obtained is used to produce Sherry. The remaining ten to fifteen percent is distilled to make brandy for the purpose of fortifying the Sherry wine later.

Sherry is one of the great expressions of the blender's art; it is made through a fractional blending system according to the solera system. This intricate blending system involves a network of several wines of varying maturity levels ranging in from the oldest (maybe ten to fifteen years) to the most recently produced. The solera method ensures a continuity of style each time the wines are released for sale. The blending system consists of drawing off one-quarter of the contents of the oldest barrels, from the bottom, for bottling. Wine is then emptied into the bottom barrel from the level of barrels above, and so on through the levels of the solera. With this process, the old wines incorporate character into the younger wines.

After fermentation is complete, Sherry is fortified with brandy for preservation and stylistic preferences. Because the fortification takes place after fermentation, Sherry is initially dry, with any sweetness being added later in the production process through sun-dried grape juice. Therefore, all Sherry begins its life as a dry wine—which is in contrast to Port Wine, for example, which is fortified prior to the completion of its fermentation process, which halts the yeast leaving considerable residual sugar.

Sherry can be broadly placed into two categories, ranging from dry, light versions such as *Fino* (FEE-noh) Sherry to much darker and less dry to sweeter versions known as *Oloroso* (oh-loh-ROAS-ohs) Sherry.

Fino Sherry Category

The first category or Fino Sherry produces light-colored, dry, and tangy fortified wines. Three styles of Fino include Fino, Manzanilla, and Amontillado. Fino Sherry are wines that have been affected and preserved by the flor, which refers to the presence of the yeast, and often are referred to as having been biologically aged. This type of aging means that the yeast has influenced the wine and protected it from the effects of oxygen. These wines have intentionally encouraged the development native airborne yeasts as air space is intentionally left in the barrels throughout the solera. This air space allows the airborne yeasts to develop and thrive as it protects the wine from oxygen and imparts its unique characteristic to the wine.

The aromas and flavors of Fino and Oloroso Sherry veer toward bakeshop (yeasty), vegetable (olive), and fruit (apple press). Fino type Sherry can be served at room temperature or slightly chilled. Their savory aroma and flavor profiles with their corresponding acidity and dryness are ideal for aperitifs, appetizers, soups, and even some types of entrées.

- *Fino Sherry* (FEE-no): This is the driest Sherry, with live yeast cells present in the wine during aging. To be classified as a Fino Sherry, a thick yeast (flor) blanket has to form within the barrel. Since the yeast never dies, its presence acts as a preservative against oxygen. Fino is a light, pale, dry, and delicate style protected by a layer of flor that grows spontaneously on the surface of the wine. Generally, it contains about 15–15.5 percent alcohol content; any higher alcohol would kill the desirable yeast formation. Fino Sherry is intended to be consumed in its youth and will decline with age rather than improve.
- *Manzanilla Sherry* (man-zah-NEE-yah): This type of Sherry is pale, delicate, and one of the most pungent of the Fino-style Sherries. Manzanilla is a type of Fino that is matured in the cool seaside-influenced atmosphere of Sanlucar de Barrameda. Within this production area, the flor grows with great abundance to give wines notable, fresh, crisp, and fragrant aroma with a slight salty tang. Manzanilla is classified as a Fino because of the predominant growth of its flor. Like the basic Fino Sherry, Manzanilla is best served chilled and makes an excellent primer as an apéritif.
- *Amontillado Sherry* (ah-mone-tee-YAH-doe): This type of Sherry starts as a Fino, but the flor that develops is thinner, less stable, and begins to die. Amontillado Sherry has been slightly affected by the passage of oxygen through the barrel-aging process, which yields a light-brown-colored, nutty, and complex wine. The aromas and flavors for an Amontillado Sherry tend to be notes of coffee shop (coffee, root beer, and butterscotch).

Oloroso Sherry Category

The second category or Oloroso Sherry produces dark-colored, less dry to sweet, and tangy fortified wines. Three styles of Oloroso include Oloroso, Cream, and PX Sherry. The category of Oloroso Sherry are wines that have NOT been affected and preserved by the yeasty flor, instead causing Oloroso types to gain oxidative and barrel-aging qualities that significantly alter the personality of the wine. The Oloroso styles maintain a deep orange-brown color, a rich flavor, and an overall less dry style compared to the Fino Sherry category.

The aromas and flavors of Oloroso-type Sherries often contain bakeshop (maple syrup, caramel, butterscotch, and brown sugar) and dried fruit (raisin). Oloroso types also have varying levels of sweetness through the addition of a sweetening agent (the juice of sun- or air-dried PX grapes) after fermentation has been completed. These Sherries generally maintain 18 percent or more alcohol. Oloroso-type Sherries can pair well with dessert items because they contain varying levels of sweetness, as well as high alcohol, which helps cut through the richness. In addition, the bakeshop aromas and flavors of the wine help to bridge the flavors in the desserts.

- *Oloroso Sherry* (oh-low-ROAZ-oh): This type of Sherry is typically off-dry to taste because a small portion of sun-dried (for twelve to twenty-four hours) grape juice is added to a fortified dry Sherry. Figure 7.30 shows a bottle of Oloroso Sherry.
- *Cream Sherry:* This Sherry is sweeter to the taste; because a greater portion of sun-dried (for ten to fourteen days) grape juice is added to a fortified dry Sherry.
- *Pedro Ximenez (PX) Sherry:* This style of Sherry is made mostly or completely from the Pedro Ximenez grape, otherwise known as PX. A portion of the grapes has been air-or sun-dried and then pressed to extract the concentrated intense juice. These wines are incredibly sweet and viscous and remain one of the few wines that can truly serve as a compatible partner with ice cream. Figure 7.31 shows a bottle of PX Sherry.

FIGURE 7.30
Oloroso Sherry.
Courtesy of Erika Cespedes.

FIGURE 7.31
PX Sherry.
Courtesy of Erika Cespedes.

Marsala

Marsala (MAHR-sahl-lah) is the youngest and less popular of the major fortified wines first produced in the 1760s. Even though it is an Italian wine made with a blend of grapes indigenous to Sicily, it is the brainchild of an Englishman, John Woodhouse. The name of the wine is taken from the port city of Marsala, found on the western tip of Sicily. Sicily is said to come from Arabic, meaning harbor of God or Marsah-el-Allah. Marsala wines are classified according to three manifestations; therefore, these characteristics become a "Triple Trinity." The first characteristic concerns the level of sweetness, the second concerns color, and the third concerns class, or ranking. Marsala has three levels of sweetness: secco, semisecco, and dolce.

- *Secco:* This designation indicates a dry Marsala, containing no more than 4 percent residual sugar.
- *Semisecco:* This designation indicates a semi-dry Marsala, containing no more than 10 percent residual sugar.
- *Dolce:* This designation indicates a sweet Marsala, containing more than 10 percent residual sugar.

Marsala's three color classifications are oro, ambra, and rubino.

- *Oro:* Oro translates as gold. White grapes such as Catarratto, Inzolia, Grillo, and Damaschino are used for this classification of Marsala.
- *Ambra:* Ambra means amber. The same white grapes are used for ambra as for oro. Unlike oro Marasala, however, a catto or musto catto is added to ambra Marsala. Catto is a reduction of wine. The wine is reduced to one-third of its original volume, which gives the Marsala its cooked taste. A sifone (a mixture of semidried grapes and alcohol) is also added to the wine; the sifone is responsible for the wine's sweetness.
- *Rubino:* This term means ruby. Red grapes are used to make this wine and provide its color. Grapes such as Perricone, Calabrese, and Nerello are used for rubino Marsala. These grapes can be mixed with white grapes, but the white grapes cannot exceed 30 percent of the total grapes used.

The third part of the trinity refers to Marsala's quality ranking. The three classifications are Marsala Fine, Marsala Superiore, and Marsala Vergine (also known as Vergine Soleras).

- *Marsala Fine:* This Marsala can have any of the sweetness or color rankings. It must be aged at least one year in wood and have an alcohol content of at least 17 percent. Sweet Marsala Fine wines are good to serve as dessert wines while the drier ones are good as aperitifs.
- *Marsala Superiore:* This Marsala can come in any color and level of sweetness. The wine must be aged for at least two years in wood and have at least 18 percent alcohol. If the wine is aged for a total of four years in wood, it can add the word riserva to the label. In addition to this, Marsala Superiore may have one of the following designations on the label:
  - L.P. London Particular
  - S.O.M. Superior Old Marsala
  - G.D. Garidaldi Dolce. The sweet G.D. Marsala Superiore wines are good to serve as dessert wines, while the drier ones are good as aperitifs.
- *Marsala Vergine* (Vergine Soleras): This Marsala can be found in any color, but it is limited to secco in its level of sweetness. This is considered one of the best Marsala made. It is aged in wood for five years and has at least 18 percent alcohol. In addition, this level of Marsala cannot have any catto or sifone added.

If the Marsala is aged for ten years, it can add the word stravechhio or riserva to the label. This level is made using the solera system (see the discussion on sherry). It should always be served as an aperitif.

Dessert Wines

Learning Objective 9
Identify some famous dessert wines found throughout the wine-producing world

Delectable ... Juicy ... Voluptuous

Dessert wines have become a general category for any wines that are rich, potent, and concentrated, with considerable levels of sweetness. These wines come from all over the world, and they can be made from many different types of grapes by various production methods. The various types and styles of dessert wines can be used to partner with desserts or they can be appreciated alone. Dessert wines often are made from grapes that have been concentrated—as a result produce high levels of sugar content, less juice, and overall reduced yield. They frequently are labor intensive—as they demand more care and attention to detail in the harvesting and production processes. Therefore, it takes more grapes and manpower to produce a dessert wine, which often translates to a higher selling price. Because of both price and concentrated rich flavor, it is a common practice to serve dessert wine in a small two-oz. portion in an undersized glass.

Production Processes

There are several methods of creating a dessert wine. Each technique removes water content and concentrates flavors and sugars, while maintaining high acid levels and potentially high alcohol levels in some cases, to prevent overpowering sweetness. The combination of alcohol and concentrated sugar levels are large contributing factors to a dessert wine's body, weight, or overall intensity. These wines can range from a consistency of light juice to heavy syrup. There are six techniques for producing dessert wines: (1) late harvest wine, (2) ice wine, (3) rot wine, (4) dried grape wine, (5) fortification of wine, and (6) enrichment wines. Each technique is unique to the country or region of origin of the wine.

Late Harvest Dessert Wines

Late harvest wine begins with leaving the grapes on the vine past the normal harvest. Through the extra hang-time, the grapes begin to reduce in water content and increase in sugar content and in weight. The sugar content is measured in brix, which can equate to a desired level of alcohol content. Grapes used for table wine normally are harvested around thirty brix (depending upon grape type and other factors) or higher. Through a late harvest, aromas and flavors also become more concentrated. Eventually, the grapes dry out and become raisins. During production, the fermentation process may stop naturally or intentionally before the yeast can consume all the sugar, leaving varying amounts of residual sugar in the wine. Late harvest wines are created in many environments around the world—produced in both warm and cool wine regions.

Some famous late harvest wines include those from Germany and Austria identified as Auselese, Beerenauslese (BA), and Trockenbeerenauslese (TBA). These wines are considered three of Germany's greatest wines (mostly made from the Riesling grape) in general, and late harvest wines in particular, in the world. These wines are produced from late harvest grapes and may even have varying amounts of noble rot.

Late harvest wines can be produced almost anywhere; it largely depends upon whether or not it's possible to create demand for them beyond the traditional versions from Germany. Late Harvest Zinfandel, produced in various areas in California, and Late Harvest Shiraz, produced in various areas in Australia, are other late harvest wines that are available in the marketplace.

Eiswein Dessert Wines

Ice wine (or Eiswein in German) is created in cold climates where the grapes are left on the vine into the late fall and winter to freeze. This process gives the wine its name, Eis means "ice," and wein means "wine" (ICE-vyne). Once the grapes are sufficiently frozen, they are handpicked in the early morning or late evening and pressed while still frozen. Since the grapes have been left on the vine for a longer period, water content is decreased and sugar content is increased. Any remaining water is frozen, leaving a sweet, concentrated juice. Most authentic ice wines are made in areas where the weather is cold enough to thoroughly freeze the grapes. Ice wines often are low in alcohol (9–11 percent) leaving considerable residual sugar. These wines are incredibly rich and viscous, with a good balance of sugar and lively acidity to maintain the integrity of such a sweetened wine. Figure 7.32 displays a bottle of Inniskillin, Canada's most famous and largest producer of "Ice wine."

FIGURE 7.32
Glass of Ice wine. Courtesy of Erika Cespedes.

Germany is the original ice wine producer in regions such as the Mosel and Rhine where they are created with the Riesling and Gewürztraminer grape varietals. Canada has become the world's leading producer in terms of both quantity and increasing quality. Canada's Ontario region accounts for about 90 percent of the country's production with grapes that include Vidal Blanc, Riesling, and even a red ice wine from Cabernet Franc.

Ice wine can be made virtually anywhere by an alternative method known as cryoextraction in which the grapes are placed in a mechanical freezing device. This method is used to artificially create the same effect as freezing on the vine. Cryoextraction can produce less expensive versions than the traditional approach to making these wines; however, the market is not overly persuaded in purchasing these ice wine imposters.

Noble Rot Dessert Wine

Noble rot, often noted by the more appealing Latin term, *Botrytis Cinerea* (boh-TRI-tis sihn-EAR-ee-uh), is produced from a so-called "friendly" fungus. This fungus can grow on certain grapes, given the appropriate climatic conditions. Rot wines are classically produced from relatively thin-skinned grapes such as Semillon, Sauvignon Blanc, Chenin Blanc, and Riesling varietals.

Noble rot causes the grape skins to break, allowing the juice and pulp to become affected by the mold, and through time, it extracts the water content by about one-third. This allows the remaining juice to concentrate into luscious, syrupy nectar consistency, maintaining natural acidity while imparting new flavors of honey and apricot with higher sugar content. The ideal environment that encourages the growth of the fungus is cool evenings and moist, foggy mornings, followed by sunny days. If the climate lacks adequate sun, the mold will turn into the undesirable gray rot. If the climate is too warm or lacks moisture, the *Botrytis* will never develop.

One of the oldest and most renowned noble rot dessert wine is produced in the Tokaj-Hegyalja wine region in northern Hungary. The variance of spelling *Tokaji* (TOKE-eye) references the wine (as opposed to the name of the region) that is made from any or all of the three permitted white grapes: Furmint (FOOR-mint), Hárslevelű (harsh-la-Velooh), and Muscat Blanc à Petits Grains. Figure 7.33 is a bottle shot of Tokaj.

Tokaji Aszú (TOKE-eye AH-soo) was crowned as the "king of wines and wine of kings" by Louis XIV. Tokaji Aszú has a distinguished history that dates back to the 17th century. It is believed that the first Aszú was made in 1630 and quickly became a vice among royal households throughout Europe. Tokaji Aszú is made from the *Botrytis*-affected harvests where the berries are literally harvested grape by grape in late October and November and sorted according to their degree of infection from *Botrytis*. The *Botrytis*-affected berries (called Aszú by Hungarians) are placed aside, and the uninfected berries are made into a dry base wine. The Aszú berries are mashed into a sweetened paste and placed into a basket called a *puttonyo* (PUH-tohn-yo). The puttonyo is capable of holding about fifty-five pounds and is used as a measure of sugar content. The sweet paste is now added to the dry white base wine in a cask called a *gönc* (gahn-ts), a 136-liter wood container. The exact number of *puttonyos* (PUH-tohn-yosh) determines the grade of the Tokaji Aszú, which can range from three to six, identified on the wine's label. Afterward, the new Aszú wine is transferred into casks and matured in cool, damp cellars. The wine will be aged a minimum of two years with barrel maturation, and an additional year in bottle for Aszú wines. The cellars maintain constant levels of temperature and high humidity, which provide ideal conditions for storing and aging Tokaji wines.

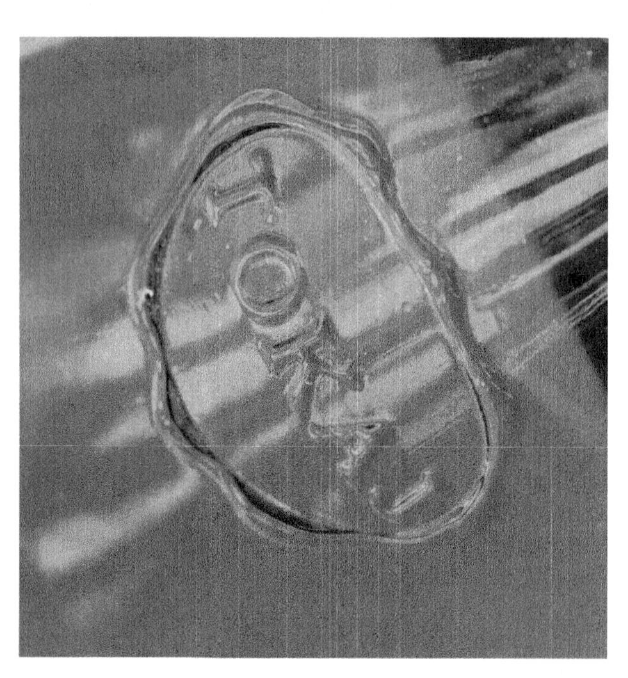

FIGURE 7.33
Bottle shot of Tokaj. Courtesy of Erika Cespedes.

Tokaji Eszencia (TOKE-eye EHS-sen-tsee-uh) is the rarest of all Tokaji wines, made purely from pressed, unblended Aszú grapes, with little or no base wine. It is the equivalent of seven puttonyos. Eszencia is the first-run juice of the Aszú grapes, which seeps from the press under the grapes' own weight. The sugar content is extremely high, and the wine will ferment at a very slow rate, often over many years. This wine is even more concentrated and sweeter than Tokaji Aszú. It is rarely made, because the small amount of juice extracted out of the dried, shriveled berries takes several years to ferment due to its high sugar concentration. Sometimes a small amount is bottled separately, after which it will develop for years in the bottle; more often, Eszencia enriches an Aszú blend. Figure 7.34 identifies a bottle of Tokaji Aszú with five puttonyos.

FIGURE 7.34
Tokaji Aszú. Courtesy of Erika Cespedes.

Noble rot is arguably more famously known in France's Bordeaux region specifically in the appellations of Sauternes (saw-TEHRN) and Barsac (bahr-SACK), both of which also name their wines by their respective location. These small tracts of land are located across from each along the Ciron River where it merges with the Garonne River. As the

Garonne River is affected by the cooling influence of the Ciron, creating ample humidity, it causes the grapes to be damp in the morning ideal for the spread of *botrytis*. Sauternes is arguably one of the best and most famous of all "rot"-based dessert wines in the world. Sauternes is not a grape, but a blend of two primary grapes in varying percentages of Sauvignon Blanc and Semillon.

Other famous *Botrytis*-affected wines made in France are *Coteaux du Layon* (koh-toh deu leh-YAWN), *Quarts de Chaume* (kahr duh SHOHM), and *Bonnezeaux* (bawn-ZOH), all produced from the Chenin Blanc grape in the Anjou area of the Loire Valley.

FIGURE 7.35
Grapes drying on racks. Courtesy of John Peter Laloganes.

Dried Grape Dessert Wine

Dried grape wines often are produced in climates that allow grapes (either red or white) to be sun- or air-dried. The grapes are harvested and allowed to dry or raisin under controlled conditions, either hanging off rafters or layered on straw mats. This process evaporates water content, concentrates flavors and sugar content, and yields a rich, viscous wine. This process of drying the grapes is known as *passerillage* (pah-seh-ree-LAHZH) in France and passimento or *passito* (pah-SEE-toh) in Italy. Dried grape wines are produced around the world with such wines as Vin Santo, Recioto, and Amarone from Italy; *Strohwein* (SHTROH-vine) from Germany; and *Vin de Paille* (van-duh-PIE) from France. Figure 7.35 shows racks of dried grapes awaiting fermentation.

Vin Santo (vin-SAHN-toe)

Vin Santo is a white, rosè or red dessert wine produced primarily in Tuscany, Italy, but can be found in other areas throughout the country. It often is made from white-wine grapes *Trebbiano* (treb-ee-AH-noh) and *Malvasia* (mal-vah-SEE-ah) as well as the red Sangiovese grapes. The grapes are dried by being either placed on straw mats or hung from rafters in the winery. The grapes are dried until they shrivel, which concentrates the grapes' sugar and flavors. The very sweet grapes are then fermented in small barrels that allow oxygen in, which causes the wine to maderize or oxidize. While in barrel, the wines must be aged for a minimum period of three years. The result is a slightly amber-brown wine that can be either sweet or dry.

FIGURE 7.36
Dried grapes. Courtesy of John Peter Laloganes.

Recioto (reh-CHAW-toh) and Amarone (ah-mah-ROH-neh)

These are dried grape wines produced in the Veneto region of Italy. When the grapes are fermented, the wine becomes either an Amarone or a Recioto, depending upon whether the yeast has consumed all the grape sugar or not. If the wine is left with residual sugar, it is known as Recioto; if it is fermented dry, it is Amarone. Figure 7.36 shows a rack of dried grapes awaiting fermentation. Figure 7.37 shows a cluster of Corvina grapes—regarded as the most important grapes that go into Amarone and Recioto wines.

FIGURE 7.37
Dried cluster of Corvina grapes. Courtesy of John Peter Laloganes.

FIGURE 7.38
Taylor Fladgate,forty-year-aged Tawny. Courtesy of Erika Cespedes.

Fortified Dessert Wines

Fortification is the process of adding a distillate (often, unaged brandy) during or toward the end of fermentation. The act of adding alcohol to the fermentation process kills the yeast leaving varying amounts of residual sugar in the wine. Wines subjected to this process often are produced in hot areas where the original purpose of the added alcohol serves to preserve the wines while in transport. Port, Madeira, and Sherry are three of the most famous versions of fortified wine that were discussed earlier in this chapter. Figure 7.38 displays a bottle and glass of a forty-year-aged Tawny Port being paired with a dessert.

Another example of a fortified dessert wine category that is lesser known than Port, Sherry, and Madeira but equally delectable is *Vin Doux Naturel* (VDN) (van doo nah-tew-REHL). These are France's versions of sweetened fortified dessert wines. VDNs can be produced from either red wine based on Grenache or white wine based on Muscat grape varietal. VDNs are produced in southern France, primarily in *Banyuls* (bahn-YOOLS) in Languedoc and *Rastau* (rah-STOW) in Rhône Valley.

Enrichment Dessert Wines

Enrichment wines can consist of table or sparkling wines that have been created by adding sugar after fermentation. The purpose of enrichment is to produce a wine with some varying degree of sweetness. The most famous enrichment wines are known by their family name, Champagne/sparkling wine. Most sparkling wine is fermented completely dry; then a dosage, or sugar mixture, is added to achieve the desired level of sweetness. When a high dosage is added, the resulting wine is sweet enough to be considered appropriate for dessert. Sparkling wines are versatile in pairing with a wide range of desserts. The better options for pairing with desserts include Demi-Sec (sweet), and Doux (very sweet). Some sparkling wines maintain incredibly fruit-forward aromas and flavors with significant amounts of residual sugar either through enrichment or stopping the fermentation process such as Moscato, Asti, Lambrusco, Sparkling Shiraz, and Brachetto.

Check Your Knowledge

Directions: Use these questions to test your knowledge and understanding of the concepts presented in the chapter.

I. MULTIPLE CHOICE: Select the best possible answer from the options available.

1. What distinguishes Champagne from other sparkling wines?
   a. A specific set of grape varietals
   b. The geographical location (climate and soil type)
   c. Production method
   d. All of the above

2. The soils in the best Champagne vineyards contain
   a. clay
   b. gravel
   c. chalk
   d. sand
3. What does the term autolysis refer to?
   a. Ripening of the grapes in the fall time
   b. Decomposition of yeast cells
   c. Sur lie aging
   d. None of the above
4. Brut style of sparkling wine is
   a. dry
   b. very dry
   c. slightly dry
   d. sweet
5. The process of collecting the yeast in the neck of the bottle during the Champagne production process is known as
   a. disgorging
   b. remuage
   c. dosage
   d. fermentation
6. After the dead yeast is removed in the Champagne process, sugar is added to the wine to adjust dryness/sweetness levels. This is known as
   a. disgorging
   b. remuage
   c. dosage
   d. riddling
7. Champagne can use any or all of the permitted three grape varietals, including Chardonnay, Pinot Noir, and
   a. Pinot Gris
   b. Pinot Blanc
   c. Pinot Grigio
   d. Pinot Meunier
8. Disgorgement is
   a. the process of cleaning the bottle before use
   b. the process of cleaning the barrels before use
   c. the process of removing dead yeast cells from a bottle of sparkling wine
   d. the process of removing the grape skins from the wine after fermentation
9. Blanc de Blanc is made from
   a. Chardonnay
   b. Pinot Noir
   c. Pinot Meunier
   d. None of the above
10. A Sherry always begins its life as a
    a. dry wine
    b. sweet wine
    c. dry or sweet wine, depending on the style
    d. medium-sweet wine
11. Two main categories of Sherry include
    a. barrel aged and bottle aged
    b. Ruby and Tawny
    c. Fino and Oloroso
    d. Fino and Manzanilla

12. Two main categories of Port include
    a. barrel aged and bottle aged
    b. Ruby and Vintage
    c. Fino and Oloroso
    d. Fino and Manzanilla
13. The solera method is
    a. a style of Sherry
    b. a style of Port
    c. an intricate blending system used to produce a Sherry
    d. none of the above
14. In making Port wine, the fortification of alcohol is added
    a. before fermentation begins
    b. after fermentation is completed
    c. any time during fermentation, depending on the style of Port being produced
    d. never. No supplemental alcohol is added to Port
15. Aged tawny Ports are often labeled as 10-, 20-, 30-, or 40-year-old wines; these years indicate the
    a. average age of wine in the blend
    b. youngest age of wine in the blend
    c. vintage date
    d. oldest age of wine in the blend
16. Malmsey Madeira is a
    a. dry style
    b. semi dry style
    c. semi sweet style
    d. sweet style
17. *Botrytis cinerea* is also known as
    a. sparkling wine
    b. ice wine
    c. noble rot
    d. both b and c
18. Which item is not associated with Eiswein (ice wine)?
    a. The grapes are frozen
    b. An extremely late harvest wine
    c. Dry-tasting dessert wine
    d. Produced in Canada

II. DISCUSSION QUESTIONS

19. List the six methods used to produce a dessert wine.
20. What is the difference between barrel-aged and bottle-aged port? List several styles within each category.

CHAPTER 8

The Brewery: Beer Production and Service

CHAPTER 8 LEARNING OBJECTIVES

After reading this chapter, the learner will be able to:

1. Identify some key historical points in the evolution of beer
2. Briefly identify the significant beer-producing countries
3. Discuss the four major ingredients used to make beer
4. Explain the brewing process and the fundamental differences between lager and ale beer categories
5. Explain the steps of a properly poured bottled beer
6. Identify some issues that may occur with draft beer

This is grain, which any fool can eat, but for which the Lord intended a more divine means of consumption . . . Beer!

—Robin Hood, Prince of Thieves, Friar Tuck

The Origins of Beer

Learning Objective 1
Identify some key historical points in the evolution of beer

The story of beer stretches back some 8,000 years—evolving in a parallel progression with that of civilization. It has served in many favorable capacities—as liquid refreshment, source of nourishment, and employed as a means of barter in exchange for goods and services. Beer is quite possibly one of the oldest alcoholic beverages known to man; where it is likely to have been independently invented by multiple cultures, but first documented in ancient civilizations of Egypt and Mesopotamia. The historical roots of brewing are also intertwined with those of early bread baking; both products were a necessary part of daily life in ancient society. Beer's existence is supported with archeological evidence of brewing as discovered on clay tablets along with the residues of clay pottery. It is likely that the first brewers were the farmers of the land back in the Middle East. With the sciences not understood at this point, beer production was likely replicated through a step-by-step trial and error approach. Figure 8.1 shows a brilliant glass of beer.

Since beer was being made in different parts of the world, various grains were used by the range of cultures. As the cultivation of cereal grains expanded, people to the south (in Africa) grew grains like millet and the lands to the east (in Asia) were more suitable for rice, but northern and western Europe favored wheat and barley. As the cultivation of barley spread north and west, certain cultures became specialists in the production of beer and certain beer styles. Ultimately, Europe became an iconic producer largely because they had an abundance of good water from snowy mountain ranges and a climate that was excellent for growing various grains and hops. The significance of this drink was well understood at an early point—its economic value was held in high regard and was critical in forging new or existing business and trading relationships.

FIGURE 8.1
Brilliant glass of beer. Courtesy of John Peter Laloganes.

The Middle Ages and Europe

The early years of beer production were led by many groups—the church was one of the most significant contributors as beer was symbolic of joy and spiritual triumph. Beer was the daily drink of the community, both because available water supplies were often polluted and because beer was an inexpensive source of nourishment, particularly for the "working people." The monks brewed as a means of revenue for sustaining the church. Beer became such an important component of life during this period that the Catholic Church named Bishop Arnold of Metz (d. 640 A.D.), the "Patron Saint of Brewers." He was as cherished among his disciples as he frequently touted the virtues of beer such as, "From man's sweat and God's love, beer came into the world." Upon the notice of his death, the people of his town went to retrieve his body and carry him back home. When his supporters stopped to rest in a nearby village, they were told there was only one goblet of beer left to feed them. Ironically, regardless of how much they drank, they never ran out of beer. The group declared a miracle, and eventually Bishop Arnold was elevated to the status of the "Patron Saint of Brewers"—St. Arnulf (Arnold).

The first guild of brewers was formed in Belgium during the reign of Duke Jean (John) I (1251–1294) and was known as the "knights of the mashing fork." Beer had become so important at this point—England's signing of the Magna Carta in 1215 contains a clause about "ale" that alludes to its significance in everyday life. It was the Germans, however, who enacted the first laws or regulations to produce beer. In 1487, Duke Albert IV set forth the first set of regulations, which was the basis for William VI's famous *Reinheitsgebot* (rhine-HITES-gah-bote) of 1516, also known as the "German Purity Law." These regulations are still followed by many German beer producers today. They stated that the only ingredients allowed in beer production were malted grain, hops, and water—later, they recognized yeast being the fourth important component.

Early America

Beer was a significant part of early American colonial life as it was the preferred beverage aboard sailing ships, including the Mayflower. Since water did not always make the voyage without turning bad, beer which contained large amounts of alcohol and hops would act as a preservative to ensure the liquid nourishment would be safe for consumption. Demand for beer increased, and by 1680, there were dozens of government-licensed taverns. Breweries were owned and operated by some iconic Generals from the Revolutionary era, including George Washington, Charles Sumner, Ethan Allen, and Israel Putnam. Perhaps one of the most famous early brewery owners was Samuel Adams, even though he did not reach the height of his fame as a brewer until recent times, when a widely marketed brewery, the Boston Beer Company crafted a beer in 1984 with his name on the label. On December 16, 1773, Samuel Adams and John Hancock, along with other members of the Sons of Liberty, dressed in Native American garb and drank pints of beer before raiding an English ship in an act of rebellion that would become known as the Boston Tea Party. Later, the Continental Congress established a daily ration of 1 quart of beer or cider per soldier each day. At the time, the daily ration of milk was only 1 pint.

Taverns also came under ownership from some of our country's founding fathers, including our second president, John Adams, who owned and managed a tavern. He also enjoyed "a large tankard of cider every morning," according to his grandson, Charles Francis Adams. The oldest tavern in America was founded in 1762 by Samuel Fraunces and is still operating in New York City's lower Manhattan. The tavern played a prominent role in the history of the American Revolution. At the end of the Revolutionary War, General George Washington said farewell to his officers at Fraunces Tavern.

At first, it was the English brewing traditions that formed the underpinning of beer styles in the United States. But eventually, the emergence of German immigration and their prolific lager-style beers in the 1800s became iconic for their contributions to the American brewing industry. The Germans formed small breweries that were ultimately transformed into the famous powerhouse of American and international companies such as Blatz, Schlitz, Miller, and Anheuser-Busch. As the years proceeded, and in part because of Prohibition (1920–1933), the lager style became less of a reflection of its high-quality German heritage and more of a mass-produced diluted impostor throughout much of the 20th century. Once Prohibition was lifted in 1933, only a couple of dozen breweries remained and succeeded, partly because of their acquisition of smaller companies, but also because they acquired expertise at refining the science and efficiency of production to benefit from the economies of scale, and additionally diverting large expenditures to marketing and advertising their products. Ultimately, brewers and consumers lost touch with America's beer heritage, leading to a culture of uniformity, where the only distinctions in products were in their labels and advertising slogans.

Up through the 1970s, Americans had very few options or alternative styles of beer readily available; people enjoyed either the singular homogenous style of beer, brewed their own, or drank something else. The mass industrialization that was prevalent in the brewing industry in the mid-1900s had a profound impact on how consumers understood beer to exist. Even today, when people dismiss beer as their beverage of choice, it's often due to their antiquated paradigm of what beer "used to be" along with limited accessibility and exposure to stimulating and unique beer styles. Therefore, beer style and quality were manifestations of the economics of attempting to deliver a product that was inexpensive, yet efficient—quite like the fast food industry. The large breweries (macro brewers) fueled the way that beer was supposed to be—leading American beer to lose its soul—and portraying a diluted and uninspired character of what it once was. By the end of the 1970s, fewer than fifty brewers operated in America. National (and eventually international) brewers became more powerful as regional and local brewers either were acquired or went out of business—the beer industry had seen constant consolidation.

Modern Times: A Turning Point

High-quality beer—which includes import, craft, and specialty beers—has emerged from being nonexistent in the 1970s to now comprising up to 23 percent of the marketplace (Beer Institute, 2013). With only a couple of dozen breweries existing at the end of America's Prohibition in 1933, beer has become one of the most widely consumed beverages in the world with well over 5,000 U.S. breweries in operation as of 2017.

Encouraged by the Jimmy Carter administration of the 1970s, Americans were legally allowed to home brew up to 100 gallons of beer per year for personal use. This law was instrumental in encouraging a nation of home brewers to explore and replicate the classic styles of Europe and the UK that had all but disappeared from the American landscape. Home brewers were encouraged to experiment with new creations that have continued to push the edge of the beer world, inspiring America to finally become one of the world's pre-eminent brewing nations just in the past forty years. Since then, the American beer industry has experienced transformational change, initially with the rise of the baby boomer's generation in the mid-1980s and then again in more recent times with the twenty–thirty-year-olds, aptly known as the "Millennials." The mass migration of consumers going from mainstream beer to craft alternatives has continued to accelerate, leading to major shifts in the U.S. beer market. Since then, customers have become more quality oriented, and the smaller quality-focused microbreweries (small production breweries) and brewpubs (restaurants that brew their own beer on-site) have sprung up throughout many major cities in the United States. With all the excitement that has surrounded smaller production breweries, there had been three, and now two multi-national brewers: Anheuser-Busch InBev (NYSE: BUD) Molson Coors (NYSE: TAP) SABMiller (NASDAQOTH: SBMRY) that control the marketplace. With a 200 billion-plus merger of AB InBev and SABMiller in 2016, they now operate the largest brewing company in the United States in volume with over 40% share of beer sales. Worldwide, this company brews well over 400 brands around the world.

Craft Beer: The 21st Century Uprising

The brewing industry originally began (and has recently been rediscovered) as a local business. Beer companies and brands are deeply rooted in local communities—something craft brewers have been able to grasp from their early stage of development. The brewing industry has not agreed on a consistent definition of the term "craft beer," nor is it legally defined. According to the Beer Institute, craft beer was generally considered to be made with an annual production of less than two million barrels, but was raised to six million barrels in 2011. This change was due to a large part in attempting to be inclusive of some higher volume brewers that are most "supportive" to the craft beer segment. Regardless of size, craft brewers can arguably be any size of production if there is more importantly, a definite devotion to the integrity of their product. The market demand for these types of beers have multiplied; even large national macro brewers have recognized the "craft" trend and have purchased smaller craft breweries and or created "faux" craft beers in their current portfolios that meet this current demand. Some examples include Molson Coors creating the *Blue Moon* brand or AB InBev creating the *Shock Top* brand and acquiring the highly successful Goose Island Brewery based in Chicago. Craft beers are not a trend; instead, they have become a way of life as craft brewers have responded to customer needs by offering intense and unique flavor profiles: "Style (instead of brand) is becoming the more important aspect of consumer choice, especially among Millennials," Demeter Group, 2013. Another upcoming movement has been the growing number of "nano breweries"; super small scale production of beer having been born out of the growing home brewing movement.

The top craft breweries are well over the original guideline of two million barrels of production to be considered craft. It's believed that the support of these iconic producers are why the craft segment of the industry decided to increase the guideline of barrel production. Many of these breweries operate multiple locations in order to be located closer to their primary markets. According to the Brewers Association, the six largest craft breweries in order of sales volume are:

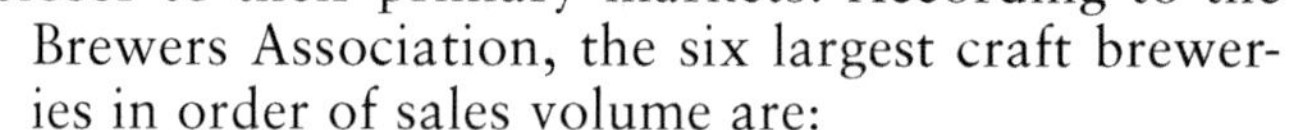

1. D.G. Yuengling & Son, the oldest operating brewer in the U.S. out of Pottsville, Pennsylvania
2. Boston Beer Company (Samuel Adams) of Boston, Massachusetts
3. Sierra Nevada Brewing Company of Chico, California
4. New Belgium Brewing Company of Ft. Collins, Colorado
5. The Gambrinus Company (Shiner Bock) of San Antonio, Texas
6. Deschutes Brewery of Bend, Oregon

Figure 8.2 is an image of the iconic Pale Ale from Sierra Nevada Brewing Company.

FIGURE 8.2
Sierra Nevada. Courtesy of Erika Cespedes.

Significant Beer-Producing Countries

Learning Objective 2
Briefly identify the significant beer-producing countries

A beverage manager should be familiar with the classic brewing countries where the traditions of beer making are often centuries old. These countries continue to thrive and showcase their classic beer styles that have originated from areas largely in northern parts of Western, Central, and Eastern Europe. Germany alone has well over 1,300 breweries, while in countries such as Belgium, the United Kingdom, the Czech Republic, Poland, Austria, the Netherlands, France, Lithuania, and Romania, the brewing sector landscape is highly diverse with forty-five or more breweries per country. Pictured in Figure 8.3 is A La Mort Subite, a café in Brussels that specializes in classic Gueuze beers.

German Beer

Beer (even more so than wine) is the most significant alcohol beverage of Germany. In comparison to the rest of Europe, Germany is the leading country for the production and consumption of beer. According to the German Beer Institute, Germany also has around 1,274 breweries, the largest number in Europe. Roughly 50 percent of all German breweries are located in the southern half of the country such as in Bavaria, with more than a hundred breweries in Baden-Württemberg and Nordrhein-Westfalen. Figure 8.4 illustrates Radeberger Pilsner from Germany which began in 1872.

Some popular German beer styles include Alt, Kölsch, and Weissbier (Hefeweizen and Dunkelweizen) along with some famous lager styles, including Pilsner, Marzen, Dunkel, Bock, Doppelbock, Vienna, Helles, and several other regional specialties.

FIGURE 8.3
A La Mort Subite located in Brussels, Belgium. Courtesy of John Peter Laloganes.

Reinheitsgebot

German beer traditionally was brewed according to the Reinheitsgebot laws of 1516. The Reinheitsgebot laws translate literally to "purity order" and is sometimes called the "German Beer Purity Law"—originally created to regulate the production of beer in Bavaria which eventually spread throughout Germany. The laws stated that only water, hops, and barley were allowed to be ingredients and later, after its discovery, yeast became the fourth legal element for beer production. The limitation of barley grains was believed to ensure the availability of sufficient amounts of affordable bread, as the more precious wheat and rye were reserved for use by bakers. In modern times, it is easy to comprehend how these laws could be limiting in terms of producing alternative styles of beer. The Reinheitsgebot laws were officially repealed in 1998, though many brewers still apply this tradition and it has become more of a critical marketing piece for Germany and for beers around the world.

FIGURE 8.4
Radeberger Pilsner. Courtesy of Erika Cespedes.

Oktoberfest

The world-famous Oktoberfest is a 16-day festival running from late September to the first weekend in October, held annually in Munich, Germany and celebrated around the world. Oktoberfest is

Germany's premier tourist attraction, with more than six million attendees every year. The first Oktoberfest took place in 1810 to celebrate the marriage of Crown Prince Ludwig with Princess Theresa Von Sachsen-Hildburghausen. The event was so successful that it was decided the celebration should occur every year.

In keeping with tradition, beers designated as Oktoberfest must follow specific production standards. The beer must be brewed within the city limits of Munich, conform to the historic Reinheitsgebot and contain a minimum of 6 percent alcohol by volume. The beers are usually made after a special Vienna style called Marzen style beer, where the beer is brewed in March (Marzen) and then cellared, clarified, and packaged by September in anticipation for Oktoberfest.

Belgium/Belgian Beer

FIGURE 8.5
La Brouette located in Brussels, Belgium. Courtesy of John Peter Laloganes.

Belgian beer production has origins that go back to the Middle Ages when the production of beer came under the watchful eye of the Roman Catholic Church. Initially, beer production was used to share with those seeking refuge on their pilgrimages; later, beer was used as a means of financing their communities. Through time, the Abbeys became the heart of agricultural knowledge and science. During this period, numerous styles of beer were discovered and the techniques of brewing were refined.

Belgian brewers have a long tradition in creating a range of beer types. It is arguable that any other beer-producing country has more diverse, native beer styles than Belgium. Their beers exhibit predominantly malt flavors; hops are generally in the background. An interesting distinction is their reliance on wild (or affectionately known as native yeast) strains such as *Brettanomyces bruxellensis* that yields an earthiness in several Belgium beer styles.

The respect and integrity that goes along with pouring Belgium beer is that each style and often the brand has its own dedicated glass. The late Michael Jackson (the celebrated beer critic and author) is noted for saying, "Belgian beers have become fashionable, yet the pleasures they offer have been truly explored by only a discerning minority of drinkers." Pictured in Figure 8.5 is the famous Belgian tavern La Brouette that specializes in classic Belgium cuisine and beers.

Some popular Belgian beer examples include Abbey Ales, Golden Ale, Trappist Beers, Wit (white) Beer, Gueuze, Lambics, Dubel, and Triples, and several other regional specialties.

FIGURE 8.6
Chimay Trappist Ale. Courtesy of John Peter Laloganes.

Trappist Beers

The name "Trappist" originates from the Cistercian abbey, La Trappe, in Normandy, France. Today, there are close to 170 Trappist monasteries around the world, but only eleven are qualified to brew beer and be awarded the "Authentic Trappist Product" (ATP) designation as defined by the International Trappist Association (ITA). Figure 8.6 identifies one of the most widely recognized Trappist beers, Chimay Ale.

Not all Trappist beers are Belgium, but the six of the eleven are possibly the most recognized out of all Trappist beers. The fundamental belief of Trappist monasteries is the need to be self-sufficient, normally accomplished through manual labor—in this case—brewing and selling beer! To be labeled an Authentic Trappist Product, a brewer must conform to the following restrictions:

- The beer must be brewed within the walls of a Trappist monastery
- The brewery must be secondary to monastic life
- The brewery must not be for profit; any surplus must be used for charity
- The beer must be constantly monitored to assure high quality

United Kingdom Beer

Beer from the UK (includes Britain, England, and Ireland) has a long history of developing distinct traditions and contributions to the many styles of beer. England is one of the few countries where cask conditioning, or the maturation of beer, is still fairly common-place to be carried out in the cellar of a pub rather than at the brewery. It was the early 18th century that saw the development of new styles throughout England, including the now popular Porter (the ancestor of Stout Ale), and India Pale Ale, which was often referred to as a bitter or extra special bitter.

Eastern European Beer

Eastern European beer has a long-standing tradition, but it is often overshadowed by the pedigree of Czech Beer. The Czech Republic is home of the lager style of beer called Pilsner. This beer is still brewed as per the Bohemian tradition and has been produced for centuries in Pilsen, Budejovice (Budvar), and other towns. Figure 8.7 shows a can of Pilsner Urquell, founded and brewed the first Czech style Pilsner in 1847.

FIGURE 8.7
Pilsner Urquell from the Czech Republic.
Courtesy of Erika Cespedes.

An important moment in history for many Eastern European beer-producing countries was the fall of communism in 1988. Since then, many breweries have been privatized and/or bought by foreign brewing companies. This has allowed the breweries to refocus on tradition and quality as well as to increase production substantially to compete in the global marketplace.

The beers throughout the Baltics, Poland, Romania, Bulgaria, and Croatia generally combine elements of Czech, German, and British brewing traditions. Most beers found in Eastern Europe are lighter-styled lager options, although there are some dark ones as well.

North American Beer

Over the past decade, there has been something just short of a "beer revolution" in North America. It's unfortunate that for decades, many Americans had become accustomed to poor quality, mass-produced beer that was indistinguishable from another. In more modern times, beer makers have admired the classic beer styles and have made concerted attempts to replicate the best versions. Through their sheer passion, brewers have opted to forge ahead and experiment, bringing about a renewed excitement of different styles that haven't been seen or experienced prior. The craft brewers have been taken more seriously as their conviction for exploration has created an entire niche in the marketplace that demands high quality. Along their journey, the brewers have occasionally collaborated with one another to funnel ideas back and forth. Never has there been so much interest and appreciation for fine beer in the United States and Canada. Coupled with the

FIGURE 8.8
American craft beers. Courtesy of John Peter Laloganes.

availability and accessibility of quality ingredients, such as hops from the Pacific Northwest and barley and wheat from the Midwest, this continent has created a new generation of beer drinkers who appreciate quality beer. More recently, America is becoming a nation of beer intellectuals. Over the past decade, beer-centric establishments have marched forward in pursuit of offering quality-oriented classical beer styles and hyper creative alternative versions that are truly original. Despite craft beer's unprecedented growth, high-quality beer is not a fad but a fundamental change in how people view and enjoy the beer category. Figure 8.8 displays a collection of craft beers that showcases the creative range that tends to represent modern American beer culture.

The Core Ingredients in Beer

Learning Objective 3
Discuss the four major ingredients used to make beer

The study of beer and its production process provides an excellent illustrative example of showcasing the sciences. Apart from ethanol (the common denominator among all alcoholic beverages), beer contains a multitude of substances undergoing chemical processes that collectively determine its distinctive personality. The core ingredients used to create beer are quite basic, yet their infinite combinations and transformations through production can yield numerous distinctive styles and types of beer. Beer consists of water, a starch source (malted barley), hops, and yeast.

Beer is produced through the initial brewing and fermentation of starches mainly derived from grains, most commonly malted barley, although wheat, rye, and corn are used as well. Beer is enhanced with hops and occasionally fruits, which adds flavor, acidity, and bittering qualities. Beer may undergo aging in barrels or stainless steel tanks for a period of weeks to years. The strength of beer usually hovers around 4–6 percent alcohol by volume (commonly referred to as "abv") but more modern styles are reaching the tipping point of 8 and 9 percent abv and even 20 percent in some astonishing—rare versions. The abv is typically dependent on the style of beer being produced and/or local traditions.

Water "The Purifier of Beer"

Beer is composed mostly of water; it comprises about 85–95 percent of a beer's ingredients. Water contains a wide range of chemical properties and mineral components that establish it as an integral ingredient as it plays an obvious role in creating aromas and flavors and structural components in a beer. Water is a molecule string of two hydrogen molecules and one oxygen molecule (H_2O). In one sense, water is a simple molecule, but it is also complex because it carries minerals that can alter the characteristics of the final product. The types of minerals or flavors that water carries are dependent on the water's source. Water with a high mineral content is referred to as "hard," while water with low mineral content is called "soft." The composition of water could have a profound impact on the finished beer. However, in modern times, almost any water can be "adjusted" to replicate an intended style of beer.

Yeast "The Fermenter of Beer"

Yeasts are single-celled microorganisms that are the necessary catalyst for the fermented grains (the sugar source) to create alcohol. They are biologically classified as fungi and are responsible for converting fermentable sugars into alcohol, CO_2 and other desirable by-products.

There are literally hundreds of varieties and strains of yeast and any number of them may be introduced into the brewing process. The brewer must first decide whether to use native or wild yeast—cultured or cultivated yeast—or a combination of both categories. Native yeast strains vary greatly from one location to another. The distinctions in these regional yeasts may seem slight, yet they can create considerable variation in the final product. The yeast used to make the original Pilsner comes from Eastern Europe, but brewers in the United States can procure a cultured version of this yeast to reproduce this style of beer. The result will be a beer that has many of the characteristics of a classic Pilsner originating from the town of Pilsen (Plzeň), Czech Republic.

Spontaneous fermentation is an age old inspired tradition commonly associated with Belgian Lambic beers—creating a slightly soured, non-filtered brew. This spontaneous fermentation can occur when the wort (unfermented beer juice) is exposed to the surrounding environment through an open vessel inviting natural/wild yeast and bacteria to literally infect the beer. One of the typical yeasts acquired through open fermentation is somewhat unpredictable, yet highly distinguishable, *Brettanomyces* (breh-tan-UH-my-sees) yeast strain, which adds a certain complex, funky flavor and potential for soured characteristics in the beer. Figure 8.9 shows a collection of wild yeast, barrel aged, sour ales from Goose Island Beer Co.

FIGURE 8.9
Goose Island sour ales. Courtesy of John Peter Laloganes.

In the current era, it has become more common for the production of both beer and wine to incorporate cultured yeasts that have been replicated in a laboratory. These cultured yeasts benefit from being more predictable in terms of the personality variations of the finished product. Regardless of native versus cultured yeast, there are two broad categories of yeast that are utilized in beer production: *Saccharomyces cerevisiae*, known as top-fermenting yeast and *Saccharomyces uvarum* known as bottom-fermenting yeast. Each category of yeast responds differently throughout their productive stages of fermentation.

Ales rely on "top-fermenting" yeasts which means that the beer ferments toward the top of the fermenting vessel. These yeasts generally ferment at warmer temperatures (55–75°F) and more quickly (3–5 days) than lagers. Ales convert less sugar into alcohol, leaving a more noticeable body with some slight residual sugar upon completion of fermentation. At these relatively warmer fermentation temperatures, beers produce higher esters (noticeable aroma and flavor by-products), providing a distinctive character associated with ale beers. Esters are a class of compounds that contribute specific odors due to the fermentation process. Some odor compounds created include: *isoamyl* (ahy-soh-am-mil) acetate, the odor associated with a banana, or ethyl *phenylacetate* (fee-nehl-ass-ih-tate), the odor associated with honey.

Lagers incorporate "bottom-fermenting" yeasts which means that they ferment toward the bottom of the fermenting vessel. Lagers typically undergo a lengthier (7–15 or more days) primary fermentation at colder temperatures (between 45–55°F) than ales. Lagers beers are typically then given a longer resting period (ranging from 32–39°F) throughout what is referred as the lagering phase. The name "lager" derives from the German "lagern" meaning to store. Classically, during this stage of lagering—the brewers in the Bavarian region of Germany would store their beer in cool cellars and caves during the warm summer months. The brewers recognized that the beers continued their fermentation during this time, and the natural formation of esters and other by-products would not occur. Through time, lager beers would begin to clarify themselves when stored in cool conditions—the lagering phase ultimately produced a "cleaner" and lighter style of beer as broadly compared to ales.

Malted Barley "The Soul of Beer"

Barley is the most common grain used in beer production; however, any grain such as wheat, oat, or rice can be added in smaller doses for aroma, flavor and mouthfeel characteristics. Malted grains provide the color, sweetness, body, and roasted flavors and aromas in beer. The malt also works to counterbalance the hop's acidity and bitter characteristics.

The range of malt styles and varying combinations are somewhat endless. They can be used to replicate a classic beer style or applied creatively to create a new style as per the vision of the brewer. The names used to identify specific types of malts can be based on somewhat inconsistent terms. Most commonly, they are identified by their various roasting levels that generate a range of specialty malts called caramel, chocolate, and black malts. Contrary to popular belief, darker beers (use of black malts) are not necessarily fuller in body or stronger in alcohol; instead, they are more robust in their aroma and flavor profile. Alternatively, malts can be named after a city where the malt style originated. Malts can also contribute to the name of a beer, such as in Britain, where beers that use mild-pale malt gave way to a style of beer known as the Pale Ale. Some beers may include only one style of malt; others may contain as many as half a dozen or more malts (and/or grains) to provide an added uniqueness to the finished beer. Figure 8.10 displays a trio of specialty malts: pale malt, caramel malt, and dark or chocolate malt.

FIGURE 8.10
Trio of specialty malts. Courtesy of Erika Cespedes.

Use of Adjuncts

Malts may also be named according to the other grains such as rye, wheat, or oats used in the brewing process. If these other grains are incorporated, they are referred to as adjuncts—simply, other agents that provide alternative characteristics to the beer. The use of adjuncts doesn't necessarily equate to a good or bad scenario, it's simply an alternative to add various grains and/or other ingredients for purpose of diversity and uniqueness. Although, with the addition to the universal grain of barley, many of the inexpensive American lager beers incorporate brewing adjuncts such as corn and rice to act as "filler." The incorporation of these ingredients is used to allow for lesser amounts of more expensive grains, making it less costly to produce the beer. Greater quantities of adjuncts such as corn and rice can also serve to incorporate lighter color, body, and flavor—this is ideal for the mainstream beer consuming public—such as America's large-scale brewer "Budweiser" adding as much as 30 percent rice to lighten their product.

The Malting Process

For the yeast to easily digest the grain, the malting process must be activated for the complex starches in the grains to be transformed into simple sugars. Malting is a three-step process that includes (1) steeping, (2) germination, and then (3) drying/roasting. The grain is initially steeped in water for a period of hours to days in order to stimulate germination (or sprouting) while being aerated for roughly one week. During this germination process, the grains will bud and put forth shoots, while an enzyme known as *diastase* (die-AH-stay-see) converts the complex starch molecules into simple fermentable sugar, allowing the yeast to easily convert it into alcohol. Along with the starch, seven main enzymes will be formed via the malting process—most notably alpha and beta *amylase* (am-ah-lase). Once the grains have germinated to the desired degree, they are placed in a kiln to dry. This drying process not

FIGURE 8.11
Trio of malts and their corresponding beer. Courtesy of Erika Cespedes.

only stops the germination but additionally has a significant influence on the beer's final personality characteristics. At this point, the germinated grains become malt as they are dried and eventually ground. The variety of grain, extent to which it can germinate, and the temperature at which it is dried can all influence the color and overall characteristics of any given beer. Figure 8.11 showcases types of malts and their possible influence on color of the finished beer.

Beyond the obvious need of malt as a source of sugar for yeast consumption, it also provides many additional characteristics to beer: visual aesthetics, aroma and flavor compounds, and density or body for the overall beer. The malting process illustrates one of many chemical responses, such as the Maillard reaction, which most influence the malt character and color of a beer. This reaction is simply known as "browning" and is essentially how amino acids react with a reducing sugar. Between that and the caramelization that occurs during this process, the various malty aromas and flavors as well as color hues in beer are created.

Hops "The Spice of Beer"

Hops are a cone-shaped perennial plant deriving from a relative of the cannabis family. They contain dozens of essential oils belonging to a group of liquid hydrocarbons called *terpenes* that are expressed in a beer's personality. Like grapes or apples, hops come in many varieties—each one offering a different taste and feature to the brew. Hops provide many aromatic and structural dimensions to beer—initially contributing unique aroma/flavor (grass, pine, citrus, floral, and herbal); then once tasted, they add distinctive structural components (dryness, spiciness, and bitterness), which help contrast with or offset any sweetness leftover from the malt. Many of famous "hop" growing areas are located in Central Europe, England, northwest U.S. (Washington state), and Australia/New Zealand. According to *USA Hops: Hop Growers of America*, Washington state produces some 71 percent of commercial hop production followed by Oregon, Idaho, and other states. Figure 8.12 shows a handful of hops, one of the major ingredients used in beer production.

Hops are divided in two main groups: (1) boiling, or bittering, hops and (2) finishing, or aromatic, hops. These two types of hops are applied at different times during the production process, depending on what type of beer the brewer is trying to produce.

FIGURE 8.12
Hops. Courtesy of Erika Cespedes.

- *Bittering Hops:* The boiling, or bittering, hops are initially added during the first boil of the wort (unfermented beer) in which the brewer attempts to extract the alpha acids, soft resins, and essential oils giving beers their characteristic bittering agent.
- *Aromatic Hops:* The finishing, or aromatic, hops may be added or layered at the end of the boil when the wort is still warm or added throughout the fermentation process. This addition of hops (often referred to as dry hopping) will dramatically increase the distinctive hop aromatic characteristics in the finished beer.

Additional to their aromatic contributions, hops also aid in head retention—the foam collected at the top of a freshly poured glass of beer. They also have anti-bacterial qualities that act to preserve the beer during its production and transportation processes, allowing it to survive many months beyond its natural shelf life.

The International Bitterness Units or IBU is a universal system that has been established to rank the degree of bitterness that hops can provide in a finished beer. Many mass produced American lagers hover around 9 IBUs which is below the average threshold for detection of hops. Some beers can be as high as 120 IBUs. Highly hopped beers have an intense aroma of pine, herbal, and citrus, along with a greater drying and slight bittering sensation. Some general approximations include:

- Pilsners and Lagers 30–40 IBUs at 4–5 percent alcohol
- American Pale Ales 30–40 IBUs at 5 percent alcohol
- India Pale Ales (IPAs) 50–70 IBUs at 5–7 percent alcohol
- Double India Pale Ales (IIPAs) 80–100 IBUs at 8–9 percent alcohol

The Brewing Process

Learning Objective 4
Explain the brewing process and the fundamental differences between lager and ale beer categories

Types of Breweries

The process of making beer is known as brewing, which on a commercial scale takes place in a dedicated facility called a brewery. The popularity of beer in this country has led many passionate people to pursue this craft of brewing as a hobby (known as home brewing) out of one's home or garage.

Breweries are often classified as per their size, based on production in barrels. According to the Brewer's Association, they define a typical large-scale brewery exceeding production of 6,000,000 barrels yearly. Some well-known examples of large-scale macro (and multinational) breweries include Anheuser-Busch InBev (AB In Bev), SABMiller Brewing Company, and Coors Brewing Company.

There are four distinct craft beer industry market segments: microbreweries, brewpubs, regional craft breweries and contract brewing companies.

The Brewer's Association defines a *microbrewery* as a brewery that produces less than 15,000 barrels of beer per year with 75 percent or more of its beer sold off-site. Microbreweries became very popular in the late 20th century—which originally brought beer back to an art form—helping to rediscover and replicate high-quality versions of classic styles. Fritz Maytag (of the Maytag appliance company), who took over Anchor Brewing Company in San Francisco back in 1969 began to brew high-quality beer for non-mainstream tastes, providing one of the first commercial examples.

A *brewpub* is a restaurant-brewery that sells 25 percent or more of its beer on site. The beer is brewed primarily for sale and consumption in the restaurant and bar. The beer is often dispensed directly from the brewery's storage tanks. Brewpubs often sell beer "to go" and /or distribute to off site accounts. Bert Grant was noted for opening the first modern-day brewpub (in 1982) located in Yakima, Washington.

Contract brewing is when a business hires a brewery to produce its beer. It can also be a brewery that hires another brewery to produce additional beer.

A *regional brewery* is one that typically produces between 15,000 and 500,000 barrels yearly.

Malting

As previously discussed, most beers are constructed with a large quantity of malted grain—as the malt provides the necessary enzymes needed for fermentation to occur. Malted barley accounts for the most prevalent and significant quantity of grain utilized—even in wheat beers. During the malting process, a raw grain is converted into malt through a process of germination and then drying. The dried grain is milled and called grist, similar to dry cereal.

FIGURE 8.13
Brewing kettle. Courtesy of John Peter Laloganes.

Wort and Mashing

One of the first stages of the beer production process involves the creation of wort—a sugary liquid that is formed ultimately for consumption by the yeast. First, the malted grain (or grist) needs to be hydrated—along with the addition of hops—is added to water and brought to a boil through a process referred to as mashing. The mashing process takes around 1–2 hours, during which the sugars are drawn out of the grist. The sweet wort liquid is then drained and strained off the grist (known as lautering) and cooled down in preparation for fermentation. Hops may be added at this point or can be added later during the stage of fermentation. Figure 8.13 identifies a brewing kettle used for beer production.

Fermentation

As the wort is cooled from the previous stage, a strain of yeasts is carefully measured and added to the mixture. This mixture is now allowed to ferment roughly between three and seven days. During this period, the yeast converts the wort's fermentable sugars into alcohol and carbon dioxide (CO_2). Once the conversion of sugar has reached the right stage, the temperature is reduced, and the yeast begins to settle to the bottom of the fermenting vessel. The settled yeast is removed and the beer is transferred to a storage vessel. Additional hops may be added (process known as dry hopping) at the end of the fermentation process to add more hoppy character.

FIGURE 8.14
Barrels used for aging. Courtesy of John Peter Laloganes.

Maturation/Aging

Once the conversion of sugar has reached the right stage, the temperature is reduced and the yeast begins to settle to the bottom of the fermenting vessel. The settled yeast is removed and the beer is transferred to a storage vessel. At this stage, the beer, which is known as green beer (or unaged), is placed in an aging tank, barrel, or bottle, and it is left to evolve for typically up to two months. Figure 8.14 shows a stack of barrels used for aging beer. There, close to freezing point, the beer matures for about two weeks (though many emerging craft beers are being aged for periods of years). Figure 8.15 shows a stainless steel tank used for allowing wine to mature prior to clarification and bottling. During maturation, lager-style beers will have residual yeast and insoluble malt proteins settled, and the beer gradually becomes clearer. This process is known as lagering—allowing the newly fermented beer to mature and develop smooth, desirable aromas/flavors.

Clarification

Fining and clarifying agents are often used in brewing to remove some of the dead yeast cells and residues from the grains, which

FIGURE 8.15
Stainless steel tank. Courtesy of John Peter Laloganes.

can otherwise leave a beer to appear cloudy. This cloudiness does not affect the wholesomeness of the beer, but it does impact the appearance. Much of the drinking public would expect to have a clear-looking beer without any remaining sediment from the production process. A clarifying or fining agent is used to remove this sediment, and a purified beer is the result.

Traditional clarifying agents such as Irish Moss (a seaweed that grows along the rocky coast lines of the Atlantic Ocean) and gelatin are still used today, but modern filters are now used to clarify commercially produced beer.

Carbonation

Beer will then be given a dose of carbon dioxide (CO_2) prior to bottling. There are varying degrees of pressure, and will be adjusted as per the style being produced. Beer can be packaged in a variety of ways: barrel, bottle, or can.

Bottle-Conditioned Beer

Bottle-conditioning is when a beer is given a small dose of yeast and sugar just prior to bottling—like the technique of incorporating the bubbles into Champagne. Bottle-conditioned beers are unfiltered so the final conditioning (referring to a beer's secondary fermentation) takes place partially or entirely in its original bottle from which it will be served. This technique is quite different from most other, more common beers, which are carbonated using high pressure injection of CO_2. Bottle-conditioned beers will vary in clarity. Since the beer undergoes some level of fermentation within the bottle, it is likely that yeast solids (referred to as sediment) will remain suspended or settled as a thin layer at the bottom of the bottle. Upon opening a bottle for consumption, it is recommended to slowly pour the beer, leaving any sediment at the bottom half of the bottle. Purists will typically prefer the yeast to be poured within their glass. For example, German Hefeweizen is an unfiltered beer that contains varying amounts of sediment. Once 90 percent of the beer is poured into the glass, the remainder of the contents is swirled in the bottle to suspend the sediment before pouring the residue into the glass.

Cask-Conditioned Beer

Cask-conditioned beers are very similar to bottle-conditioned beers but the location of the secondary fermentation is the wood barrel from which it is dispensed. The casks will contain live yeast in suspension that will remain in contact with the beer for a period of time.

Both bottle and cask-conditioned beers remain in contact with yeast cells so that over a period of maturation the beer will achieve a natural, characteristically gentle carbonation and will develop more complexity of flavor. With the popularity of craft beers, both conditioning techniques have been experiencing a revival in the marketplace.

Packaging

Beer is primarily purchased in bottles and kegs with a growing selection available in canned options. More recently, local brewers and brewpubs are using growlers to offer draft beer "to-go" for the consumer to transport back home. Each vessel has its own advantages and disadvantages; most likely, the best option is based on customer preference and the operation's availability of refrigerated and non-refrigerated storage space.

Draft Beer

Beer in a keg (otherwise known as draft beer) is dispensed from a barrel rather than a bottle or can. Barrels (also known as kegs) are nothing more than very large cans of beer. The beer in a keg can be pasteurized or non-pasteurized and will be *tapped*, or opened by attaching a contraption (connected to lines of CO_2 and/or nitrogen) with a spout, and are manually controlled at the touch of a button or the pull of a handle. Kegs are very popular with large gatherings as they are more cost effective as opposed to purchasing individual beer bottles or cans. Kegs are dispensed with CO_2 though with certain brands of draft beer (Guinness and Boddingtons), nitrogen and CO_2 are combined in order to obtain a denser head and creamy mouthfeel.

Draught/draft beers from literally tens of thousands of breweries and numerous brands are available; it's simply not possible for distributors to offer every possible beer. Availability of each size of keg will vary by brand of beer and location. Buyers should ask distributors about availability of the various keg sizes. The chart below offers a breakdown of the two most common-sized kegs: The full-sized kegs are known as ½ barrels, and half-sized kegs are known as ¼ barrels that are utilized in beverage-based establishments.

THE ANATOMY OF A KEG

Measurement	Keg (½ Barrel)	Half Keg (¼ barrel)
Height in inches	23¼	14¾
Diameter in inches	17	17
Quantity in ounces	1984	992
Quantity in gallons	15.5	7.75
Weight of a full keg in pounds	161	82
Weight of an empty keg in pounds	30	17
# of 12 oz. beers	165	82
# of 16 oz. beers	124	62
Freezing point	29°F	29°F

*These dimensions and weights may vary slightly because of differences between keg manufacturers.
*Half kegs may also be referred to as a pony.
*The # of beers do not account for foam or beer in the lines, so the actual number will be a bit smaller depending on these variables.

Growlers

The growler is another option, used solely for dispensing directly from a tap, usually at brewpubs. A growler is a container usually made of glass that comes in various sizes (often 64 oz.), shapes, and colors. Ideally, it's colored in dark amber or brown glass to greatly reduce UV light from spoiling the beer. With the onset of local craft beers, home brewing, and brewpubs, the popularity of the growler has taken hold. This allows consumers to take out their favorite draft beers in a glass container that is filled on demand directly from the draft system.

Bottles and Cans

For the beverage industry and the occasional consumer, bottled and canned beers are purchased by the case. Each case typically contains twenty-four bottles or cans (12 oz. each) of beer (total 2.25 gallons). Beer cans look like soda cans, which have an identifiable pull tab on top. While the standard vessel size is 12 ounces, there are some alternatives that include 16-ounce and imperial pints (20 ounces) among them. Buying bottled beer is appropriate for establishments that do not forecast large sales of certain brands or styles of beers but would still like to offer them as an option for their customers. If bottled beer is stored properly, it maintains 100 percent yield. Most beers come in a case containing twenty-four beers, but take note that some imported or micro brews come by the twelve pack or different amounts in each case.

Canned beers have been in production since the 1930s; yet over this time, they have come to represent poor quality, very similar to the screw cap closures used for wine. More recently, there has been renewed interest in canning from craft brewers, who are realizing that cans (like the screw cap closures used for wine) can provide a perceived purity and freshness. Additionally, canned beer provides an assurance of integrity by preventing light damage, which is otherwise not possible with bottled beers. The only "real" difference in the taste associated with canned beer versus bottled beer is a perceived one. As beer companies (both small and large) search for that innovative point of difference, it is expected that consumers will be exposed to more canned beer in the near future.

Beer Service

Learning Objective 5
Explain the steps of a properly poured bottled beer

All too often, "professional" beer service is relegated to the lowest form of delivery. In many respects, proper service of beer has an uphill battle to garner the equivalent respect as wine service. Beer service should be given as much attention as the wine and food, both of which are commonly grounded in ritual. Too many beverage servers are afflicted with the "it's just beer" mentality. It is the common and half-hearted effort that draft beer is served flat and bottled beer is served without a glass—both examples speak of the ignorance about the significance that visual and aroma elements contribute in the overall quality of the product. Call it snotty or overly fussy, but it's time that beer receives the appropriate attention it deserves. Certainly, different expectations may arise if one is drinking a beer at the ball park versus having beer in a restaurant—but in some instances, paying $6 or $7 for a beer gets less attention and service than a $2 coffee at the Dunkin Donuts. Too often, draft beer is being poured into warm glassware, dirty or mismatched glassware—the worst scenario is observing it being dispensed into a glass from a tap while the copper spout is submerged in the beer—as if one is leaving a garden hose unattended to fill up a bucket of water. It is accurate that in many areas of operating a successful business, paying attention to the little details leads to increased customer satisfaction and repeat business. Paying proper attention to the service of beer is yet another source to differentiate an establishment from its competition. While some consumers may not notice, the more intelligent customer will recognize and appreciate the attention devoted to professional beer service.

Regardless of the packaging, beer is ideally consumed from a glass so the drinker can appreciate both visual and aromatic nuances of beer. In addition, when drinking from a glass, the beer is least likely to be disturbed for greater preservation of the beer's carbonation.

Serving Temperature for Beers

Commonly, beers are broadly categorized as per the major types of yeast used to produce these—ales and lagers. This arrangement provides some broad guidance as to how beer can be best enjoyed.

Ales

The ale category of beer employs yeast that ferments at the "top" of the fermentation vessel, and typically at higher temperatures around 60–75°F. Thus, ale yeast makes for a quicker fermentation period of 7–8 days, or even less. Ale yeasts are known to produce by-products called *esters*, which are "flowery" and "fruity" aromas ranging from apple, pear, pineapple, grass, hay, plum, prune, and so on. Therefore, the ale category of beers generally express themselves better when served a bit warmer (than most beers Americans are familiar with) at an ideal range of 45–55°F for optimal point of consumption.

Lagers

The lager category of beer is named after the word *lager* which comes from the German word *lagern* and means, "to store." Lager beers are brewed with bottom-fermenting yeast that works slowly at around 34°F, and are often further stored at cool temperature to mature. Lager yeast produces a beer that is often leaner and "cleaner" with fewer by-product characteristics than ale yeast. Lager yeasts permit highly aromatic grassy and piney aromas and flavors as well as allowing the invigorating crisp acidity of the hops to accentuate the palate. Lager beers are generally served colder than ales in order to accentuate their refreshing character in an optimal temperature range between 35 and 45°F.

The Perceptual Visual Experience

Properly poured beer in a clean beer glass presents an incredible perceptual experience that is captured through a beer's color, carbonation, foam, and head.

Color

The color associated with a beer is a direct reflection of its addition of specialty malt. The types of malt used in combination with the degree of its clarification will determine a beer's color hue and clarity. For determining the color in each beer, modern brewers use the standard reference measurement (SRM), expressed on a scale of 4 to 80+ when referring to its hue.

Carbonation

The carbonation (or more simply the bubbles) found in a glass of beer can be perceived as quite sensual as they dance across the palate. While the unopened bottle cap (or other closure) of a beer remains in place, an equilibrium exists between the dissolved gas and any bubbles that may be formed. Upon opening the bottle, the equilibrium is disrupted due to the reduction of pressure (Boyle's Law) and out flows the beer along with the CO_2 from the solution as it's transformed into foam and bubbles, known as diffusion.

Before a bubble can be released, it must form a nucleate from the imperfections or scratches on the surface of the beer glass. The inside of a glass will contain a nucleation site—some areas with a special etching pattern or scratch that provide a spot to encourage carbonation to form. After a bubble is released from its nucleation site, it enlarges as it makes its way to the surface. Bubble enlargement during ascent is caused by a continuous diffusion of dissolved carbon dioxide through the bubble's gas/liquid interface. Each one of the 1.5 million tiny bubbles are scented with mini aroma compounds. When bubbles flow to the top of a glass, they collect in the head where they burst and spray a miniscule amount of liquid into the air. To accentuate this, specific glassware with a tapered head works to concentrate the aroma.

Foam

Foam is a major characteristic of beer that is caused by the sudden decrease of pressure when the beer is released from its storage vessel, whether it's a bottle, can, or keg.

The carbon dioxide dissolved in the beer turns to gas and forms bubbles that rise to the surface. The beer foam consists predominantly of a dispersion of CO_2 in beer. The CO_2 bubbles rising through the liquid to the surface accumulate molecular proteins. Gas diffuses from smaller bubbles to bigger ones ("coarsening"). In small bubbles, carbon dioxide will exert a greater pressure than is found in big bubbles. If a small bubble is next to a big bubble, the gas contained within the bubbles will try to reach equilibrium. Bear in mind that the nature or style of beer can dictates its foaming qualities.

Head

Head is the technical term for the layer of creamy foam that forms at the top of a beer once it has been poured into a drinking vessel. For quality-oriented drinkers, a 1-inch head is a general standard. The head is not only a definite psychological aesthetic, it's also a textural (mouth feel) benefit. It acts as a catalyst for dispersing the aromatic nuances associated with the beer's personality. The head contains minute bubbles that are condensed—each individual bubble is a pocket of gas surrounded by a thin wall of beer. The head works toward preventing the carbon dioxide from escaping the liquid too quickly. It partly acts as a safety net—slowing the release of the beer's aromatic compounds, contributing additional dimensions of texture to the palate throughout the drinking experience. The mouthfeel and the creamy sensation of the beer are preserved. Once the head forms, it begins to disappear—sometimes slower than other times.

The volume and quality of the head formation in any given glass of beer depends largely on the volume of CO_2 gas liberated at the moment of pouring. It will be the result of three highly controllable factors of serving temperature, cleanliness of the beer glass, and pouring technique that combine to influence the head's formation, stability, and the degree of persistence.

How to Professionally Pour Beer

Whether pouring beer from a bottle or the tap of a keg, the angle of the glass is vital to the success of a visually aesthetic and professionally poured beer. The angle of glass will help control the beer's formation of head and ensure that it will remain foamy just as the higher powers of the world intended.

The first step of a properly poured beer is to begin with a clean glass. This step may appear obvious; however, many dishwashers leave residuals of cleaning detergents, oils, dirt, and remains from a previous beer. These factors may inhibit head or foam creation and introduce some undesirable aromas and flavors.

FIGURE 8.16
Pouring bottled beer (step 1). Courtesy of Erika Cespedes.

1. Hold the glass at an angle of 45°. Figure 8.16 illustrates step one of pouring beer into a glass.
2. Pour the beer, targeting the middle of the slope of the glass. Don't be afraid to pour hard or add some air space between the bottle and glass for showcasing the attractive appearance, to provide a bit of showmanship, and to assist in inspiring a good *head* for the beer.
3. At the half-way point of pouring, adjust the glass to a 90° angle and continue to pour the stream toward the middle of the glass. This will induce optimal head formation (about 1" minimum) as it's a necessary component to a well-poured beer. Figure 8.17 shows the straightening of the glass in order to build head formation. Figure 8.18 shows the final beer being poured into an upright glass.

FIGURE 8.17
Pouring bottled beer (step 2). Courtesy of Erika Cespedes.

FIGURE 8.18
Pouring bottled beer (step 3). Courtesy of Erika Cespedes.

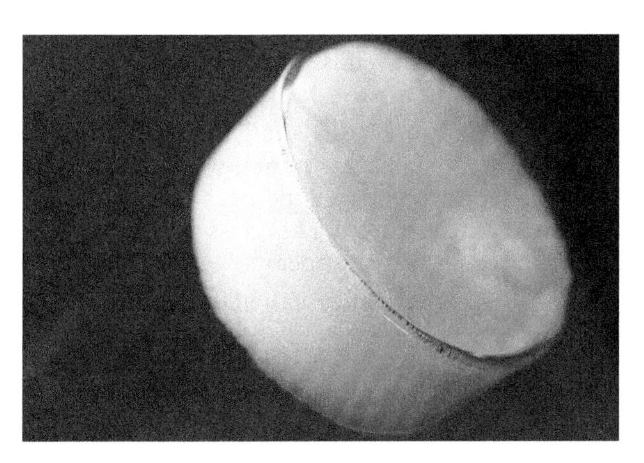

FIGURE 8.19
Head formation. Courtesy of Erika Cespedes.

The head plays as such an important role. It not just provides an aesthetic element, but also acts as a catalyst for transporting the beer's aromatic characteristics. The head, like the crema on a properly extracted espresso shot, assists in allowing full enjoyment of the aromas as it slows and preserves the release of aromatics. Figure 8.19 illustrates a nice head formation on a glass of beer.

Bottle-conditioned beers (beers that have been unfiltered prior to bottling) may contain a considerable amount of yeast sediment. The yeast is intended to be served in the beer and is often contained in the bottom and last $^1/_5$ of the beer bottle. Prior to pouring out the remaining $^1/_5$ of the beer, swirl the bottle a bit to integrate the yeast and beer together. Finally, proceed to pour the remaining contents of the beer into the glass.

Beer Glassware

Upon first sight, beer has a brilliant aesthetic that is illuminated by the combination of its color, carbonation, foam, and head; these work collectively to imprint the perceptual experience into the drinker's memory. The combination of these visual attributes is the most striking trait recognizable in a beer, signifying the importance of consuming beer from a glass as opposed to a bottle or can. Though glassware can be interchangeable depending upon the preference of the establishment, the drinking vessel truly plays a role in the beer's integrity and maximum enjoyment; it has become a greater part of the customer's expectation. When drinking from a glass, not only does one get to enjoy all the gustatory elements that beer has to offer, but the drink is also minimally disrupted allowing the carbonation to dissipate at a slower rate than it would if the beer were consumed out of its bottle.

Scientific studies illustrate that different shapes of glassware will impact head development and retention. Varying levels of head retention and presentation are desired with different styles of beers, different styles of glassware should be used accordingly.

1. **Pint Glass** One of the most universal of all beer glasses throughout America is the pint glass. This glass has become one of the most prolific of beer glasses mainly because it is durable and stackable, making it easy to store behind the bar and easy to carry. The pint glass is characterized as nearly cylindrical, with a slight taper and wide mouth. There are two standard sizes: the 16-oz. (common in the United States; the poor man's pint glass) and the 20-oz. Imperial, and has a slight ridge toward the top. When a pint of beer is ordered in the United Kingdom, an *imperial pint* of beer must be served. Glasses for imperial pints are clearly marked with a "fill line" of 20 oz. In the United Kingdom, if beer is not poured to this line, it cannot be served. As of January 2011, the United Kingdom is moving to break from tradition to potentially allow for smaller portion sized beers in addition to pints and half-pints.
2. **Beer Stein** The beer stein or mug can be very ornate and can be designed with or without a lid. The mug is designed to be very durable which lends itself quite user-friendly for everyday use behind a bar.
3. **Pilsner Glass** The pilsner glass is a tall, slender beer glass that is shaped somewhat like a funnel, with a larger top than bottom. This shape allows the beer to "show off" and accentuate its carbonation. This glass provides a very aesthetic option for customer appeal.
4. **Goblet** The goblet (sometimes called the chalice) is a bowl-shaped glass with a stem. The large surface area maintains a healthy thick head on the top of the beer. These glasses are commonly used for French and Belgian beers.
5. **Yard** The yard glass is a very tall, thin glass that is about a yard in length. The traditional yard glass holds 42 oz. These days, half-yard glassware can be found as well. In some cases, there are specially designed holders for ease of consuming a beer contained in the yard glass. Figure 8.20 showcases a half-yard glass.
6. **Snifter** The snifter is commonly used for spirits such as brandy and cognac. However, these wide-bowled and stemmed glasses with their tapered mouths are perfect for capturing the aromas of strong ales such as barley wine or a Bourbon stout. Size options exist, but they all provide room to swirl and enhance and capture the aromatics while allowing time to savor the beverage.
7. **Weizen** Weizen glass is a tall glass to showcase the beer's color with a curvature at the top to allow for headspace for development of the white, creamy head. Used for wheat-based beers known for producing a healthy, stable head. Figure 8.21 showcases a weizen glass.

FIGURE 8.20
Half yard glass. Courtesy of John Peter Laloganes.

FIGURE 8.21
Weizen glass. Courtesy of Erika Cespedes.

Troubleshooting Draft Beer

Learning Objective 6
Identify some issues that may occur with draft beer

Draft Beer (often interchangeable with the term draught) derives from a pressurized container called a keg (or a cask in some cases). This is one of the most common methods used for dispensing beer in beverage establishments, particularly bars, around the world. The keg is pressurized with CO_2 which drives the beer toward the dispensing device (or tap). Some beers include the addition of nitrogen gas as a method of incorporating a denser head and creamy mouthfeel. Occasionally, the draft system can be compromised—below are several scenarios that identify common draft beer problems. Each problem includes a likely cause complete with possible remedies. Kegs of beer are extremely profitable but come with some occasional challenges. Figure 8.22 identifies a draft beer being poured into a nonic pint glass.

FIGURE 8.22
Pouring draft into a nonic glass. Courtesy of Erika Cespedes.

PROBLEM—HAZY OR CLOUDY BEER

Likely Cause	Possible Remedy
Yeast or bacteriological infection	Clean beer dispensing equipment
Old beer	Rotate stock
Cooler is too cold	Store at approximately 45°F
Failure to utilize beer fast enough	Trade out beer and clean lines

PROBLEM—UNPALATABLE BEER

Likely Cause	Possible Remedy
Yeast or bacteriological infection	Clean beer dispensing equipment
High storage temperature	Store in cool room (approximately 45°F)
Old beer	Rotate stock
Use of non-approved detergents	Use only brewery approved detergents
Dirty lines, equipment, or glasses	Clean with approved detergent

PROBLEM—FLAT BEER OR POOR HEAD

Likely Cause	Possible Remedy
Excessive or insufficient CO_2 pressure	Decrease or increase pressure
Beer too cold	Check and adjust temperature
Faulty CO_2 regulator	Service regulator
Pressure too low	Adjust pressure
Dirty glassware	Properly wash glassware

PROBLEM—HEADY BEER

Likely Cause	Possible Remedy
Insufficient or excessive pressure	Increase or decrease pressure
Out of gas	Change cylinder
Faulty CO_2 regulator	Repair CO_2 regulator
Beer too warm	Check temperature
Faulty bar tap	Service tap
Poor dispensing techniques	Train staff
Warm glasses	Chill glasses

Check Your Knowledge

Directions: Use these questions to test your knowledge and understanding of the concepts presented in the chapter.

I. MULTIPLE CHOICE: Select the best possible answer from the options available.

1. Beer is most often made from which of the following grains?
   a. Wheat
   b. Rye
   c. Oats
   d. Barley
2. Hops add which component to beer?
   a. Aroma/flavor
   b. Bitterness
   c. Preservation
   d. All of the above
3. Dry hopping is the process of adding hops to the beer
   a. during the boiling of the wort
   b. during fermentation
   c. at the end of the wort boil
   d. all of the above
4. Malt can be measured at different levels of gravity which can indicate
   a. the acidity of the beer
   b. the sweetness of the beer
   c. the level of body or weight of the beer
   d. None of the above
5. A form of packaging beer includes
   a. kegs
   b. bottles
   c. cans
   d. all of the above

6. Bottle- and barrel-conditioned beer indicates
   a. unfermented beer
   b. secondary fermentation in the bottle or barrel
   c. the beer is aged in contact with the yeast cells
   d. both b and c
7. A full keg (½ barrel) of beer contains
   a. 1894 ounces
   b. 1948 ounces
   c. 1984 ounces
   d. none of the above
8. Reinheitsgebot (otherwise known as the Purity Law)
   a. states only the basic four ingredients (water, yeast, hops, and barley malt) can be used in a beer
   b. are the only beers that are allowed to be served at the German Oktoberfest
   c. originally didn't include the yeast in the law as one of the ingredients
   d. all of the above
9. In beer production, the degree of toasted/roasted grain can determine a beer's
   a. color
   b. aromas/flavors
   c. body
   d. both a and b
10. A beer's malt and hops both work to
    a. add sweetness
    b. add bitterness
    c. counterbalance one another
    d. add alcohol
11. Pouring a beer properly will allow for what level of head?
    a. Approximately 1"
    b. ¼ inch
    c. 2"
    d. As little as possible
12. When pouring beer, it's essential to begin with a clean glass and
    a. tilt the glass at an angle of 45°
    b. keep the glass straight
    c. allow the beer to gently enter the glass
    d. tilt the glass at an angle of 15°
13. The most universal serving glass used for beer is the
    a. Stein
    b. Pilsner
    c. Goblet
    d. Pint
14. During the germination process of producing malt, the grains bud and then put forth shoots. What is the name of the enzyme that converts the starch into fermentable sugar?
    a. Diastase
    b. Amylase
    c. Esters
    d. Carbon dioxide
15. How many Trappist breweries are currently licensed to brew beer?
    a. 2
    b. 4
    c. 11
    d. 0

16. The use of wheat, corn, rice, or oats as an additional ingredient in the production of beer is technically considered an
    a. additive
    b. adjunct
    c. booster
    d. unacceptable agent
17. Which country did the original Pilsner derive from?
    a. Belgium
    b. Germany
    c. Czech Republic
    d. England

II. DISCUSSION QUESTIONS

18. Identify the two categories of beer. What are some significant distinctions between them?
19. Briefly explain the brewing process.
20. Explain how beer obtains its color.

CHAPTER 9

Beer Styles: Ales and Lagers of the World

CHAPTER 9 LEARNING OBJECTIVES

After reading this chapter, the learner will be able to:

1. Identify what a beer style is and its importance to the beverage manager
2. Explain an ale category and identify several styles
3. Explain a lager category and identify several styles
4. Explain a beer cocktail and provide one example
5. Explain the concept of novelty beers

Nothing ever tasted better than a cold beer on a beautiful afternoon with nothing to look forward to than more of the same.

—Hugh Hood

Beer Styles

Learning Objective 1
Identify what a beer style is and its importance to the beverage manager

The beverage professional strives to learn the vast assortment of beer styles—beers that offer a dizzying display of color, an assortment of aroma and flavors combined with a range of mouthfeel characteristics. A "beer style" is the phrase used to identify a beer that describes its overall character and oftentimes its origin. A beer style has often been replicated over many decades or even centuries of brewing, trial and error, marketing, and consumer acceptance. At the same time, beer is an evolving beverage with modifications of its original styles made to adapt to an ever-changing and constantly evolving marketplace. While there are many types and styles of beer brewed throughout the world, the basics of brewing are shared across national and cultural boundaries. Figure 9.1 showcases a trio of beer styles.

Beer is a "sensory" science that incorporates an incredible sum of chemistry and biology not only into its conception but also into its vast world of stylistic variations. Each beer style is a collaboration composed of numerous decisions throughout its stages of production. One approach to making sense of the world around us—whether it's understanding music, art, or beer—is to find some means of broad categorization for recognition and communication. Many can relate to the idea of organizing taste in music; someone who may enjoy the sounds of the Pixies may also appreciate the sounds of Nirvana or Radiohead, all of which can be broadly categorized as alternative or garage rock. Not unlike in music, beer professionals can determine meaningful ways to communicate their favorite beer by

FIGURE 9.1
Selection of beer. Courtesy of Erika Cespedes.

FIGURE 9.2
Spiteful Brewing "The Hot Dog Meets the Bush" Saison beer. Courtesy of John Peter Laloganes.

referencing a style recognizable by others. The traditional European brewing regions—Germany, Belgium, the United Kingdom, Ireland, Poland, and the Czech Republic—have local classic styles of beer. Some countries, notably the United States, Canada, Japan, and Australia have embraced European styles and in the process allowed creativity to flourish and go beyond these classic styles to such an extent that they have effectively created their own indigenous types.

Beer styles presented in this chapter are ones that are, for the most part, widely available in bars and restaurants across the United States. The styles offer a well-defined description based primarily on an expected sensory analysis experience. The style guidelines offered are suggested, typical characteristics when one is presented with the given style. Keeping in mind the vast world of beer (and beverages in general) there is a lot of room for interpretation allowed by each individual producer and/or geographic location. Figure 9.2 pictures a creative beer that takes a classic Saison style and reinterprets it by collaborator brewers—Spiteful Brewing and Denali Brewing. The beer is called, *The Hot Dog Meets the Bush*, due to the intent of creating a beer with ingredients unique to each brewery's home. Spiteful Brewing based in Chicago used mustard seeds, due to the city's obsession with the Chicago hot dog. To complement that, Denali Brewing of Alaska, contributed Alaskan blueberries to blend for a refreshing Saison.

For a further, in-depth understanding into beer styles, the Beer Judging Certification Program (BJCP) is an organization that was founded in 1985 as a means of assessing and certifying beer judges. The BJCP offers a comprehensive book of style guidelines that not only acts as a study tool for beer judges, but is also used by brewers looking for some guidelines when crafting a beer or looking to enter a beer into a competition. There are many styles of beer available in today's market place; in the BJCP guidebook, they detail well over 100 styles of beers with additional dozen or so style guidelines for specialty types of beers.

Categories of Beer

In the beer family, there are two main branches or broad categories based on types of yeasts used and fermentation characteristics: Those beers that are produced by *top-fermenting yeasts* or ales, and those that are made with *bottom-fermenting yeasts* or lagers. Within these two major classifications of beer, there are several styles—and some styles even have derivatives or sub-styles. Each beer style contains specific personality traits that deem the beer unique and consequently distinctive. There are certain classical examples within each beer style—if a brewer specifically has the intention of reproducing a classical beer, then he or she is working toward an established style. In some cases, there is enough variation of a classic style, ultimately leading to the creation of a new style or sub-style of beer. Figure 9.3 identifies a chart that showcases how beer can be broken down from a category, to a style, and then a possible sub-style.

One category of beer that doesn't neatly fall under any traditional style is that of a *Session Beer*. This "style" of beer can be formulated as almost any beer style—but

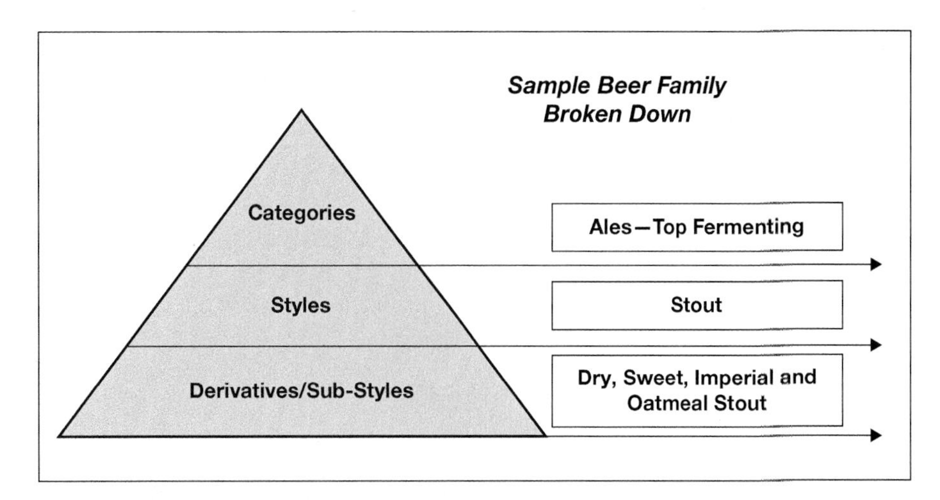

FIGURE 9.3
Beer family.

it's fashioned to be highly drinkable, containing no higher than 5 percent abv (alcohol by volume). The cutoff of abv level is a debatable one, but certainly these beers are designed to be an option for consumers who enjoy big flavor, without all the added alcohol content.

Session beers typically feature a balance between malt and hop characters and likely maintain a clean finish. The purpose of a session beer is to allow a beer drinker to have multiple beers, within a reasonable time period or session, without overwhelming the senses or reaching levels of intoxication. Some respected industry examples include:

- Founders Brewing – All Day IPA (Grand Rapids, Michigan)
- Evil Twin – Bikini Beer IPA (Brooklyn, New York)
- Bell's – Oarsman Berliner Weissbier (Kalamazoo, Michigan)
- Lagunitas Brewing – DayTime IPA (Petaluma, California)

Top-Fermented Beers

Learning Objective 2
Explain an ale category and identify several styles

Ale Style Beers

These beers often incorporate the term "ale" in their name (Pale Ale, Amber Ale, and so on) and, in addition, Porters, Stouts, Belgian specialty beers, wheat beers, and many German specialty beers fall within this category. The oldest English term for beer is *ale*; it is a generic term for English-style top-fermented beers. Ales are usually copper in color but sometimes darker, and they use a top-fermenting style yeast strain (*Saccharomyces cerevisiae*) that is generally fermented at warmer temperatures and relatively quickly in about 3–5 days. Ales convert less sugar into alcohol leaving a more noticeable body with some slight residual sugar after fermentation. At these relatively warmer fermentation temperatures, ales produce by-products called *esters* that are more evident in taste and aroma—which many regard as a distinctive character of ale beers. Ales generally have a more robust and complex aroma and flavor profile versus lager beers. They are best consumed at cool temperatures of 55–65°F rather than stereotypically frigid ones. Figure 9.4 illustrates an Amber Ale from New Belgium Brewery.

Some common styles of the beer that fall into the Ale category include: Abbey Ale, Altbier, Barley Wine, Brown Ale, Cream Ale, Golden/Blond Ale, Kölsch, Lambic,

FIGURE 9.4
New Belgium "Fat Tire" Amber Ale. Courtesy of Erika Cespedes.

Gueuze, Fruit Lambic, Pale Ale, India Pale Ale, Imperial Pale Ale, Porter, Saison, Scotch Ale, Dry Stout, Sweet Stout, Oatmeal Stout, Imperial Russian Stout, Trappist Ale, Dubbel, Trippel, Weizenbier, Hefeweizen, Dunkelweizen, Witbier.

1. **Abbey Ale** Abbey beers are made in commercial breweries around the world—intended to emulate the styles of the classic Trappist beers of Belgium. Sometimes, these beers are referred to as Belgian-styled beers that may or may not actually come from Belgium. These beers vary largely from producer to producer, but they often share an ample degree of aroma and flavor concentration with higher amounts of abv.
   - **Appearance** These beers can range dramatically in color variation, from deep yellow or gold through light amber to copper.
   - **Aroma/Flavor** These beers range dramatically in aromas and flavor, but are often found to have more of a malt emphasis.
   - **Structural Components** These beers range in structure, and therefore are quite dependent on the producer and the style they are trying to emulate.
   - **Industry Examples** Brewery Ommegang-Abbey Ale (Cooperstown, New York), The Lost Abbey Brewing Company-Lost and Found Abbey Ale (San Marcos, California), Goose Island Beer Co.-Pere Jacques (Chicago, Illinois), New Belgium Abbey-Belgium Style Dubbel (Fort Collins, Colorado).
2. **Altbier** (ullt-beer) "Alt" translates to "old" in German, and traditionally, Altbiers are conditioned for a longer than normal period of time. This style of ale was being made in Dusseldorf, Germany quite possibly 3,000 years ago. Altbier utilizes top-fermenting yeast (ale)—but is fermented at cold "lager beer" temperatures—and then aged for 1–2 months. This method was utilized by the Germans prior to the discovery of bottom-fermenting lager yeasts.
   - **Appearance** Altbier varies from a copper to reddish-brown color.
   - **Aroma/Flavor** Altbier contains medium-to-high malty flavor, often with biscuit, nutty, and toasty nuances. Hop aroma is typically low to moderate.
   - **Structural Components** Altbier has a medium-light to medium body that is fairly balanced in hops and barley, yet has a slight bitter flavor. The beer contains moderate to highly moderate carbonation and typically contains about 4 to 5 percent abv.
   - **Industry Examples** Rush River Brewing Company-Über Alt (River Falls, Wisconsin), Widmer Brothers Brewing Company-Okoto Festival (Portland, Oregon), Off-color Brewing-Scurry (Chicago, Illinois).

FIGURE 9.5
Anchor's "Old Foghorn" Barley Wine. Courtesy of Erika Cespedes.

3. **Barley Wine** Barley Wines despite their name are brewed from grain and not grapes. These beers are very strong and intense—usually the strongest ale offered by a brewery that can almost rival a wine in terms of alcohol content. Barley Wines typically range between 10 and 14 percent alcohol. Barley Wine is highly complex, often hoppy, intense in maltiness with a full, slightly viscous body, and are generally designed to age for an extended period of time. Pictured in Figure 9.5 is Old Foghorn Barley Wine from Anchor Brewing Company.

- **Appearance** Barley Wine has great depth in both color intensity (fairly deep) and color hue (ranges from light amber to copper). High alcohol and viscosity may be visible in the obvious "legs" or "tears" on the sides of the glass.
- **Aroma/Flavor** These beers offer strong and intense malt flavor with noticeable bitterness. Aromas are reminiscent of dried fruits with caramel and butterscotch. While this beer is strongly malted, it can also contain the occasional counterbalance of hoppy assertiveness.
- **Structural Components** These beers are full-bodied and chewy, with a velvety, luscious texture (although the body may decline with long conditioning). Barley Wine is, in general, lightly to moderately carbonated, with low head retention.
- **Industry Examples** Sierra Nevada Brewing Co.-Bigfoot Barley Wine (California), Rogue-Old Crustacean Barley Wine (Oregon), Bell's Brewery-Third Coast Old Ale (Michigan), Anchor Brewing-Old Foghorn Barley Wine (San Francisco), Three Floyds-Behemoth (Indiana), Flying Dog Brewery-Horn Dog Barley Wine (Maryland).

4. **Brown Ale** Brown Ales can range from sweet to bittersweet beers with undertones of malt. As its name suggests, this ale has a dark brown color and has become known as *Nut Brown Ale* by some producers. Some producers have included additions of nuts or coffee into the production process. Spawned from the traditional English Brown Ale, the American versions simply use American ingredients, yielding a more intense expression. Figure 9.6 identifies an American Nut Brown Ale.
   - **Appearance** This beer yields a dark reddish-brown to brown color.
   - **Aroma/Flavor** Brown Ales contain medium-to-high malty aromas and flavors with varying nuances of caramel, nutty, toasty, chocolate, licorice, and raisin. Hop aroma is typically low to moderate.
   - **Structural Components** Brown Ales tend to be fairly malty and slightly sweet and rich on the palate. The beer maintains a medium-high to full body with low hop bitterness.
   - **Industry Examples** Samuel Smith's-Nut Brown Ale (England), New Castle-Brown Ale (England), Dogfish Head Craft Brewery-Indian Brown Ale (Milton, Delaware), Goose Island Beer Co.-Hex Nut Brown Ale (Chicago, Illinois), Abita Beer-Pecan Harvest Ale (Louisiana).

FIGURE 9.6
AleSmith Nut Brown Ale. Courtesy of Erika Cespedes.

5. **Cream Ale** Cream Ales are clean and brilliant in appearance with subtle aromas and flavors and light-bodied—all of which are geared toward an inherent mainstream appeal to the consuming public. This style originated from the American light lager style where additions of adjuncts such as corn or rice are used to lighten the body. It is not uncommon for smaller craft brewers to brew all malt Cream Ales. In recent times, this beer style has diminished in availability as other styles have garnered greater popularity.
   - **Appearance** These beers are often pale straw to gold in color. Cream Ales maintain a low-to-medium head with medium-to-high carbonation. Head retention is low due to the extensive use of adjuncts.
   - **Aroma/Flavor** Cream Ales are low-to-medium aromatic. It is possible to detect a slight citrus element as well as corn-like aroma (due to the possible use of 20 percent adjuncts). The hop aromas range from low or even non-existent—yet, neither malt nor hops prevail in the taste.

FIGURE 9.7
Duvel Belgian Golden Ale. Courtesy of John Peter Laloganes.

- **Structural Components** The beer can vary from somewhat dry to faintly sweet. Cream Ales are generally light and crisp, with a smooth mouthfeel and relatively high carbonation levels.
- **Commercial Examples** New Glarus Brewing-Spotted Cow (Wisconsin), Sixpoint Brewery-Sweet Action (New York).

6. **Golden/Blond Ale** Golden Ales are effervescent beers that were originally created in Belgium to compete with Pilsner beer coming out of Eastern Europe. Traditionally, the beers are bottle-conditioned or simply refermented in the bottle. The most famous versions are the very strong ones like the deceptively drinkable classic *Duvel* (Doov'l) at 8.5 percent abv. References to the devil are included in the names of many commercial examples of this style, referring to their potent alcoholic strength and as a tribute to the original example, Duvel. Figure 9.7 pictures a bottle of Duvel.
   - **Appearance** Golden Ales use very pale malts that yield a straw yellow to golden yellow color shade. They offer good clarity with a long-lasting, white head.
   - **Aroma/Flavor** These beers are lightly to fairly aromatic with a moderate spiciness and low-to-moderate malt. Golden Ales have a low-to-moderate yet distinctive floral hop and fresh dough-like character.
   - **Structural Components** Golden Ales are light-to-medium bodied with substantial carbonation levels that lead to a dry, refreshing liveliness. Golden Ales have a low to moderately bitter aftertaste. Pictured in Figure 9.8 is Canada's *Unibroue* La Fin Du Monde, a triple fermented "Golden Ale."
   - **Industry Examples** Duvel (Belgium), Leffe (Belgium), La Chouffe (Belgium), Redhook-Blonde Ale (Portsmouth, New Hampshire), Flying Dog Brewery-Tire Bite Golden Ale (Maryland), Two Brothers Brewing Company-Prairie Path Golden Ale (Illinois), Unibroue La Fin Du Monde (Canada). Pictured in Figure 9.9 is La Chouffe, a Belgium unfiltered Blonde Ale, which is refermented in the bottle.

FIGURE 9.8
La Fin Du Monde. Courtesy of Erika Cespedes.

7. **Kölsch (Koollsh)** Kölsch is the local brew of the city of Cologne, "Köln" in German. This beer is one of the palest German beers made, similar to Britain's Pale Ale style. Many Cologne brewers ferment on the cooler side of typical ale temperatures; then the beer is "lagered" or stored between two and four weeks. Classically, this beer is served in a tall, narrow glass.
   - **Appearance** Kölsch can offer a very pale straw yellow-to-light golden color. Authentic versions are filtered to a brilliant clarity. These beers maintain a delicate white head with poor retention.
   - **Aroma/Flavor** This beer is often muted-to-lightly aromatic with very low malt aroma. Kölsch maintains subtle hop and mineral character with citrus and tree fruit notes.
   - **Structural Components** These beers are medium-light to medium in body with a smooth, clean, and crisp mouthfeel. Kölsch has medium to medium-low bitterness, dry, and medium acidity. The beer maintains moderate levels of carbonation.

FIGURE 9.9
La Chouffe. Courtesy of John Peter Laloganes.

- **Commercial Examples** Goose Island Beer Co.-Summertime (Chicago), Alaskan Brewing-Summer Ale (Alaska), Harpoon-Summer Beer (New England), New Holland Brewing-Lucid (Michigan), Shiner-Kölsch (Texas).

8. **Lambic Beers** Lambic is a very distinctive style of Belgian beer that relies on wild yeast for fermentation. Straight "unblended" Lambic beers are often a true representation of the "house character" of a brewery that will be used as the base beer for many of its derivative styles. Lambic beers often have more in common and reminiscent of white wine than many of the popular beer styles.

   Lambic beers often consist of 70 percent barley and 30 percent unmalted wheat as well as seasoned with a small dose of hops. The Lambic is exposed to the native airborne wild yeasts and bacteria that are present in the brewery—the beer then undergoes spontaneous fermentation. It is this process that gives the beer its distinctive sour—cider-like personality. After fermentation, the Lambic is siphoned into old wood barrels and left to mature for one to three years.

   - **Appearance** Lambic beers range depending upon age, from a pale yellow (in younger beers) to deep golden (in more mature beers) in color. Clarity may be slightly hazy with poor head retention.
   - **Aroma/Flavor** Young Lambics are noticeably sour, but aging can bring this character more in balance with the malt, wheat, and barnyard/earthy characteristics.
   - **Structural Components** Lambic beer is typically light to medium-light body with a dry, tart, and lightly bitter mouthfeel. Lambic beers are very lightly, if at all, carbonated.
   - **Industry Examples** Some of the few bottled versions readily available in the marketplace are Cantillon Grand Cru Bruocsella (Belgium) and Brouwerij Lindeman's Cuvée René Grand Cru (Belgium). It is possible to obtain greater options in one of the many bars in and around Belgium.

8A. **Gueuze or Geuze** (GOO-za) This is a type of Belgian Beer that is derived from the Lambic, which acts as the base for the many of its variations. Some Lambic beers are enhanced with fresh fruits such as raspberries or cherries; although the prestigious choice will be Gueuze, considered by many to be the noblest of Lambic styles.

Gueuze is produced through spontaneous fermentation and becomes naturally sour due to the wild microorganisms encouraged to infiltrate this beer during the production process. The art of making Gueuze is more about the blending process than a brewing one. Typically, Gueuze will be a blend of younger 1–2-year-old beer with an older 2–3 years old beer from wood barrels. This mixture is then bottled and prepared for secondary bottle fermentation. "Young" Lambic contains fermentable sugars while "old" Lambic has the characteristic complexity. Due to the beer's secondary bottle fermentation, Gueuze is sometimes referred to as "Brussels Champagne." Figure 9.10 pictures "A la Mort Subite," a Belgium beer bar that specializes in Gueuze styles.

FIGURE 9.10
A la Mort Subite. Courtesy of John Peter Laloganes.

- **Appearance** These beers are golden in color with excellent clarity. Gueuze maintains a thick, long-lasting, white head.
- **Aroma/Flavor** A moderately sour/acidic aroma balanced with wheat, malt, and barnyard/earthy characteristics with tree and citrus fruit with undertones of honey.
- **Structural Components** A traditional dry Gueuze has no fruit flavoring and will be tart, sour, and naturally effervescent. They are light to medium-light body. While some may be more dominantly sour and acidic, balance is the key and denotes a better quality Gueuze. Gueuze beers are often highly carbonated.
- **Industry Examples** Cantillon-Gueuze (Belgium) and Brouwerij Lindemans-Gueuze Cuvée René (Belgium).

8B. **Fruit Lambic** Fruit-based Lambic beers are often produced similar to Gueuze—through the addition of younger and older Lambic beers blended together. Fruit was traditionally added to Lambic or Gueuze to increase the variety of beers available in local bars and cafes. The choice of fruit is commonly added half-way through the fermentation process that allows the yeast to consume much of the fruit's sugar content.

FIGURE 9.11
Lindeman's Framboise Lambic. Courtesy of Erika Cespedes.

Fruit Lambic beers are *not* just—fruit beer—instead it is the most natural and high quality expression of beer that has been enhanced with fruit. The most traditional styles of fruit Lambic beers include:

- *Kriek* (**creek**) made with sour cherries
- *Framboise* (**frahm-bwahz**) made with raspberries
- *Pêche* (**pesh-eh**) made with peach
- *Pomme* (**pohm**) made with apple
- *Cassis* (**cah-sees**) made with black currants

Pictured in Figure 9.11 is Brouwerij Lindeman's Framboise Fruit Lambic.

- **Appearance** The variety of fruit generally determines the color, though lighter-colored fruit may have little effect on the color. Fruit Lambic often contain a thick head that showcases the shade of fruit used within production. Fruit Lambics are intended for early consumption as extended aging will dissipate their definable color hue.
- **Aroma/Flavor** The choice of fruit added to the beer will provide the dominant aroma and flavor. The beer will have undertones of sourness and slight barnyard/earthiness.
- **Structural Components** Fruit Lambics have light to medium-light body and range from having a moderate-to-high tart, and a common acidic sensation. Fruit Lambic can be dry, containing varying degrees of residual sugar. The beers are often highly carbonated to assist in providing a certain freshness and liveliness to counterbalance the fruit elements. Pictured in Figure 9.12 is Brouwerij Lindemans Framboise side label that identifies its production notes.
- **Industry Examples** Cantillon (Belgium), Mort Subite (Belgium), and Brouwerij Lindemans (Belgium).

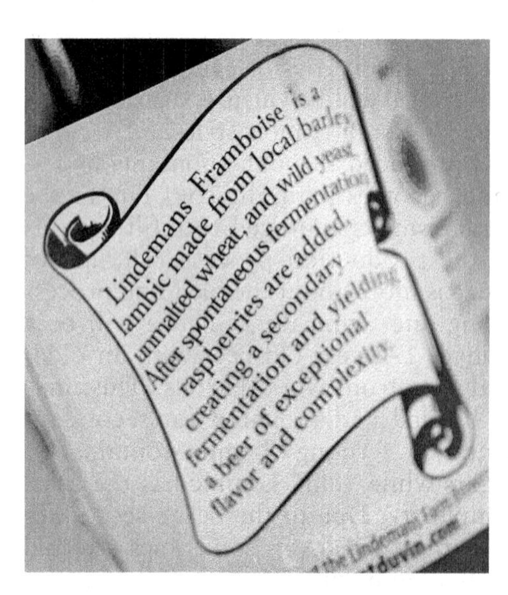

FIGURE 9.12
Ingredients in Framboise Lambic. Courtesy of Erika Cespedes.

9. **Pale Ale** Pale Ale, sometimes referred to as a "Bitter" in England, is originally an English-style beer with numerous

FIGURE 9.13
Half Acre "Daisy Cutter" Pale Ale. Courtesy of Erika Cespedes.

American adaptations, which reflects the use of indigenous ingredients (hops, malt, yeast, and water). Bitter is a very generic term use to describe a wide variety of ales. They can be lightly or highly hopped. An "Extra Special Bitter" may suggest a varying version that is slightly higher in alcohol and/or hop content. Pale Ale beers tend to be one of the more prolific styles brewed in modern day. Pictured in Figure 9.13 is Half Acre Beer Company's Daisy Cutter Pale Ale.

- **Appearance** Pale Ales can range from golden to deep amber in color. They are generally quite clear, although dry-hopped versions may be slightly hazy. Pale Ales maintain a moderately large white to off-white head.
- **Aroma/Flavor** This beer style contains moderate to strong hop aroma due to the application of the dry hopping technique. The prevalent aromas of citrus, grass, and pine identify hop's assertive use. The often low to moderate malt aromas and flavors such as bread, toast, and biscuit are often necessary to balance the presence of hops.
- **Structural Components** Pale Ales are medium-light to medium body with a moderate-to-high hop bitterness and dry finish. Overall, Pale Ales have a refreshing and smooth sensation with a slight lingering bitterness. The carbonation is typically moderate to high. Pictured in Figures 9.14 and 9.15 is Sierra Nevada's flagship Pale Ale beer.
- **Industry Examples** Sierra Nevada Brewing Co.-Pale Ale (California), Deschutes-Mirror Pond (Oregon), Three Floyds Brewing-X-Tra Pale Ale (Indiana), Half Acre-Daisy Cutter Pale Ale (Illinois).

9A. **India Pale Ale** India Pale Ales (or IPAs) can be thought of as a Pale Ale hyped on steroids—these beers are brewed in a manner that gains an increased gravity and hop dominance. They were originally an English beer that was intentionally given large doses of hops in order to survive their long voyage and temperature extremes in route from England to India. Along with Pale Ale, IPAs are extremely popular beer styles that are replicated by many breweries around the world. Figure 9.16 identifies a bottle of Hopothesis IPA.

FIGURE 9.14
Sierra Nevada Brewing Pale Ale. Courtesy of Erika Cespedes.

FIGURE 9.15
Sierra Nevada Pale Ale. Courtesy of Erika Cespedes.

FIGURE 9.16
Hopothesis India Pale Ale (IPA). Courtesy of John Peter Laloganes.

- **Appearance** IPAs can range from golden amber to light orange-copper in color. They should appear relatively clear, although unfiltered dry-hopped versions may be a bit hazy. IPAs maintain a strong head with off-white color.
- **Aroma/Flavor** This beer style contains moderate to strong hop aroma from dry hopping such as citrus, grassy, and pine hop character. A moderate to moderately-high hop aroma of floral, earthy, fruity, and/or slightly grassy tend to be slightly balanced from the moderate caramel-like or toasty malt presence. Despite the substantial hop character typical of these beers—sufficient malt flavor, body, and complexity help to provide balance.
- **Structural Components** IPAs have medium-light to medium body with moderate to medium-high carbonation. The heavy inclusion of hops can combine with the carbonation to render an overall dry bitter sensation. Figure 9.17 identifies a collection of IPAs, with Syndicate Ale's IPA pictured upfront.
- **Industry Examples** Fuller's-IPA (England), Summit Brewing Co.-India Pale Ale (Minnesota), Samuel Smith's-India Ale (England), Goose Island Beer Co.-IPA (Illinois), Bell's Brewery-Two-Hearted Ale (Michigan), The Great Divide-Titan India Pale Ale (Colorado), Three Floyds Brewing-Alpha King (Indiana), Sierra Nevada Brewing Co.-Celebration Ale (California), Dogfish Head-60 Minute IPA (Delaware), Anchor Brewing-Liberty Ale (California).

9B. **Imperial IPA** Imperial IPA is a recent innovation reflecting the trend of American craft brewers' "novelty beers" for increasingly extreme products. The adjective "imperial" simply implies a stronger version of an IPA; "double," "extra," "extreme," would be equally suitable.

- **Appearance** The color range for Imperial IPAs can vary from golden amber to medium orange-red copper. These beers are typically a bit cloudy or hazy due to often being unfiltered and dry-hopped. Imperial IPAs maintain a strong head stand with off-white color.
- **Aroma/Flavor** This beer style contains an intense hop aroma from dry hopping that contributes highly aromatic elements of citrus, grass, and pine character. Slight malt notes may be detected but certainly aren't apparent—these beers truly showcase the hops.
- **Structural Components** Imperial IPAs are medium-light to medium body and offer moderate to medium-high carbonation. These two elements combine to render an overall dry, lingering bitter sensation.
- **Industry Examples** Russian River-Pliny the Elder (California), Three Floyds Brewing-Dreadnaught (Indiana), Sierra Nevada Brewing Co.-Torpedo Extra Pale Ale (California), Bell's Brewing-Hop Slam (Michigan), Great Divide Brewing-Hercules Double IPA (Colorado), Dogfish Head-90-minute IPA (Delaware).

FIGURE 9.17
Collection of IPA. Courtesy of John Peter Laloganes.

FIGURE 9.18
Fuller's London Porter. Courtesy of Erika Cespedes.

FIGURE 9.19
Fuller's London Porter (2). Courtesy of Erika Cespedes.

10. **Porter** Porters are dark—substantially malty ales with a complexity of roasted aromas and flavors. This style of ale became popular in England, back in the 18th century. Porters and Stouts are rather broad styles that are somewhat similar and open to brewer interpretation. The history of both Porters and Stouts are intertwined—at some point in history they have become two distinct styles of beer. Porter may be distinguished from Stout because it often lacks a strong roasted barley character. Porters tend to provide more smokiness and dark fruit aromas and flavors as compared to Stout beer. Porters are brewed with dark malts and almost always aged for a period of time in oak barrels. Figures 9.18 and 9.19 showcase Fuller's London Porter.
    - **Appearance** Porter beer can range in color shade from medium-to-dark brown and deep-to-opaque in color intensity. This beer will retain off-white to light tan head.
    - **Aroma/Flavor** This beer is fairly to highly aromatic as it contains mild to moderate roasted quality as evident in its aromas and flavors of chocolate, caramel, nutty, and toffee elements. Porters may also contain secondary aromas and flavors such as coffee, licorice, or biscuits.
    - **Structural Components** Porters are medium-to-full body with a medium-low to medium hop bitterness that will serve to balance the somewhat sweet element from the roasted malt. These beers have moderate-to-moderately high levels of carbonation.
    - **Industry Examples** Fuller's London Porter (England), Samuel Smith-Taddy Porter (England), Anchor Brewing-Porter (California), Sierra Nevada Brewing-Porter (California), Deschutes Brewery-Black Butte Porter (Oregon), Rogue-Mocha Porter (Oregon).
11. **Saison** (seh-ZOHN) *Saison* is French for "season" named after the original intention of these beers being traditionally brewed in the winter for consumption throughout the summer season. These beers are also known as *farmhouse ales* since they originally were made in farmhouses in the French-speaking region of Belgium.
    - **Appearance** Saison beers are very pale straw to light golden in color. The beer will be slightly cloudy from starch haze and/or yeast sediment with a dense, white, and lingering head. Figure 9.20 identifies Brewery Ommegang's Farmhouse Saison Ale.
    - **Aroma/Flavor** These beers are often lightly aromatic yet highly complex with notes of wheat, citrus, orange, coriander elements with hints of herbal. Saison may have *Brettanomyces* characters of slight earth and barnyard-like. This style of beer is occasionally dry-hopped for more assertive floral and/or piney character.
    - **Structural Components** Saison beers are typically light-to-medium body. They are often bottle-conditioned with some yeast character and high carbonation. Hop bitterness is moderate to slightly assertive. This style contributes a slight sourness quality with medium-to-high alcohol content ranging from 4.5 to 9 percent abv.

FIGURE 9.20
Brewery Ommegang Farmhouse Ale. Courtesy of John Peter Laloganes.

- **Industry Examples** Brasserie Dupont-Saison Farmhouse Ale (Belgium), Goose Island Beer Co.-Sophie (Chicago, Illinois), Brewery Ommegang-Hennepin Farmhouse Saison (New York).

12. **Scotch Ale** Scotch (or Scottish) Ales originated in Scotland where they have been also known as *wee heavy*—an obvious reference towards the beer's potentially high alcohol content. This powerfully rich—highly malted—full-bodied ale has alcohol levels ranging between 7 and 10 percent abv.

    In the 19th century Scotland, a classification, based on the now out-of-date shilling currency, was devised in order to distinguish each one on the basis of the beer's alcohol level—light (60 schilling), heavy (70 schilling), and export (90 to 160 schilling).

    - **Appearance** Scottish Ales range from light copper to dark brown in color shade with a deep color intensity. The beer maintains a relatively dense tan head and evident "legs" or "tears" (in medium and high alcohol versions) on the side of the glass. Figure 9.21 showcases a Scotch Ale in a snifter glass, allowing one to slowly sip and savor the beer.
    - **Aroma/Flavor** These beers have a fairly to highly aromatic intense malt aroma and flavor with nut, smoke, peat, and earth notes. Scottish Ales traditionally undergo a lengthy boil in the kettle for caramelization of the wort. These ales may also have secondary notes of dried fruit such as plums and raisins. Overall hop character is low with light floral or herbal aromas, allowing the beer's signature malt aromas and flavors to be evident.
    - **Structural Components** Scottish Ales can range from medium-full to full-bodied. These beers maintain a high level of unfermentable sugars and in combination with its abundance of malt—they tend to create a thick, viscous mouthfeel. The slight presence of hops and alcohol both assist in balancing the sweetness from the malt. These beers maintain moderate levels of carbonation.

FIGURE 9.21
Scotch Ale. Courtesy of Erika Cespedes.

- **Industry Examples** McEwan's-Scotch Ale (United Kingdom-Scotland), Arcadia Brewing Company-Loch Down Scotch Ale (Michigan). Figure 9.22 pictures Arcadia's Loch Down Scotch Ale.

13. **Stout** Stout beer is a term used for very dark ales that utilize roasted barley for their intense character and dark color. These beers are produced in varying degrees of alcohol content, sweetness, and bitterness with aromas and flavors of unsweetened chocolate, coffee, and espresso.

13A. **Dry Stout** Dry Stouts (otherwise known as *Irish Stouts*) are dark, roasted, and bitter ales. The Dry Stout style evolved from attempts to capitalize on the success of London Porters, but instead would reflect a fuller, creamier, more "stout" body, and strength. When a brewery offered a Stout and a Porter, the Stout was always the stronger beer (it was originally called a "Stout Porter"). Dry Stouts, if particularly strong in alcohol, can be designated as a "Foreign Extra Stout." Figure 9.23 pictures a can of Guinness Draught Stout.

FIGURE 9.22
Arcadia "Loch Down" Scotch Ale. Courtesy of John Peter Laloganes.

FIGURE 9.23
Guinness Draught Stout. Courtesy of John Peter Laloganes.

- **Appearance** Dry Stout styles are dark brown to black in color shade and remain opaque in intensity. These beers maintain a thick, creamy, long-lasting, tan- to brown-colored head.
- **Aroma/Flavor** Stout beers are often fairly to highly aromatic with coffee, roasted barley, and chocolate/cocoa aromas and flavors.
- **Structural Components** Dry Stouts are medium-light to medium body—somewhat misleading due to the beer's intense color, aromas, and flavors. Due to the hops and roasted levels of barley, these beers have slight astringency and moderate levels of bitterness similar to bittersweet or unsweetened chocolate. The dryness apparent from this style of Stout derives from the addition of unmalted roasted barley. Dry Stout has low to moderate carbonation and low to moderate alcohol (4.2 percent) content.
- **Commercial Examples** Guinness-Draught Stout (Ireland), Murphy's-Stout (Ireland), Three Floyds Brewing-Black Sun Stout (Indiana), Goose Island Brewing Co.-Dublin Stout (Illinois).

13B. **Sweet Stout** Sweet Stouts (otherwise known as *Milk* or *Cream Stouts*) are very dark, sweet, full-bodied ales with modest roasted aromas and flavors. The use of the term "milk" or "cream" (neither of which are terms that are permitted in England) name is derived from the use of lactose, or milk sugar, as a sweetening agent. Since the beer is high in lactose—thus the name milk stout is used and can be compared to a sweetened cappuccino or caramel latte.

- **Appearance** Sweet Stouts range from dark brown to black in color shade and are opaque in color intensity. These beers maintain a creamy tan- to brown-colored head.
- **Aroma/Flavor** These beers are fairly to highly aromatic with mild roasted grain aromas and flavors of coffee, caramel, and/or chocolate notes as well as a cream, custard-like aroma and flavor from the presence of the milk sugar.
- **Structural Components** Sweet Stouts contain large amounts of residual dextrin and additions of unfermented lactose sugars. Both of these components give the beer a fuller, creamier body and a medium-to-high sweetness that counterbalances the bitter roasted character. Variations exist, with the level of residual sweetness and the intensity of the roasted character—the balancing of these two elements are the most variable components in this beer style. The hop bitterness is moderate, yet lower than in dry stout. This beer has low-to-moderate levels of carbonation with a typically 3–6 percent abv.
- **Industry Examples** Hitachino-Nest Sweet Stout "Lacto" (Japan), Samuel Adams-Cream Stout (Boston), Left Hand Brewing Company-Milk Stout (Longmont, Colorado).

13C. **Oatmeal Stout** Oatmeal Stout is very dark, full-bodied, roasted malt focused ale with complementary aromas and flavors of oatmeal. Approximately 5–10 percent of oats are

FIGURE 9.24
Samuel Smith's Oatmeal Stout. Courtesy of Erika Cespedes.

added to the roasted malt throughout the production of this beer to provide increased body and richness. Traditionally, this variation of Stout was an English seasonal style of sweet Stout that relied on oatmeal for body and complexity rather than lactose for body and sweetness. Figure 9.24 shows a bottle of Samuel Smith's Oatmeal Stout.

- **Appearance** Oatmeal Stout is dark brown to black in color shade with opaque intensity. This beer maintains a thick, creamy, and persistent tan- to brown-colored head.
- **Aroma/Flavor** This style of beer contains intense roasted grain aromas, often with a coffee-like character. Oats can contribute a nutty, grainy, or earthy element. Dark grains can combine with malt sweetness to give the impression of milk chocolate or coffee with cream. Light-to-medium hops with the balance toward malt.
- **Structural Components** These Stout beers offer medium-to-full body with a smooth and silky mouthfeel and medium to medium-high carbonation. Variations exist, but Oatmeal Stout can range from fairly sweet to quite dry. The level of bitterness also varies, as does the oatmeal impression. Light use of oatmeal may give a certain silkiness of body, while heavy use of oatmeal can be fairly intense with an almost oily mouthfeel.
- **Industry Examples** Samuel Smith-Oatmeal Stout (England), Young's-Oatmeal Stout (England), Goose Island Brewing Co.-Oatmeal Stout (Illinois).

13D. **Russian Imperial Stout** The Imperial Russian Stout is rich and complex, with variable amounts of roasted grains, malt, hops, and alcohol. This beer was first brewed in England during the 18th century. The Russian Imperial Stout was brewed to high gravity and hop level in England for export to the Baltic States and Russia. This beer gained "imperial" status because it was supposedly a favorite of Catherine the Great, the Empress of Russia in the 18th century. Imperial Russian Stouts tend to be rather high in alcohol (8–11 percent abv). Pictured in Figure 9.25 is the Yeti Imperial Stout from Great Divide Brewing Company.

FIGURE 9.25
Great Divide Brewing Co. Yeti Imperial Stout. Courtesy of Erika Cespedes.

- **Appearance** This beer ranges from very dark reddish-brown to black in color shade with opaque color intensity. Russian Imperial Stouts generally form a head that is tan to dark brown in color but low-to-moderate in retention. High alcohol and viscosity may be visible as "legs" or "tears" on the side of the glass.
- **Aroma/Flavor** This beer is fairly to highly aromatic with rich and intense elements. The roasted malt character contributes coffee, dark chocolate, or slightly burnt toasted notes. Secondary aromas and flavors consist of dark fruit (plums, prunes, and raisins) elements. Hop aroma can vary from very low to quite assertive. Aged versions of Russian Imperial Stout tend to be slight vinous, or wine-like such as Port.

FIGURE 9.26
Orval Trappist Ale. Courtesy of John Peter Laloganes.

FIGURE 9.27
Rochefort 10 Trappist Ale. Courtesy of John Peter Laloganes.

- **Structural Components** The Russian Imperial Stout is full-bodied and intense, although the body may decline with long bottle aging. Carbonation may be low to moderate as it declines with age. Alcohol strength will be evident, but should be balanced and not hot or spicy. The beer will range from relatively dry to moderately sweet, usually with some lingering finish of roasted qualities and bitterness.
- **Commercial Examples** Three Floyds Brewing-Dark Lord (Indiana), Bell's Brewing-Expedition Stout (Michigan), North Coast Brewing-Old Rasputin Imperial Stout (California), Samuel Smith-Imperial Stout (England), Deschutes Brewery-The Abyss (Oregon), Rogue-Imperial Stout (Oregon), Great Divide Brewing-Yeti Imperial Stout (Colorado).

14. **Trappist Ales** Trappist Ales can only be legally applied to a brewery operated by the Trappist monks. Trappist Ales are strong beers with a high amount of alcohol (7–10 percent) and many are bottle-conditioned. This style of beer originated and is still currently made in Belgium and Dutch Monasteries. The Trappist order originated in the Cistercian monastery of *La Trappe*, located in Normandy, France. In the mid-17th century—the Abbot of La Trappe felt that the Cistercians were becoming too liberal and in response—introduced reform in the abbey and the Order of Cistercians of the Strict Observance began. Pictured in Figure 9.26 is an Orval Trappist Ale from Belgium.

    Breweries were only introduced as a means of providing sustenance for the monks (and community as a whole) during lent and eventually became a means of income for the monasteries. One of the fundamental tenets is that monasteries should be self-supporting—beermaking, cheesemaking, or some other means of generating income was necessary. Trappist beers are brewed under the approval and control of the Trappist monks. The International Trappist Association (ITA) was created to ensure the Trappist name wasn't being abused. To bear the official seal awarded by the ITA, the monasteries must follow certain regulations: the beer must be brewed within the walls of the monastery and under control of the monks, the existence of the brewery must be directed towards sustaining the monastery and not towards financial profitability, it must be secondary to the monastic way of life, and it is to be constantly monitored for "irreproachable quality." Pictured in Figure 9.27 is Rochefort 10 Trappist Ale.

There are eleven Trappist breweries in the world, six in Belgium. They are:

1. *Achel* (ay-shell) – Belgium
2. *Chimay* (shee-may) – Belgium
3. *Stift Engelszell* (stif-ayn-gehl-zel) – Austria
4. *La Trappe* (lah-trahp) – Netherlands
5. *Orval* (ohr-vaehl) – Belgium
6. *Rochefort* (rowsh-fehr) – Belgium
7. *Spencer* – United States

8. *Tre-Fontane* (tray-fohn-tane) – Italy
9. *Westmalle* (west-mahl) – Belgium
10. *Westvleteren* (west-vlee-tehr) – Belgium
11. *Zundert* (zoon-dehr) – Netherlands

These beers tend to be complex and spicy with a high amount of alcohol (about 7 to 10 percent). There are several spin-offs of the Trappist Ale—the two most famous examples are the dubbel and tripel in reference to their alcohol content being two or three times the amount of a standard beer. Many of the Trappist Ales are so stylistically varied. The Trappist style and its derivatives (dubbel and tripel) have also been popularized by many craft brewers.

14A. **Dubbel** The dubbel (or double) is a Belgian Trappist beer that originated at monasteries in the Middle Ages. The "double" reference is in regard to the alcohol content that typically represents twice the standard to equate between 7 and 8 percent abv. These beers are traditionally bottle-conditioned (refermented in the bottle) and express a deep copper-red color with a moderately strong, malty, complex aroma, and flavor. While this style originates from the Trappists, breweries around the world are allowed to replicate the dubbel style.

- **Appearance** Dubbel beers showcase dark amber to copper color, with an attractive reddish hue. The creamy off-white head is large, dense, and long lasting.
- **Aroma/Flavor** These beers are incredibly complex—showcasing rich malty sweetness that may provide hints of chocolate, caramel, and/or toast as well as raisins, plums, dried cherries, and cloves.
- **Structural Components** Dubbel beers are medium-full body with medium-high carbonation. The beers balance edges toward the malt with medium-low bitterness.
- **Industry Examples** Westmalle-Dubbel (Belgium), La Trappe-Dubbel (Belgium), Chimay-Premiere Red (Belgium), Lost Abbey-Lost and Found Abbey Ale (California), and Allagash-Double (New England).

Pictured in Figure 9.28 is a bottle of Westmalle Dubbel Trappist Ale.

FIGURE 9.28
Westmalle Dubbel Trappist Ale. Courtesy of John Peter Laloganes.

14B. **Tripel** The Tripel or Triple is a Belgian Trappist beer that originated at monasteries in the Middle Ages. The "triple" reference is in regard to the alcohol content that typically represents three times the standard—to an equivalent of 8–10 percent abv. These beers are traditionally bottle-conditioned (refermented in the bottle) and express some serious complexity. Triple is more potent than the double though its delicate body and somewhat more pale color suggest the contrary. Similar to the dubbel, this style originates from the Trappist beers, however, has been replicated in high-quality versions around the world.

- **Appearance** Tripels are deep yellow to deep gold in color and contain ample effervescence with a long-lasting, creamy head.
- **Aroma/Flavor** These beers are complex with moderate to significant spiciness, light malt character, and low alcohol yet fairly to highly aromatic of hops. Alcohols are soft, spicy, and low in intensity.

FIGURE 9.29
Westmalle Tripel Trappist Ale. Courtesy of John Peter Laloganes.

- **Structural Components** Tripel beers are medium-light to medium body, although lighter than the substantial gravity would suggest (thanks to sugar and high carbonation). High alcohol content adds a pleasant creaminess. The beer edges toward medium-to-high bitterness (typically at least 30 IBUs) and is very dry, with substantial carbonation leading to a dry finish. The Tripel beer strongly resembles a strong golden ale but slightly darker and somewhat fuller-bodied.
- **Industry Examples** Westmalle-Tripel (Belgium), Chimay-Cinq Cents (White) (Belgium), Unibroue-La Fin du Monde (Canada), Allagash Brewing Company-Tripel Reserve (New England).

Pictured in Figure 9.29 is a bottle of Westmalle Tripel Trappist Ale.

15. **Weizen/Weissbier** (vice-beer) There exist two primary types of Weisse beer: Hefeweizen (light, unfiltered wheat beer) and Dunkelweizen (dark, unfiltered wheat beer).

15A. **Dunkelweizen** (doonn-kel vite-sen) Dunkelweizen is an older and darker-styled Bavarian wheat beer. This beer grew out of favor for the preferred lighter and fresher hefeweizen. This moderately dark, spicy, fruity, malty, refreshing wheat-based ale reflects the best yeast and wheat characteristics of a hefeweizen blended with malty caramel richness. By German law, at least 50 percent of the grist must be malted wheat, although some versions use up to 70 percent.

Since the beer is commonly unfiltered, the yeast is usually decanted into the glass. The proper way to pour wheat beer is to leave 10 percent in the bottle to swirl and dissolve the sediment (yeast) before pouring the remainder into the glass.

- **Appearance** This beer style contains a light copper to mahogany brown color. A very thick, mousse with a long-lasting, off-white head is characteristic of Dunkelweizen. The high protein content of wheat impairs clarity in this traditionally unfiltered style, although the level of haze is somewhat variable. The suspended yeast sediment, which should be roused before drinking, also contributes to the cloudiness.
- **Aroma/Flavor** Weisse beers are fairly to highly aromatic. They have a refreshing, lightly hopped aroma with noticeable banana, citrus, clove, vanilla, bubblegum, bread, and yeast character. The presence of darker-type barley malts gives this style a deep, rich barley malt character. A light to moderate wheat aroma (which might be perceived as bready or grainy) may be present and is often accompanied by a caramel, bread crust, or richer malt aroma.
- **Structural Components** Dunkelweizen offers a medium to medium-plus body. The texture of wheat as well as yeast in suspension imparts the sensation of a fluffy, creamy fullness. The malty presence provides an additional sense of richness and fullness. The malt can be low to medium-high, but shouldn't overpower the yeast character. This beer is moderate in effervescence.
- **Industry Examples** Weihenstephaner-Hefeweissbier Dunkel (Germany), Hacker-Pschorr-Weisse Dark (Germany).

15B. **Hefeweizen** (hay-fuh-veit-senn) This beer is pale and unfiltered wheat-based refreshing ale. This beer has origins in southern Germany and remains one of the most famous summer beers around the world. *Weizen* means "wheat" in German: its name is indicative of the beer's typical large portion of malted wheat used during the brewing process. By German law, at least 50 percent of the grist must be malted wheat, although some versions use up to 70 percent.

Since the beer is commonly unfiltered, the yeast is usually decanted into the glass. The proper way to pour wheat beer is to leave 10 percent in the bottle to swirl and dissolve the sediment (yeast) before pouring the remainder into the glass.

- **Appearance** These beers are pale straw to dark gold in color. *Weisse* means "white" in German which is representative in these beers as pale to golden color with predominately thick and long-lasting, white head. The white and cloudy appearance is derived from a combination of several aspects: (1) pale wheat malt used in production yields a pale to golden color, (2) the high protein content of wheat impairs clarity in an unfiltered beer, and (3) the remaining suspended yeast sediment that remains in the unfiltered beer. Figure 9.30 pictures a tall glass of Ayinger Bräu-Weisse, a German Hefeweizen.
- **Aroma/Flavor** Weisse beers are fairly to highly aromatic. They have a refreshing, lightly hopped aroma with noticeable banana, citrus, clove, vanilla, bubblegum, bread, and yeast character. These beers often don't age well and are best enjoyed while young and fresh.
- **Structural Components** These beers have a medium-light to medium body. The quantity of suspended yeast may increase the perception of body. The texture of wheat imparts the sensation of a fluffy, creamy fullness with a tart, citrusy character. Weisse beer contains high levels of carbonation with a dry finish.
- **Industry Examples** Paulaner-Hefe-Weizen (Germany), Ayinger Bräu-Weisse (Grmany), Hacker-Pschorr Weisse (Germany), Sierra Nevada Brewing Co.-Kellerweis Hefeweizen (California).

15C. **Witbier** *Witbier* (sometimes referred to as white beer) is a refreshing, moderate-strength wheat-based ale deriving from Belgium about 500 years ago. This style is usually made from equal portions of unmalted wheat and malted pale barley, spiced with ground coriander seeds, orange peels, and other spices and herbs. Figure 9.31 shows a classic Wit Ale from Wittekerke.

FIGURE 9.30
Ayinger Bräu-Weisse. Courtesy of Erika Cespedes.

FIGURE 9.31
Wittekerke Wit Ale. Courtesy of Erika Cespedes.

- **Appearance** These beers illustrate a very pale straw to light gold in color. The designation "white" beer may refer to the pale head formed during fermentation, or to the fact that these beers are often unfiltered, and therefore hazy that gives it a whitish-yellow appearance. These beers are dense with a white head and good retention.
- **Aroma/Flavor** Lightly to fairly aromatic, this beer has a pleasant fruitiness often with a honey and/or vanilla character and zesty ginger, orange, coriander, and lemon notes.
- **Structural Components** Witbier is medium-light to medium body, often having a smoothness and light creaminess from the unmalted wheat. The beer offers low to medium-low bitterness, medium acidity with slight sourness. Witbier finishes dry and often a bit tart. It is fairly to highly carbonated. The alcohol is moderate at 4.5–5.5 percent abv.
- **Commercial Examples** Hoegaarden-Wit (Belgium), Wittekerke (Belgium), Allagash Brewing Company-White (New England), Brewery Ommegang-Witte (New York), Hitachino Nest-White Ale (Japan).

Bottom-Fermented Beers

Learning Objective 3
Explain a lager category and identify several styles

Lager Style Beers

These beers were originally found largely in Germany and Eastern Europe (more precisely in the Czech Republic and Poland). Lager category of beers utilizes "bottom-fermenting" yeast strain (*Saccharomyces uvarum* or *carlsbergensis*) that means they ferment toward the bottom of the fermenting vessel. Lagers typically undergo a lengthy 7–15 days primary fermentation period at cold temperatures. Lagers style beers are typically given a long secondary fermentation (ranging from 32 to 39°F) throughout what is referred as the *lagering phase*.

The name "lager" derives from the German *lagern* meaning *to store*. Classically, during this second stage of lagering—the brewers in the Bavarian region of Germany would store their beer in cool cellars and caves during the warm summer months. During this period, the brewers recognized that these beers needed to continue their fermentation and in the process, the slow cool fermentation prohibited the natural production of by-products. Through this lagering period, the beers also began to clarify themselves. This cooler environment inhibits the natural production of esters, creating a crisper neutral tasting product. When stored in cool conditions, the lagering phase ultimately produces a "cleaner and lighter" style of beer as broadly compared to the ale category of beers.

Lagers are best served cold between 35 and 45°F. These beers were introduced to the United States in the 1840s by the German immigrants who originally developed them back in the 7th century. Today, lager style beers consist of more than 90 percent of the beer produced and consumed in the United States. Unfortunately, many of the large-scale American breweries produce infamous diluted versions of the German classics.

Some common styles of the beer that fall into this category include: American Lager, Bock Beer, Eisbock, Maibock, Doppelbock, Dortmunder, Lager, Light Beer, Märzen, Pilsner, and Rauchbier.

16. **American Lager** The standard lager is a mass-market beer that is produced in most countries. They are intended to be well-chilled in order to accentuate their refreshing and thirst-quenching effects—very similar to ice cold water. This highly popular style is the typical preference for non-craft beer consumers. Often these beers are branded as Pilsner styles, but should not be confused with traditional German and Czech styles. Additionally, there are alternative lager style beers called Ambers which contain slightly more concentrated aromas, flavors, and mouthfeel than typical lager versions.

- **Appearance** These beers appear very pale straw in color shade and are highly carbonated. They are brewed with high percentage (up to 40 percent) of rice or corn as adjuncts in order to lighten the style. These beers are filtered for the visual appeal of clarity and to showcase their bubbles.
- **Aroma/Flavor** Lager beers are generally muted-to-lightly aromatic with subtle malt aroma and flavor—essentially a neutral profile. The hop aroma and flavor may range from none to a light, spicy, or floral presence and some noticeable light levels of yeast character. Occasionally, it's possible to pick up a "cooked corn" or grainy characteristic; Amber versions yield more caramel and slight nutty characteristics.
- **Structural Components** The structural components of lager consist of a light body—made possible from addition of a high percentage of adjuncts such as rice or corn. The standard lager is highly carbonated that assists in providing some freshness and a perception of liveliness. Amber lagers provide greater mouthfeel with a persistent finish.
- **Industry Examples** Pabst Blue Ribbon, Miller High Life, Budweiser, Molson Golden, Labatt Blue, Coors Original, Foster's Lager, Flying Dog Amber Lager (Maryland).

16A. **American Light Lager** The standard American light lager is designed to appeal to the mass-market, especially among non-craft beer drinkers. They are intended to be well-chilled in order to accentuate their refreshing and thirst-quenching effects. Light beer has a lighter taste and is lower in calories than "traditional beer." This beer can be made by two methods:

  - By adding enzymes that lower the calories and alcohol content of the beer
  - By diluting regular beer that was fermented dry in order to obtain the desired abv

- **Aroma/Flavor** This beer style offers a nearly flavorless profile. It typically contains a low to no malt and hop aroma, although it can be perceived as grainy, or corn-like with a subtle amount of yeast and apple characteristics.
- **Structural Components** These are very light (sometimes watery) in body, very highly carbonated with a slight carbonic bite on the tongue.
- **Industry Examples** Bud Light, Coors Light, Keystone Light, Michelob Light, Miller Lite.

17. **Bock Beer** Bock Beer originated in the Northern German city of Einbeck, and was recreated in Munich starting in the 17th century. The name "bock" or "billy-goat" in German is often used in logos and advertisements. Bock Beers are a dark, strong, malty lager beer with alcohol levels between 6.5 and 10 percent. Figure 9.32 pictures a glass of Anchor Brewing's American Bock Beer.

- **Appearance** Bock Beers are light copper-to-brown in color shade with good clarity despite the dark color. It maintains a large, creamy, persistent, off-white head.
- **Aroma/Flavor** These beers have a strong malt aroma with subtle toasty and caramel characteristics. Bock Beer has virtually no hop aroma. Some alcohol may be noticeable.
- **Structural Components** Bock Beers are medium to medium-full bodied with moderate to moderately low carbonation. Some alcohol warmth may be detected.

FIGURE 9.32
Anchor Brewing American Bock beer. Courtesy of John Peter Laloganes.

17A. **Eisbock** (ice bock) Eisbock is otherwise known as an *ice beer*. The beers gain their potent strength from the unique production process of being frozen near the end of their maturation period. Because water freezes before alcohol, the chilled brew can be drained off the ice crystals that form in the tank. During this process, the beer loses about 7–10 percent of its water content. As a result, the alcohol concentration in the beer increases, usually to about 10 percent abv, about twice as much as the 4.5–5.5 percent of a regular German lager.

- **Appearance** These beers range from deep copper-to-dark brown in color. The lagering phase of production provides good clarity. Head retention may be impaired by higher-than-average alcohol content and low levels of carbonation. Eisbock maintain an off-white to deep ivory colored head.
- **Aroma/Flavor** Eisbock beers are dominated by a balance of rich, intense malt aromas and flavors with slight toast and caramel qualities. The presence of alcohol is evident, but not overly sharp.
- **Structural Components** These beers tend to be full-bodied with significant warmth deriving from the alcohol content. Eisbock beers have low levels of carbonation. Hop bitterness just offsets the malt sweetness enough to avoid a cloying character.

17B. **Maibock** (my-BOCK) Maibocks (or *Helles* "hell-ess" Bock) are a Bavarian favorite being produced in anticipation of springtime and the month of "May." These beers are in essence a pale version of a traditional bock. While quite malty, this beer typically has less dark and rich malt compared to the traditional bock. Maibocks tend to be more dry, hoppy, and bitter than a traditional bock.

- **Appearance** The color shade of Maibock beer is straw yellow to golden yellow with a watery to pale color intensity. The lagering process provides excellent clarity. Maibock beers are known to sustain a large, creamy, white head.
- **Aroma/Flavor** Maibock beers are lightly to fairly aromatic. They contain moderate to strong malt aroma, often with a lightly toasted quality. Moderately low hop aroma, often with a slight spicy quality.
- **Structural Components** These beers are medium-bodied with moderate to moderately high carbonation levels. They are smooth and clean with no harshness or astringency, despite the increased hop bitterness.
- **Commercial Examples** Ayinger-Maibock (Germany), Hacker-Pschorr-Hubertus Bock (Germany), Summit Brewing Company-Maibock (Minnesota).

17C. **Doppelbock** (duh-ppel) Doppelbock (literally "double bock") is a stronger and usually darker version of the traditional Bavarian Bockbier. It is exceptionally malty, with very little bitterness. Standard Doppelbocks may have as much as 7 percent abv yet Doppelbocks can hover near 10–13 percent abv. Many Doppelbocks have names ending in "-ator," often as a tribute to the classical Paulaner Salvator.

- **Appearance** Doppelbock beers are deep gold-to-dark brown in color shade. The lagering phase should provide good clarity. These beers maintain a large, creamy, persistent head with varying color from white to off-white.
- **Aroma/Flavor** Doppelbock beers are very rich and malty with some slight caramel, chocolate, and toast aroma, and flavor with hints of prune or plum. There is virtually no hop aroma. Moderate alcohol aroma may be present.
- **Structural Components** These beers are often medium-to-full body. They are smooth, without harshness, and maintain moderate to moderately low

carbonation levels. Hop bitterness varies from moderate to moderately low but always allows malt to dominate. Many Doppelbock versions are fairly sweet, but decline in maturation.

- **Industry Examples** Paulaner-Salvator (Germany), Ayinger-Celebrator (Germany), Spaten-Optimator (Germany), Samuel Adams-Double Bock (Boston).

18. **Dortmunder** (dort-moon-dehr) Dortmunder is a traditional German style indigenous to the Dortmund industrial region of Germany. It is brewed to contain a slightly fuller body than other light lagers, providing a firm malty body to complement the hop bitterness. Balance and smoothness are the hallmarks of this style. It has the malt characteristics of a *Maibock* and the hop character of a *Pils*, yet Dortmunder is slightly stronger than both with an alcohol level at about 5.5 percent abv.
    - **Appearance** Dortmunder beers are clear with a light gold-to-deep golden color. They maintain a persistent white head.
    - **Aroma/Flavor** This beer style is lightly to fairly aromatic with low-to-medium in malt and hop aromas and flavors with some slight mineral characteristics in the background.
    - **Structural Components** Dortmunder has a medium body with moderate carbonation levels that provide a smooth yet crisply refreshing beer. The hop bitterness lingers slightly in aftertaste.
    - **Industry Examples** DAB-Dortmunder Kronen (Germany), Ayinger-Jahrhundert (Germany), Great Lakes Brewing-Dortmunder Gold (Ohio), Bell's Brewery-Lager (Michigan), Old Dominion Brewing-Lager (Delaware).
19. **Märzen** (Mair-tsen) Otherwise known as the famous German Oktoberfestbier or Märzen-Oktoberfestbier. *Märzen* is German for "March," indicative of the traditional month in which the beer was brewed. In the Middle-Ages, brewers had a difficult time brewing quality beer during the summer months because the beers would easily spoil with bacteria. To have an ample supply of drinkable beer during the summer months, brewers would create an extra malty and hopped beer in March that would be preserved throughout the summer in cool caves until fall time and the new brewing season would begin. Most modern Märzen beers are aged for several months for greater depth and complexity.
    - **Appearance** This beer can range quite extreme from dark golden to deep orange-red to copper in color.
    - **Aroma/Flavor** Märzen beers are fairly aromatic with a light hop aroma and a moderate to moderate-high toasted malt aroma with a subtle toasty character.
    - **Structural Components** They often offer a medium body, smooth, medium acidity, and a slight to moderate bitterness. The beer has a creamy texture and medium carbonation. "Fest" type beers are special occasion beers that are usually stronger than their everyday counterparts.
    - **Industry Examples** Paulaner-Oktoberfest (Germany), Hacker Pschorr-Original Oktoberfest (Germany), Hofbräu-Oktoberfest (Germany), Summit Brewing Company-Oktoberfest Märzen Style (Minnesota), Goose Island-Oktoberfest (Illinois), Samuel Adams-Oktoberfest (Boston).
20. **Pilsner** Pilsner or "Pils" was originally produced in Bohemia (now the Czech Republic) in 1842. The same brewery still exists and produces *Pilsner Urquell* (uhr-KVEL). The Pilsner style of beer has become quite possibly—one of the most popular styles of beer in the world. Pilsner beers are pale, dry, and crisp with a bitterness that tends to linger. They are light in body and color with

high level of carbonation. Figure 9.33 features a trio of canned Pilsners from the original Pilsner Urquell.

- **Appearance** Pilsner beers are straw yellow to light gold in color shade. They are very clear and transparent in color intensity. Pilsners maintain a creamy, long-lasting, white head.
- **Aroma/Flavor** The Pilsner is typically muted-to-lightly aromatic with an expression of citrus, pine, spice, and slight bready component. It contains moderate levels of hops to provide some citrus, pine, and hay aromas and flavors.
- **Structural Components** Pilsner beers are dry and crisp—clean and refreshing. They prominently feature the hop's bitterness. Pilsners are medium-light body and medium to high in carbonation levels. Generally, Pilsner beers contain about 5 percent abv.
- **Industry Examples** Pilsner Urquell (Czech Republic), Warsteiner (Germany), Heineken (Netherlands), Goose Island Beer Co.-Pils (Chicago).

21. **Rauchbier** (r-ow-x-beer) Rauchbier is an incredibly unique and rare style of beer—identified through its seriously intense—beechwood smoked malt. These traditional beers of Germany are called Rauchbier, where *rauch* means "smoke." Most Rauchbiers are brewed with a bit more hops in order to counterbalance the otherwise assertive smokiness.
    - **Appearance** These beers vary in color—though most appear amber-to-brown in color with a large, creamy, rich, tan- to cream-colored head.
    - **Aroma/Flavor** The concentration and character of the smoke imparted by the use of smoked malts may vary from low to assertive. The different sources used to smoke malt are reflective in their unique aroma and flavor characteristics. Some woods used—beechwood, for example, which is most common in Rauchbier; peat, or other hardwoods like oak, maple, mesquite, alder, pecan, apple, or cherry are examples of smoked malts. The various woods may be reminiscent of certain smoked products due to their food association (e.g., hickory with ribs, maple with bacon or sausage, and alder with salmon). Smoky flavors may range from woody to somewhat bacon-like depending on the type of malts used. Figure 9.34 pictures a bottle of Rauchbeer made in the Marzen style.
    - **Structural Components** The mouthfeel varies with this beer style. The beer is dry, ranging from medium to full-bodied with some moderate

FIGURE 9.33
Pilsner Urquell. Courtesy of Erika Cespedes.

FIGURE 9.34
Rauchbeer (Bier). Courtesy of Erika Cespedes.

bitterness deriving from hops and medium to medium-high levels of carbonation.

- **Industry Examples** Alaskan Brewing Company-Smoked Porter (Alaska), O'Fallon-Smoked Porter (Missouri), Schlenkerla-Weizen Rauchbier (Germany), Rogue-Smoke (Oregon), Left Hand-Smoke Jumper (Colorado).

22. **Vienna Style** Named after the city of Vienna, traditionally this style of lager is brewed using a process known as decoction. The decoction process extends the boiling process of the wort to allow it to contain caramelization aromatic flavor components.

    - **Appearance** This beer can range from a medium red-brown to copper in color.
    - **Aroma/Flavor** This beer offers a predominant malt aroma that is rich in "bread" characteristics that is somewhat toasty with subtle influence of hops.
    - **Structural Components** Moderate-strength lager with a medium body that is soft and smooth with moderate bitterness, yet finishing relatively dry with slight malt sweetness.
    - **Industry Examples** Boston Beer Company (Sam Adams)-Boston Lager (Boston, Massachusetts), Great Lakes Brewing Company-Eliot Ness Vienna Lager (Cleveland, Ohio).

Beer Drinks/Cocktails

Learning Objective 4
Explain a beer cocktail and provide one example

The differing densities of liquids in a beer cocktail cause them to remain largely in separate layers, just as in a *pousse-café* (poos-cah-fay)—the spirit-based, layered cocktail. The effect is best achieved by pouring over a spoon turned upside down over the top of the glass so that the liquid runs gently down the sides rather than splashing into the lower layer and mixing with the other beverage.

- *Black and Tan* is a drink made from Bass Ale with a layer of Guinness Stout over the upper half. To prepare a Black and Tan, fill a pint glass halfway with Pale Ale then add the Stout. The top layer is best poured slowly over an upside-down tablespoon placed over the glass to avoid mixing the layers. There is a specially designed black-and-tan spoon, bent in the middle so that it can sit on the edge of the pint glass for easier pouring.
- *Half and Half* or *Black and White* is a beer drink made from a Pale Ale such as Harp Lager with a layer of Guinness Stout over the upper half.
- *Black and Blue* or, dark side of the moon, is Blue Moon (Belgian-styled white beer) with a layer of Guinness Stout over the upper half.
- *Black Velvet* is a beer cocktail made from Guinness Stout or any other Stout beer with a sparkling wine, traditionally Champagne, poured slowly over the upper half of the glass. It is traditionally served in a Champagne Flute.

Novelty Beers

Learning Objective 5
Explain the concept of novelty beers

The intention of novelty beers (or as the industry uses the phrase *extreme beer*) is to appease a small group of the consuming marketplace, yet garner a larger amount of interest and contention along the way with the broader mainstream public. The novelty beers are something original and interesting or exciting, though often for only a limited time. The range of novelty can be quite mild and/or extreme—in some examples even shocking.

Industry Example of a Mild Novelty Beer

Moody Tongue Brewery (from Chicago) created the Shaved Black Truffle Pilsner, a one-time limited (500 bottles) German Pilsner brewed with premium Australian black truffles. The beer was released on November 14, 2014 through an online lottery and limited to one beer per person at $120 per 22-ounce bottle. Figures 9.35 and 9.36 feature Moody Tongue's Black Truffle Pilsner.

Another well-known Chicago producer, Goose Island Beer Co. released *Bourbon County Coffee Stout* in March 2010. The beer was a collaboration/partnership with *Goose Island Beer Co.* and the local coffee roaster *Intelligensia Coffee and Tea Company*. This beer contained high gravity, which contains three times the amount of grain than their flagship beer, was aged for one hundred days in Bourbon barrels, and then infused it with Intelligensia's black cat espresso. Pictured in Figure 9.37 is Goose Island Bourbon County Coffee Stout.

Industry Example of an Extreme Novelty Beer

Scottish brewery *Brew Dog* created a Belgian-style ale, *The End of History* with an alcohol content of 55 percent. They accomplished this high alcohol content through the Eisbock method of freezing the beer and gradually removing the ice crystals thereby extracting water content. Directly from their website, Brew Dog is "a beacon of nonconformity in an increasingly monotone corporate desert." They state, "We are proud to be an intrepid David in a desperate ocean of insipid Goliaths." The same brewery subsequently created an even more shocking beer. Released in July 2010, they produced a special bottling of *The End of History* while covering the bottles with a dead animal. The special bottling could be covered in either a dead weasel, a squirrel, or a hare. Brew Dog's co-founder James Watt said: "We want to show people there is an alternative to monolithic corporate beers, introduce them to a completely new approach to beer and elevate the status of beer in our culture."

FIGURE 9.35
Moody Tongue's Black Truffle Pilsner. Courtesy of John Peter Laloganes.

FIGURE 9.36
Moody Tongue's Black Truffle Pilsner (2). Courtesy of John Peter Laloganes.

FIGURE 9.37
Goose Island Bourbon County Coffee Stout. Courtesy of John Peter Laloganes.

Check Your Knowledge

Directions: Use these questions to test your knowledge and understanding of the concepts presented in the chapter.

I. MULTIPLE CHOICE: Select the best possible answer from the options available.

1. This ale-styled beer is usually made from equal portions of unmalted wheat and malted pale barley, spiced with ground coriander seeds, orange peels, and other spices and herbs.
   a. Porter
   b. Pale Ale
   c. Hefeweizen
   d. White/Witbier
2. These beers can be thought of as a Pale Ale hyped on steroids—these beers are brewed in a manner that gains an increased gravity and hop predominance.
   a. Porter
   b. IPA
   c. Pale Ale
   d. Hefeweizen
3. Despite the name of this beer style, it is brewed from grain and not grapes. These beers are very strong and intense—usually the strongest ale offered by a brewery that almost rivals wine in alcohol content.
   a. Barley Wine
   b. IPA
   c. Saison
   d. White/Witbier
4. French for "season" named after the original intention of these beers being traditionally brewed in the winter for consumption throughout the summer season.
   a. Saison
   b. IPA
   c. Lambic beer
   d. Golden ale
5. Historically, this was a local brew of the city of Cologne ("Köln" in German). It is one of the palest German beers made—similar to Britain's pale ale style.
   a. Kölsch
   b. Pilsner
   c. Hefeweizen
   d. Stout
6. These ales are sweet to bittersweet ales with undertones of malt. As its name suggests, this ale has a dark brown color and has become known as Nut Brown Ale.
   a. Stout
   b. Porter
   c. Brown Ale
   d. Alt Beer
7. This style of ale contains an intense hop aroma from dry hopping that contributes highly aromatic elements of citrus, grass, and pine character.
   a. Saison
   b. Pilsner
   c. Imperial IPA
   d. Stout

8. The original version was created in Pilsen Czech Republic, often abbreviated as Pils.
   a. Lager
   b. Pilsner
   c. Lambic
   d. Marzen
9. This is a very distinctive style of Belgian beer that relies on wild yeast for fermentation. Straight "unblended" beers are often a true representation of the "house character" of a brewery.
   a. Lambic
   b. Gueuze
   c. Saison
   d. Golden ale
10. These are an effervescent beer that was originally created in Belgium in order to compete with Pilsner beer styles. Traditionally, the beers are bottle-conditioned.
    a. Golden ale
    b. Lambic
    c. Pale Ale
    d. Rauchbier
11. This word means "wheat" in German—therefore, its name is indicative of the beer's typical large portion of malted wheat used during the brewing process.
    a. Hefeweizen
    b. White/Witbier
    c. Pale Ale
    d. Lambic
12. An incredibly unique and rare style of beer—identified through its seriously intense beechwood smoked malt.
    a. Golden ale
    b. Lambic
    c. Rauchbier
    d. Pale Ale
13. The name means "billy-goat" in German and is often used in logos and advertisements.
    a. Dortmunder
    b. Bock beer
    c. American light lager
    d. Golden ale
14. These beers are made in commercial breweries around the world—made to emulate the styles of Trappist beers. Sometimes these beers are referred to as Belgian-styled beers.
    a. Abbey ale
    b. Trappist ale
    c. Stout
    d. Marzen
15. These beers are a blend of younger (1–2 years old) with older (2–3 years old) Lambic beers.
    a. Marzen
    b. Rauchbier
    c. Pilsner
    d. Gueuze
16. This style of beer is dark, smoky, and substantially malty ale with complex aromas and flavors of roasted character. It is the predecessor to the Stout style of beer.
    a. Cream stout
    b. Imperial stout

   c. Porter
   d. Barley wine

17. This beer was originally brewed to high gravity and hop level in England for Stout export to the Baltic States and Russia.
   a. Cream stout
   b. Imperial stout
   c. Porter
   d. Barley wine

18. This lager styled beer has a watery appearance and is light in aroma and flavor—it is also lower in calories than its "traditional version."
   a. Pale ale
   b. American light lager
   c. Alt beer
   d. Hefeweizen

II. DISCUSSION QUESTIONS

19. Identify your favorite style of beer. Be prepared to justify your response.
20. The concept of novelty beers (or as the industry uses the phrase *extreme beers*) is for what purpose?

CHAPTER 10

Sake, Cider, and Mead

CHAPTER 10 LEARNING OBJECTIVES

After reading this chapter, the learner will be able to:

1. Explain how sake is unique as its own beverage category
2. Distinguish how seimaibuai influences the grade and the finished characteristics of a sake
3. Distinguish between a Junmai and a Honjozo sake category
4. Explain the major distinctions of cider from United States, Spain, France, and England
5. Briefly explain a mead

It is the man who drinks the first bottle of saké; then the second bottle drinks the first, and finally it is the saké that drinks the man.

—Japanese proverb

The Essentials of Sake

Learning Objective 1
Explain how sake is unique as its own beverage category

Sake (pronounced sah-keh, not sak-ee) is a rice-based brewed beverage that has been documented to be more than two thousand years old. In Japanese, the word sake means "the essence of the spirit of rice" because it exists as a clear liquid (with one exception) made from rice, yeast, water, and koji (a mold). Sake generally hovers between 10 and 16 percent alcohol by volume, largely determined by the category of a given sake style. While sake has remained an important part of the Shinto religion and Japanese culture for thousands of years, other similar rice-based alcoholic beverages are produced by other Asian countries, such as China, Thailand, Korea, India, Malaysia, and the Philippines. Figure 10.1 identifies a Korean Rice Fermented Beverage.

Sake is no longer confined to the shelves of sushi bars and Japanese restaurants, but is now becoming an ever-present feature on beverage menus around the country. Consumers, most notably Millennials (those roughly under 35 years of age), are gravitating toward sake more than generations in the past.

Overall, consumers are becoming more educated about sake and embracing it in other non-Asian dining settings. As the sake category grows in the United States, sake is being incorporated by mixologists as an enhancement agent behind the bar, as many consumers are now opting for sake-based cocktails. On the horizon for the future growth in sake is a greater emphasis on *terroir-oriented* versions, which will further promote and emphasize regional differences.

Sake as Its Own Category

"Rice wine" is a term often used to classify sake; however, this is a definite misnomer. Rather than argue whether sake is a wine or a beer—recognize that sake is

FIGURE 10.1
Korean Rice Fermented Beverage. Courtesy of Erika Cespedes.

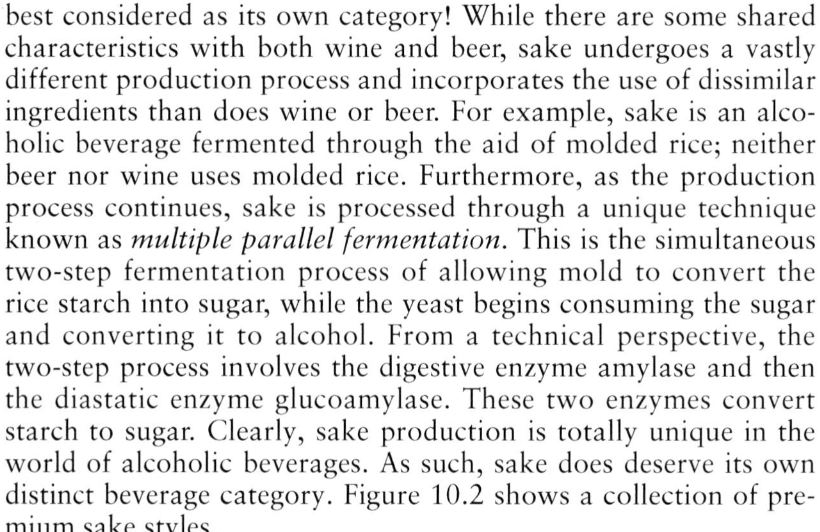

best considered as its own category! While there are some shared characteristics with both wine and beer, sake undergoes a vastly different production process and incorporates the use of dissimilar ingredients than does wine or beer. For example, sake is an alcoholic beverage fermented through the aid of molded rice; neither beer nor wine uses molded rice. Furthermore, as the production process continues, sake is processed through a unique technique known as *multiple parallel fermentation*. This is the simultaneous two-step fermentation process of allowing mold to convert the rice starch into sugar, while the yeast begins consuming the sugar and converting it to alcohol. From a technical perspective, the two-step process involves the digestive enzyme amylase and then the diastatic enzyme glucoamylase. These two enzymes convert starch to sugar. Clearly, sake production is totally unique in the world of alcoholic beverages. As such, sake does deserve its own distinct beverage category. Figure 10.2 shows a collection of premium sake styles.

Ingredients Used to Produce Sake

Sake is brewed from four basic ingredients: water, rice, *koji* (KOH-jee), and yeast. On the surface, it seems fairly uncomplicated, yet inherently things become more complex when the multitude of variations and strains of rice and yeast and types of water come into consideration. All ingredients are crucial elements in their own way that influence the production and final characteristics of sake.

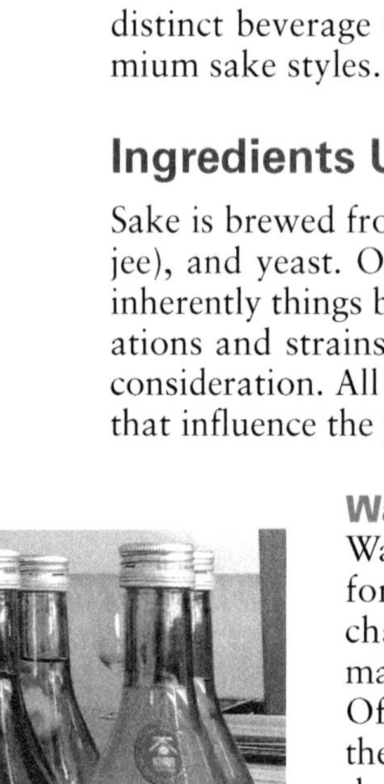

FIGURE 10.2
Selection of premium sake. Courtesy of John Peter Laloganes.

Water

Water forms 80 percent of the base of sake, therefore, it plays a significant influence in the finished character. Water in sake brewing can come from many sources, including wells, rivers, and streams. Of course, water can be chemically altered and synthetically produced according to specifications and desired results. Other producers can simply use local tap water, and filter it or alter it as is necessary to fit their needs. As a general rule, the harder (more minerals) water can create a more robust, savory style of sake. And in reverse, the softer (less minerals) water can create a gentler, cleaner style of sake. It is largely believed that Japan produces some of the best sake largely because most water across Japan (from the north to the south) is much softer in comparison to the rest of the world.

Rice Variety (or Sakamai)

There are about 100 different strains or varieties of sake rice, each type yields specific aroma and flavor profiles. This multitude of variety is one of the main reasons for modernizing the sake beverage category and aiding the surge in popularity in sake consumption. Premium sake is brewed with special rice that is very different than table rice, as it is about 20–30 percent larger, offering a harder exterior (so it doesn't easily crumble during the polishing stage). With these larger grains of rice, there is a concentrated white opaque core of starch at its center. The center core of starch is called the *shinpaku* (shin-pah-koo) or white heart. Surrounding the shinpaku, closer to the surface of the rice grains, are the fats and proteins that adversely influence fermentation and may lead to off-flavors— strange and generally unwanted components in the profile. This goes to address the importance of premium sakes that have an increase in polishing (or milling), which removes much of these exterior components leading to a higher grade of sake.

Yeast

Yeast influences the many elements of sake characteristics. The main job of the yeast is to convert sugar to alcohol and carbon dioxide. It is the heart of the creation of all alcoholic beverages—and no different when it comes to sake production. But different yeast strains will produce different characteristics such as esters, alcohols, acids, and other chemical compounds that affect the nuances of fragrance and flavor.

Over the last ten years or so, dozens of new yeast strains have been developed and brought into use; the process of coming up with a new, specialized yeast strain usually takes about three years. This has been one of those great technical advances in the sake world as each yeast will give rise to its own specific array of chemical compounds and final style characteristics. In modern premium sake, fruity and wine-like aromas from esters are given off by these different strains of sake yeasts.

Koji (KOH-jee)

The official name for the koji mold is *Aspergillus oryzae*, responsible for breaking the starch content contained in the rice into sugar. Without koji, there can be no sake—since rice is a complex starch and unable to be digested by the yeast. Koji is a mold that is sprayed on the steamed white rice where it breaks down the starch molecules and converts them to sugar within about 48 hours. Koji is ready for use when it looks like the rice contains a small amount of white frosting on each grain.

The Production Process

Learning Objective 2
Distinguish how seimaibuai influences the grade and the finished characteristics of a sake

Milling and Polishing

Once proper sake rice (in the case of premium sake) has been obtained, the rice is then milled, or polished, to prepare it for brewing sake. The milling process removes anywhere from 30 to 50 percent of the starch. The amount of milling will greatly influence the taste and ultimately the sake classification. Finally, sake is classified by its rice milling rate, or *seimaibuai*, the percentage of rice grain remaining after milling and polishing.

Washing and Soaking

Next, the white powder (called nuka) left on the rice after polishing is washed away, as this makes a significant difference in the final quality of the steamed rice. The rice is now soaked to attain a certain water content deemed optimum for steaming that particular rice. The degree to which the rice has been milled in the previous step determines what its pre-steaming water content should be. The more a rice has been polished, the faster it absorbs water and the shorter the soaking time. Often it is done for as little as a minute, sometimes it is done overnight.

Multiple Parallel Fermentation

Koji mold (koji-kin) is sprayed on the steamed white rice where it breaks down the starch molecules and converts them to sugar within about 48 hours. Once yeast has been added, the whole mix is then allowed to ferment over the next two to four weeks. This mash (or moromi) is allowed to sit from 18 to 40 days, after which it is pressed, filtered, and blended.

Filtration and Pasteurization

Most sake is then pasteurized once. This is done by heating it quickly by passing it through a pipe immersed in hot water. This process kills off bacteria and deactivates enzymes that would likely adverse flavor and color later on. Sake that is not pasteurized is called *namazake*, and maintains a certain freshness of flavor, although it must be kept refrigerated to protect it.

Aging

Finally, most sake is left to age for about six months, rounding out the flavor, before shipping. Before shipping, it is mixed with a bit of pure water for a slight dilution to bring the near 20 percent alcohol down to 6–16 percent or so, and blended to ensure consistency. At this point the sake may be pasteurized a second time.

Style Categories of Sake

Learning Objective 3
Distinguish between a Junmai and a Honjozo sake category

Sake falls into one of two main categories. The first category is the "pure" rice style (called "Junmai-shu" in Japanese) and the second category is the alcohol-added "fortified" style (called "aruten" in Japanese or sometimes just referred to as Honjozo-shu). Figure 10.3 illustrates the two main categories of sake along with several styles under each category. The Junmai-shu and Honjozo-shu are not only categories, but also the entry level styles within each category. The "shu" suffix, by the way, simply means "sake," and is often dropped when discussing sake. Hence, Junmai-shu is sometimes called simply Junmai; Honjozo-shu is very often called Honjozo.

The categories of sake are further broken down by style per their milling rates. The percent is referring to what has been milled from the exterior of the rice. The Japanese word *seimaibuai* (say-my-boo-eye) is the actual term for how much a grain of brewing rice is polished or milled. For example, a seimaibuai of 50 percent means that each grain of rice that makes up that particular sake has been milled or polished 50 percent leaving a remaining value of 50 percent (Daiginjo style or grade). Ginjo is a special type of Junmai or Honjozo, and considered the highest achievement of the brewer's art, due to its high rate of milling away the exterior of the rice grains.

The shinpaku or "white heart" is concentrated at the center of the grain—with proteins, fats, and amino acids located toward the outside. The rice that is less

Sake Style Categories

	Seimaibuai (Rice Milling %)	Junmai (Pure Rice Style)	Honjozo (Fortified Rice Style)
Milling Rate % (↑)	50% milled away	Junmai Daiginjo	Daiginjo
	40% milled away	Junmai Ginjo	Ginjo
	30% milled away	Junmai	Honjozo
-------	No minimum milling required	------	Futsu-shu

FIGURE 10.3
Sake style categories.

polished will yield a sake that is bolder, full-bodied, and robust. With increased milling, one can remove more of the exterior that would otherwise lead to unwanted flavors and aromas in the brewing process. Greater milling leads to cleaner, more elegant, and more refined sake. It also allows more lively aromatics to come about. The more you polish the rice, the higher the grade of sake.

Junmai Sake

Junmai (joon my) sake is a category and a style. As a category, it is referencing pure rice premium sake, also meaning that it is unadulterated without any distilled alcohol. Figure 10.4 shows a picture of Junmai sake from Niigata, Japan.

FIGURE 10.4
Junmai sake. Courtesy of Erika Cespedes.

- *Junmai Daiginjo* (die gheen joe): A style of Junmai, brewed with very highly polished rice (to at least 50 percent, with 50 percent milled away). *Daiginjo* means "ultra-premium" and refers to sake of the highest grade.
- *Junmai Ginjo* (gheen-joe): A style of Junmai, using highly polished rice (at least 60 percent remaining, with at least 40 percent milled away) and fermented at colder temperatures for longer periods of time.
- *Junmai:* The entry level style to this category of premium, pure sakes. Made with rice polished to at least 70 percent of its original size, with minimum 30 percent milled away.

Honjozo Sake

Honjozo (hone joe zoe) is sake to which a very small amount of distilled ethyl alcohol (called brewer's alcohol) has been added to the fermenting sake in the final stages of production. Slight amount of water is added later, in order to balance out the brewer's alcohol. The addition of distilled alcohol (made from pure sugar cane), makes the sake lighter, more fragrant, sometimes a bit drier, and in the opinion of many, easier to drink. Figure 10.5 shows a bottle of Ginjo style sake from Japan.

FIGURE 10.5
Ginjo sake. Courtesy of Erika Cespedes.

- *Daiginjo:* A style of the fortified category, brewed with very highly polished rice with a minimum of 50 percent or more milled away. Given a very small amount of pure distilled alcohol.
- *Ginjo:* A style of the fortified category, brewed with a highly polished rice, at least 40 percent is milled away and fermented at colder temperatures for longer periods of time. Given a small amount of pure distilled alcohol.
- *Honjozo:* The entry level style to this category of premium, fortified sakes. Made with rice polished to at least 70 percent of its original size, with minimum 30 percent milled away and varying amounts of pure distilled alcohol.

Note that most run-of-the-mill cheap sake has an excessive amount of brewer's alcohol added to it. However, Honjozo sake contains only small amounts of this added alcohol. Adding alcohol does not make a sake lower grade; it is part of one manner of brewing that produces specific results (like lighter, more fragrant sake with a more robust structure and perhaps longer shelf life). In "Futsuu-shu," an ungraded and non-premium sake, the brewer's alcohol is added for the purpose of increasing yields.

FIGURE 10.6
Nigori sake. Courtesy of Erika Cespedes.

Alternative Styles

- *Futsuu-shu* (foot soo shoo): It is an "ungraded" everyday type of sake; the vast majority of sake available in the marketplace. It is a style of sake that has no minimum milling percentage and therefore has higher levels of impurities. This type of sake is produced with copious amounts of pure distilled alcohol to increase yields.

  The overall quality level of Futsuu-shu sake ranges from poor to acceptable allowing for an accessible price point for those that place price over high quality.
- *Genshu:* Genshu style sake is at full strength and undiluted. It is typically bottled at 18–20 percent alcohol. As most sake is best served lightly diluted with water, this style of sake, however, is best served on ice, or on the rocks.
- *Nama* (nah-mah): Nama means "draft" and is sold either on draft or in a can. In this sake, fresh sake is microfiltered instead of other sake which is pasteurized twice, once before aging and once in the process of bottling. Nama has a fruity and fresh taste with pleasant aroma.
- *Nigori* (nee-gohr-ee): Nigori style sake tends to have a cloudy appearance because of lack of filtering as compared to other styles of sake. This style is often served after meals; it is lightly sweet and may contain light carbonation. Figure 10.6 displays a bottle of Nigori style of sake.
- *Kimoto* (kee-moh-toe): Traditionally made via pole-ramming (yama-oroshi), which is depicted in much artwork, also known colloquially as "Nama." This is similar to *batonnage*, as the moto is stirred with a long, wooden pole. This process continues throughout fermentation for about 45 days.
- *Yamahai* (yah-mah-hi): This style is brewed with native lactic bacteria and no stirring. Yamahai usually contains a detectable level of sweetness and acidity, with deep, more pronounced flavors.

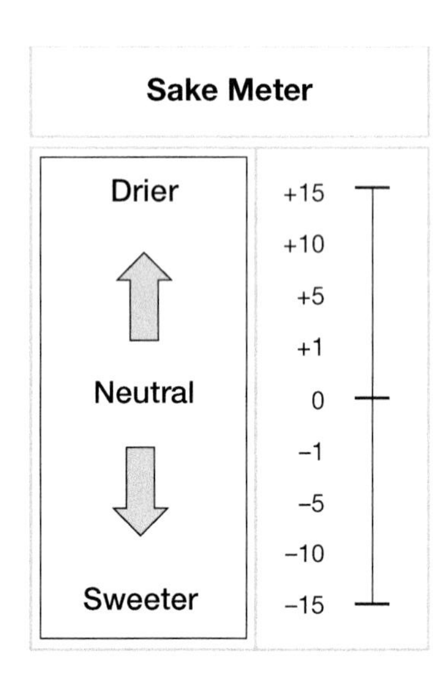

FIGURE 10.7
Sake meter.

The Sake Meter

This is a useful tool that is often made available for most premium sake. In some cases, the meter or rating of dry-to-sweetness will be identified on the back label of a bottle of sake. Or alternatively, the meter rating can be found when searching for the technical notes on the internet, for a given sake. The sake meter is a scale that ranges from −15 to +15. The higher the rating, particularly at 0 or above, indicates a drier style of sake. Conversely, a sake meter reading that is lower than 0 indicates a sweeter style of sake. Figure 10.7 depicts a sake meter that measures the dryness or sweetness levels.

FIGURE 10.8
Sakazuki. Courtesy of Erika Cespedes.

FIGURE 10.9
Nigori sake being chilled. Courtesy of John Peter Laloganes.

Serving Vessels

When serving sake, one can choose a traditional or contemporary vessel. Traditionally, it is accustomed to drinking sake (particularly if it is warmed) in small porcelain cups called *sakazuki* (sah-kah-zoo-kee). Certainly, this choice can work to convey the Asian theme, but it is not necessary to enjoy the delight of sake. For a more contemporary approach, it has become common practice to serve sake in smaller white wine or sherry glasses. Figure 10.8 shows a sakazuki being filled with Junmai sake.

Serving Temperatures

Serving sake that has been warmed or heated is considered classic, particularly for "everyday" types of sake. Serving sake this customary way can add a special, traditional atmosphere to your enjoyment. To serve warm, pour sake into a small open-mouthed carafe (Tokkuri). Warming it in a hot water bath over very low heat to approximately 110°F is the best and gentlest method. Never boil sake, and usage of a microwave is not recommended. Warming up sake is known to take away any impurities or undesirable nuances in the aromas and flavors. Futsuu-shu sake is one type of sake that is often heated, as it can mask the impurities of this ungraded style. Premium sakes such as Junmai are best served at room temperature, slightly chilled, or even over ice.

For a more contemporary approach to serving sake, chilling is becoming increasingly popular. This tends to promote sake as a more versatile drink before, during, and after meals. Ginjo type sake, were specifically developed for chilling. Chill sake to about 50°F. Figure 10.9 displays a Nigori sake being chilled on ice.

The Essentials of Cider

Learning Objective 4
Explain the major distinctions of cider from United States, Spain, France, and England

Cider is defined as a beverage made from fermented apples (or pears) with not more than 8 percent alcohol. The term "hard cider" is often used to reference a drink from fermented apple juice as opposed to simply apple juice. Cider is a clear (but not always) beverage and lightly carbonated showing different degrees of sweetness. Apples are to cider as to what grapes are to wine and barley is to beer. Similar to wine, ciders can be made in a variety of styles—different apples, yeast, and barrels are just some variables that can provide uniqueness and complexity.

Humans first discovered apples in the early Paleolithic period, some 750,000 years ago. The carbonized remains of apples have been dated at 7500 BCE and records describing the extensive planting of apple trees in elevated gardens along the Nile River Delta have been dated at 1300 BCE.

In a sense, the history of the apple is a history of our agrarian roots. There is a long history in America, dating back to the Colonial period where hard cider was the alcoholic drink of choice. Cider made its way via the English settlers, who soon learned that apples were much easier to grow in the New England soil than the barley and grains for beer. A young man by the name of Johnny Chapman, better known as Johnny Appleseed, was largely, if indirectly, responsible for spreading the production of cider. Sweet or hard cider was the drink of America's past. It was so culturally ingrained, there was even a weakened version called Ciderkin for children. Cider was once Americans' alcoholic drink of choice, but consumption started to diminish in the mid-to-late 1800s with the influence of the German immigrants and their passion for brewing beer.

In recent years, hard cider has been gaining in popularity, particularly with craft beer drinkers; also offering a viable alcohol-based alternative for those with gluten sensitivities or celiac disease. The surge in popularity is countering a nearly 200-year-old trend of declining interest in the beverage. According to Greg Hall (founder of Virtue Cider), "Craft cider is at the intersection of craft beer and the local farm to table food movement." Cider is produced all around the world, however, there is a cider renaissance in the United States. Consumers are searching for artisan ciders out of Midwest and East Coast. They are producing a collage of different styles, ranging from bone-dry versions to super sweet ice cider versions where the apples have been frozen to extract out water content.

Cider Production

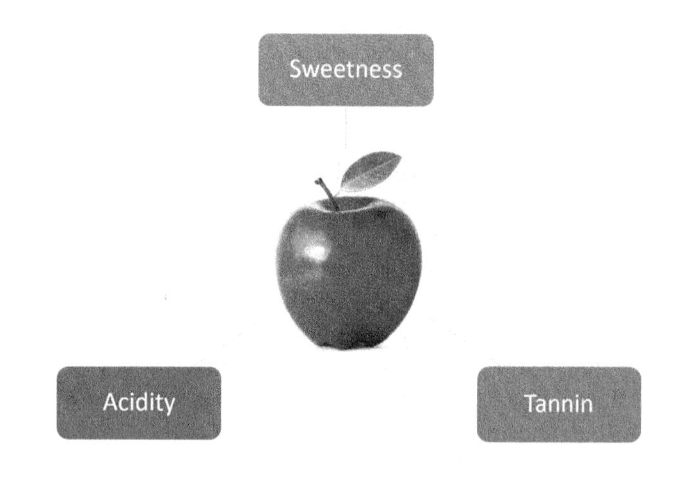

FIGURE 10.10
Components needed in quality cider. Courtesy of John Peter Laloganes.

Apple varieties grown in the United States are over 2,500; the critical step in the manufacture of hard cider is the apple selection. Apples used for cider production are often not the ones purchased in the grocery store for purposes of eating; instead, cider apples have more noticeable characteristics of sourness, sweetness, and bitterness. Most ciders are typically blends of a dozen or so different apple varieties to achieve balance since it is nearly impossible to find a single apple that will be able to provide the complexity needed for a craft oriented cider. Figure 10.10 illustrates the need for sourcing a blend of apples that can collectively offer a mix of sweetness, sourness (from acidity), and tannin.

Green or immature apples make a poor cider; however, some may be added to increase the acid content (tartness). Rotten apples should not be mixed up with the good quality cider apples; some apples harbor *acetobacter*, the active bacteria in vinegar making and would deem the cider undrinkable.

Prior to sealing the bottle, the cider has either seen a reintroduction of some low levels of CO_2, or has been given a dosage (similar to Champagne) where a "dose" of sugar has been added into the bottle. This dose will induce a secondary fermentation carried out in the actual bottle for a creation of natural carbon dioxide.

United States and Canadian Ciders

Unites States producers are making cider in thirty-three states, but mostly center around the Midwest and East coast. Generally, the larger producers (Woodchuck,

Angry Orchard, and Crispin) dominate the market with their off-dry to more often sweeter versions that appeal to the masses of consumers. In contrast, the craft producers (Argus, Vander Mill, and Virtue) focus on dryer styles that often replicate classic versions of French, English, and Spanish ciders. United States ciders tend to be:

- Styles that are all diverse, ranging from dry to sweet and fruity to funky
- Higher in alcohol than other countries, ranging from 4 to 8 percent
- Experimental with bottle fermentation and aging in old calvados, wine, bourbon, or rum barrels

English Ciders

Cider has been a popular drink in the UK for centuries, particularly Somerset cider. In Somerset, apple orchards grow in abundance and their soil and climate are vital ingredients in cider making, in the same way as vineyards are to wine. Traditional farmhouse cider making in Somerset owes its origins not just to the historical apple orchards of monasteries but to generations of small farmers throughout the county. English ciders tend to be:

- More often dry and tannic compared to French ciders
- Typically complex from the use of native yeasts
- Higher in alcohol percent compared to French ciders

French Ciders

French cider (*cidre* in French) is produced predominantly in northern France in the regions of Normandy and Brittany. They vary in dryness/sweetness as well as strength, typically ranging from 3 to 6 percent alcohol. Figures 10.11 and 10.12 show Eric Bordelet cider from Normandy, France. French ciders tend to be:

- Lighter, subtle, and fruitier compared to English and Spanish cider
- Range of dry-to-sweet styles
- Lower in alcohol (ranging from 3 to 6 percent) compared to English and Spanish ciders

FIGURE 10.11
Eric Bordelet cider. Courtesy of Erika Cespedes.

FIGURE 10.12
Eric Bordelet cider (2). Courtesy of Erika Cespedes.

Spanish Ciders

The overwhelming majority of Spanish ciders come from the Asturian and Basque regions of northwest (Green) Spain. Asturian cider is called *sidra* while the Basques opt for the term *sagardoa.* Spanish ciders tend to be:

- Higher in alcohol than other ciders, typically ranging between 5 and 6 percent
- Most often extremely dry, tart, and tannic
- Fermented with only native yeasts
- Earthy, rustic, and musty in aromas and flavors
- Cloudy in appearance due to bottle fermentation and unfiltering

The ciders characteristically are relatively cloudy and with sediment. Therefore, during service, the bottles should be shaken before opening to activate the natural carbonation before pouring. It is a tradition for Asturian bartenders to pour *Sidra* from overhead (*escanciar*) to a glass held at the waist in order for the resulting aeration to release the full flavors. The small amount of cider should be poured into a narrow glass from a height of around 3 feet. This aerates the cider, enhancing the bouquet and the natural carbonation, and is called "throwing" the cider, which produces the gas Spaniards call estrella.

The new generation of Asturian producers introduced a sweet, dessert style cider by adapting the process of Canadian ice cider. Since an apple never freezes on the branch (they fall to the ground well before) in Asturias, "frost cider" is made by freezing the freshly pressed juice to minus 20°C. The juice is then thawed gradually to extract a concentration of flavor with five times the sugar allowing fermentation to higher alcohol levels while arresting fermentation with ample residual sugars.

Pear Cider

Pear cider, or poiré in French and perry by the English is often an additional product offering in the portfolio of apple cider producers. For marketing purposes in recent years, producers across the globe have been using the name perry cider interchangeably with pear cider. The use of these terms are debatable, because technically speaking, pears used for perry cider are of a special cultivar that are higher in tannin and acid than standard pears used for most other pear cider.

Both cider and perry have early roots in Normandy, France, and then traveled north to England by way of the Norman Invasion in the 11th century. While some perries finish completely dry, most perries are slightly sweet. The principal differences between perry and apple-based cider are that pears must be left for a critical period to mature after picking, and the pomace must be left to stand after initial crushing to lose some of its tannins, a process analogous to wine maceration. After initial fermentation, the drink undergoes a secondary malolactic fermentation (MLF) while maturing, allowing some of its sharp acids to round and slightly soften. Most perries are naturally carbonated in the bottle.

Just a "Taste" of Mead

Learning Objective 5
Briefly explain a mead

Mead is honey wine (or a beverage fermented from honey water) that is brewed just like any other fermented beverage. Honey is full of simple sugars allowing for an easy and quick fermenting process. Honey, which takes on the aromatic characteristics of the flowers that the bee pollinates, is the primary ingredient being fermented—and mead can easily incorporate fruit or spices. They can be fermented in different ways, ranging from dry to sweet and still to sparkling. Mead usually consists of fairly high alcohol content (in comparison to the average beer), between 8 and 12 percent in alcohol, but can be as high as 14 percent. The color can range from clear to golden, and flavors range from dry to sweet depending upon the length of fermentation. This drink dates back to the ancient Greek, Egyptian, Inca, and Aztec cultures: Mead could have been one of the first fermented beverages produced by man.

Mead was an important and popular drink during the 5th and 6th centuries. According to Charlie Papazian in *The New Complete Joy of Home Brewing*, during this time, the custom of the honeymoon was started. The newlywed couple would be offered a month's—or a moon's—supply of mead. It was believed that if the couple consumed the mead for a month after the wedding, it would help produce a male child, which back in the day was an essential factor in carrying on the family name and the transfer of property and noble title.

Check Your Knowledge

Directions: Use these questions to test your knowledge and understanding of the concepts presented in the chapter.

I. MULTIPLE CHOICE: Select the best possible answer from the options available.

1. Junmai-shu sake
   a. is the purest and highest grade
   b. includes a small amount of brewer's alcohol (lightly fortified) to bring out its flavor
   c. is an ungraded sake without any milling regulations
   d. is commonly used in cooking
2. Sake is unique in that is undergoes multiple parallel fermentation. Which response best explains this process?
   a. Sugar is converted to alcohol by yeast, and then sake is fortified with a light spirit
   b. Molded rice starch is converted to sugar, and then sugars are converted to alcohol by yeast
   c. Rice is allowed to germinate, and then sugars are converted to alcohol by yeast
   d. Both a and c
3. Sake is brewed from four basic ingredients: rice, water, yeast, and
   a. Honjozo
   b. Junmai
   c. Ginjo
   d. Koji
4. Which one of the following is NOT considered a "pure" sake?
   a. Junmai
   b. Junmai Ginjo
   c. Junmai Daiginjo
   d. Honjozo
5. Nigori sake is best described as
   a. cloudy, unfiltered sake
   b. one of the most common brewing method with cultured lactic acid added
   c. aged in wooden barrels
   d. dry, bright, floral, and light-bodied
6. The koji mold acts to
   a. break down the starch molecules
   b. convert the starch molecules to sugar for fermentation
   c. contribute aroma and flavors to the sake
   d. all of the above

7. The more the sake rice is milled or polished down
   a. it's considered everyday sake with high impurities
   b. the more smooth, elegant, and refined is the sake
   c. the more flavorful and rich is the sake
   d. it has more alcohol because of the high sugar content
8. Mead is made from a base of
   a. koji mold
   b. honey
   c. pears
   d. apples
9. The vast majority of Spanish ciders come from
   a. Somerset
   b. Normandy and Brittany
   c. Bordeaux and Burgundy
   d. Asturia and Basque
10. The sake meter is a scale that ranges from –15 to +15. It measures the
    a. quality level
    b. quantity of brewer's alcohol
    c. dryness/sweetness level
    d. amount of koji mold used
11. Futsuu sake is best described as
    a. a premium pure sake
    b. a premium fortified sake
    c. a cloudy, unfiltered sake
    d. an "ungraded" everyday type of sake
12. The tradition for Asturian bartenders, pouring *Sidra* from overhead is known as
    a. escanciar
    b. estrella
    c. sagardoa
    d. all of the above
13. Which is the correct style of pure sake using highly polished rice (at least 60 percent remaining, with at least 40 percent milled away)?
    a. Junmai Daiginjo
    b. Junmai
    c. Honjozo
    d. Junmai Ginjo
14. Sake is classified by its rice milling rate, or
    a. seimaibuai
    b. shinpaku
    c. sakazuki
    d. sakamai
15. Which country produces cider that is almost always dry, tart, and bitter?
    a. France
    b. United States
    c. England
    d. Spain
16. What is the term often used to distinguish that a cider has alcohol?
    a. Soft
    b. Hard
    c. ABV
    d. Alcohol added

17. Honjozo sake has a small amount of what agent added to it?
    a. Carbonation
    b. Alcohol
    c. Coloring
    d. Aged sake

II. DISCUSSION QUESTIONS

18. Distinguish how seimaibuai influences the grade and also the finished characteristics of a sake.
19. Explain the impact of adding the brewer's alcohol to premium sake.
20. Explain the major distinctions of cider from United States, Spain, France, and England.

CHAPTER 11

The Distillery: Spirits and Liqueurs of the World

CHAPTER 11 LEARNING OBJECTIVES

After reading this chapter, the learner will be able to:

1. Explain the process and science of distillation
2. Explain the different methods of distillation
3. Identify all the popular clear spirits and their main fermentable source
4. Identify all the popular brown spirits and their main fermentable source
5. Identify at least four of the categories for liqueurs
6. Briefly explain the craft spirit movement

Alcohol is the anesthesia by which we endure the operation of life.

—George Bernard Shaw

Primer on Spirits and Distillation

Learning Objective 1
Explain the process and science of distillation

The term *spirit* originates from Middle Eastern alchemy as a reference to the vapor given off and collected during the production process known as distillation. The process of distillation hasn't changed much since its discovery in early civilization. Historically, alchemists were the ones often involved in creating special elixirs and potions that can have some transformative effects on people. In 1250, Arnaud de Villeneuve is said to be the first distiller in France—creating a liquid he called *eau-de-vie* (ohh-duh-vee) or in Medieval Latin, *aqua vitae* (ah-kwah-vee-tay)—both meaning "water of life" as the alcohol was often thought to contribute to the virtues of living a long and fruitful life. Figure 11.1 shows vodka, considered one of the purest types of spirits, possibly one of the earliest produced.

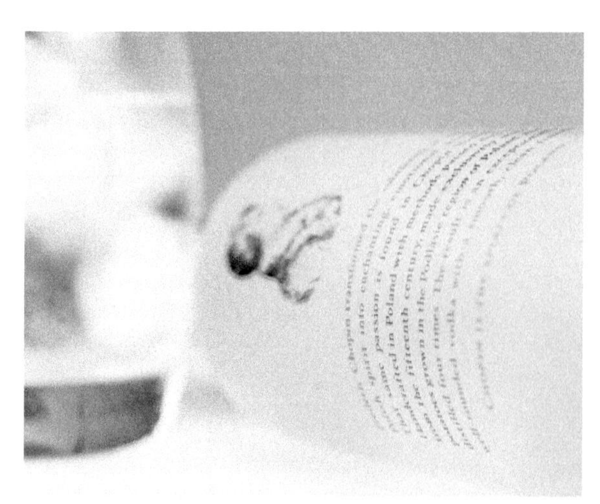

FIGURE 11.1
Chopin vodka. Courtesy of Erika Cespedes.

Spirits and liqueurs are potent drinks with a concentration of large levels of alcohol. These beverages are produced via the distillation process—a method of separating two mixtures based on the differing volatility levels of their boiling points. Initially, a fermented beverage, known as a wash or *mash*, is heated gradually (1–2 hours) to the boiling point within a still (the device used to distill the beverage). This part of production relies on the laws of fractional distillation—a difference in boiling points that allows for the separation of ethanol from water. Since the two significant variants of the wash will vaporize at different temperatures

FIGURE 11.2
Copper column still. Courtesy of John Peter Laloganes.

(water at 212°F and ethanol at 173°F)—the alcohol will boil out of the wash first. For the next 2–3 hours, a mix of alcohol, water, and other flavor enhancing agents boil and vaporize out of the wash. The vapors will recondense as a liquid called a distillate. The distillate comes out of the still in stages, starting with the intensely and undesirably flavored "heads," followed by a long phase of desirable "hearts," and finishing with lower proof and less desirable "tails." The hearts of the run are collected and set aside while the heads and tails are added back to the next distillation. The new mixture can be further aged in barrels and/or flavored by the distiller. Therefore, spirits and liqueurs are not "created" by distillation; instead, they are merely concentrated and purified from the initial alcohol base. Pictured in Figure 11.2 is a column copper still. Copper has been known as the superior material used for stills; they offer excellent conductivity and even distribution of heat ensuring a more consistent product while also diminishing the possibility of undesirable volatile compounds throughout the distillation process.

All distilled spirits begin as a clear concentration of alcohol from a fermented liquid such as wine or beer. The finished distillate may remain clear, or it may become gold to golden-brown in color, depending on how it is aged or treated after distillation. Clear spirits, such as vodka and rum, can be bottled right after distillation. Spirits with gold to golden-brown hue will further benefit from being aged in charred wood barrels.

Multiple Distillations

During the first distillation, the alcohol level of the liquid will double at the very least. If, for example, a fermented beverage has 12 percent alcohol before it is distilled, the distillation concentrates the alcohol to approximately 24 percent. Sometimes, it takes more than one distillation to achieve the desired alcohol content of a specific spirit. Therefore, the process can be repeated many times to achieve greater levels of alcohol along with further clarification and purity as the alcohol is passed through a series of activated charcoal. If a spirit goes through two or three distillations, the alcohol content can increase to 40, 50, and even 95 percent alcohol by volume (abv). It is possible for producers to identify and promote their product as having undergone *double* or *triple distilled* as an added marketing advantage.

Scotch, a type of whisky from Scotland, is double distilled whereas their neighbors in Ireland distill their Irish whiskey three times. There are some select spirits that are marketed as *pure grain alcohol*, which need to be distilled many more times before reaching their intended 190 proof (95 percent abv). The 190-proof grain alcohol is the highest alcohol level available on the market and has very limited use because of the dangers and potency involved with a drink that is so concentrated. A 100 percent pure alcohol level can be achieved in a laboratory or industrial conditions; however, it has very little application in the beverage industry.

Distilling: An Environmentalist's Dream

The products and by-products of fermentation and distillation are environmentally clean, nontoxic, and pure. For example, the carbon dioxide produced during fermentation is natural and does not harm the environment. It can be collected and recycled for use to carbonate drinking water or soda. Leftover mash can be sold for cattle feed. Grape skins, stems, and seeds can be used as mulch. Even the excess yeast can be used as an ingredient in various products. The only product that might be harmful is the alcohol itself, but alcohol is only harmful when it is consumed in excess.

Differentiating Between a Spirit and a Liqueur

The term *spirit* is typically used generically to reference any product that has been distilled, although there is an obvious distinction between a spirit and a liqueur. The term *spirit* refers to a distilled beverage that contains no added sugar and contains at least 20 percent abv—versus a liqueur, which refers to a distilled beverage that contains added levels of sugar, cream, and/or flavorings. The six basic categories of spirits include: vodka, rum, gin, Tequila, whiskey, and brandy.

NEED PROOF ... HERE IT IS ...

The term **proof** (used universally with the term alcohol percent) dates to 16th century England where it was intended to be used as a "test or demonstration". Originally, the term came about when a buyer or seller wanted to "prove" the amount of alcohol in a distilled beverage. First, they would mix the beverage with gunpowder. Once lit, the gunpowder/alcohol mixture would burn a slow blue flame—indicative of 50 percent abv, or 100 proof. If the alcohol content was lower, the gunpowder would have trouble burning. If the alcohol content was higher, the gunpowder would flame up. The scale of proof tops out at 200 proof.

Methods of Distillation

Learning Objective 2
Explain the different methods of distillation

The basic distillation procedure is similar regardless of the spirit being produced. There are predominately two distinctly different types of stills—the pot still and continuous still—they perform the same basic function but create vastly differing results.

Pot Still

With pot distillation, a batch of fermented liquid is placed into a capped and sealed copper pot. As the liquid mash heats up (to 173.1°F), the alcohol content present in the liquid begins to boil (alcohol boils at a lower temperature than water) and then turns into a vapor. The alcohol vapors rise upward into the head of the still; then they're drawn off into an arm and then to a cooling coil. The coil allows the collected vapors to condense back into a liquid, then being allowed to drain off into a collection vessel. The newly condensed liquid (at the point it is a distilled beverage) is a mix of alcohol and congeners. Congeners are a collection of impurities to some degree, but they are also responsible for providing desirable complex layers of flavor compounds. A pot still is considered inefficient as it allows a distiller to make only one batch of spirits at a time, prior to cleaning the still and producing another batch. The pot still was exclusively used for distillation until the invention of the continuous still in the early 1800s.

Column/Continuous Still

The continuous still was created in the early 1800s to avoid the inefficient process of draining and cleaning the pot still between batches. The continuous still is an ongoing distillation which means it has the added benefit of operating continuously without interruption and without need for maintenance between batches of alcohol.

FIGURE 11.3
Stainless steel column still. Courtesy of John Peter Laloganes.

Throughout the distillation process, the fermented mixture is continuously fed into the still and the fractional derivatives of spirits and water are removed throughout its operation. The mash enters the top of the column, near the coolest part of the still where it immediately starts to sink to the bottom. Since the still is heated from the bottom, steam begins to rise, allowing the liquids to interact while the heat vaporizes the mash and forces the alcohol and other volatile molecules up the still. Simultaneously, the water and solid matter in the mash fall back to the bottom of the still. Since column stills contain partitions, or perforated plates that create separate chambers within the still, each time the vapors hit a plate, they start to condense again, leaving the congeners behind in the condensation. As the vapors rise from chamber to chamber, and from plate to plate, they reduce in congeners while maintaining more of the ethanol. Ultimately, the alcohol vapors are diverted out of the top of the still and into a condenser, where they transform back into a liquid. Figure 11.3 shows a continuous stainless steel still.

Column stills can distill proofs as high as the 190s, or 95 percent abv, however, most spirits will be watered back down to a reasonable alcohol content of around 40–45 percent.

Popular Clear Spirits

Learning Objective 3
Identify all the popular clear spirits and their main fermentable source

Non-aged or clear spirits are fermented, distilled, and bottled without being exposed to aging in wood barrels. They are—with some key exceptions—less expensive than their aged counterparts because the producer does not incur the added expense of additional barrel and storage costs. Clear spirits include most gin, vodka, silver rum, silver Tequila, and grappa with lesser known ones such as aquavit and arak. With very few exceptions, clear spirits are distilled and sold at a minimum of 80 proof (40 percent abv). This high proof allows the spirit, if desired, to be chilled below 32°F without freezing. In fact, many of the clear spirits, arguably, are better if they are served this cold.

Vodka

The name "vodka" comes from the Slavic word *voda*, meaning water. Vodka is believed to have originated sometime in the 1400s and was most likely originally used for medicinal purposes. It is identified, at least historically, with Poland and Russia. Though in modern day, this spirit is made in Denmark, Finland, France, and even the United States. Vodka is America's (if not the entire world's) "number one" most popular distilled spirit—largely because of its adaptability and versatility for drink creations. Vodka can be made from just about any fermentable agent—though most common sources include corn, wheat, rye, or barley as well as potatoes, beets, grapes, or sugarcane on occasion. The spirit is composed of ethanol, water, traces of impurities, and any remaining flavoring agents from the sugar source and production process.

As per standards in United States, vodka must be void of any "distinct" taste, color, aroma, or character. Setting the law aside, for anyone who has consumed a flight of unadulterated vodkas side-by-side, it becomes clear that subtle, yet certainly noticeable aroma differences do exist. Its subtle flavor reflects not only where it's made, but also the base of fermentable sugar source. Vodka's "lack" of distinct

FIGURE 11.4
Vodka on the rocks. Courtesy of Erika Cespedes.

characteristics makes it suitable to mix with juices like orange, tomato, and cranberry juice, sodas like 7-up, ginger-ale, cola, and other alcohol agents. Vodka offers a unique base to create the likes of cocktails such as the martini, bloody Mary, cosmopolitan, and the Moscow mule. Pictured in Figure 11.4 is a bottle of Chopin Vodka, one of the few premium vodkas that use potatoes as the base in their mash.

Purification of Vodka

Congeners (cahn-jen-ehrs) are chemical substances that are naturally produced through the fermentation process contributing complex aromas and flavors to a spirit. Regarding vodka, less congeners is generally considered to equal better vodka. Therefore, extensive use of purification during the production process is a common practice for vodka. This spirit may be double or triple distilled, yielding further purity each time it undergoes the process. Purification is additionally conducted as the distilled vodka progresses through activated charcoal filters to heighten its purity and absorb trace amounts of congeners that would otherwise contribute undesirable aromas and flavors. Studies conducted have also shown that congeners are the contributing factor to the effects and symptoms of hangovers. Of course, without conducting any research, truly the number one factor in any hangover . . . is drinking!

Styles of Vodka

Apart from the fermentable agent and/or geographical origin, vodkas may be classified into two main groups—clear vodkas and flavored vodkas. Clear vodkas have always been the staple of success and popularity. Just when vodka sales may have seemingly hit a plateau, the spirit industry created a premium and super premium movement with the customer responding, "Yes, I will drink less, but I will drink better." While most vodka is "unflavored," a large part of the increased surge in popularity has been the additional drink creations brought on with the flavored vodkas. Figure 11.5 displays a collection of Grey Goose, flavored vodka.

FIGURE 11.5
Flavored vodka. Courtesy of Erika Cespedes.

Clear Vodka

Vodka is ethyl alcohol in its purest form, especially when it's distilled at higher proof and left unaged. These spirits can be fermented and distilled from virtually any grain or vegetable material ranging from cereal grains, potatoes, molasses, beets, grapes, and any number of other products.

Modern day vodkas from Russia and Scandinavia are generally wheat based, while many Polish versions are made from rye, with some potato devotees. In the United States, vodkas are a "mixed bag" as they are created from all different sources, dependent on the preference of the distiller. Typically, an unaged spirit made from grapes have generally been called *eau de vie* but when distilled at high-enough proof they can also fall under the vodka category.

However, any vodka that is not made from the traditional use of grain or potatoes must list all ingredients on their label.

Flavored Vodka

Originally, flavored vodkas were used to mask the flavor of the more primitive ones. In recent years, these flavored vodkas have surged in popularity, particularly in higher volume bars and night clubs. Distillers have added flavors such as mandarin orange, lemon, pepper, and not to forget the novelty of cotton candy or green tea flavor. The flavor options are endless, and vodka producers continue to bring more to the marketplace every year. In addition, vodka has seen an increase of small-batch vodkas that are marketed as ultra-premium alternatives.

Gin

Gin is made from a neutral spirit base, flavored with more than forty botanicals—most notably the juniper berry, which has been used for its medicinal properties for centuries. The name *gin* is derived from either the French word *genièvre* or the Dutch word *jenever*—both mean "juniper,"—referring to the defining flavor agent. Gin can be affectionately regarded as flavored vodka.

Gin is created when the distiller macerates the botanicals in a neutral spirit and then distills the product another time. Beyond the reliance of the juniper berry, other botanicals are included in the production of gin, such as grains of paradise from Africa, orange peel, lemon peel, star anise, licorice root, fennel, caraway seeds, ginger, nutmeg, and coriander.

Gin is the main component in some of the world's most famous cocktails. It offers a unique base to create the likes of cocktails such as Tom Collins, Aviation, French 75, Gimlet, Gin Fizz, Negroni, Singapore Sling, and the Classic "Gin" Martini. Figure 11.6 displays a collection of gin.

FIGURE 11.6
Selection of gin. Courtesy of Erika Cespedes.

Styles of Gin

Gin has a long history closely associated with Holland and England. Yet, the consumption of gin in the United States dates way back to colonial times. Its most infamous period in American history was during Prohibition (1920–1933) when "bathtub" gin became a catch-all phrase for illicit manufacturing of all different types of alcohol. But in recent years, there is a renaissance underway with small-batch artisan producers out of Portland, Oregon; Seattle, Washington; Middleton, Wisconsin; Chicago, Illinois; and more.

London Dry Gin

London Dry gin is the most recognized style consumed around the world. Its distinctive profile incorporates a predominant juniper berry characteristic and is blended with the botanicals during distillation process as opposed to having them being added later. This style of gin is regarded as a highly aromatic and well-integrated version.

Old Tom Gin

This is an old style of gin dating back to the 1800s, but faded out with the success of other styles, notably London Dry. Old Tom gin resurfaced in 2007 when Hayman's Old Tom gin was created. Old Tom gin tends to be more viscous and fuller-bodied than all other styles, with its sweetness derived from naturally sweet botanicals, malts, or added sugar.

Plymouth Gin

Plymouth gin was created in 1793 from the historical city of Plymouth in England. This is one of the more unique styles. It is name protected, and maintains geographical orientation of sorts where it is defined by its origin. This style is not legitimately allowed to be duplicated outside of Plymouth. Its characteristics resemble the aromatics of a London dry with the mouthfeel of an Old Tom gin.

Jenever or Dutch gin

Even though gin is associated more frequently with England, it was first invented in Holland by *Franciscus Sylvius* (b.1642–d.1672), a Dutch physician and scientist. He originally used gin to treat people with kidney and bladder ailments. Holland increased the production of gin as it gained considerable popularity during times of war throughout the 17th century. As the British soldiers were fighting alongside the Dutch, the British were introduced to the new drink that quickly became known as *Dutch courage* as they would imbibe prior to engaging into battle. In 1689, the Dutch born William of Orange became King of England; soon after everything that represented Dutch culture quickly came into vogue—gin consumption throughout England tripled.

Modern and Western Gin

Modern gins are a newer style of gin. These are gins with an intention of softening the focus on Juniper, and allowing other botanicals to be recognized. This allows for a greater inclusion of balanced flavors. Some examples of modern gin include Tanqueray No. Ten, Aviation Gin, and Hendricks.

Rum

Originally a product of the Caribbean and Latin America, rum is a distilled beverage made from sugarcane, or its derivative, molasses. Initially, the spirit is clear upon completion of distillation. Less common, but arguably more complex and intriguing are the ones that are aged in oak barrels for a period of months to years.

Rum owes a great debt to a simple plant: sugarcane. Hundreds of years ago, there began a love affair with sugar throughout Europe as the colonies became established around the Caribbean to make the sweet commodity. The by-product of sugar production—namely, molasses became a source of concern. At the time, there wasn't much use for the thick, sticky, sweet substance until it was soon discovered that molasses could be fermented and then distilled.

Rum has a sordid history closely tied to the infamous slave trade in the 17th and 18th centuries around the Caribbean and several Central and South American countries. Eventually the drink's popularity spread to Colonial North America and the demand for sugar, a necessity for rum production, became widely apparent and the need for labor source was an essential. A triangular trade was established between Africa, the Caribbean, and the English-controlled Colonies. Rum, sugar, and slaves were often a medium of exchange and eventually providing the impetus for the American Revolution. Pictured in Figure 11.7 is a bottle of Kraken Black spiced rum.

FIGURE 11.7
Spiced, dark rum. Courtesy of Erika Cespedes.

Styles of Rum

Rum has numerous regional variations within the Caribbean, the epicenter of world rum production. Virtually every major Caribbean island produces its own distinct rum style based on their innate sense of uniqueness. Due to the tropical climate commonly associated in most rum-producing areas, rum matures at a much faster rate than is typical for whisky or brandy. An indication of this higher rate is the angel's share, or amount of product lost to evaporation.

The various styles of rum can be grouped according to their color or through the primary language traditionally spoken in their production origin. The color of the spirit is primarily determined by the amount of time the spirit has spent aging in oak barrels. The longer it's been aged, the more color and flavor it picks up from the wood.

While the rules for rum production vary greatly from country to country, there are several main types: light, gold, spiced, flavored, rhum agricole, cachaca, overproof, age dated, and dark rums. Rum is the main component in some of the world's most famous cocktails. It offers a unique base to create the likes of cocktails such as Mai Tai, Mojito, and Piña Colada.

Light Rum

Light rum (may also be called silver, white, or clear rum) is clear and maintains a fairly mild aroma and flavor with a lighter body as compared to golden or dark rums. They are minimally aged (maybe a year at most) and act as great base for cocktails as they blend well with the numerous mixers and liqueurs. Most light rum is 80 proof (40 percent abv). Most Spanish-speaking islands and countries (Cuba, Puerto Rico, Venezuela, Colombia, and the U.S. Virgin Islands) traditionally produce this style of rum.

Gold Rum

Gold rum, also called Amber rum, has an amber or darker color gained through the application of barrel aging, or in some low-quality versions, through the addition of caramel coloring. Over time, aged rums will lose some of their water content due to evaporation. This missing liquid has long been called the "angel's share." The aging process allows the rum to acquire a greater assertive aroma and flavor of vanilla and other bakeshop qualities while gaining greater viscosity. Most English-speaking islands and countries (Barbados, Jamaica, Belize, and Trinidad) are traditionally known for darker rums that retain a greater amount of molasses flavor.

Spiced Rum

Spiced rum obtains their distinctive and spicy characteristics through the addition of spices (cinnamon, fennel, allspice, and rosemary) and possibly caramel. Lower-quality versions may be derived from light rums with a larger addition of caramel color and flavor; better-quality versions will use amber rum allowing the product to be more natural.

Flavored Rum

Flavored rums often consist of a light rum base that has been infused with fruits, herbs, and/or nuts. These rums can be partnered with light rum or stand alone with the addition of fruit and juices to create one of the many—tropical style cocktails.

Dark Rum

Dark rum generally undergoes extended aging from barrels that were heavily charred and/or have been given some dose of caramel coloring. The color, aromas, and flavor are very assertive with intense spice and bakeshop characteristics. Many dark rums are full-bodied. Most French-speaking islands (Martinique and Haiti) are famous for their agricultural rums being produced primarily from sugarcane juice.

Overproof Rums

These contain much higher than the standard 40 percent abv (80 proof), with many containing as high as 75 percent (150 proof) or more. These types of rums are used as floaters over the top of a finished cocktail. One of the most common examples is Bacardi 151.

Añejo and Age-Dated Rums

These rums are aged for a lengthy period of time and then blended from different vintages or batches to ensure a continuity of flavor from year to year. Some aged-dated rums will provide age statements identifying the youngest rum in the blend. For example, a 10-year-old rum contains a blend of rums that are minimum ten years old. These rums are commonly consumed like Cognac—intended for slow sipping and savoring of the drink.

Rhum Agricole

The French name for cane juice rum, a style of rum originally distilled in the French Caribbean islands from freshly pressed sugarcane juice rather than molasses.

Cachaça

Cachaça (kuh-shah-suh) is a Brazilian white rum—distilled from sugarcane juice. Outside Brazil, cachaça is used almost exclusively as an ingredient in tropical drinks, with the Caipirinha being its most famous cocktail.

Tequila and Mezcal

Tequila is a Mexican spirit, essentially a type or subset of mezcal made from the fermented and distilled sap of the blue Weber agave plant. Tequila can only be made in 124 municipalities within the state of Jalisco (including the town of Tequila and most of today's production) and some surrounding areas. The official Mexican standard or NOM (Norma Oficial Mexicana) requires Tequila to be produced from a minimum of 51 percent blue Weber agave and allows for the addition of up to 49 percent other sugar sources.

NOM defines 100 percent agave Tequila as containing sugars exclusively from the blue Weber agave plant and it must be bottled at the distillery. If the label does not say 100 percent agave it is a mixto. Pictured in Figure 11.8 is a label stating 100 percent agave.

FIGURE 11.8
Tequila made from 100 percent Blue Weber agave. Courtesy of Erika Cespedes.

The blue Weber agave plant is considered superior to other agave varieties. It appears like a giant aloe vera plant with spikes on the tips. Blue Weber agave takes between eight and twelve years to grow before it can be harvested—this obviously impacts both the availability and price of Tequila.

In the ground, the plant produces a large bulb called a piña, which has the look similar to a pineapple. The agave's leaves are removed and the piñas are quartered and slowly steamed until all starches are converted to sugars. The piñas are then crushed in order to extract its sweet juices, yeast is added, and finally it's fermented. In the production process, Tequila is twice distilled in either copper stills or continuous stills with an alcohol content range from 70 to 110 proof.

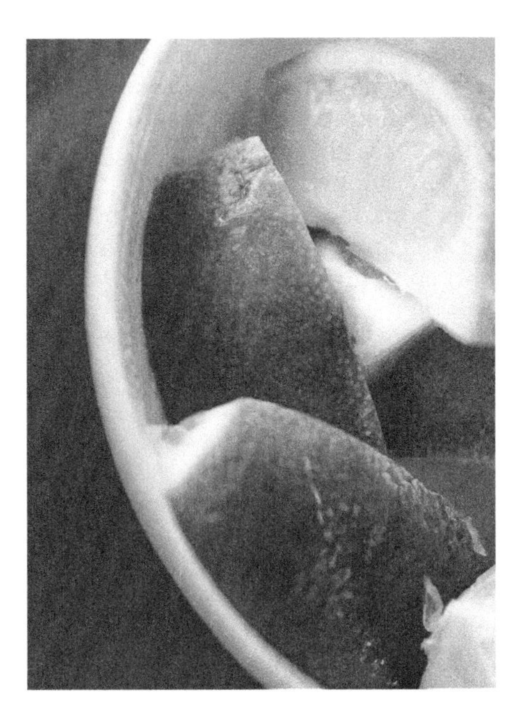

FIGURE 11.9
Lime wedges. Courtesy of Erika Cespedes.

Tequila is an incredibly popular North American spirit that can be consumed both neat (with no other ingredients) or as a base spirit in the numerous variety of cocktails such as the Margarita and a Tequila Sunrise. When blanco Tequila is consumed neat, it is often served in a narrow shot glass and may be served as Tequila cruda. *Tequila cruda* consists of taking a single shot of blanco or reposado Tequila accompanied with salt and a slice of lime. The other Tequila styles—if intending to be drunk neat—will often be served in a low-ball glass or snifter, very similar to drinking and appreciating Cognac. Figure 11.9 shows a plate of lime wedges, an essential ingredient in a Tequila cruda.

Styles of Tequila

Each of Tequila's five categories is based on a combination of its age and corresponding color. As with wine and other spirits that are aged in oak casks, Tequila is enhanced with aging in oak barrels (mostly American oak, but some producers use French oak), while the harshness of the alcohol mellows and the drink becomes more complex.

Blanco (BLAHN-ko)

Blanco or white Tequila or called the occasional *plata* (PLAH-tah) or silver Tequila is a clear spirit that has not been aged, or aged for at most two months spent in either stainless steel tanks or previously used oak barrels prior to being bottled.

Joven (HOE-ven)

Joven or young Tequila or called the occasional *oro* (OHR-oh) or gold Tequila is the result of blending silver Tequila with some varying amounts of Reposado and/or Añejo Tequila in the higher quality versions. But the vast majority of joven Tequilas are unaged that are typically mixtos (51 percent blue Weber Agave with the addition of sugars, caramel coloring, oak extract, glycerin, syrup, and other additives). This type of Tequila is often blended with mixers or consumed as a shooter.

Reposado (RAY-po-sah-doe)

Reposado or rested Tequila is aged a minimum of two months, but less than a year in oak barrels. The barrels begin to mellow the aromas and flavors to the agave and impart subtle oak characteristic, yielding the Tequila with a light straw color.

Añejo (AN-yeay-ho)

Añejo or aged or the prestigious vintage Tequila is aged for a minimum of one year, but less than three years in oak barrels. These Tequilas emphasize greater influence of vanilla, butterscotch, and baking spices from the barrel.

Extra Añejo

Extra Añejo (extra aged or ultra-aged) Tequila is aged for a minimum of three years in oak barrels, emphasizing a significant blend of agave character with the influence of the barrel. This is a newer category that was established in 2006. This type of Tequila is often served neat with at most, a splash of water.

Mezcal

Mezcal (mezz-KEHL), like its cousin Tequila is made from agave. While Tequila must be made with a minimum of 51 percent blue Weber agave, any lesser percentage of blue Weber agave would be labeled as mezcal. Unlike Tequila, mezcal can use any of the eight approved varieties of the agave plant. It's made in eight specific regions of Mexico, although Oaxaca is the center of the mezcal world with 80–90 percent of production. Mezcal distillers traditionally slow-roast the agave by burying it in pits with hot rocks, this works to infuse its signature smoky characteristic.

It is a common misconception that Tequilas contain a "worm in the bottle." Only certain mezcal (usually from the state of *Oaxaca*, wah-hahk-ah) are ever sold in this manner, which initially began as a marketing gimmick in the 1940s. The worm is actually the larva from a moth that lives on the agave plant.

Marc and Grappa

Marc (mahrk) is a French spirit produced from the remaining residues of winemaking—skins, seeds, stems, and any leftover grape juice. These remains are set aside to ferment and then distilled into Marc. This spirit is also known as *grappa* (grahp-pah) when it authentically comes from Italy but can be identified according to either name when it comes from elsewhere. Most of this spirit, regardless of the name, is non-aged and clear. If it is aged, it will appear golden to golden brown. In most scenarios, this spirit is consumed *neat*, or without ice or mixers, after a meal to aid digestion.

Lesser Known Clear Spirits

Fruit brandy, or the French term *eau de vie*, is a spirit that can be made from almost any fruit—commonly apples, pears, berries, or plums. The specific name of the spirit is determined by which fruit the spirit is produced from. For example, a brandy made from raspberries is called *framboise*. If the drink is made from strawberries, it is called *fraise*; if made from pears, it is called *poire*, and cherry brandy is known as *kirsch*.

Akvavit or *Aquavit* (AW-kwuh-veet) is a potato-based spirit traditionally from Scandinavia. This spirit is predominantly flavored with caraway and varying amounts of aniseed, fennel, cumin, dill, and/or bitter orange. Aquavit may be referred to as "drinking snaps," which comes from a Nordic verb, *snappen,* which describes how aquavit is traditionally drunk. It is not in reference to the sweet-flavored liqueur *schnapps* we may normally associate with the term. The Aquavit is snatched or seized and "thrown back" in one gulp; in essence, it is drunk as a shot.

Arak (air-RAK) and *Raki* (reh-key) are strong Eastern European and Middle-Eastern spirits with up to 50 percent alcohol. Bottles labeled with either Arak or Raki terms may contain fennel, figs, dates, raisins, or plums. Because of their potent alcohol strength, this spirit is typically chilled and/or served with water and ice. Arak or Raki, or some similar type of drink may have been the first distilled spirit—quite possibly originating from India or China between 800 and 1000 BCE.

Popular Aged (or Brown) Spirits

Learning Objective 4
Identify all the popular brown spirits and their main fermentable source

After the distillation process, a spirit is clear and watery in appearance. If this clear spirit is aged (often in a wood barrel) for any length of time, the alcohol will imbibe some noticeable color—enhanced aroma and flavor—and increased mouthfeel. Figures 11.10 and 11.11 show a barrel and a bottle of KOVAL's 4-grain whiskey. The slow passage of oxygen through the wood barrels enhances the overall spirit by providing complexity and mellowing the alcohol. The aging process for distilled beverages usually takes at least two years (but the process can last as many as

FIGURE 11.10
Barrel of KOVAL's 4-Grain Whiskey. Courtesy of John Peter Laloganes.

FIGURE 11.11
Bottle of KOVAL's 4-Grain Whiskey. Courtesy of Erika Cespedes.

twenty years or more) for the spirit to gain significant benefits. The bottle is usually identified by its length of aging as opposed to wine which is identified by date of production. For example, twelve-year-old bourbon spent twelve years in an oak barrel, though it may be older because it likely spent more time stored in the bottle prior to being consumed. It is likely that aged spirits cost more than non-aged beverages. The additional time these spirits spend evolving in the barrel costs money because the producer is unable to recoup the costs on its investment during that time. In addition, the producer incurs storage costs and product loss caused by evaporation.

Brandy

Brandy is an aged spirit produced from the distillation of wine. The word brandy derives from the Dutch word *brandewijn*, or "burnt wine," which is how the clever Dutch traders introduced their "burnt" or boiled wine to Europe. Through distilling the wine, they made it easier to preserve and transport the product, yet unknowingly changing the fundamental character and irreversible aspects of the drink. Originally, the Dutch thought it was possible just to add the water back into the brandy, though eventually an appreciation for the newly created spirit flourished.

Brandy is a broad term used to indicate any wine deriving from grapes that has subsequently been distilled into a spirit. Since most brandies are distilled from grapes, the greatest brandy-producing areas of the world have roughly paralleled those areas also producing grapes for winemaking. Brandy is often best consumed as a sipping drink, one that is usually served in a short-stemmed, bowl-shaped glass called a *snifter*. The bowl of this glass is meant to be cradled in the hand, which

FIGURE 11.12
Brandy snifter. Courtesy of Erika Cespedes.

slightly warms the brandy. The snifter can also be gently swirled to assist in releasing the brandy's volatile aromas. Figure 11.12 shows a snifter of brandy.

Most countries produce some form of brandy—either clear or the more popular and well known colored version. Some examples of clear brandy include *Pisco* (pee-skoh) from Peru, Brazil, and Chile; and *Aguardente* (ah-gwahr-dehn-tay) from Spain and Portugal. Both brandies are aged in old, neutral wood so they lend no color to the spirit. Most other brandies have the recognizable golden-brown color hue. Brandy is often consumed neat, or with a splash of water, but it's also the main component in some of the world's most famous cocktails. It offers a unique base to create the likes of cocktails such as Brandy Alexander, Pisco Sour, and the Sidecar.

Some brandies produced around the world are given an aging classification and/or appellation system that allows the consumer to better understand the brandy's origin and its characteristics. Spain (known for Brandy de Jerez) and France (known for Cognac, Armagnac, and Calvados) are two countries that have implemented a grading and appellation system to control the production of brandy.

Cognac and Armagnac

Cognac (cohn-YAK) and *Armagnac* (ar-muhn-YAK) are the world's most renowned brandies, both produced within specific regions of France. They are the only two of three officially designated brandies in all of Europe–Jerez in Spain is the third. Both Cognac and Armagnac begin their lives as humble, insipid white wine made primarily from the white grapes, *Ugni Blanc* (oo-nee blahwn–also known as *Trebbiano*) and/or French Colombard, and then begins a process of being transformed into the two most reputable brandies in the world.

Cognac is the world's most renowned brandy from the Charente region of France situated just north of Bordeaux along the Atlantic coast. Armagnac is Cognac's under-appreciated, yet older land-locked relative located in the Gascony region of France just along the foothills of the Pyrenees Mountains in southwest France. Armagnac has the distinction of being the oldest wine distilled in France; production dates to the 14th century. Cognac and Armagnac are two distinct but similar types of French brandy.

Production of Cognac

The Cognac-production area is divided into six districts or crus that cover much of the *Charente* (sheh-rahnt) and all of the Charente-Maritime (close to the Atlantic Ocean) departments north of Bordeaux. Each district maintains its own unique characteristics, based on desired geological elements, largely the degree of chalky soil. The defining chalky soil contributes to a grape's high levels of acidity, ultimately, a highly important aspect in the finished spirit. The crus form rough concentric circles around the town of Cognac, beginning with Grande Champagne, which is the most respected cru, not to be confused with the Champagne region of northern France. Circling outward, the other crus are Petit Champagne, Borderies, Fin Bois, Bons Bois, and Bois Ordinaire. Figure 11.13 shows a map of Cognac.

Another Cognac label designation—"Fine (feen) Champagne" Cognac is the result of an assembly of Grande and Petite Champagne spirits with a minimum of 50 percent deriving from Grande Champagne.

Cognac is double-distilled in a copper pot still, created through two stages known as the *chauffe* (showf). The first chauffe produces a lighter spirit called a *brouillis* (broo-yee) with an alcohol content of 24 to 30 percent abv. The brouillis is then

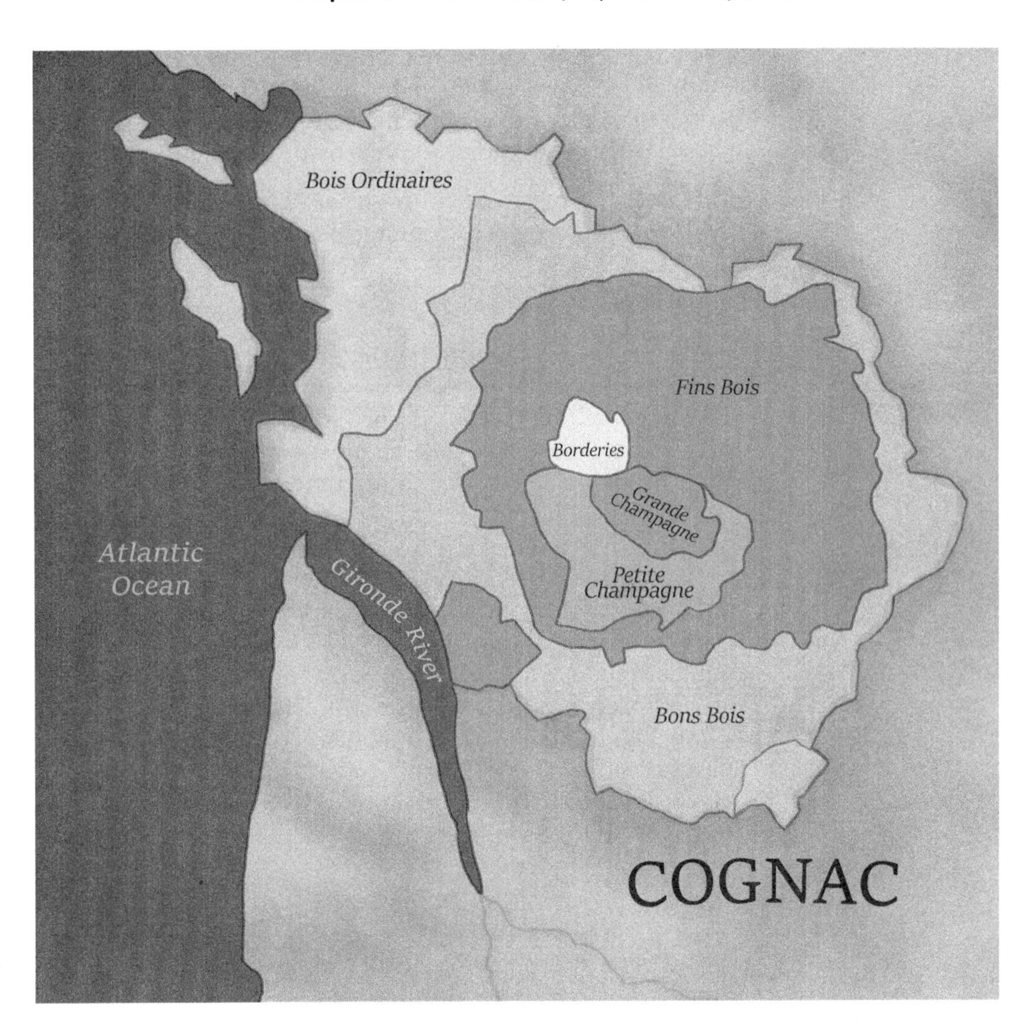

FIGURE 11.13
Map of Cognac. Courtesy of Thomas Moore.

redistilled in a second heating called the *la bonne chauffe* (lah-bohn-showf) where the spirit undergoes a "cutting process" or separation by the distiller. The "cutting" of the spirit has three segments—the head (the first, higher-alcohol vapors), the heart (the middle, clear spirit) and last, the tail (the remaining weaker vapors). Only the best part of the cutting—the "heart" of the distillation—is reserved for aging to eventually become Cognac. The left-over head and tail segments of the spirit will be mixed with the next batch of wine or brouillis to be redistilled. At this stage, the heart is a clear spirit that ranges between 68 and 72 percent abv and will be slightly diluted with water to cut and mellow some of the alcohol prior to aging.

Styles of Cognac

Styles of Cognac are based on its aging designations. The age of Cognac is determined solely by the number of years that it has matured in oak barrels—these clearly labeled designations are marketed and increasingly priced accordingly to their ascending order of age. The coding is designated based on the length of aging of the youngest spirit in the blend. Since most Cognac is blended, the spirit will likely be older than its legal minimum. There are three official age designations for these spirits:

- "VS" or Very Superior, which must be aged a minimum of two years
- "VSOP" or Very Superior Old Pale, which must be aged a minimum of four years

- "XO" or Extra Old, which must be aged a minimum of ten years, though it is likely to average well over twenty years in age. XO Cognac also can be labeled with some of the following terms: extra, Napoleon, VVOSP, cordon bleu, vieille reserve, grande reserve, royal, and vieux

Producers have created non-legal marketing terms like "Hors d'Age" (meaning ageless) to denote special Cognac and Armagnac that are older than the typical XO.

Production of Armagnac

The Armagnac-production area is divided into three districts along the foothills of the Pyrenees Mountains found in southern France. The Armagnac region consists of Bas Armagnac in the west that produces fruity, complex types of brandy, the Grand Bas Armagnac that produces the most prized Armagnac, and La Tenareze in the center that produces a coarser brandy that ages well.

Unlike Cognac, Armagnac is distilled only once and retains more of a fruity, yet rustic mouthfeel. This approach produces a somewhat less refined spirit with approximately 40 percent alcohol (therefore, no need for dilution of the final product). Ultimately, the clear spirits will be transferred into oak barrels and left to age in cellars for a minimum period of two years. The coopers have traditionally used wood from the *Limousin* (lee-moo-zahn) and the *Tronçais* (traohn-kay) forests. The Tronçais forest provides wood with soft, fine grains, which is particularly porous to alcohol. The Limousin forest produces medium grained wood, which is harder and even more porous. As the spirit ages, its aromas and flavors concentrate and the color darkens to a warm shade of amber. Throughout the aging process, the spirit will mellow and become less aggressive while the color darkens and the aromas and flavors become more complex.

Styles of Armagnac

Styles of Armagnac are based on its aging designations. The age of Armagnac is determined solely by the number of years that it has matured in oak barrels—these clearly labeled designations are marketed and increasingly priced accordingly to their ascending order of age. The coding is designated based on the length of aging regarding the youngest spirit in the blend—since most Armagnac is blended—the spirit will likely be older than its legal minimum. Pictured in Figure 11.14 is a bottle of VS Armagnac. There are four official age designations for Armagnac:

- "VS" or Very Superior, which must be aged a minimum of three years
- "VSOP" or Very Superior Old Pale, which must be aged a minimum of four years
- "XO" or Extra Old, which must be aged a minimum of six years
- "Hors d'Age," which must be aged a minimum of ten years

Extra, Napoleon, and vieille reserve are also terms that can be used on the label. Single vintage year/age statements may also be used; either a single vintage or an age statement such as 10 or 20 years old.

FIGURE 11.14
Armagnac VS. Courtesy of Erika Cespedes.

Brandy de Jerez

The Moors (a group from northern Africa) settled in southern Spain, very near *Jerez* (hehr-eth) in the year 711. The tribe is recognized for introducing the distillation technique to this part of the world. Being devout Muslim, the Moors opted not to drink the spirit; instead they distilled the local fortified wine, Sherry, in order to make perfumes, antiseptic, and for medicinal purposes. The spirit became known as *Brandy de Jerez*—required to be aged in American oak which previously has contained Sherry, and utilize the traditional Spanish blending system (the solera method). Brandy de Jerez must be aged exclusively in the province of Cádiz, more famously known as the *Sherry triangle*. The Spanish government declared a *denominacion especifica* (DE) for Brandy de Jerez in 1987.

Styles of Brandy de Jerez

Just like Cognac and Armagnac, the age of Brandy de Jerez is determined solely by the number of years that it has matured in oak barrels—these clearly labeled designations are marketed and increasingly priced accordingly to their ascending order of age. There are three designations of Brandy de Jerez:

- Brandy de Jerez Solera, aged on average for one year
- Brandy de Jerez Solera Reserva, aged on average for three years
- Brandy de Jerez Solera Gran Reserva, aged on average for ten years

Calvados Brandy

Calvados (kehl-vuh-dose) is an apple-based brandy from the Normandy region of France. Calvados is distilled from specially grown and selected apples where it is common for a producer to use well over 100 specific varieties. Each variety can add a different dimension of complexity—some are selected for their tartness and bitterness, while others may be selected for their sweetness and fruit. It can take between 8 and 16 pounds of apples to make a single bottle of Calvados.

The rules for Calvados vary according to their designation awarded by the French Appellation d' Origine Contrôlée System (AOC). *AOC Calvados* contains the broadest requirements with the more restrictive and prestigious appellation of *AOC Calvados Pays d'Auge*. Another designation is the *AOC Calvados Domfrontais* which requires at least 30 percent of pears in the initial cider prior to distillation.

The fruit is harvested by hand in the fall-time and then initially fermented into a dry cider when it then undergoes two distillations, which traditionally occur in copper Alembic pot-stills.

Styles of Calvados

After two years of minimum French oak aging requirements, the wine can be labeled with the Calvados designation only after passing a blind tasting by a committee of local experts. Most Calvados are a blend of multiple years, though some producers produce vintage Calvados. The aging system references the youngest one in the blend. There are five official Calvados aging designations:

- Fine or Trois étoiles or pommes, which must be aged a minimum of two years
- Réserve or Vieux, which must be aged a minimum of three years
- VO or Vieille Réserve, which must be aged a minimum of four years
- VSOP or Grande Reserve, which must be aged a minimum of five years
- XO, Napoléon, or Hors d'Age, which must be aged a minimum of six years

Whiskey (or Whisky)

Whiskey is a general term referring to a distilled spirit derived from grain—mostly from barley or corn, but whiskey can also come from wheat and rye or a combination of these grains. Essentially, whiskey is distilled beer that has been aged in oak barrels.

FIGURE 11.15
Selection of Whiskey. Courtesy of Erika Cespedes.

Most of the time it is distilled twice before it is placed in barrels where the aging process can last as little as two years, but it can also extend for several decades. Figure 11.15 displays a collection of whiskey.

Whiskey includes Scotch (from Scotland), Irish whiskey (from Ireland), Canadian whisky (from Canada), and lastly, Bourbon and Tennessee whiskey (from the United States). Each of these types of whiskey has some unique qualities beyond the obvious geographical origin that make them unique.

The word *whiskey* can be spelled two different ways. *Whiskey* (ey) refers to American and Irish whiskey. *Whisky* (y) refers to Scotch and Canadian whisky. Some exceptions to this rule do exist, however. For example, Maker's Mark and Old Forester bourbon use both the Scotch and the Canadian spelling.

Whiskey is often consumed neat, or with a splash of water, but it's also the main component in some of the world's most famous cocktails. It offers a unique base to create the likes of cocktails such as Vieux Carré, Manhattan, Mint Julep, Sazerac, and the Whiskey Sour.

Scotch Whisky

Scotch derives from Scotland and comprises of two main categories: single malt and blended. Most Scotch whisky contains the characteristic malted barley that is smoked and dried over peat. Peat is compost that is comprised of a soft carbon fuel made from vegetable matter that occurs naturally and is harvested from the land. When peat is burned, it has a very pungent complex aroma that is eventually imparted into the malted barley. Corn, wheat, and other cereal grains can be used for blended Scotch, but only barley is allowed for single malt whisky. Regardless of Scotch category, the law requires Scotch to be aged for at least three years in either a used American white oak bourbon barrel or a used Sherry barrel. Pictured in Figure 11.16 is a bottle of Glenfiddich Single Malt Scotch.

Single malt scotch can derive from any of five areas: (1) Lowland, (2) Highland, (3) Campbeltown, (4) Islay (eye-lah), and (5) Speyside. In comparing the five different single malt Scotches, one might find that Lowland Scotch is mild, gentle, and sweet with a little smoke. Highland Scotch, on the other hand, is more full-bodied—intense smoke flavors, but is balanced. Campbeltown is similar to Highland Scotch, but it is more peat-flavored. Islay also is full-bodied, but it has salty and intense "peat" characteristics due to the area's proximity to the ocean and reliance on peat as a source of fuel. Speyside is quite similar to Lowland Scotch, but it has some reminiscent qualities of Sherry. Figure 11.17 shows two Single Malt Scotches–the Macallan is a Highland Scotch and the Balvenie is from Speyside.

FIGURE 11.16
Glenfiddich Single Malt Scotch. Courtesy of Erika Cespedes.

Blended Scotch originally became vogue as an alternative to the intense, smoky flavor that is characteristic of single malt Scotch. As its name suggests, blended Scotch does not come from a single source—the Scotch can consist of several types of grains and/or the same grains from different locations blended

FIGURE 11.17
Selection of Single Malt Scotch. Courtesy of Erika Cespedes.

FIGURE 11.18
Johnnie Walker Black Label Blended Scotch. Courtesy of Erika Cespedes.

together to construct this classic drink. Pictured in Figure 11.18 is a bottle and snifter of Johnnie Walker Black Label Blended Scotch.

Irish Whiskey

Irish whiskey and Scotch are quite similar, but Irish whiskey lacks the smoky flavor of Scotch. Unlike Scotch, peat is not as prevalent in Irish whiskey, therefore the barley of Irish whiskey is not exposed to smoke as it is being dried. In addition, Irish whiskey is rarely made from single malt. Many grain combinations are used in the production of Irish whiskey, including corn, rye, wheat, and oats. In addition, Irish whiskey usually is triple-distilled and must be aged for a minimum of three years. In most cases, it is not shipped out of the country unless it has been aged 5–8 years. Like Scotch, Irish whiskey is aged in used bourbon barrels or Sherry casks. Pictured in Figure 11.19 is a bottle and snifter glass of Jameson Irish whiskey.

FIGURE 11.19
Jameson Irish Whiskey. Courtesy of Erika Cespedes.

Bourbon Whisky

Not unlike many of the world's great spirits, wines, and beers—bourbon and religion are undoubtedly linked throughout history. Bourbon is named after the famous, "Bourbon County," Kentucky. Legend has it that Baptist minister Elijah Craig made the first bourbon.

According to a resolution passed by the U.S. Senate, bourbon can be made anywhere in the United States, but it cannot be made outside of the United States. Pictured in Figure 11.20 is a bottle of Basil Hayden's Kentucky Small Batch Bourbon.

There are four defining agents that distinguish bourbon from other American whiskeys. Bourbon must be:

- made from a mash that is at least 51 percent corn
- made from no other additives, other than water
- distilled no more than 160 proof, but enters the barrel at no more than 125 proof
- aged for a minimum of two years in charred new white oak barrels

FIGURE 11.20
Basil Hayden's Small Batch Bourbon. Courtesy of Erika Cespedes.

Technically, bourbon can be made anywhere in the United States as long as these four rules are followed, but the only state that can be listed on the label is Kentucky. Most bourbon is now made in or near Louisville, Lexington, and Bardstown, Kentucky.

Tennessee Whiskey

Geographically, Tennessee is very close to Kentucky; therefore, it makes sense that the whiskey made in the two states would maintain similar characteristics. The main distinction between the two spirits—Tennessee whiskey is maple charcoal filtered—this contributes a maple aroma and flavor that bourbon does not contain. There are two famous distilleries in Tennessee that make this type of whiskey: Jack Daniel's and George Dickel.

Canadian Whisky

Canadians spell their "whisky" in the Scottish fashion, without an "e" as opposed to "whiskey" from the United States or Ireland. Canadian whisky is traditionally made from large amounts of rye but now is more often made from multi-grains and in some cases containing a large percentage of corn. Although no single type of grain can be more than 49 percent each of mash. Canadian whisky must be aged in white oak barrels for a minimum of three years.

Rye Whiskey

Rye whiskey, or rye malt whiskey, is produced from a minimum of 51 percent rye. Rye whiskey can be made from 100 percent rye, but that is very rare. Like bourbon, rye whiskey must be aged for a minimum of two years by law (four years is more common, however) in new white oak barrels. This whiskey has characteristic aroma and flavor nuances of caraway seeds; otherwise it is likened to spicy rich bourbon.

Apéritifs and Digestifs

This is a broad and expansive category of alcoholic beverages. They include a wide range of styles but most can fall within two major types: those typically served before (apéritifs) the meal and those consumed after (digestifs) the meal.

An apéritif, deriving from the Latin *aperire*, "to open" is used in the sense of opening up the appetite. Many apéritifs are made from a wine base that has been fortified with herbs, spices, and other botanicals with the most recognizable example being vermouth.

A digestif is a bittersweet liqueur typically served after a meal, in theory, to aid in digestion. Digestifs have become more appealing when incorporated into cocktails or simply called, *amaros* (Italian for "bitter") like Cynar, Fernet-Branca, and Campari.

Vermouth

Vermouth is an aromatized wine that has been infused with botanicals. It is a type of aperitif used either as a component in cocktails, or consumed on its own, possibly over ice. The word "vermouth" derives from "wormwood," and is inherited from earlier Hungarian and German wormwood-infused wines of the same name.

FIGURE 11.21
Carpano Antica Sweet Vermouth. Courtesy of Erika Cespedes.

Wormwood remains vermouth's principal defining botanical. Most vermouths carry varying levels of bittersweet character that works to stimulate the production of gastric juices and promote appetite. These products are relatively low in alcohol content when compared to other spirits, but slightly higher than still wine.

Vermouths have either a white (dryer versions) or a red (sweeter versions) wine base. Largely due to their application of use, these products are often discussed more in the spirit world as opposed to the wine world, even though they are a type of fortified wine. Modern vermouth—as a commercial product—originated in the region around Turin, Italy in the late 18th century as a moderately sweet, herbaceous beverage. The naming and labeling of these products are not always straightforward, but there are several popular brands listed below:

- *Cinzano* (chin-ZAH-no): Cinzano can trace its history back to 1757. In 1815, Cinzano relocated to Torino, Italy. The brand is one of the world's most recognizable producers of white and red versions of vermouths around the world.
- *Dolin* (doe-LEEN): Produced in *Chambéry* (shahm-bay-RHEE), France. Dolin invented and commercialized the "white vermouth" style in 1821.
- *Carpano* (car-PAH-no): In Torino Italy, Antonio Benedetto Carpano invented the commercial model for red vermouth in 1786. Figure 11.21 shows a bottle of Carpano Sweet Vermouth.

Liqueurs

Learning Objective 5
Identify at least four of the categories for liqueurs

Liqueurs (lih-CURE), also known as cordials, are spirit-based with varying levels of sweetness and flavored with the infusion of fruits, herbs, and spices. The base spirit could be any neutral grain spirit, or the liqueur can be made from a base of vodka, Tequila, and whiskey. Liqueurs can be used as part of cocktail or they can be served alone as an after dinner drink.

The word *liqueur* comes from the Latin word *liquefacere*, which means "to melt or dissolve." Liqueurs are made by one of three methods: maceration, percolation, or distillation. To make liqueur, a variety of selected ingredients are dissolved into a neutral distilled spirit. The maceration method allows the flavoring ingredient to soak into the distilled spirit, bleeding its flavors into the spirit. The percolation method works the same way, but the spirit is sprayed over the flavoring ingredient until the spirit takes on its flavor. In the distillation method, the flavoring components are distilled with the spirit. This method is used for seeds and other ingredients that can withstand high heat. Pictured in Figure 11.22 is a bottle of Chambord liqueur—one of the most famous liqueurs known throughout the world.

FIGURE 11.22
Chambord black raspberry liqueur. Courtesy of Erika Cespedes.

The sugar content of liqueurs can range from 2.5 to 35 percent of the total weight, and the consistency of the beverage can be a thick, syrup-like substance or it can be the viscosity similar to other distilled spirits. The alcohol levels of liqueurs can range from 34 to 100 proof, yet most liqueurs do not exceed 60 proof.

Coffee and Chocolate-Based Liqueurs

- *Crème de Cacao* (krem de ca-COW): Crème de Cacao is a chocolate and vanilla bean-based liqueur. This liqueur is available in both white and brown varieties.
- *Cream Liqueurs:* Coffee and chocolate-cream liqueurs are consumed by themselves and often used to fortify coffee. They tend to be at the low end of the alcohol range, with levels between 30 and 40 proof. These liqueurs are composed of cream, a spirit, and the main flavoring agent. An example of a cream liqueur is Bailey's Irish Cream and Carolans Irish Cream.
- *Godiva:* Godiva is a chocolate liqueur with varying options of white chocolate, dark chocolate, milk chocolate, mocha chocolate, and caramel milk chocolate. They are produced from the renowned Godiva chocolatier.
- *Kahlua®:* Kahlua® is a coffee-flavored liqueur that derives from Mexico. Kahlua® has an alcohol level of 53 proof and is used in baking and in making candies. This liqueur is often served in coffee beverages, cocktails, or over ice.
- *Patrón XO Cafe:* Patrón XO Café is a blend of premium Tequila and coffee. The taste is dry, not sweet as with most low proof coffee liqueurs. The high level of alcohol (70 proof) brings out the essence of pure coffee and Tequila.
- *Tia Maria®:* Tia Maria® is a coffee liqueur made in Jamaica. It is very similar to Kahlua®, and some would argue that they can be used interchangeably. Tia Maria® is sweeter than Kahlua®, and it has an alcohol level of 53 proof.

Fruit-Based Liqueurs

- *Fruit Brandy Liqueur:* Fruit-based brandy liqueurs include mostly fruit-flavored grape spirits. Generally, they come in three flavors: apricot, cherry, and peach. They typically contain an alcohol level of about 40 proof.
- *Chambord* (Sham-BOARD): Chambord is a French black raspberry liqueur that has an alcohol level of 33 proof. This liqueur was supposedly inspired by a visit from King Louis XIV when he visited Château Chambord in the late 1600s.
- *Crème de Cassis* (KREM de kah-CEASE): Crème de Cassis is a sweetened, dark red liqueur made from French blackcurrants. While crème de cassis is a specialty of Burgundy, France—it is also made in Loire Valley, France, and Canada. This liqueur is partnered with white wine to create the famous French cocktail—*Kir*—or a *Kir Royal* when the white wine is substituted with Champagne.
- *Kirsch* (KEERSH): Kirsch is a colorless cherry brandy made primarily in Germany.
- *Malibu:* Malibu is a clear, coconut-flavored liqueur. It has an alcohol level of 56 proof.
- *Maraschino* (mahr-ah-SKEE-no): This is a clear liqueur that is flavored with cherries. Maraschino is aged for several years in ash wood barrels or glass after it is sweetened with simple syrup. The alcohol level can range from 50 to 100 proof.
- *Midori* (mih-DOOR-ee): Midori derives from the Japanese word for "green" and is in fact a green-colored liqueur that has an evident mélange of sweet

melon aroma and flavor with 46 proof. It was originally created in Japan in the early 1980s.

- *Poire Williams:* Poire (PWAR) Williams is a clear liqueur made in Switzerland and eastern France. It is made from the William pear and has an alcohol level of about 60 proof.
- *Sloe Gin:* Sloe gin is made from the tart sloe plum, so it is not really gin at all. Sloe gin is red and has an alcohol level that ranges from 42 to 60 proof.
- *Southern Comfort:* Southern Comfort, a liqueur with two different proof levels, 70 proof and 100 proof, has an amber color and a peach flavor. It is very popular in the United States. On its label, this liqueur claims creation in the Big Easy, or New Orleans, Louisiana, in 1874. This liqueur was created by bartender Martin Wilkes Heron.

FIGURE 11.23
Cointreau orange liqueur. Courtesy of Erika Cespedes.

FIGURE 11.24
Grand Marnier orange liqueur. Courtesy of Erika Cespedes.

Orange-Based Liqueurs

- *Aperol* (ap-err-ohl): Aperol is an Italian aperitif that was originally created in 1919—it didn't become popular until the mid-20th century. Aperol is made from an infusion of bitter and sweet oranges along with other herbs and roots. Aperol is bright orange in color and contains 11 percent alcohol. Aperol is the main ingredient in Spritz—the common aperitif drink consumed in Northeastern Italy's Veneto region.
- *Aurum* (ohr-room): Aurum is an orange liqueur with a brandy base. It is triple-distilled in a special way such that the orange flavor is not added until the last distillation. This Italian liqueur boasts an alcohol content of 80 proof.
- *Campari* (cam-pahr-ee): Campari is a bitter Italian apèritif made with a unique blend of herbs and spices with orange being the dominant flavor.
- *Cointreau* (kwahn-TROW): Cointreau is a French orange liqueur. It is double-distilled and infused with orange peel and some secret ingredients. Cointreau's alcohol level is 80 proof. Pictured in Figure 11.23 is a bottle of Cointreau.
- *Curaçao* (CURE-uh-soh): Curaçao is a rum-based, orange-flavored liqueur that is available in three colors: clear, blue, and orange. The clear version is also known as *triple sec*. The alcohol ranges from 50 to 80 proof.
- *Grand Marnier* (GRAN mahr-nYAY): Grand Marnier has the reputation of being the king of the orange liqueurs. This reputation is only fitting because the base of the liqueur is top-quality Cognac. Grand Marnier has an amber color and its aroma hints at the Cognac, oranges, and barrel aging that make up this famous liqueur. It has an alcohol level of 80 proof and should be served in a brandy snifter. Pictured in Figure 11.24 is a bottle of Grand Marnier.

- *Mandarine Napoleon:* This is an orange-colored liqueur with the flavor of tangerines. Mandarine Napoleon is made from a base of French brandy. It takes its name from Napoleon I, who liked to drink a similar beverage. It has an alcohol level of 76 proof.
- *Pimm's:* Pimm's is the British equivalent of an orange-flavored liqueur. However, it is based on London gin, so it also has a flavor of assorted herbs. It has an alcohol level of 50 proof.
- *Van der Hum:* This is South Africa's answer to an orange-flavored liqueur. It is made with a special orange that is indigenous to South Africa, the *naartjie*, which is like a tangerine. It has a low alcohol level of only 50 proof.

Licorice-Based Liqueurs

- *Absinthe* (ab-sinth): Absinthe is crafted through a direct distillation of macerated whole herbs and botanicals in neutral alcohol and water. For lack of a better category, it falls under licorice-based liqueurs—though it doesn't contain added sugar content. Absinthe must contain grand wormwood which, rumor has it, can cause hallucinations in high doses. Absinthe also contains anise and fennel; other whole herbs and botanicals are permitted but the primary flavor of distilled absinthe is anise. White- and green-colored versions of absinthe are available—regardless, the color typically clouds (like Sambuca and Ouzo) upon the addition of water.

  Absinthe is the spirit that has more mystery and intrigue than quite possibly any other spirit in the world. Absinthe was popular in the 1800s among the bohemians (like our hippies or hipsters) but this also led to its downfall. This liqueur was so popular in Paris in the 1860s that 5:00 pm was called the "green hour" referencing the drink's *verte* or green color. In the early 1900s, there was a connection with drinking absinthe, supposedly going insane, and subsequently committing violent crimes. The problem was most likely due to drinkers easily becoming intoxicated based on absinthe containing 90–148 proof in alcohol. Eventually, countries banned absinthe until the restrictions were loosened some 100 years later. There has been a modern revival as of 2007: new regulations allowed absinthe to be produced, shipped, purchased, and consumed in the United States.
- *Anis/Anisette* (ah-nees/ann-uh-SET): Anis is a clear, licorice-flavored cordial. A similar liqueur known as *anisette* is French in origin. Anisette is very sweet and has an alcohol level higher than anis, which ranges from 42 to 96 proof.
- *Galliano* (gal-YAH-noh): Galliano is a yellow liqueur with licorice and vanilla flavors. Galliano is Italian and has an alcohol level of about 70 proof.
- *Goldwasser* (gold-VAY-suhr): This liqueur has gold flakes floating in a clear spirit. The spirit is flavored with aniseed, caraway seed, and oranges. With an alcohol level of 60 to 80 proof, Goldwasser stands alone most of the time so that the gold flakes are not hidden.
- *Ouzo* (OO-zoe): Ouzo is an anise-flavored liqueur that is exclusively a Greek production. Ouzo is a popular apéritif when it is traditionally served over ice and mixed with water—altering its color from clear to a cloudy white. Ouzo is widely consumed straight from a shot glass. For a more sophisticated option, ouzo serves as an excellent digestif when served in a snifter and garnished with a few floating coffee beans.
- *Pastis* (pass-stee): Pastis is the French version of anise. It is a licorice-flavored liqueur and has an alcohol level of 90 proof.

FIGURE 11.25
Sambuca. Courtesy of Erika Cespedes.

- *Pernod* (pear-NOH): Pernod is a popular brand of pastis, or licorice-flavored liqueur.
- *Sambuca* (sam-BOO-kah): Sambuca is a popular Italian anise-flavored liqueur. Its most popular version is clear or white Sambuca, though another version, a black sambuca is also available. Sambuca can be served in various ways: poured neat, over ice or mixed with water or coffee. In Italy, the popular *Caffè corretto* incorporates the liqueur into coffee or espresso as a sweetener to substitute sugar. It is also common to serve sambuca in a snifter with some floating coffee beans or *Sambuca con la mosca* (literally, Sambuca with fly). Commonly three coffee beans are used to represent health, happiness, and prosperity. Pictured in Figure 11.25 is a bottle of Sambuca liqueur.

Nut-Based Liqueurs

Nut-based liqueurs are a general term used to describe sweetened spirits flavored with nuts such as almonds, walnuts, or hazelnuts. The alcohol levels of these liqueurs range from the high 40 to about 80 proof.

- *Amaretto* (ah-muh-ret-oh): Amaretto has the flavors of almonds and apricots. One of the base ingredients of the liqueur is apricot stones, the seed from inside the apricot. Amaretto is served with coffee and chocolate desserts. The alcohol proof is in the mid-50s and its color is a deep orange brown.
- *Crème de Noyeaux* (krem de noy-YOH): This is an almond-flavored liqueur made from apricot kernels. The name comes from the French *noyau*: "kernel, pit, or core."
- *Frangelico* (fran-JELL-ih-koh): This is one of the most famous "nut" based liqueurs. Produced in northern Italy, its origins date back more than 300 years to the presence of early Christian monks living in the hills of the area. The *Frangelico* name is part of a local legend–an abbreviation of "Fra. Angelico," a hermit monk believed to have inhabited the Piedmont region during the 17th century. The Frangelico bottle is distinctively shaped like a monk's garment, with a traditional rope belt around its waist. This liqueur is infused with local hazelnuts and small amounts of cocoa and vanilla. Pictured in Figure 11.26 shows a bottle of Frangelico liqueur.

FIGURE 11.26
Frangelico liqueur. Courtesy of Erika Cespedes.

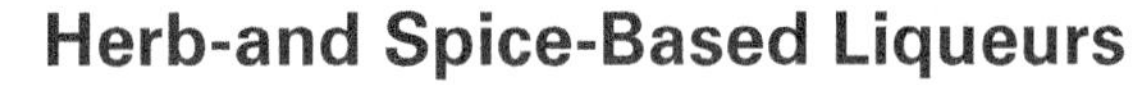

Herb-and Spice-Based Liqueurs

- *Benedictine:* Benedictine is a herbal liqueur with a Cognac base. It is named for an order of Christian monks. After being invented in the 1500s, Benedictine was not made for almost 80 years (between 1789 and the 1860s) because the French banned its production after the French Revolution. Benedictine started being produced by a descendant of the monk's lawyer. The label identifies the letters "D.O.M." that stand for "Deo Optimo Maximo" (To God, the best and greatest), which is the Bénédictine motto. The liqueur has an alcohol level of 80 proof. Pictured in Figure 11.27 is a bottle of Bénédictine liqueur.
- *Crème de Menthe:* Crème de Menthe is a highly sweetened liqueur flavored with mint leaves. The liqueur is available in both green and clear options—necessary for the appropriate cocktails.

FIGURE 11.27
Bénédictine liqueur. Courtesy of Erika Cespedes.

- *Drambuie* (dram-BOO-ee): Drambuie is an amber liqueur made from Scotch whiskey, honey, and herbs. Drambuie has an alcohol level of 70 proof.
- *Fireball Cinnamon Whisky:* It is a cinnamon-flavored whisky-based liqueur.
- *Glayva* (glah-VAH): Glayva is an amber liqueur made with Scotch whiskey, honey, and herbs, but unlike Drambuie, the producer also uses oranges. Glayva has an alcohol content of 80 proof.
- *Jägermeister* (YAY-gher-my-ster): Jägermeister is an herbal-based liqueur—made with a blend of over fifty herbs, fruits, and spices. It is produced in Germany and is commonly relegated for shooters.
- *Kummel* (kim-uhl): Kummel holds the distinction of being one of the oldest liqueurs. Its major flavoring component is caraway seed. Kummel has a vodka base and is produced in Germany, as well as many other Eastern European countries. It has an alcohol level ranging between 54 and 70 proof.
- *Strega* (strey-guh): Strega is an Italian herb, orange, and spice liqueur with a yellow color. It has an alcohol level of about 80 proof.

Other Liqueurs

- *Advocaat* (ad-voh-kaht): Advocaat is in essence Dutch eggnog that can be added to coffee or enjoyed alone. This creamy, yellow-orange liqueur is used as both an aperitif and a digestive; the alcohol level ranges between 30 and 40 proof.
- *Chartreuse* (sharh-TROOZ): Chartreuse is green or yellow colored, herb liqueur. The production method for chartreuse is highly guarded by the silent Carthusian Order of monks. Chartreuse has ties to both France and Spain; the production has bounced between the two countries during the past two centuries because of political unrest in the two countries. Yellow chartreuse has an alcohol level of 80 proof, while its green sibling touts an alcohol level of 110 proof.
- *Crème Liqueurs:* This class of liqueurs should not be confused with cream liqueurs. Crème liqueurs have no cream and are very sweet and infused with many fruit options such as banana, raspberry, plum, and strawberry. These liqueurs may also incorporate nut flavors such as almond and hazelnut, or floral and garden agents such as rose petal, mint, celery, tea, and violet. They usually feature a picture indicating their flavor on the label. The alcohol level ranges from the high 40s to about 80 proof.
- *Parfait Amour:* This liqueur gets its vivid bluish-purple color from a vegetable dye. Parfait amour gets its flavor from violets, cinnamon, cloves, coriander seeds, and citrus fruit. The name means "perfect love" in French, but it is Dutch in origin.

The Demand for Craft Spirits

Learning Objective 6
Briefly explain the craft spirit movement

With the explosion of the craft distilling movement in the United States, the popularity often encompasses local and small batch producers. A small yet passionate group of artisanal distillers are redefining the spirit world by introducing new products for the local U.S. markets, ones that can even rival their worldly counterparts. Figure 11.28 shows a sign prominently displayed in the front of Chicago Distilling Company.

Among the spirits being crafted by these microdistillers include Pinot Noir brandy, London Style gin, Elderflower liqueur, single grain whiskeys, gin aged in

FIGURE 11.28
Drink Local Booze. Courtesy of Erika Cespedes.

FIGURE 11.29
Selection of local whiskey from KOVAL distillery. Courtesy of Erika Cespedes.

FIGURE 11.30
Chicago Distilling Company. Courtesy of Erika Cespedes.

neutral Pinot Noir barrels, Gewürztraminer grappa, Eastern-style gin, Hazelnut-spiced rum and vodka being infused with various agents from hot peppers, saffron, tarragon, and chocolate to basil.

Some examples of microdistilleries in the United States include:

- *Clear Creek Distillery* in Portland, Oregon—one of the nation's first microdistilleries.
- *Tuthilltown Spirits* in Hudson Valley, New York—New York's first whiskey distillery since Prohibition.
- *KOVAL* in Chicago, Illinois—Chicago's first distillery since the mid-1800s. Figure 11.29 displays a collection of KOVAL's whiskey.
- *North Shore Distillery* in Chicago, Illinois.
- *CH Distilling* in Chicago, Illinois—Chicago's first distillery bar.
- *Ransom Distilling* in Portland, Oregon—Producer of artisan gin, grappa, and wine. One of the first distilleries to age their gin in older wine barrels.
- *The Chicago Distilling Company* in Chicago, Illinois. Figure 11.30 shows a sign for Chicago Distilling Company.
- *Anchor Brewing and Distilling* in San Francisco, California—America's first single-rye malt whiskey called Old Portrero.

Check Your Knowledge

Directions: Use these questions to test your knowledge and understanding of the concepts presented in the chapter.

I. MULTIPLE CHOICE: Select the best possible answer from the options available.

1. Bourbon must be made
   a. with at least 51 percent corn
   b. aged for minimum two years in new charred oak barrels
   c. with no additives other than water
   d. all of the above

2. Ouzo is a ______ flavored liqueur from Greece.
   a. coffee
   b. anise
   c. nut
   d. feta cheese
3. Chambord is a ______ flavored liqueur from France.
   a. raspberry
   b. blackberry
   c. black raspberry
   d. orange
4. While a liqueur is a spirit, it is distinctive in that it contains varying amounts of
   a. fruit
   b. nuts and spices
   c. sugar and flavorings
   d. both a and b
5. Which location is NOT one of the production areas for single malt Scotch?
   a. Lowland
   b. Highland
   c. Westside
   d. Campbeltown
   e. Islay
6. Which part of the "cutting" of the spirit is most desired?
   a. the head
   b. the heart
   c. the feet
   d. the tail
7. Which of the following is true of the distillation process?
   a. It creates the initial alcohol content
   b. It removes water content
   c. It concentrates the alcohol content
   d. Both b and c
8. Spirits can be broadly classified into
   a. sweet and dry
   b. black and white
   c. clears (non-aged) and browns (aged)
   d. vodka and whiskeys
9. Which drink is NOT a popular clear (or non-aged) spirit?
   a. Vodka
   b. Gin
   c. Rum
   d. Brandy
10. Which drink is NOT a popular brown (or aged) spirit?
    a. Scotch
    b. Brandy
    c. Bourbon
    d. Gin
11. Tequila must be made
    a. from fermented grains
    b. from at least 51 percent agave
    c. from at least 51 percent blue Weber agave
    d. in Spain

12. Añejo Tequila is
    a. often used for shooters
    b. otherwise known as white or silver Tequila
    c. aged for a minimum of two months in oak barrels
    d. aged for a minimum of one year and is often drunk neat
13. Marc and grappa are spirits made from
    a. wine
    b. fruit
    c. the remains of wine
    d. sugarcane
14. A significant difference between Cognac and Armagnac is that
    a. Cognac is single-distilled while Armagnac is double-distilled
    b. they derive from separate regions in France
    c. Cognac is double-distilled while Armagnac is single-distilled
    d. both b and c
15. Gin is essentially
    a. only made in England or in Holland
    b. a flavored vodka
    c. aged in oak for complexity
    d. a flavored liqueur
16. For each consecutive distillation, the spirit becomes
    a. more clarified and pure
    b. higher in alcohol percent
    c. cleaner with the removal of congeners
    d. all of the above
17. The *proof* of a given spirit is
    a. half of the alcohol percent
    b. twice the alcohol percent
    c. it depends
    d. all of the above
18. Bourbon must be made
    a. with at least 51 percent corn
    b. distilled to no more than 160 proof
    c. aged for minimum two years in new American charred white oak barrels
    d. with no additives, other than water
    e. both a and d
    f. all of the above

II. DISCUSSION QUESTIONS

19. Explain the distillation process.
20. What is the difference between Tequila and mezcal?

CHAPTER 12

Mixology: The Art and Science of the Cocktail

CHAPTER 12 LEARNING OBJECTIVES

After reading this chapter, the learner will be able to:

1. Explain the characteristics of a proficient bartender
2. Identify the foundational elements needed in the production of a cocktail
3. Recognize several bartender tools of the trade
4. Familiarize oneself with common terminology used at the bar
5. Provide some characteristics that define the contemporary bar
6. Explain each of the drink-making techniques
7. Provide some examples of molecular mixology
8. Provide some examples for ensuring that a cocktail is produced consistently
9. Identify several types of garnishes for drink making
10. Identify all 22 classic cocktails

There are two reasons for drinking: one is, when you are thirsty, to cure it; the other, when you are not thirsty, to prevent it. . . . Prevention is better than cure.

—Thomas Love Peacock, Melincourt, 1817

The Essential Primer on Mixology

Mixology combines both the art and science of integrating compatible liquids and other ingredients to create a libation known as the cocktail. The act of combining specific types and amounts of alcohol and mixers (such as water, soda, juice, or milk) is done in such a way that it provides a visual, gustatory, and revitalizing aesthetic for the consumer and a profitable activity for the beverage operation. Figure 12.1 illustrates a visually appealing and potentially profit generating cocktail. The art of mixology is expressed in the drink's presentation and the bartender's flair at showmanship and entertainment. The bartender or barkeep has been the traditional title of the highly-trained expert who creates and reproduces cocktails. Recently, the term *mixologist*, meaning one who conducts mixology has become increasingly popular as this vocation has established itself as a credible and respectable career path. The usage of these often interchangeable terms of bartender and mixologist

FIGURE 12.1
Craft cocktail. Courtesy of Erika Cespedes.

FIGURE 12.2
Bartender behind the bar. Courtesy of Erika Cespedes.

can be argued for their rightful placement of job title. Pictured in Figure 12.2 is a bartender skillfully pouring from a strainer into a cocktail glass.

When a bartender produces a drink, it may appear as if the pouring of liquids is done without regard—in reality, the bartender is a highly-trained professional who applies specific drink-making techniques with the use of carefully measured ingredients. It is imperative for bartenders to follow recipes that have been previously tested for expected taste and ones that have been cost-out for achieving some profitable objective. If the bartender does not follow the recipes, or if the establishment does not have standardized drink recipes, then quality, consistency, and profit cannot be assured and the integrity of the beverage program can be compromised.

Service at the Bar

Learning Objective 1
Explain the characteristics of a proficient bartender

Bartenders pour and serve wine, beer, spirits, and practice *mixology*—the practice of creating and replicating cocktails. It takes a complex skill-set that involves study, patience, and experimentation to mix different liquids with differing chemical properties in order to achieve the "perfect drink." As a beverage manager, it is critical to be selective when recruiting and hiring bartenders. To be successful at building a beverage program, a bartender must not only be proficient at drink making, but also in providing excellent customer service. Giving good customer service includes how bartenders interact with the customers in both slow and hectic moments. Regardless of how busy, any guest walking up to the bar or initially sitting down should at least be acknowledged, and then when time permits, the bartender can provide a prompt and proper introduction which is polite, spirited (pun intended), professional and unhurried. The sign of a great bartender is to convey a certain "calm-under-fire" quality with precise bursts of movement.

In many bars and restaurants, the bartender employs a *bar-back,* an individual who serves like an assistant to the bartender. They don't reproduce or serve drinks, but instead act as a support to the bartender by performing a variety of functions such as stocking, cleaning, running drinks, changing kegs, and so on. While the bar-back may not be directly involved with drink making, they free up the bartender to spend more time in being attentive and directly servicing the customer.

In some beverage establishments, there is a heightened and unique approach to service delivery. Having an element of theatrics through providing spirits with flair or flame and perhaps tableside service can add a unique distinction.

Serving Spirits with Flair or a Flame

Molecular mixology is applying scientific analysis and techniques to drink making. This new approach to mixology involves working with physical properties of drink by making foams, gels, mists, and applying heat to caramelize sugars and ultimately using an appropriate glass. Some of these drink creations waver between food and drink and clearly provide an element of distinction as they are served to the customer.

Even though most spirits are served at room temperature or cooler, some of these drinks are served with a flame. The nature of distilled alcohol allows the liquid to be lit easily. With the proper safety measures, lighting a drink can be a very dramatic manner for a bartender to present a drink. Lighting a drink can also sell more beverages since surrounding enthusiasts are likely to order the same. Huber's Café was established in 1879, and remains Portland Oregon's oldest restaurant. They are partly famous for their signature cocktail—*Spanish Coffee* that consists of Bacardi 151 rum, Bols Triple Sec, Kahlua, coffee, fresh whipped cream, and a touch of nutmeg prepared tableside with grand flair as it's lit on fire just before it's served.

Tableside Cocktail Service

Tableside cocktail service has gained more popularity by innovative operators looking to differentiate themselves in a crowded marketplace. Tableside cocktail service offers an added value and an element of drama. This approach works to heighten the visually appealing aspect. Some such suggestions include Bloody Mary and Mimosa carts for Saturday and Sunday brunch service or Martini and Manhattan carts for evening times. One other option can be shaking Margarita cocktails tableside and then straining them into the customer's glass.

The Foundation of a Cocktail

Learning Objective 2
Identify the foundational elements needed in the production of a cocktail

Distilled spirits are the base ingredients for the clear majority of cocktails. It becomes imperative for the beverage manager to know and understand the differences between the various spirits and liqueurs. This allows them to make remarkable cocktails or to more effectively manage those who are reproducing them. This book is not intending to provide a comprehensive list of all the possible cocktails or every variation known to man. Nor is it intending to serve as the "be-all know-all" guide to bartending. For that, look to *Mr. Boston*, the pre-eminent bartender's recipe and drink guide: It has been in print since 1935, with its most recent edition in 2011 containing more than 1,500 cocktail recipes. *The Beverage Manager's Guide to Wines, Beers, and Spirits* is focused on providing a foundation for the beverage manager, not the bartender. It is not necessarily expected of a beverage manager to be able to perform at the same level as his/her employees; however, one should be competent in bar terminology, basic drink-making techniques, and having a solid understanding of some classic cocktails. Included in this book are 22 classic cocktail recipes, selected and tested by Kai Wilson, famed Chicago mixologist, previously of the Drake Hotel and most currently, at Fountainhead, a craft beer and whiskey emporium. The selection of 22 classic cocktails represents drinks that have stood the test of time and therefore are important for the beverage manager. Mr. Wilson points out that "It is not a comprehensive list, but the selections were made to balance out the base spirit and to include a variety of drinks categories (sours, savory, and so on) and cocktail production techniques." Kai mentions that while he has striven to accurately document the cocktail recipes and their corresponding background details, any errors are strictly his own.

When creating (or reproducing) a cocktail, there is a foundation upon which it is built—it is often comprised of anywhere from two to three to four elements. All elements collectively assist to solidify the finished taste and appearance of the cocktail. The four significant components to cocktail production are ice, base spirit, modifiers, and garnish.

Ice

For the unsuspecting consumer, ice is simply frozen water used to maintain a chilled beverage. However, from the mixologist's perspective, ice is a fundamental component to any cocktail, and arguably it's as important as the base spirit used. The standard ice used throughout the beverage industry is sometimes referred to as "cheater" ice—made from a machine that produces a cylindrical ice with a hollow center. It is designed to consume a large proportion of the glass's volume as to "cheat" the drinker out of a portion of the product (think excessive use of breadcrumbs in the production of crab cakes). The major disadvantage of this ice is the rapid rate of dilution transforming an otherwise quality-oriented cocktail into a dull, flat, and watered-down version. The focus of beverage program must be on the quality of ice and on balancing the function of chilling a drink while simultaneously maintaining the integrity of the cocktail.

FIGURE 12.3
Ice. Courtesy of Erika Cespedes.

Without doubt, ice is one of the most important cocktail agents that drink makers take for granted. Surprising to some, ice is available in many shapes and sizes—all of which respond differently within the drink. Once ice is in contact with alcohol (as opposed to fruit, water, or any other agent), it begins a rapid rate of dilution. A drink with a larger and thicker ice cube will dilute less than a smaller and thinner cube. Due to less surface area, using an ice ball works to slow the dilution rate allowing the spirit to remain less adulterated with water. Based on the ice cubes available, the drink recipe may or may not need to be adjusted to adapt to the level of dilution. It is interesting to note that ready-made ice is a relatively new addition to the bar. In the past, bartenders had to crack their own ice off large blocks. Pictured in Figures 12.3, 12.4, and 12.5 is an ice ball being used with a bourbon whiskey to lessen the rate of dilution.

Base Spirit

The base spirit or liqueur is considered the largest quantity of volume in the drink. It tends to be the key spirit or liqueur that provides the essential foundation to the cocktail. The most common base spirits include: vodka, gin, rum, tequila, brandy, and whisky(ey). In many recipes, there is one main spirit; however, there are recipes that also call for the addition of two or more spirits in the same cocktail.

FIGURE 12.4
Ice with bourbon. Courtesy of Erika Cespedes.

FIGURE 12.5
Ice with bourbon (2). Courtesy of Erika Cespedes.

FIGURE 12.6
Seasonal bar ingredients. Courtesy of Erika Cespedes.

Modifiers/Mixers

The modifiers/mixers are the additional liquid agents added to a cocktail that work to enhance the base spirit or overall drink. This could be a secondary spirit or liqueur—or bitters, syrups, some herbs, spices, fruit juice, or soda agent.

Garnish

The garnish is the decorative agent whether it is a rimming procedure or as simple as a lemon wedge that adds a subtle enhancement yet highly visual appeal to the cocktail. Garnishes assist in solidifying the drink. Pictured in Figure 12.6 is a garnish tray that has been assembled for use throughout the evening shift.

Bartender's Tools of the Trade

Learning Objective 3
Recognize several bartender tools of the trade

Bartenders have a few special tools necessary to perform their vocation. Figure 12.7 displays several tools that are necessary for the bartender to perform the job. Having the appropriate tools are essential for replicating many of the popular cocktails and for assisting the bartender with pouring drinks more accurately and efficiently each time they are produced. The following is a list of some of the more common tools of the trade:

Cocktail Shaker Both Boston and standard shakers fall into this category. The Boston shaker consists of two containers; usually at least one is stainless steel, and the other is glass (often a pint glass) that allows one to overlap the other. The standard shaker is a stainless-steel tin with a removable strainer at the top. Figure 12.8 illustrates a bartender using a cocktail shaker with a pint glass.

FIGURE 12.7
Bar tools. Courtesy of Erika Cespedes.

FIGURE 12.8
Bartender shaking a cocktail. Courtesy of Erika Cespedes.

FIGURE 12.9
Hand-held juicer. Courtesy of Erika Cespedes.

Bar Spoon Varying from the average spoon, the bar spoon contains a long spiral handle ideal for reaching the bottom of tall glasses. This type of spoon is essential for stirring and layering drink-making techniques as well as tedious tasks like fishing cherries or olives out of a jar.

Jigger This metal double-sided measuring device is used for producing consistent alcohol-based drinks. A jigger usually holds 1½ or 1¼ oz. of alcohol on the larger side and some fraction of that ¾ or ½ on the smaller side.

Juicer This is a handheld device used for juicing citrus fruits as needed. Figure 12.9 displays a handheld juicer.

Muddler Similar to the appearance of a miniature bat-like device, the muddler is a thick stick made of wood or stainless steel. It's used to crush ice, mash fruit, and express the essential oils from herbs. Figure 12.10 displays a muddler.

Blender Blender is an essential machine used to blend drinks and crush ice for making frozen cocktails such as frozen margaritas or strawberry daiquiris.

Speed Pourers Speed pourers are placed in the opening of a spirit bottle. They are very useful for free pouring as they reduce spills by slowing the flow of alcohol from the bottle and into a glass. Figure 12.11 displays a set of speed pourers that will be inserted into the neck of a bottle in order to gain maximum control of the flow of spirits.

Cocktail "Hawthorne" Strainer These strainers are a circular metal tool with a handle and metal spring over the top. The cocktail "Hawthorne" strainers are specially designed to block unwanted ice when pouring a drink into a glass after it has been shaken or stirred. Figure 12.12 displays a Hawthorne strainer being used to hold back the ice that was used to make the cocktail, while being strained into a glass with fresh ice.

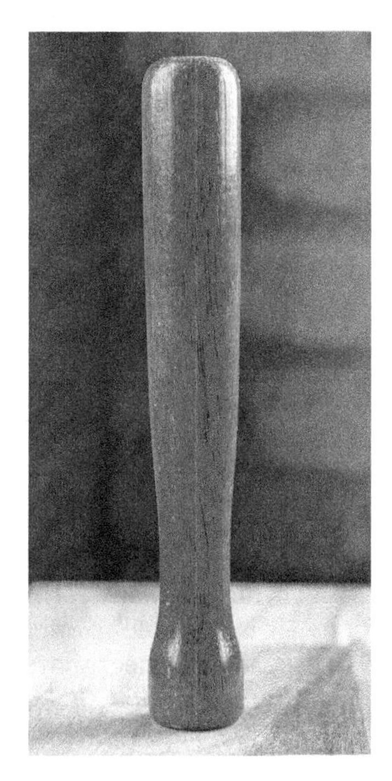

FIGURE 12.10
Muddler. Courtesy of Erika Cespedes.

FIGURE 12.11
Pour spouts. Courtesy of Erika Cespedes.

FIGURE 12.12
Straining a cocktail into fresh ice. Courtesy of Erika Cespedes.

Essential Drink-Making Terminology and Ingredients

Learning Objective 4
Familiarize oneself with common terminology used at the bar

Every vocation, from the medical field to having a specialty in law, has their own specialized lexicon. They share a language that is often considered foreign or technical to the outsider; the beverage industry is no different. Below is a list of terms, ingredients, and concepts that have special meaning for those working in a bar setting.

- *Bitters* is a distilled spirit used in the same manner as vermouth, as an enhancement to the drink. Bitters are flavored with botanicals and gain their spicy, bitter sensation from aromatic roots, herbs, spices and/or fruit. These can be thought of as the bartender's "spice rack." Bitters were originally prescribed by physicians as medicinal tonics in the 1700s. In the subsequent century, bitters became known as aromatic or cocktail bitters and were an increasingly popular agent used in drops and dashes, bringing subtle complexity to cocktails. Angostura and Peychaud's bitters were considered the classic and popular brands. Although recent times have brought an explosion of brands with different aroma and flavor profiles such as grapefruit, chocolate, and peach. Additionally, many mixologists have begun producing their own bitters in-house as a point of differentiation.
- *Call Drink* consists of a requested brand name of a spirit combined with some form of a mixer. Sometimes a call drink is referred to as a premium drink.
- *Carbonated Water* (club soda, soda water, sparkling water, or seltzer) is water that has been injected with carbon dioxide causing the water to become effervescent.
- *Chaser* refers to a mixer served separately from the "other" drink and often consumed immediately after drinking a straight shot.
- *Fizz* is a term that describes any beverage that has been carbonated or which emits small bubbles.
- *Flavored Syrups* are an easy way to sweeten a drink where it can provide a means to balance other agents such as sourness (from acidity) or spiciness (from alcohol). Flavored syrups are made ahead of time, as part of a bartender prep either in the beginning of their shift or even made a day in advance. These syrups are not only a time saver but also an excellent way to provide a depth of flavor. Through infusing agents such as herbs, spices, vegetables, or other agents, the production time for drink making is reduced at the bar. Some examples of flavored syrups include: cardamom bourbon syrup, ginger, lemon thyme syrup, and coffee and cinnamon syrup.
- *Flip* is a chilled, creamy drink made of eggs, sugar, and a wine or spirit. Brandy and sherry flips are two popular ones.
- *Frappé* is the name for a drink consisting of a liqueur or partially frozen fruit drink served over crushed ice.
- *Grenadine* is a classic pomegranate-based sweetener integral in many popular cocktails. Grenadine is an essential ingredient though most bars purchase artificially sweetened and flavored products that lack any form of authenticity. It is possible to mix equal parts of POM (pomegranate juice) and sugar into a jar and shake vigorously until the sugar is dissolved. The mixture can be brought to a brief boil to more effectively dissolve the sugar into the substance.
- *Highball* is spirit served with a mixer in a medium to tall highball glass.
- *Lace* is the last ingredient poured on top; also known as float.
- *Lowball* is a short drink made of spirits served with ice, water, or soda in a small glass.
- *Mist* is a liqueur served over crushed ice, often served as an after-dinner drink.
- *Mixer* refers to any addition to a drink other than alcohol and ice, such as juice and soda.

- *Neat* is a manner of serving a single unmixed spirit or liqueur, unadulterated, without any water, ice, or another mixer. Neat drinks are often served in a rocks glass (or lowball) or upon request, can be served in a snifter.
- *Nightcap* involves consuming an alcoholic beverage (often wine, spirits, or liqueur) prior to bedtime or at the closing time at a bar.
- *On the Rocks* refers to any drink served over ice.
- *Shooter* also known as *a shot* when it is consumed all at once.
- *Shot* is generally equivalent to 1¼ to 1½ ounce, though can also be considered 2 oz. in some beverage establishments. A "pony" shot is 1 oz.
- *Simple Syrup* (equal parts of sugar and water brought to a boil until the sugar dissolves) is a saturated mixture of sugar and water. As a liquid, it easily dissolves in drinks when a bit of sweetness may be needed.
- *Sling* is a tall brandy, gin, or whiskey drink with lemon juice, sugar, and soda water—hot or cold.
- *Smash* is a short julep made of a spirit, sugar, and mint, served in a small glass.
- *Sour* is a liquid concoction made from lemon/lime juice and sugar. Many bars will buy pre-made sour mix but some make their own with a mixture of sugar, water, and citrus juices (such as lemon, lime, and/or orange juice).
- *Splash* is the term used to inform a bartender to add a small amount of liquid, often to a base spirit; for example, a whiskey with a splash of water. Adding a splash of water to a whiskey (or other base spirit) will not only slightly dilute the alcohol by volume, but it also works to unlock the aroma molecules allowing them to be easier to identify.
- *Straight-Up* (or up) is a manner of preparing a drink as shaken or stirred with ice, strained, and served in a glass. "Straight up" means "chilled and served without ice in a glass."
- *Super-Call* (also known as top shelf or super premium) is either a higher proof spirit or a super-aged version.
- *Swizzle* is a tall rum-based drink served over cracked ice.
- *Tonic Water* is carbonated water with the addition of dissolved quinine. It contains a distinctly bitter taste.
- *Toddy* is a sweetened drink of a spirit (especially whiskey) and hot water, often with spices, and served in a tall glass.
- *Vermouth* (available in both dry and sweet versions) is an aromatized wine that has been spiced and fortified. It is used in small doses in various cocktails, especially the Martini and Manhattan. The flavors of vermouth can come from allspice, anise, bitter almond, bitter orange, cinnamon, clove, fennel, ginger, nutmeg, saffron, thyme, and vanilla. Figure 12.13 displays a Gin Martini made with an addition of dry vermouth.
- *Virgin* drinks are made without alcohol; a nonalcoholic drink.
- *Well-Drink* is a generic spirit (undefined by a brand name) and a mixer. For example: gin and tonic or rum and coke.

FIGURE 12.13
Martini. Courtesy of Erika Cespedes.

The Contemporary Bar

Learning Objective 5
Provide some characteristics that define the contemporary bar

One of the bigger cocktail trends in recent years has been the use of the term, *culinary cocktails*. These are specialized alcoholic beverages that not only integrate and merge bar and kitchen ingredients but also carry a chef mindset to the production of drinks. The modern mixologist prefers to bring an application of freshness and handcrafted products to the front-of-house not much different from what the chef has been doing for decades within the kitchen. Mixology has begun to incorporate house-made bitters, infusions, flavored syrups, and even barrel aging cocktails within the restaurant. Modern mixology tends to incorporate local and seasonal culinary ingredients (herbs, fruits, and vegetables) that enhance the overall appeal and drinking experience. Figure 12.14 displays a bourbon thyme cocktail with house made lemon honey syrup. Culinary cocktails are changing the way people think about their drinks and due to our continually evolving palates this trend is likely to continue. Another example of a culinary cocktail is offering the *Caprese Martini* in the late summer months of August and September. It's crafted by adding some vodka that was previously infused with local heirloom tomatoes and cracked pepper into a shaker filled with ice along with some muddled basil and a few drops of balsamic syrup. Shake until well chilled and strain the cocktail into a martini glass. Garnish with a skewer of fresh ball of mozzarella and cherry tomato with basil leaf.

FIGURE 12.14
Bourbon cocktail. Courtesy of Erika Cespedes.

Drink-Making Techniques

Learning Objective 6
Explain each of the drink-making techniques

There are many drink-making techniques that can be used when producing cocktails. These techniques are universal and it is imperative that every bartender master them—certainly at the very least these techniques should be theoretically understood and added to the beverage manager's repertoire.

Building

The "building" technique is the oldest and simplistic preparation technique. When building a cocktail, start by adding ice to the serving glass, then pour the ingredients into the same glass one by one. Usually, the ingredients are floated on top of one another with the alcohol initially poured in the glass prior to any mixers. Occasionally, a swizzle stick is placed in the glass, allowing the ingredients to be mixed as desired. Caution should be taken to ensure the drink is never built to the rim of the glass.

SCREWDRIVER

The orange juice in a Screwdriver makes this cocktail a great companion with brunch. This cocktail tastes even better when made with freshly squeezed orange juice.

Ingredients:

– 2 ounces of vodka
– 4 ounces of orange juice

Preparation:

1. Build the ingredients in a highball glass filled with ice.
2. Serve with a tall straw and a slice of an orange.

Variation:

Tequila Sunrise Replace the vodka with tequila and float ½ ounce of grenadine over the top. Garnish with an orange slice with a cherry on a pick. Serve with a tall straw.

Stirring

The "stirring" technique is quite simple for drink making. Special equipment is not needed other than a mixing glass and a bar spoon. Figure 12.15 shows a special bar spoon resting inside a mixing glass. With numerous classic cocktails making a comeback, this drink-making technique has once again become popular. There are two approaches for using this technique. Some bartenders prefer to stir their drinks without adding ice to ensure that the ingredients do not become diluted. In this instance, the ingredients are mixed first in one glass and then poured over ice in another glass to chill the drink. However, some people enjoy stirring a drink together with ice, and can easily do so without fear of diluting the drink.

When using the stirring technique to make a drink, avoid the use of crushed ice. Crushed ice can melt rather quickly, easily diluting the flavors of the drink. Cubed ice will not melt as easily, and will effectively chill the drink before releasing too much water into the ingredients.

FIGURE 12.15
Mixing glass and bar spoon. Courtesy of Erika Cespedes.

- To stir a drink without the addition of ice, the ingredients should be added to a mixing glass. A bar spoon, straw, or mixing rod is then used to stir the ingredients. It is best to stir slowly, in a clockwise motion, with six or so complete rotations made. The stirred drink can then be added to a chilled glass or a new glass filled with fresh ice.
- To stir a drink with ice, add the ingredients to a mixing glass that has been filled ¾th of the way with ice. A bar spoon, straw, or other mixing device is then used to stir the ingredients. The most effective stirring method is to stir slowly, in a clockwise motion, with six or so complete rotations made. The stirred drink can then be strained and added to a chilled glass or into a new glass filled with fresh ice.

The highly popular and classic Old-Fashioned and Manhattan cocktails offer excellent examples of demonstrating the stirring drink-making technique. Figure 12.16 displays a classic Old-Fashioned cocktail. This is considered one of the oldest American cocktails mentioned as far back as the early 1800s. The Old-Fashioned cocktail is made from predominantly a base of whiskey (rye or bourbon) with a bit of sugar and aromatic bitters (possibly muddled together), with a slight dilution in some form, from a splash of either water or club soda, and then garnished with a cherry and orange rind. The Manhattan recipe is another stirred cocktail with the recipe located in a further section within this chapter titled 22 Classic Cocktails.

FIGURE 12.16
Old-Fashioned cocktail. Courtesy of Erika Cespedes.

Muddling

The "muddling" technique is used to extract maximum oils and flavors from fresh ingredients such as citrus fruits and herbs. The ingredient(s) can be crushed directly in a mixing glass with the use of a muddler or the back end of the bar spoon. Muddling is usually done directly in the mixing glass that is being used to create and serve the drink. Muddling is completed just prior to most of the liquid ingredients being added to the glass. Some people find that smashing the ingredients using an up-and-down motion is acceptable, while others apply a less aggressive approach by twisting the muddler to combine the ingredients. The effects of muddling may not seem obvious, but they can certainly make the difference between a so-so drink and one that is extraordinary.

The very popular cocktail, the Cuban Mojito is an excellent example of demonstrating muddling drink-making technique. The technique involves limes being muddled with sugar and mint in the bottom of a glass. The recipe for this cocktail is included in a further section within this chapter titled 22 Classic Cocktails.

Shaking

The "shaking" technique is part entertainment for the visual appeal and part functional to produce the drink. The use of a cocktail shaker simultaneously chills and mixes ingredients.

Normally, the shaker is filled two-third with ice cubes, then any liquid ingredients are poured in the shaker, and finally the lid is placed tightly on the shaker. (Note that some bartenders prefer to add the liquid ingredients first and then add the ice, limiting the rate of dilution.) Hold the shaker in both hands, with one hand on top and the other supporting the base, and give a short, firm shaking motion for 10–20 seconds or until frost begins to form on the exterior of the shaker. Once the cocktail has been sufficiently chilled, it is ready to be strained. Remove the shaker lid and strain into a chilled glass or a glass filled with fresh ice. Pictured in Figures 12.17, 12.18, and 12.19 is a cocktail being strained into a glass. Figure 12.20 illustrates a classic Gin Martini with olives.

FIGURE 12.17
Straining a cocktail. Courtesy of Erika Cespedes.

FIGURE 12.18
Straining a cocktail (2). Courtesy of Erika Cespedes.

FIGURE 12.19
Straining a cocktail (3). Courtesy of Erika Cespedes.

FIGURE 12.20
Martini with an olive garnish. Courtesy of Erika Cespedes.

MARTINI

The classic Martini is one of the most standard drinks offered at every beverage establishment. There are numerous methods for making the Martini: gin or vodka; up or on the rocks; olive or lemon twist.

Ingredients:

- 2 ounces of gin or vodka
- ¼ to ½ ounce of dry vermouth
- 1 green olive or lemon twist for garnish

Preparation:

1. Pour the ingredients into a cocktail shaker (if shaken) or mixing glass (if stirred) filled with ice cubes.
2. Shake for 10–20 seconds OR stir for six cycles.
3. Strain into a chilled cocktail glass.
4. Garnish with the olive or lemon twist.

Variations:

There are many variations of the classic martini that are a matter of personal preference:

- Dry Martini–Traditionally uses little or no vermouth.
- Gibson–Garnish with a cocktail onion.
- Dirty Martini–Add about 1 ounce of olive brine into the cocktail shaker.

Shaking Egg Whites

Egg white—the albumen part of the egg will work to enrich the texture and appearance of certain cocktails. When the egg whites are shaken vigorously, the egg's proteins loosen while benefiting from an incorporation of tiny air bubbles to create foam. Some of the foaminess is integrated into the body of the drink, while providing an additionally visual and textured cap of foam on top of the cocktail.

There are two methods for creating egg white foam: dry shake and reverse dry shake. The dry shake is simple: place all the ingredients except for ice into a shaker tin, cover it tightly, and shake vigorously until a nice froth forms. Then add ice for a second round of shaking to properly chill the drink. The reverse dry shake is similar, but provides for less dilution. All ingredients, including ice but excluding the egg white, are placed into the shaker and given a vigorous shake. Then strain out the ice and add the egg white into the shaker, shaking vigorously to incorporate the ingredients. Pour the cocktail into a glass.

Blending

Blending is a vigorous technique that uses an electric blender in order to combine and "froth" fruit and/or other ingredients, such as ice, that don't break down well through shaking. This technique creates a smooth, often frozen mixture such as frozen Daiquiris, Margaritas, and Piña Coladas.

The drink's ingredients should be added into the blender first, then the ice. Although typical ice cubes can be used, crushed ice blends quicker. Begin by blending on a low speed and slowly progress to medium speed. There is no time limit given for blending drinks, since it can take different amounts of time to reach a smooth consistency. Be careful to not over blend as it is possible to dilute the ice and the drink will taste watered down. When blending, if there is an apparent hole in the center of the vortex as the drink is blending—the mixture is too thin and requires more ice. On the other hand, if the drink is moving slowly or not at all while blending—the mixture is too thick and needs more liquid.

STRAWBERRY DAIQUIRI

This is a popular frozen, blended cocktail ideal for summer time. The variations are endless with substituting any seasonal fruit in place of the strawberries.

Ingredients:

- 2 ounce light rum
- ½ ounce of triple sec
- 1 ounce lime juice
- ½ ounce of sugar
- 1 cup ice
- 5 strawberries

Preparation:

1. Combine all the ingredients in a blender.
2. Blend well at high speed.
3. Pour into a pint glass or daiquiri glass.
4. Serve with a tall straw and a strawberry garnish.

Layering

To layer or float a liquid (i.e., cream, liqueurs) on top of one another is one of the more difficult techniques applied by bartenders. Learning to master the art of floating layers of alcohol upon one another can lead to amazing visually aesthetic rainbow-colored drinks. Layering drinks were most popular in the late 1800s and early 1900s, especially in Europe with "pousse-cafés"—that would be comprised of layer upon layer of liquids.

When layering, slowly pour the liquid on the rounded or back side of a spoon as it rests against the inside of the glass. The ingredient should run down the inside of the glass and remain separated from the liquid below it. Each liquid has its own specific gravity or viscosity—learning the approximate gravity of each liquid allows this technique to occur easily. Higher density alcohols are said to sink below those that are of lighter densities. If the specific gravity of the liquid is unknown, read and compare the proofs of the bottles. Lower proofs of liqueurs generally mean there is more sugar and that the liqueur is thicker and heavier (for example, 151 proof rum can be floated on top to make flaming shooters). Keep in mind, the same types of liqueurs made from different companies can occasionally have different proofs and amounts of sugar content, thereby altering the viscosity or gravity of the liquid. When learning to layer, using a specific gravity chart may be useful. Very talented bartenders can layer drinks with more than eleven layers. The late *Max Allen*, Bartender Emeritus of the "Seelbach Hilton" in Louisville, Kentucky, was touted for being able to layer drinks—thirty-two to thirty-three levels deep.

B-52

This remains one of the most popular shooter (pousse-café) drinks.

Ingredients:

- 0.5 ounce coffee liqueur
- 0.5 ounce Irish cream
- 0.5 ounce triple sec

Preparation:

1. Layer the ingredients (one by one) into a pousse-café glass or a shot glass.
2. An adventurous person will light the top layer of the drink on fire. *Caution* ... Don't serve the drink until the flame has been extinguished.

Variations:

The options are endless. Utilize the gravity chart below and begin to experiment.

The following table lists the gravities of several popular spirits and liqueurs listed from heaviest to lightest. As a rule, the greater the difference in gravities, the easier it is to keep two alcohols from mixing. The table is not complete, but can allow one to begin layering with some common options that are available widely in the marketplace. The liquids with the lowest value (0.94) would be the lightest in the table versus liquids with the highest value (1.18) would be the heaviest.

Spirit or Liqueur	Gravity (In Order of Heaviest to Lightest)
Grenadine/Crème de Cassis	1.18
Anisette	1.175
Crème de Noyaux	1.165
Crème de Almond	1.16
Coffee Liqueur Crème de Banana Crème de Cacao White Crème de Cacao Goldwasser	1.14
Coffee Liquor Parfait d'Amour	1.13
Cherry Liqueur Crème de Menthe Strawberry Liqueur	1.12
Blue Curaçao Galliano	1.11
Amaretto Blackberry Liquor	1.10
Apricot Liquor Tia Maria Triple sec	1.09
Amaretto di Saronno Drambuie Frangelico Orange Curacao	1.08
Benedictine D.O.M.	1.07
Campari Fruit Brandy (apricot, blackberry, cherry, peach) Yellow Chartreuse	1.06
Midori	1.05
Cointreau Peach and Cherry Liqueur Brandy Benedictine Peppermint Schnapps Sloe gin	1.04
Green Chartreuse	1.01
Water	1
Tuaca	0.98
Southern Comfort Almost any 80 or 100 proof spirit	0.97
Kirsch	0.94

Flaming

Flaming is a technique that lights a drink on fire, normally to enhance the flavor but also to add visual appeal or flair for the consumer. The basic rule of thumb is anything 80 proof or above will light; the higher the proof, the easier the alcohol will be to ignite. The key to flaming a drink is to heat both the glass and the alcohol until they are very warm. The alcohol is then ignited with a lighter—some spirits will ignite quite easily if their proof is high. Flaming a beverage can be an impressive way to spark conversation among guests, but for obvious reasons this technique should only be attempted with caution. Always extinguish a flaming drink before serving it. (*Note:* Some jurisdictions outlaw tableside flambé entirely.)

Molecular Mixology

Learning Objective 7
Provide some examples of molecular mixology

Molecular mixology takes its inspiration from molecular gastronomy, the food equivalent. The essential concept underlying this modern movement is to apply advanced and unconventional scientific practices to the traditional production techniques of drink making. The practices and techniques of molecular mixology is a sophisticated approach as it propels drink making to its highest level of presentation—one that represents a work of movable and consumable piece of art. However, Alton Brown (the famous celebrity Chef) states an important truth that "Molecular gastronomy is not bad ... but without sound, basic culinary technique, it is useless." Therefore, it is equally important to stress the basics when it comes to drink making.

Incorporation of molecular practices not only enhances the visual appeal but also increases the flavors and overall guest experience. Molecular mixology manipulates the states of matter to create unexpected visual, flavor, or textural elements in the drink. Some popular techniques include the use of spherification, emulsification, and gels. For some bars and restaurants, the incorporation of such avant-garde techniques can be considered gimmicky—with the absence of discretion, the molecular mixology approach can be inappropriate and disastrous given the type of beverage program and target market.

Spherification

This is the process of transforming liquid into a small sphere or shape that is coated with an extremely delicate gel membrane. Regardless of the internal ingredients, spheres have been more favorably coined as ravioli, caviar, eggs, and gnocchi due to their visual and textural resemblance. The delicate spheres break very easily when placed in one's mouth, releasing an explosion of flavor.

Emulsification

This adds a layer of flavored foam to a cocktail with similar consistency of the foam on a cappuccino. The texture of foam varies depending on what it is made from, how much liquid is incorporated, and the size of air bubbles. Mixologists combine flavors like raspberry juice with a stabilizer such as lecithin using foaming equipment. Example of application: Captain Mocha Joe—an aged rum cocktail that is infused with Madagascar vanilla, featuring espresso caviar and a cocoa foam.

Gels (or Fancy Jello Shots)

This technique is the jellifying of different liquids. Unlike those vodka-spiked jelly shots associated with many college parties, the gelatin cocktails of molecular mixologists are made with perfectly crafted cocktails.

The Pour Station

Learning Objective 8
Provide some examples for ensuring that a cocktail is produced consistently

The work area of a bar should be arranged so that every ingredient is easily accessible to the bartender. If the bar is organized correctly, the bartender can efficiently produce any drink when one is ordered by the customer. Quality bartenders will know the layout of the bar so they can easily obtain ingredients and equipment without necessarily looking. Efficient bartenders stay organized so that drink production keeps pace with customer demand. Common strategy is to place the most frequently used ingredients and tools nearest to the bartender.

Designated pour station(s) should be arranged in select locations in the bar area where most or all the drinks will be produced. This station becomes the central place for drink production—requiring appropriate ingredients and equipment to be accessible. Additionally, the pour station acts as a location for service staff to obtain drinks for their customers sitting elsewhere in the establishment.

The pour station should be arranged in a way that allows a minor amount of empty space on the counter for any immediate glassware to sit while drinks are being prepared. Backup glasses should be in a nearby place where they are easily accessible. The "under bar" in the pour station—the major (well-type) spirits should be arranged in the *speed rail* (a shelf that holds the most frequently used spirits). Common garnishes such as sliced lemons, limes, oranges, and cherries should be easily accessible. Additionally, ice bins should be near as well as any juice or mixes such as orange, cranberry, tomato, pineapple juice and other soda type mixers such as soda, cola, and 7-up should be kept at hand and chilled.

Measuring the Spirits and Liqueurs

Ensuring that drink ingredients are measured accurately is critical for several reasons. The beverage manager wants to ensure that the guest has a personalized, yet standardized standard experience and is not going to be overserved. Also, the beverage manager is concerned about cost control. The amount of alcohol in a drink can be measured using several methods. The first method requires the bartender to use a small double-sided measuring device called a *jigger*. A standard jigger has two sides: The small side holds three-fourth (¾) portion and the larger side holds a one and one-fourth ounce (1¼) portion. Figure 12.21 displays a double-sided jigger. Another method to measure alcohol is the application of *free pouring*—the bartender pours a drink with a silent count synchronized to pour the correct amount into the vessel. This method allows a bartender to showcase bit more flair and exercise greater speed in the drink production process. Though without the use of the jigger, the possibility of over or under pouring becomes more concerning. The free pouring approach can work well, assuming that the bartender has practiced and developed a good sense of timing to ensure consistency.

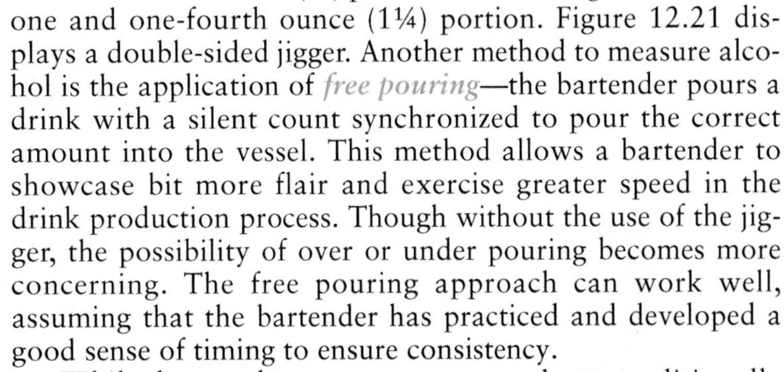

FIGURE 12.21
Double-sided jigger. Courtesy of Erika Cespedes.

While bar and restaurant owners have traditionally focused on point-of-sale (POS) systems for food orders, auto-pour spouts can now interface with the POS as well. Liquor control technologies can assist in inventory purposes resulting in improved profitability.

The automatic pouring system is another approach to creating drinks using a mechanized dispensing system. The machine can pour specified amounts of alcohol—though it is very expensive. Theoretically, the system may save money in the long run through eliminating accidental or intentional over/under pours and can assist with the overall consistent production of drinks. There are certainly many pros and cons to consider when moving to an automated system—there always seems to be a sense of flair and dramatics that can't be replaced when a bartender conducts free pouring.

Selecting the Correct Glassware

As with wine and beer, glassware for spirits also comes in various shapes and sizes all suitable for specific drinks. It is important to know which glass belongs to the appropriate or suggested cocktail. Serving a drink in improper glassware is a sign of lacking detail that can make or break the appearance and experience of a well-made cocktail. Glassware is just as important as following a proper drink recipe, using the right garnish, and charging each drink the correct selling price. Any well-written drink recipe should suggest appropriate glassware—usually deriving from tradition. Not only should shape of a glass be appropriate, but the size of a glass should be compatible with the size of the cocktail.

The numerous options of glassware can prove overwhelming but having the proper glassware is an important element of a drink for various reasons. Glassware not only adds visual appeal, but they also serve an important function in enhancing the aromas and flavors of any alcohol beverage. The shape and size of the glass can also work to emphasize a type or style of drink. The "Classic Martini," for instance, should be served in the very distinct cocktail or martini glass. If served in any other glass, the drink might still be considered a Martini, but an informed customer will know that something about the authenticity and integrity of the drink is absent. With so many variations available, it's possible to find glassware in nearly every style to accommodate any drink as well as any budget. Figure 12.22 displays an example of a cocktail/martini glass.

Regardless of type or size, ensure all glassware is cleaned spotless prior to pouring a drink into it and serving it to the customer. Glasses should be washed in warm water with a small amount of detergent, rinsing them afterward with cold water, and then polishing them with a suitable cloth. Refrain from grabbing or holding glassware on the lip or upper half of the glass, instead, all glassware should be held by the base or stem of the glass to avoid fingerprints.

Chilled glasses are a great way to serve cocktails. Glasses can be placed in a cooler prior to serving, or for a quicker method—fill the glass with ice and water for at least two minutes before preparing the drink. Either method will encourage a chilled and frosty appearance on the glass.

FIGURE 12.22
Cocktail/Martini glass. Courtesy of Erika Cespedes.

Shot Glass This type of glassware is used for customers who desire straight alcohol; however, there are some mixed drinks that call for shot glass as the preferred glass of service. Alcohol served in a shot glass is usually room temperature, but may be from a pre-chilled bottle of spirit. The glass holds between 1 and 2 oz. of alcohol.

Cordial Glass This type of glassware resembles a stemmed shot glass, but can be a bit more ornate. The glass can be used as a shot glass or for cordials such as Irish cream, Chambord, or Kahlua. The *pousse-café* (POOSE-cah-fay) glass is a straight, narrow (occasionally slightly flared toward the top) glass used for layered drinks. Oftentimes, the bartender may utilize a shot or cordial glass if a pousse-café glass is not found.

Rocks Glass This type of glassware is used for serving alcohol either "neat" or for mixed drinks served over cubed ice (rocks). It is also known as an Old-Fashioned glass.

FIGURE 12.23
Bloody Mary with beer. Courtesy of Erika Cespedes.

Collins Glass This type of glassware is tall and thin that is used for drinks such as the Tom Collins and a Bloody Mary. Sometimes the Collins glass is referred to as the highball glass and looks very similar to the rocks glass, but it is larger in size and therefore utilized for mixed drinks served over cubed ice with the addition of juice or soda. Figures 12.23 and 12.24 show a Bloody Mary with a bountiful garnish and served in a highball glass.

Snifter The brandy glass or snifter has a large bowl and a short, stubby stem, which encourages the drinker to hold the bowl of the glass cradled in their hand. The cradling action allows the brandy to be savored and slightly warmed through body temperature, which serves to enhance the aromas and enjoyment of the drink.

Cocktail Glass The cocktail or martini glass is a cone-shaped cocktail glass designed for drinks that have been shaken or stirred with ice in another container and strained into the glass. This glass provides a nice visual presentation and has gained considerable popularity in American bars and restaurants.

Alternative Glassware *Whiskey sour glass* looks like a wine glass used specifically for whiskey sours. The *margarita glass* looks like an oversized Champagne coupe glass. It's used for margaritas on the rocks or with crushed ice. The *tulip glass* is a stemmed glass with an obvious tulip shape often used for frozen daiquiri type drinks. The *copper tin cup* is used for the resurgence of the Moscow Mule. Figure 12.25 displays a Moscow Mule copper mug.

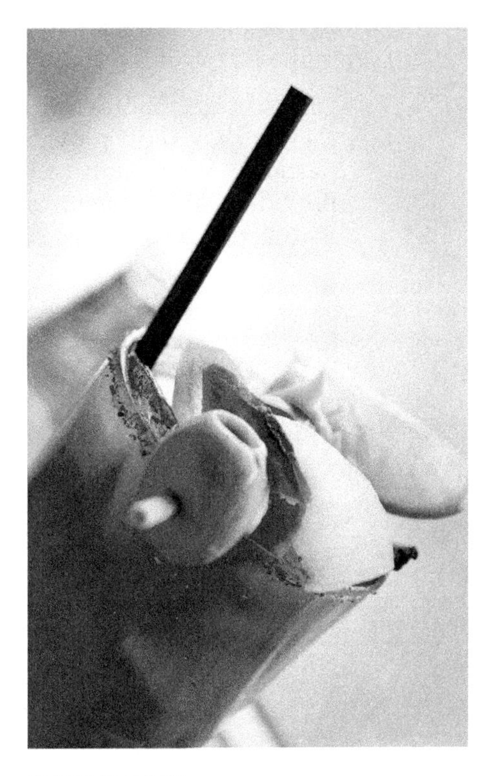

FIGURE 12.24
Bloody Mary with bountiful garnish. Courtesy of Erika Cespedes.

FIGURE 12.25
Moscow Mule copper mug. Courtesy of Erika Cespedes.

Garnishing and Rimming Drinks

Learning Objective 9
Identify several types of garnishes for drink making

Most drinks have an exact garnish used as a finishing touch to their presentation. Cocktail garnishes began in the United States during the late 1700s when Betsy Flanagan first used feathers from rooster tails to garnish drinks. Thus, the "cock tail" came into existence for the first time in Yorktown, New York. Obviously, rooster feathers are not used any longer, but the application of garnishes continues to play a small part in the enhancement of flavor and large part of the aesthetic appeal.

Sometimes, the garnish is as simple as a slice of orange, a wedge of lemon, sprig of mint, a green olive, or a cherry; however, garnishes also can be very ornate and complicated. Regardless, every garnish must begin with a selection of fresh ingredients. Figure 12.26 displays a tray of fresh garnishes used for drinks throughout the bartender's shift. One of the most common garnishes is a lemon or lime twist—a thin strip of citrus peel, without pith and without the meat of the fruit. This garnish derives its name from the "twisting" action over the surface of the drink in order for the fruit's skin to express its oils. Other simple garnishes can be a slice or wedge of a fruit. A *slice* is a round cross section of the whole fruit, whereas a *wedge* is a chunk of a section of the fruit. Figure 12.27 shows lemon slice and Figure 12.28 shows a lime wedge. These slices and wedges can be combined in a drink for more complicated garnishes.

FIGURE 12.26
Selection of garnishes. Courtesy of Erika Cespedes.

Decoration of a cocktail will normally consist of one or two fruit, herb, or cherry garnishes that either complement the flavor of the drink, or contrast with the color, or both. It is important to avoid overpowering the drink. When garnishing with a slice of fruit, be careful with the size: too thin is flimsy, while too thick can unbalance the look and even the flavor of the cocktail.

- *Citrus Zest:* The exterior of the citrus fruit is slightly shaved off providing some color from the particles of the skin (referred to as zest) while providing some potent oils from the fruit's skin.

FIGURE 12.27
Sliced lemon garnishes. Courtesy of Erika Cespedes.

FIGURE 12.28
Lime wedges. Courtesy of Erika Cespedes.

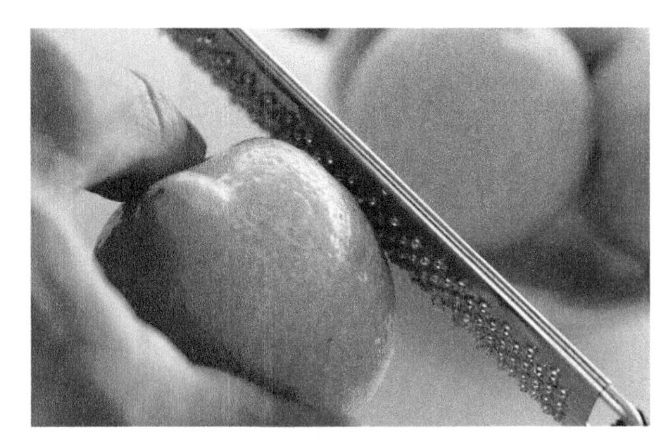

FIGURE 12.29
Lemon zest. Courtesy of Erika Cespedes.

FIGURE 12.30
Half moon oranges. Courtesy of Erika Cespedes.

Figure 12.29 displays the use of a zester shaving the exterior of the citrus fruit.

- *Citrus Half-Moons:* To make half-moons, two steps are involved. First step involves halving an orange midway between the ends and then placing each half cut-side-down on a cutting board. Now slice the semi-circle cross-section (or half-moon) every ¼ inch, discarding the end pieces. Figure 12.30 displays a plate of orange half-moons.
- *Citrus Twists:* To make a citrus twist, cut a thin slice of the citrus fruit crosswise and simply twist to serve over the top of the glass.
- *Citrus Peel Spirals:* To make a spiral of citrus peel, use a pairing or vegetable peeler to cut away the skin, working in a circular motion. Take care not to cut into the bitter pith. Figures 12.31 and 12.32 display the procedure of obtaining citrus peel spirals.
- *Citrus Peel Knots:* Using strips of peel, carefully tie each strip into a knot.
- *Citrus Flag:* Using a cocktail stick, spear a citrus peel or citrus half-moon with a cherry. Figures 12.33 and 12.34 display the procedure for making a citrus flag.
- *Cocktail Sticks:* These extremely useful wooden cocktail sticks are needed for spearing through pieces of fruits, vegetables, and cherries. Figure 12.35 displays olive garnish on a stick.
- *Maraschino Cherries:* These are preserved sweetened cherries that remain the most widely used cocktail garnish. To put their popularity into perspective, Mr. Angelo Puccineill, owner of Portland Oregon's Matador says, "Anybody that's

FIGURE 12.31
Citrus peel. Courtesy of Erika Cespedes.

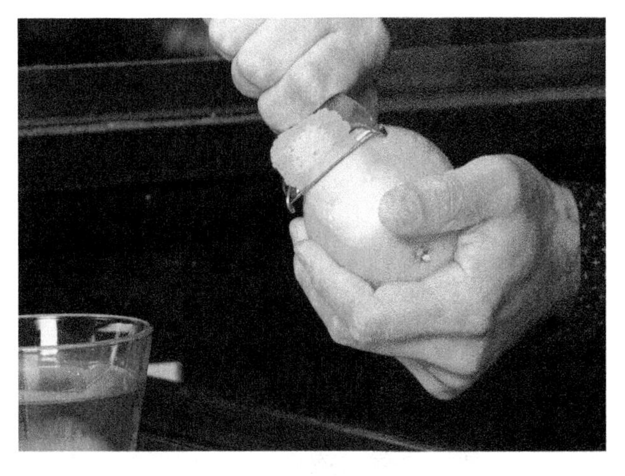

FIGURE 12.32
Citrus peel. Courtesy of Erika Cespedes.

FIGURE 12.33
Composing a citrus flag. Courtesy of Erika Cespedes.

FIGURE 12.34
Citrus flag. Courtesy of Erika Cespedes.

FIGURE 12.35
Olive garnish on a cocktail stick. Courtesy of Erika Cespedes.

FIGURE 12.36
Sprig of mint. Courtesy of Erika Cespedes.

poured a fair share of drinks in their life would never complain about a maraschino cherry . . . it's like getting mad at soda water."

- *Specialty Cherries:* Luxardo Original Maraschino Cherries are the "Cadillac" of cherry garnishes. These are proprietary (from the Luxardo family) cherries that derive from the Veneto region of Italy. Luxardo cherries are sour Marasca cherries that are candied and steeped in a sugared cherry juice.
- *Pineapple Wedge:* Pineapple wedges are commonly used for tropical cocktails. It can also be speared with a cherry.
- *Sprig of Herbs:* This involves the bartender grabbing or twisting off a bunch of herbs (typically mint) as use for a garnish. Figure 12.36 displays a bartender grabbing a sprig of mint. Figure 12.37 showcases how the sprig of mint appears in the finished cocktail.
- *Straws:* Straws are essential and go well with many cocktails. The general rule is—short glass, short straw—tall glass, tall straw.

Rimming a Glass

Rimming the top of a glass can add a decorative touch and additional flavor to cocktails. Figure 12.38 displays how a glass will look with a rimmed garnish. Most commonly, cocktails use salt or sugar to rim a glass, but some recipes also work well with powdered sugar or cocoa. If you decide to experiment with rimming different cocktails, be sure to choose an appropriate accent to the taste of the drink. The procedures for rimming a glass include:

FIGURE 12.37
Mint garnished cocktail. Courtesy of Erika Cespedes.

FIGURE 12.38
Rimmed glass. Courtesy of Erika Cespedes.

FIGURE 12.39
Chocolate martini with rimmed glass. Courtesy of Erika Cespedes.

1. Moisten the outside rim of the glass with a fresh lemon or lime wedge.
2. Fill a saucer or bowl with salt (never use iodized salt), sugar (never use superfine sugar), or other appropriate dry agent.
3. Angle the glass so that it is parallel to the table and dip the rim into the dry agent while slowly turning the glass so that only the outer edge is covered.
4. Shake off any excess salt or sugar over a sink or wastebasket.
5. Fill the glass with your mixed cocktail and garnish.

Example #1: Bloody Mary—Moisten the rim with a small amount of lemon or lime juice, then dab the rim in celery salt.

Example #2: Chocolate Martini—Moisten the rim of a glass with a small amount of Kahlua (or other coffee liqueur), then dab the glass in chocolate powder or chocolate shavings. Figure 12.39 displays a chocolate martini with a rimmed glass.

22 Classic Cocktails

Learning Objective 10
Identify all 22 classic cocktails

While new cocktails are invented every day, below is a selection of 22 essential classic cocktails for every beverage manager. Classics have staying power as they have endured the test of time. And like all classics, whether it's a car or a drink, the beauty of these cocktails comes from their simplicity. *The Beverage Manager's Guide to Wines, Beers, and Spirits* offers tested cocktail recipes by Kai Wilson. These cocktails are ones that have been in existence for quite some time and have experienced a sustainable surge in consumption and popularity over the years. The selection of 22 classic cocktails represents drinks that have stood the test of time. Below is a list of the 22 classic cocktails along with their corresponding recipe. These cocktails are arranged according to their dominant base spirit.

Vodka-Based Classic Cocktails

Bloody Mary, Cosmopolitan, Moscow Mule

BLOODY MARY

Yield: 1 Cocktail

Method: Build
Glassware: Collins/Highball

Ingredients:
1.5 oz. vodka
½ oz. lemon juice
1 dash Worcestershire sauce
Pinch of celery salt
Pinch of ground Pepper
2 dashes tabasco or other hot sauce
1 tsp horseradish (optional)
3 or more oz. tomato juice

Garnish: Celery stalk, olive, lemon or lime wedge (other garnishes as desired)

Directions:
1. Build the ingredients in a mixing cup over ice.
2. Stir well or roll the ingredients between two glasses or mixing cups.
3. Strain onto new ice in highball or pint glass.
4. Add seasonings and garnish on top.

Fernand Petiot is credited with creating the Bloody Mary at Harry's New York Bar in Paris. It was presumably named in conjunction with a newspaper reporter who hailed from Chicago who mentioned the Chicago Bar "A Bucket of Blood" and a waitress there named Mary. Another version of the story merely claims that it was named after the Queen who had received the same moniker centuries before. When Petiot moved to the United States, he could not find vodka readily available and made a version using gin called the Red Snapper.

COSMOPOLITAN

Yield: 1 Cocktail

Method: Shaken
Glassware: Cocktail glass

Ingredients:
1.5 oz. vodka (or citrus vodka)
1 oz. Cointreau
½ oz. lime juice (fresh, not Rose's)
¼ oz. cranberry juice

Garnish: Lemon peel

Directions:
1. Pour ingredients into a cocktail shaker with ice.
2. Shake well.
3. Strain into a chilled cocktail glass.
4. Garnish with a lemon peel.

The Cosmopolitan was created at The Strand in Florida and became popular in San Francisco before eventually making its way to New York where the original recipe was altered to include fresh lime juice and Cointreau.

MOSCOW MULE

Yield: 1 Cocktail

Method: Build
Glassware: Copper mug

Ingredients:
2 oz. vodka
Juice of ½ lime
Ginger beer to fill

Garnish: Lime wheel

Directions:
1. Add gin over ice in highball glass or copper mug.
2. Squeeze half a lime into glass and drop into drink.
3. Fill with ginger beer.

Created by Percy T. "Jack" Morgan, the owner of the Cock 'n Bull Restaurant on Sunset Boulevard in Los Angeles. A colorful tale that results somehow in Jack having only vodka and ginger beer on hand that leads to creation of the drink is appropriate as a "cock and bull" story. The truth is that John G. Martin of G.F. Heublein Brothers, owners to the rights of Smirnoff vodka, asked his friend Jack to create a cocktail to help promote the sales of vodka. Jack, who imported ginger beer from his native England, combined it with the vodka and for extra flair, served it in copper mugs obtained from his fiancee's place of business. The drink caught on quickly with Hollywood celebrities as Jack's restaurant was a popular spot for stars to be seen.

Gin-Based Classic Cocktails

Aviation, French 75, Martinez, Negroni, Tom Collins

AVIATION

Yield: 1 Cocktail

Method: Shaken
Glassware: Cocktail glass

Ingredients:
2 oz. Old Tom or Modern Dry gin
¼ oz. maraschino liqueur
½ oz. lemon juice
Dash of crème de violette

Garnish: Lemon peel

Directions:
1. Pour the ingredients into a cocktail shaker filled with ice.
2. Shake well.
3. Strain into a chilled cocktail glass.
4. Garnish with a lemon peel.

The cocktail is named for the light blue color that results from the use of crème de violette in the drink.

The drink is sometimes claimed to have been created to commemorate Charles Lindbergh's transatlantic flight but the drink has been documented to have predated his trip by a number of years.

FRENCH 75

Yield: 1 Cocktail

Method: Build
Glassware: Champagne Flute

Ingredients:
1 oz. gin
½ oz. simple syrup
½ oz. lemon juice
Fill remainder with sparkling wine

Garnish: Lemon peel

Directions:
1. Pour the ingredients into champagne flute.
2. Garnish with a lemon peel.

Created in 1915 by Harry McElhone at Harry's New York Bar in Paris. Named after the 75mm cannon used by the French in World War I.

MARTINEZ

Yield: 1 Cocktail

Method: Stir
Glassware: Cocktail glass

Ingredients:
2 oz. gin
¾ oz. sweet vermouth
¼ maraschino liqueur
Dash of orange bitters

Garnish: Orange peel

Directions:
1. Pour ingredients into a mixing glass.
2. Stir well.
3. Strain up or over new ice in a chilled cocktail glass.
4. Garnish with orange peel.

The Martinez is one of the predecessors to the Martini. One story of the Martinez is that it was created by Professor Jerry Thomas for a patron traveling to Martinez, California.

NEGRONI

Yield: 1 Cocktail

Method: Stir
Glassware: Cocktail glass

Ingredients:
1 oz. gin
1 oz. sweet vermouth
1 oz. Campari

Garnish: Orange wheel

Directions:
1. Pour ingredients into a mixing glass.
2. Stir well.
3. Strain up or over new ice in a chilled cocktail glass.
4. Garnish with orange wheel.

It is widely believed that the Negroni was created and named for Count Cammillo Negroni in the 1920s when he ordered an Americano with gin at Cafe Casoni in Florence, Italy.

The Negroni can also be shaken and served in a cocktail glass with a lemon twist.

TOM COLLINS

Yield: 1 Cocktail

Method: Shaken

Glassware: Collins/Highball

Ingredients:

1.5 oz gin

¾ oz. lemon juice

¾ oz. simple syrup

Club soda to fill

Garnish: Lemon wedge and cherry

Directions:

1. Pour the ingredients except club soda into a cocktail shaker filled with ice.
2. Shake well.
3. Strain over fresh ice into a collins glass.
4. Fill remainder with club soda.
5. Garnish with a lemon wedge and cherry.

Originating in the late 18th or early 19th century at Old Limmer's House in London, it is credited to either a John or James Collins. The "Tom" in Tom Collins came later as the drink frequently used Old Tom gin (a sweet style of gin as opposed to London Dry).

Rum-Based Classic Cocktails

Daiquiri, Mai Tai, Mojito, Piña Colada

DAIQUIRI

Yield: 1 Cocktail

Method: Shaken

Glassware: Cocktail glass

Ingredients:

1.5 oz. light rum

¾ oz. lime juice

¼ oz. simple syrup

Garnish: Lime wheel

Directions:

1. Pour the ingredients into a cocktail shaker filled with ice.
2. Shake well.
3. Strain into a chilled cocktail glass.
4. Garnish with a lime wheel.

It is a Taino (native Caribbean) name for a beach near Santiago, Cuba. The story is that Jennings Cox, an American mining engineer who was in charge of a mine near Daiquiri that also bore its name, wanted to make drinks for visiting friends but had only white rum, limes, and sugar. He combined them and named it the Daiquiri.

An alternative story is that William A. Chanler, a U.S. Congressman, purchased the Santiago Iron Mines in 1902 and introduced the drink to New York City, was the inventor.

In 1909, Lucius W. Johnson, Rear Admiral for the U.S. Navy introduced the drink to the Army and Navy Club in Washington, DC.

The likelihood is that these Americans merely "discovered" this drink while on the island and adopted it readily, especially given the similarity of the drink to the Gin (or Lime) Rickey that had been popularized in Washington, DC, only a few years before the Spanish-American War.

The most commonly poured Daiquiri is actually the Daiquiri No. 2 from the El Floridita Bar book published in the 1937 and created by Constantinto Ribilaigua Vert, which adds Curacao or Triple Sec.

Yield: 1 cocktail

Ingredients:

2 oz. rum

5–7 dashes Curacao

1 tablespoon orange juice

1 teaspoon superfine sugar

juice of ½ lime

cracked ice

Add ingredients to cocktail shaker; shake and pour into chilled cocktail glass.

Garnish with lime.

MAI TAI

Yield: 1 Cocktail

Method: Shaken
Glassware: Collins/Highball

Ingredients:
1 oz. light rum
1 oz. aged rum
½ oz. lime juice
½ oz. orange liqueur
½ oz. orgeat syrup

Garnish: Mint sprig

Directions:
1. Pour the ingredients into a cocktail shaker filled with ice.
2. Shake well.
3. Strain over new ice into a chilled Collins glass.
4. Garnish with a mint sprig.

From Tahitian "mai tai roa e!" meaning approximately "out of this world, the best!" There is a debate as to whether this was created by Victor Bergeron for his Trader Vic's restaurant or by Donn "The Beachcomber" Beach.

MOJITO

Yield: 1 Cocktail

Method: Muddle
Glassware: Collins/Highball

Ingredients:
2 tsps. granulated sugar or 1 oz. simple syrup
6–8 mint leaves
Half of a lime
2 oz. light rum
Club soda

Garnish: Mint sprig

Directions:
1. Place the sugar (or simple syrup) and mint leaves into a highball or Collins glass.
2. Gently muddle the mint leaves to release the oils into the sugar.
3. Squeeze the juice of half a lime into the glass.
4. Drop the spent lime into the glass.
5. Add the rum.
6. Add the ice.
7. Fill with club soda.
8. Garnish with a mint sprig.

A classic drink from Cuba using local ingredients of mint, limes, sugar, and rum. Mint comes in various varieties and it is believed that this drink was originally developed with Yerba Buena but other mints can be used.

PIÑA COLADA

Yield: 1 Cocktail

Method: Shaken
Glassware: Cocktail glass

Ingredients:
2 oz. light rum
2 oz. pineapple juice
1.5 oz. cream of coconut (Coco Lopez)

Garnish: Pineapple wedge and cherry

Directions:
1. Pour the ingredients into a cocktail shaker filled with ice.
2. Shake well.
3. Strain over new ice into a chilled cocktail glass.
4. Garnish with pineapple wedge and cherry.

The name Piña Colada literally means strained pineapple, a reference to the freshly pressed and strained pineapple juice used in the drink's preparation.

There are various claims as to the inventor of the Piña Colada. Ramón 'Monchito' Marrero Pérez claims to have first made it at the Caribe Hilton Hotel's Beachcomber Bar in San Juan on August 15, 1952, using the then newly-available Coco López cream of coconut. Coco López was developed in Puerto Rico in 1948 by Don Ramón López-Irizarry. López-Irizarry, in promoting his product, sought out the staff at the Hilton Caribe to develop a drink that would use it.

Ricardo García, who also worked at the Caribe as a bar-back, says that it was he who invented the drink, and that Monchito, the head bartender at the time, took the credit.

There is some evidence, however, that such a drink already had a long tradition among natives who would mix pineapple juice, coconut milk, coconut cream (a very intensive process to do manually), and other ingredients with the local rum.

Tequila-Based Classic Cocktails

Margarita, Tequila Sunrise

MARGARITA

Yield: 1 Cocktail

Method: Shaken
Glassware: Margarita glass

Ingredients:
1.5 oz. tequila
½ oz. orange liqueur
1 oz. lime juice
Garnish: Lime wheel

Directions:
1. Pour the ingredients into a cocktail shaker filled with ice.
2. Shake well.
3. Strain into a chilled margarita glass, optionally with a salt rim.
4. Garnish with a lime wheel.

There are many creation myths for this drink, most of which involve a bartender creating the drink for a woman named Margarita.

The truth is that from the early 1800s, a class of drinks called a "Daisy" involved a base spirit, orange liqueur, and lemon juice. Margarita means daisy in Spanish. With many U.S. citizens crossing the border during Prohibition, it is believed that the old daisy recipe was merely adapted at the Mexican border towns to use tequila and Mexican limes in place of the other spirit and the lemon.

The addition of a salted rim came from the tradition that came out of the late 1920s when doctors recommended lime juice and salt to help "treat" Spanish Flu symptoms. Salt and lime have been associated with shots and cocktails with tequila ever since.

TEQUILA SUNRISE

Yield: 1 Cocktail

Method: Build
Glassware: Collins/Highball

Ingredients:
2 oz. tequila
4 oz. orange juice
½ oz. grenadine
Garnish: Orange wheel and cherry

Directions:
1. Pour the tequila and orange juice into a highball glass filled with ice.
2. Float the grenadine in a circle on top of the glass and let it begin to sink to the bottom of the glass.
3. Garnish with orange wheel and cherry.

Created by Gene Sulit at the Arizona Biltmore Hotel sometime in the 1930s or 1940s. The original cocktail contained Crème de Cassis but later versions have substituted grenadine in place of this liqueur.

Brandy-Based Classic Cocktails

Brandy Alexander, Pisco Sour, Sidecar, Vieux Carré

BRANDY ALEXANDER

Yield: 1 Cocktail

Method: Shaken
Glassware: Cocktail glass

Ingredients:
1.5 oz. brandy
1 oz. Dark Crème de Cacao
1 oz. heavy cream (or half & half)
Garnish: Ground nutmeg

Directions:
1. Pour the ingredients into a cocktail shaker with ice cubes.
2. Shake well.
3. Strain into a chilled cocktail glass.
4. Garnish with a dusting or rimming of nutmeg.

The origin of this drink is unclear, as many take credit for its invention. The Brandy Alexander was likely named after Troy Alexander, a bartender at Rector's, a New York City restaurant, who wanted to serve a white drink at a dinner celebrating Phoebe Snow, a character in a popular advertising campaign in the early 20th century.

PISCO SOUR

Yield: 1 Cocktail

Method: Shaken

Glassware: Old-Fashioned/Rocks glass

Ingredients:

2 oz. pisco

1 oz. lime juice

½ oz. simple syrup

egg whites from 1 medium egg

Garnish: 2–3 dashes of Angostura Bitters

Directions:

1. Pour ingredients into a shaker without ice.
2. Shake well.
3. Add ice and shake until chilled.
4. Strain into a chilled Old-Fashioned glass.
5. Once the egg white foam has settled, garnish with 2–3 dashes of Angostura Bitters on top of the foam.

The inventor of this cocktail is not known but it is the national drink of Peru.

SIDECAR

Yield: 1 Cocktail

Method: Shaken

Glassware: Cocktail glass

Ingredients:

1.5 oz. Cognac or Armagnac

1 oz. orange liqueur

½ oz. lemon juice

Garnish: Sugar rim and orange peel

1. Pour the ingredients into a cocktail shaker filled with ice.
2. Shake well.
3. Strain into a sugar-rimmed chilled cocktail glass.
4. Garnish with an orange peel.

As origins of most cocktails go, there are a few versions of how the Sidecar came into being. One story says that it was developed in a Parisian bistro during World War I by a friend who rode to the favorite bar in the sidecar of a motorcycle. Which bar this was is left to speculation, but it is popularly thought to be Harry's New York Bar.

Another claim to the Sidecar invention attributes to Frank Meier who worked at the Paris Ritz. As Gaz Regan pointed out in The Joy of Mixology, this was later disputed by a man named Bertin who worked at the Ritz after Meier.

The next story moves to Buck's Club in London. In his 1922 book, *Harry's ABC of Mixing Cocktails*, Harry MacElhone credits the drink to Pat MacGarry, one of the great bartenders of the day. This was also backed up in Robert Vermeire's 1922 *Cocktails and How to Mix Them*.

Whichever theory is correct will remain a matter of debate and opinion. One thing that is agreed upon is that the Sidecar is a classic sour drink.

VIEUX CARRÉ

Yield: 1 Cocktail

Method: Stirred

Glassware: Cocktail glass

Ingredients:

¾ oz. rye whiskey

¾ oz. cognac or Armagnac

¾ oz. sweet vermouth

2 dashes Peychaud Bitters

2 dashes Angostura Bitters

1 teaspoon Benedictine

Garnish: None

Directions:

1. Pour ingredients into a mixing glass.
2. Stir well.
3. Strain up or over new ice in a chilled cocktail glass.

Named after the section of New Orleans and supposedly created at the Carousel Bar in the Hotel Monteleone.

Whiskey-Based Classic Cocktails

Manhattan, Mint Julep, Sazerac, Whiskey Sour

MANHATTAN

Yield: 1 Cocktail

Method: Stirred
Glassware: Cocktail glass

Ingredients:
2 oz. rye whiskey
½ oz. sweet vermouth
2–3 dashes Angostura Bitters

Garnish: Cherries

Directions:
1. Pour ingredients into a mixing glass.
2. Stir well.
3. Strain up or over new ice in a chilled cocktail glass.
4. Garnish with cherries.

The origin of the Manhattan cocktail is difficult to pin down. One story credits the Manhattan Club but the original story has details that don't hold up to scrutiny. It is likely that it was created in the late 1800s as the popularity of vermouth rose in the United States.

MINT JULEP

Yield: 1 Cocktail

Method: Muddled
Glassware: Julep Cup

Ingredients:
Leaves from 4–5 mint sprigs
½ oz. simple syrup
2.5 oz. bourbon whiskey

Garnish: Mint sprig

Directions:
1. Place the mint leaves and simple syrup into a julep cup.
2. Gently muddle the leaves to release the oils into the syrup.
3. Add the bourbon to the cup.
4. Fill half way with crushed ice and stir well until the outside of the cup becomes frosty.
5. Add the remaining crushed ice.
6. Make a hole with the stem of the bar spoon and insert a straw into the hole.
7. Garnish with a mint sprig next to the straw.

The word julep comes from Sanskrit through Persian and refers to a flavored syrup ("julebs" with sweetened rose-water are still widely available in any Middle Eastern market place). As a class of cocktails, these "smashes" (referring to the muddling of the herbs) have been around since at least the beginning of the 19th century.

The julep also classically utilizes pewter or silver cups and smashed (crushed) ice that keeps the cocktail very cold.

SAZERAC

Yield: 1 Cocktail

Method: Stirred
Glassware: Old-Fashioned/Rocks glass

Ingredients:
2 oz. rye whiskey
1 cube demerara sugar
3 dashes Peychaud Bitters
1/8 oz. Absinthe

Garnish: Lemon peel

Directions:
1. Fill an Old-Fashioned glass with ice and set aside.
2. In a mixing glass place the demerara sugar cube and dash the Peychaud Bitters onto the cube.
3. Muddle the sugar cube to incorporate the bitters.
4. Add the rye whiskey to the mixing cup along with ice.
5. Stir well.
6. Empty the original Old-Fashioned glass of the ice.
7. Pour a small amount of Absinthe into the chilled glass and roll it around the inside to coat.
8. Pour off any excess Absinthe from the Old-Fashioned glass.
9. From the mixing glass, strain the chilled ingredients into the Absinthe coated, chilled Old-Fashioned glass.
10. Garnish by releasing the oils of the lemon peel over top the contents of the glass but hang the lemon peel off the side of the glass (do not put into the drink).

SAZERAC (*Continued*)

The cocktail is named after the Sazerac Coffee House of New Orleans. Back in the late 18th and early 19th centuries, "coffee houses" served many libations, including alcoholic ones. The original recipe uses cognac, along with Absinthe (both showing the French heritage of New Orleans) and locally produced Peychaud's Bitters. When *Phylloxera* decimated the vineyards of France, rye whiskey was substituted (obviously for the taste as bourbon was now readily available as it made its way down the Ohio and Mississippi Rivers).

WHISKEY SOUR (WITH EGG WHITE)

Yield: 1 Cocktail

Method: Shaken
Glassware: Cocktail glass

Ingredients:
1.5 oz. whiskey
1.5 oz. lemon juice
¾ oz. simple syrup
Egg whites from 1 medium egg

Garnish: Lemon peel

Directions:
1. Pour the ingredients into a cocktail shaker without ice.
2. Shake well.
3. Add ice to shaker.
4. Shake well.
5. Strain into a chilled cocktail glass.
6. Garnish with a lemon peel.

Sours are a class of drinks invented in the early 1800s and documented by Jerry Thomas in 1862 in the first published cocktail book. Sours consist of a base spirit, sugar, and lemon juice (sometimes lime when the base spirit was from south of the border). Many recipes also include egg white as an option. This produces a silkiness that offsets the tartness of the citrus juice. It is possible that the egg white is compensating for gomme syrup, which is rarely used in cocktails any longer.

Mocktails

Mocktails are alcohol-free beverages that may look like a traditional alcohol-based adult cocktail or resemble something that looks more like a smoothie. With a new generation of mixologists, these beverages have evolved far beyond soda pop and mocktails of Piña Colada. For various reasons, customers may not be able to consume alcohol—the creation of mocktails allows bartenders to appease this customer.

- *Watermelon and Mint Blend:* Pureed and strained watermelon juice blended with Greek yogurt and ice, then garnished with a sprig of mint. Figure 12.40 displays a watermelon mint smoothie.
- *Cucumber and Lemon-Thyme Fizz:* Cucumber water flavored with a hint of lemon-thyme syrup and a splash of soda water.
- *Strawberry Chamomile Martini:* Pre-made chamomile tea shaken with some pureed strawberry or strawberry syrup and strained into a chilled cocktail glass.
- *Lemon Grass Jasmine Ice Tea:* Jasmine tea blended with some lemon grass syrup.
- *Fuzzy Lemon Fizz:* Peach nectar topped off with lemon-lime soda.
- *Lemonade:* Sweetened lemon juice. Creative versions can add lavender or raspberry house-made syrups.
- *Phil Collins:* 7-Up or Sprite, lime juice, simple syrup, and club soda served with ice.
- *Shirley Temple:* Ginger ale that has been sweetened and colored with grenadine.
- *Virgin Drinks:* Many drinks can be made without alcohol simply by replacing the alcoholic drink with something nonalcoholic, such as ginger ale or sparkling water.

FIGURE 12.40
Watermelon mint smoothie. Courtesy of Erika Cespedes.

Check Your Knowledge

Directions: Use these questions to test your knowledge and understanding of the concepts presented in the chapter.

I. MULTIPLE CHOICE: Select the best possible answer from the options available.

1. "Tonic water" can be described as
   a. sparkling water
   b. soda with ginger overtones
   c. the same as club soda
   d. bitter
2. When a spirit is asked to be *neat*, it means that it
   a. has a splash of water
   b. is served over ice
   c. is shaken with ice and then strained
   d. has nothing added to the base spirit
3. The term *rocks* indicates that the drink should be poured into a glass with
   a. ice
   b. stones
   c. sugar
   d. a mixer
4. A mixer is an additional product other than the spirit such as
   a. juice
   b. soda
   c. both a and b
   d. none of the above
5. A standard classic martini is made with
   a. cranberry juice
   b. brandy
   c. rum
   d. gin
   e. both b and d

II. TRUE/FALSE Circle the correct answer.

6. True/False The classic layered drink is called the "pousse-café."
7. True/False Grenadine is deep red-colored syrup flavored with strawberries.
8. True/False Another name for soda water is club soda.
9. True/False A nightcap is having cookies and milk prior to going to bed.
10. True/False A well drink is when a customer requests a specific brand of alcohol.

III. MATCHING: Match the correct drink-making techniques.

11. Building ______
12. Floating ______
13. Stirring ______
14. Muddling ______
15. Shaking ______

a. The ingredients and ice are poured into a cocktail shaker and shaken vigorously for 10–20 seconds and strained into a chilled or ice filled glass.
b. The spirit with the highest gravity (density) on the bottom and top it with successively less dense spirits.
c. When a high proof liqueur or spirit is topped on a drink as the last act before serving the cocktail.
d. This method is used for frozen drinks that require an electric blender.
e. A lowball or highball glass is filled with ice, the liquor is poured in, then the mixer is poured on top.

16. Blending ______

17. Layering ______

f. Often done in a lowball glass or mixing tin where fruit slices, sugar, and bitters are placed in the bottom of the glass and pressed to extract the oils from the peel and the juice from the flesh.

g. The ingredients are poured into a mixing glass filled two-third with ice. Then the bar spoon is used to stir the ingredients, then strained into a chilled glass or onto fresh ice in the serving glass.

IV. DISCUSSION QUESTIONS

18. Explain two pros and cons of free pouring versus using a jigger.
19. What is the difference between stirring and shaking a drink?
20. What is the difference between serving a drink "neat" versus "up"?

CHAPTER 13

Coffee and Tea

CHAPTER 13 LEARNING OBJECTIVES

After reading this chapter, the learner will be able to:

1. Recite the origins of coffee
2. Explain the common methods of harvesting and processing coffee
3. Describe the range of roasting levels and each of their characteristics
4. Describe each of the levels of grinding coffee beans and their recommended brewing method
5. Identify the critical variables when brewing coffee and how they can influence the quality of extraction
6. Recognize several espresso-based drinks
7. Identify several manual methods of brewing coffee
8. Identify some origins for specialty tea production
9. Identify and explain the five categories of tea
10. Understand the variables of proper brewing of tea

Like any other language, learning to speak coffee requires education and practice.

—Anonymous

Coffee Primer

There exists plenty of intellectual study within the humble cup of coffee—it offers a symphony of chemical compounds acting to awaken and captivate the drinker via the stimulating effects found within this exhilarating liquid. Coffee offers well over 1,200 chemical components, most of which contribute not only to the aroma and taste but also to its invigorating physiological effects on our bodies. Most notably it contains caffeine, an odorless, bitter alkaloid responsible for the stimulating effect of coffee and tea. The caffeine in coffee helps to promote enlightenment and creative contemplation, allowing for an elevated and energized response without even recognizing it. Figure 13.1 displays a cup of coffee on the left and a specialty coffee drink illustrating some coffee art on the right. Coffee has long been considered a virtue, serving as a protagonist for the exchange of ideas and opinions that have provoked scholarly thought throughout the centuries. It has provided a backdrop for lively debate, political and social upheaval—transforming entire civilizations from the consumption of a single cup of coffee ever since its discovery in the Ethiopian highlands. Coffee is a product (not much different than food, wine, and beer) that has evolved dramatically over the past

FIGURE 13.1
Coffee and Latte. Courtesy of John Peter Laloganes.

twenty-five to thirty years as it remains one of the most commonly traded commodities in the global economy.

The Origins of Coffee

Learning Objective 1
Recite the origins of coffee

The story of coffee has its beginnings in Ethiopia (northeast Africa), the original home of the *Arabica* coffee plant, which still grows wild in the forest of the highlands. According to legend, Kaldi, an Abyssinian goatherd, who lived around 850 A.D. discovered coffee. He noticed that after eating the bright red berries that grew on the green bushes nearby, his goats started behaving in an abnormally exuberant manner, rearing on their hind legs and bleating loudly. Kaldi reported his findings to the abbot of the local monastery, who made a drink with the berries and found that it kept him alert through the long hours of evening prayer. The abbot shared his discovery with the other monks at the monastery, and knowledge of the energizing berries began to spread. Figure 13.2 displays coffee cherries which eventually will have their outer husk removed to release the green coffee beans contained within.

FIGURE 13.2
Coffee cherries. Courtesy of Big Shoulders Coffee.

The Coffee Plant

The coffee plant grows cherries, and within the cherry is the pit—the coffee bean. Most coffee plants grow best in tropical and subtropical environments, what is known as the "coffee belt," roughly within 1,000 miles of the equator. This belt encompasses more than fifty countries around the world, providing an ideal natural climate for *Coffea* plants to thrive. Coffee plants are grown most prominently in the following points of origin: *North America* (includes: Mexico and Hawaii), *Central America* (includes: El Salvador, Guatemala, and Costa Rica), and *South America* (includes: Ecuador, Columbia, and Brazil—the world leader in production). Other significant countries that grow coffee include: *Asia* (Vietnam and Indonesia), *Africa* (primarily Kenya and Ethiopia), *Indonesia* (Sumatra, Java, India, Bali, and Papua New Guinea), Haiti, and Jamaica. All areas share some common characteristics; for the most part the areas are humid (some are dry), warm to hot with some levels of elevations. There are several species of coffee plants but the most important commercial varietals will fall within two main categories: Arabica and Robusta. These types are then subcategorized by location or origin of being grown.

- *Arabica:* The earliest cultivated species of the coffee plant and the most widely grown and highly prized for its ability to produce specialty, high quality coffee beans. It supplies approximately 70 percent of the world's coffee, and is dramatically superior in quality to Robusta coffee.

  Specialty Coffee Practice of merchandising and selling coffees by country of origin, roast, flavoring, or a special blend, rather than by brand or trademark.

- *Robusta:* Robusta produces about 30 percent of the world's coffee. It is a lower-growing, higher-yield tree that produces "robust" but bland coffee of inferior quality than Arabica. Often used as the basis for blends of instant coffee, and for less expensive blends of pre-ground commercial coffee.

Considerations in Origin

In the specialty segment of the coffee industry, the concepts of single-estate, direct trade, single-origin, and fair trade coffees have grown increasingly more important. As the consumer has gained a more sophisticated preference for coffee, they have also become aware of the potential moral considerations when purchasing their coffee. The awareness of connection and relationship from bean-to-cup has become increasingly more important in the marketing of coffee on one's beverage menu.

- *Single-Estate Coffee:* Coffee produced by a single farm, single mill, or single group of farms, and marketed without mixture with other coffees. Many specialty coffees are now identified by estate name, rather than the less specific regional or market name.
- *Direct Trade Coffee:* This is a sourcing practice when coffee beans are purchased directly from farmers. This practice involves direct communication and negotiation between the buyer and farmer without the use of an intermediary.
- *Single-Origin Coffee:* Unblended coffee from a single country, region, and crop.
- *Fair Trade Coffee:* Coffee that has been purchased from farmers (usually peasant farmers) at a "fair" price or "living" wage as defined by international agencies.

Harvesting and Processing Coffee

Learning Objective 2
Explain the common methods of harvesting and processing coffee

Harvesting coffee involves the process of picking or cultivating the coffee "cherries" through one of several methods: strip picking and selective picking. Strip picking is the understandably cheaper route, but if about 15 percent of unripe cherries are used, the resulting coffee will be of rather poor quality. Selectively handpicking only ripe cherries, an obviously more expensive approach, requires multiple passes in the fields over a period while waiting for the cherries to become ripe. Figure 13.3 identifies some coffee cherries that have just been harvested.

When processing the coffee cherries, the purpose is to remove the outermost layers of the coffee cherry or fruit, called pulping. This can be done by several methods; the two most common include the wet (or washed) method or natural (or dry) method of processing.

FIGURE 13.3
Collecting coffee cherries. Courtesy of Big Shoulders Coffee.

1. **Wet (or Washed) Processed Coffee** Most of the world's coffees are processed using this method of processing. In the traditional wet process, the coffee skins are removed as they sit in tanks of water and enzymes that loosen the exterior husk and sticky fruit pulp (or mucilage) for several days, after which it's then washed off the beans. For efficiency purposes, it is also possible for the pulp to be scrubbed from the beans by machine. Washed coffees are known for being bright, acidic, but mild in flavor. Broadly speaking, the Americas tend to use the washed method (though dry processing is occasionally used) of processing that yields a clean and crisp coffee. Its aromatics show bright, citrus (lemon and grapefruit), occasionally nutty aromas and flavors with medium body.

FIGURE 13.4
Collecting green beans. Courtesy of Big Shoulders Coffee.

2. **Natural (or Dry) Processed Coffee** This method of processing is done by removing the husk and fruit pulp after it has been dried in the sun. In some coffee regions, such as southern Ethiopia, coffee cherries are dried on patios or raised drying beds. The coffee cherry undergoes a sort of natural fermentation that allows the exterior husks to become loose in order to be easily removed. The coffee beans develop the final flavor profile of the coffee resulting in complex, deeply-dimensioned, with a more obvious fruit (berries) characteristic. Naturally processed coffees typically have lower acidity levels. It is common for dry-processed coffees to come from: Yemen, Ethiopia, and Brazil. Figure 13.4 identifies drying trays as the coffee cherries are dried and have had their husks removed to expose the raw "green" coffee beans from within.

Decaffeination Processes

Coffee is decaffeinated prior to the beans having been roasted (in its green state). There are several methods for removing caffeine component (trace amounts will remain), such as using solvents or soaking the beans in water and using activated charcoal.

Roasting Coffee Beans

Learning Objective 3
Describe the range of roasting levels and each of their characteristics

Beans are stored *green*, a state in which they can be kept without loss of quality or taste. A green bean has none of the characteristics of a roasted bean as it only contains light aromas associated with grass and hay. At this stage the beans contain little to no taste. Roasting is the process of dramatically transforming the chemical and physical properties that are locked inside the raw, green coffee beans. Figure 13.5 displays a range of coffee beans at different roasting levels from green stage through dark roast.

Understanding the progression of the roast levels helps to understand the finished characteristics once the coffee is brewed and tasted. Color and shine are the initial clues used to determine degree of roast, but a skilled roaster uses other clues that are complemented by the audible cues (first and second crack) and the aromas released throughout the roasting process. It takes years of training to become an expert roaster with the ability to "read" the beans and make decisions with split-second timing. The difference between perfectly roasted coffee and a ruined batch can be a matter of seconds. Figure 13.6 displays the coffee beans being meticulously inspected during the roasting process. Figure 13.7 displays Tim Coonan, owner of Chicago's Big Shoulders Coffee, transferring the hot coffee beans from their roasting vessel to the cooling phase of the roasting process.

FIGURE 13.5
Range of roasted coffee beans. Courtesy of Erika Cespedes.

FIGURE 13.6
Checking on roasting levels. Courtesy of Big Shoulders Coffee.

FIGURE 13.7
Cooling coffee beans with Tim. Courtesy of Big Shoulders Coffee.

Roasting involves both art and science as there are some definite chemical changes taking place, but the element of art involves the choice of the level of roast as determined by the vision of the intended finished product. Tim Coonan, proprietor of Big Shoulders, a local coffee roaster and cafe based in Chicago, believes in the coffee beans being expressed on the basis of their varietal characteristics. Therefore, this means taking great care to not let the roast dictate style but letting the type of bean and its origin dictate style. This is a divergent approach from many of the large coffee chains that insist on the roast style dictating and driving the coffee regardless of origin.

Progression of Roast Levels

Roasting causes chemical changes to take place as the beans are rapidly brought to very high temperatures. As heat is applied to the beans, sugars begin to caramelize, bound-up water escapes, the structure of the bean breaks down and oils migrate, and the bean expands in size as the roast becomes darker. The degree of roast will ultimately have a significant impact on the color of the coffee and more importantly, the aroma, flavor and taste (body, acid, and bitterness). The ideal roast level is a personal choice that is sometimes influenced by national preference or geographic location. Figure 13.8 displays a handful of perfectly roasted coffee beans at the dark stage. Within the six stages of roasting coffee beans, four of them are detailed below. Each one of the levels offers some general characteristics for each roast.

FIGURE 13.8
Handful of roasted beans. Courtesy of Big Shoulders Coffee.

- *Light Roasts:* This roast is light brown in color, and is generally preferred for milder coffee varieties where roasters want the element of origin to come through in the finished product. There will be no oil on the surface of these beans because they are not roasted long enough for the oils to break through to the exterior of the bean. These roasts may go by the names half city roast or cinnamon roast. Figure 13.9 identifies coffee beans at a light roast stage.

FIGURE 13.9
Light roast. Courtesy of Erika Cespedes.

FIGURE 13.10
Medium roast. Courtesy of Erika Cespedes.

- *Medium Roasts:* This roast produces a medium-brown bean in color with a stronger flavor and a non-oily surface. This roast, similar to the light roast, is advantageous when the roaster wants the element of origin to come through in the finished product. The finished coffee will have characteristics associated with a toasted grain and nutty profile with pronounced acidity. It's often referred to as the American roast because it is generally preferred in the United States, but may also be called a city roast or breakfast roast. Figure 13.10 identifies coffee beans at a medium roast stage.
- *Medium-Dark Roasts:* This roast produces a rich, dark color. This roast often has some oil or prominent shine on the surface of the beans and when tasted, offers a slight bittersweet aftertaste. The name of the roast may be referred to as full city roast. Figure 13.11 identifies coffee beans at a medium-dark roast stage.
- *Dark Roasts:* This roast produces shiny black beans with an oily surface and a pronounced bitterness. The darker the roast, the less the acidity. Dark roast coffees run from slightly dark to char. There are many trade names used, often interchangeably, and may include the following: European roast, espresso roast, or Viennese roast. Figure 13.12 identifies coffee beans at a dark roast stage.

The roasting process begins with procuring unroasted, green coffee beans. Figure 13.13 identifies green beans. As the yellow stage begins, the coffee beans start

FIGURE 13.11
Medium-dark roast. Courtesy of Erika Cespedes.

FIGURE 13.12
Dark roast. Courtesy of Erika Cespedes.

FIGURE 13.13
Green beans. Courtesy of Erika Cespedes.

absorbing heat, losing moisture, and begin to develop color. The cinnamon stage is next when some bean expansion is visible and the coffee begins to shed its silver skin. The steam that is being released at this point becomes quite fragrant throughout the proximity of the roasting process.

- *1st Crack:* Maillard reaction starts to take place. In this process, hundreds of different chemical aromatic compounds are created. Sugars are browning, and the bean is also beginning to expand violently, creating a very audible, cracking or popping sound. After the first crack, the roast can be considered complete any time according to the taste preference.
- *2nd Crack:* The audible "sizzling" sound that occurs during this stage is referred to as 2nd crack. The CO_2 outgassing begins. An oily sheen begins to appear on the bean. Sugars are caramelizing, and this is when a more intense roasted aroma can be detected.

When the coffee beans have reached the peak of their desired level of roast, they are quickly cooled to stop the process. Roasted beans have a highly aromatic smell and weigh less than green beans because the moisture has been roasted out. The cooled beans will now be packaged and begin a degassing process over the course of the next 24–48 hours. This is a natural process in which recently roasted coffee releases carbon dioxide gas, temporarily protecting the coffee from the staling impact of oxygen.

Grinding Coffee Beans

Learning Objective 4
Describe each of the levels of grinding coffee beans and their recommended brewing method

Just prior to brewing a batch of coffee, the beans must be ground into smaller granular pieces to increase their surface area which is necessary for extracting its personality characteristics. It is critical that coffee beans be ground freshly for each batch of coffee since they are susceptible to becoming stale. Once oxygen meets the beans, they will degrade over time allowing various flavor compounds to undesirably change and break down.

Since the grind size influences the rate of extraction and final quality of the coffee, it will need to be adjusted for different brewing methods. As a rule, the finer the grind size, the faster rate of extraction (think espresso which takes about 20 seconds to extract), and in reverse, the coarser the grind, the slower rate of extraction (think drip brew which takes about 2–4 minutes). Therefore, finding the proper grind size is critical because it will dramatically influence the final product. Below are the four types of grind size along with their recommended brewing method.

- *Coarse Grind:* This is a grind size that has a textured appearance and feel similar to kosher salt. This type of grind size is best recommended for the following brewing methods: French press, toddy makers (cold brew method), vacuum coffee maker, or percolator.
- *Medium Grind:* This is a grind size that has a textured appearance and feel of coarse sand. This type of grind size is best recommended for drip makers or pour-over methods of brewing. The drip brew (or auto drip) is possibly the most common method of brewing in the home and quite common in most restaurants.
- *Fine Grind:* This is a grind size that has a textured appearance and feel similar to table sugar or table salt. This type of grind size is recommended for stovetop espresso pots, sometimes referred to as mocha pots.
- *Superfine:* This is a grind size that has a textured appearance and feel similar to powder sugar with some slight grit. This size is best recommended for espresso machines and possibly for Turkish style coffee. Although some prefer their Turkish coffee beans to have an even finer grind that is similar to the texture of flour.

Types of Grinders

There are two categories of grinders: blade and burr grinders. Each one has their pros and cons, but the burr grinder is considered superior for the sake of quality.

Blade Grinder This is a small coffee grinder with a propeller-like blade to grind the coffee beans. It is a basic and simple process to operate this type of grinder. It's functional for all types of grind levels except for fine to superfine levels. The tradeoff with this type of grinder is that it's slightly messy and noisy. More importantly, it tends to pulverize (not grind) the coffee, therefore, it produces inconsistent grind size by yielding "boulders" and will impact the quality of the extraction.

Burr Grinder Coffee grinder with two shredding discs or burrs that can be adjusted for maximum effectiveness. This type of grinder offers precision of a consistent grind, and a good degree of versatility because the grind can be adjusted for different preferences from batch to batch. The downside of these types of grinders is the expense, and the learning curve. It takes a while to adjust grind size to determine the ideal degree of extraction when brewing.

Variables in Brewing

Learning Objective 5
Identify the critical variables when brewing coffee and how they can influence the quality of extraction

Coffee beans are transformed into a drink through its manipulation with water—referred to as brewing method. There are many ways to brew coffee, but four of the most popular are: pressure (espresso), gravity (drip brew), steeping (French press), and boiling (Turkish coffee). To make great coffee, the variables in brewing must be considered. These variables change depending upon the brewing method and desired outcome of the finished product. Ultimately, they are the building blocks that contribute the aromas, flavors, and taste (body, acid, and bitterness) in a finished cup of coffee.

- *Grind:* This depends on brewing method, personal preference, and desired end product. The grind size can range from coarse–to–superfine.
- *Heat:* The temperature of water should typically be just under the boiling point, somewhere between 195 and 205°F.
- *Ratio:* This refers to the strength of the finished coffee which will be influenced by the proportion of water used per quantity of coffee. A standard "cup" or portion of coffee is measured at 6 ounces and would need a proportion of 2 tablespoons (.36 ounces) of ground coffee. If one is weighing coffee and water (the more accurate method), a general guide is to begin with a ratio of 1:15 (or even 1:16) which is simply 1 gram of coffee per 15 grams of water, and then adjusted slightly more, or less, per one's preference of strength.
- *Time:* The time it takes to extract the right amount of desirable qualities from the coffee.
- *Turbulence:* The pressure applied to extract the liquid from the grinds. There is less pressure for a pour-over or drip brew method where they rely largely on gravity, versus a French press that is a full immersion method where a plunger will force the grinds through the coffee.

For example, using the pour-over manual method of brewing uses a coarse to medium grind (depending on preference), with water temperature between 195 and 205°F in a ratio of 1 gram (or slightly more) of coffee per 15 grams of water, and it should steep for 3.5–4.5 minutes with a slight stir once the coffee has started to "bloom," this is when the hot water has been added, the coffee initially develops a cap from the release of its CO_2.

Extraction Rates

Extraction is the process of pulling out desired characteristics from the coffee into the finished product. This happens in different ways depending upon the type of brewing method. Some methods use gravity; others manual agitation or even forced steam. Regardless of method, obtaining the right degree of extraction for the brewing method is critical. The variables identified in the previous section will dramatically influence the appropriateness of extraction.

If an over-extraction occurs, this often means the coffee has been ground too fine, or there was not enough water used in ratio to the quantity of coffee. With over-extraction, there tend to be too many unpleasant elements transferred from ground coffee—therefore the brew time is too long. The result is excessive bitterness or a harsh taste. If the espresso brewing method was used, the release of the crema will display very dark brown almost black rings.

Just as unfortunate with over-extraction, is when coffee is under-extracted. This occurs when not enough of the characteristics of the ground coffee have been transferred. The results include a watery, sour, and dilute appearance and taste. Typically, an under-extracted coffee means the grinds are too coarse, causing the hot water to run through the grinds too quickly—therefore the brew time is too short. Another possibility is the ratio of water was too high in relation to the amount of coffee. If the espresso brewing method was used, the release of the crema may display blond or white spots.

The Language of Espresso

Learning Objective 6
Recognize several espresso-based drinks

Espresso is a term that has multiple meanings. It can be used to describe a level of roast for coffee beans traditionally used for espresso brewing, and can also refer to a method of brewing in which hot water is forced under pressure through a compressed bed of finely ground coffee. Because of this high pressure, the espresso process is able to extract oils and contains a higher concentration of suspended and dissolved solids that are not normally attained through more conventional brewing methods. From an aromatic perspective, espresso is a very concentrated, rich and intense coffee beverage. Beyond drinking espresso, it is also the basis for an entire selection of drinks, many combining brewed espresso with steam and or frothed milk.

To brew espresso, it is necessary to brew it using a specialized machine that can create a pressurized brewing process. The machine uses around 12 atmospheres of pressure (atm) to force water through finely ground coffee. The modern-day espresso machine was created by an Italian, Achilles Gaggia, in 1946. Gaggia invented a high-pressure espresso machine by using a spring powered lever system.

As the espresso is extracted through "pulling" or extracting the coffee through the machine, the grind size will be important so that the espresso shot is not under or over extracted. Figures 13.14, 13.15, 13.16, and 13.17 identify the correct fine grind

FIGURE 13.14
Collecting finely ground coffee beans. Courtesy of Erika Cespedes.

FIGURE 13.15
Portafilter with ground coffee. Courtesy of Erika Cespedes.

FIGURE 13.16
Tamping down ground coffee. Courtesy of Erika Cespedes.

FIGURE 13.17
Compacted ground coffee. Courtesy of Erika Cespedes.

for espresso as well as the tamping process to compress the ground coffee in order to provide the correct level of resistance and extraction of the water. The combination of water pressure, water temperature, and proper grind size of coffee all result in soluble oils, lipids, and fats being extracted from the grounds and transferred into the brewed espresso. With a properly extracted shot of espresso, there will be a crema, a pale brown colored, cream-like foam that cascades (like a Guinness Stout) and eventually settles on the surface to form a cap. Figures 13.18 and 13.19 display the beautiful crema forming over the surface of the espresso shot.

FIGURE 13.18
Upper view of espresso shot. Courtesy of Erika Cespedes.

FIGURE 13.19
Side view of espresso shot. Courtesy of Erika Cespedes.

FIGURE 13.20
Espresso in Demitasse. Courtesy of Erika Cespedes.

FIGURE 13.21
Marshmallow Latte. Courtesy of Big Shoulders Coffee.

If the espresso shot is to be served standalone, it is often served in a *demitasse*, "half cup" in French; typically, a 3-ounce cup. Figure 13.20 displays an espresso in a demitasse cup. For the more caffeine dependent, one may select a *doppio*, or a double espresso which can range from 3 to 6 ounces of straight espresso.

Specialty Coffee Drinks

Espresso serves as the basis for many of the specialty coffee drinks that sell for $5 or more per order. It is often combined with varying amounts of steamed milk and milk froth—milk that is heated and frothed with a steam wand as a part of the espresso machine.

- *Americano; Café Americano:* This drink includes an espresso lengthened with hot water.
- *Cappuccino:* This specialty drink contains espresso as the base, with the addition of equal parts of steamed milk and froth.
- *Café Latte:* This drink contains espresso as the base, with the addition of about three times as much hot milk, then topped with froth. Figure 13.21 identifies a specialty coffee drink, *Marshmallow Latte* compliments of Big Shoulders Coffee.
- *Café au Lait:* This is a coffee drink that does not use espresso, but it does combine one-third drip brewed coffee with two-thirds hot frothed milk.
- *Macchiato:* This can either be a serving of espresso "stained" or marked with a small quantity of hot frothed milk (espresso macchiato), or a moderately tall (about 8 ounces) glass of hot frothed milk "stained" with espresso (latte macchiato).
- *Mocha:* This term has a couple of meanings. It can represent a single-origin coffee from Yemen where it acquired the name from the ancient port of Mocha; also a drink combining espresso as the base, combined with chocolate and steamed milk.

Manual Methods of Brewing

Learning Objective 7
Identify several manual methods of brewing coffee

The manual method of brewing coffee has become an ever-increasing popular option for enthusiasts, specialty coffee shops and fine dining restaurants. This approach to brewing allows one to control the variables of the brewing process—creating a cup of coffee that's exactly suited to one's preferences, and highlighting the unique characteristics of the origin and roast of the coffee beans. There are numerous manual methods of brewing coffee. Two categories of these methods include: immersion and the pour-over methods.

- *Immersion Methods:* Clever, French press, Aeropress, Siphon (or vacpot)
- *Pour-Over Methods:* V60, Chemex, and Kalita wave

Listed in the next section is a step-by-step approach to brewing with one of the methods in each category—the French press and a Chemex.

Immersion Method—French Press

This brewing method separates the spent grounds from brewed coffee by pressing them to the bottom of the brewing receptacle with a mesh plunger. This method is referred to as a full immersion method due to the full contact of the coffee grinds with the hot water. It tends to produce a full flavor with the presence of a slight grit and a muddy appearance because the filter only consists of a small wire mesh. The recipe below needs to be adjusted depending upon the desired number of portions and size of the French press. Keep in mind, French press devices come in various sizes such as 34 ounce, 51 ounce and even a smaller 16 ounce version.

1. Place pot on a flat surface, hold handle firmly and pull plunger unit straight up and out of the pot. It's recommended to pour some hot water into the device in order to preheat for 30 seconds prior to beginning the brewing process. Once done preheating, remove the water and begin step #2.
2. Place 2 tablespoon of coarse ground coffee into the pot. The proportion of coffee and water will be increased depending upon the number of portions of coffee desired.
3. Pour 6 ounces of hot (between 195 and 205°F) water into the pot.
4. Place plunger in unit. Turn lid to close off the pour spout opening. Let coffee brew for 4 minutes. At the initial 30 second mark, give the mixture a few stirs (use a wooden spoon or chopsticks) in order to ensure there are no clumps or air pockets.
5. Once the total brewing time of 4 minutes has been achieved, hold onto pot handle firmly and then use weight of hand to apply pressure on top knob and slowly press down. Minimal pressure produces best results.
6. Turn lid to open the pour spout. Then pour coffee immediately into serving cups.

FIGURE 13.22
Finished carafe of coffee. Courtesy of Big Shoulders Coffee.

Pour-Over Method—Chemex

The Chemex is one of the more popular pour-over methods in recent times. It is highly regarded for producing a clear looking (less murky) cup of coffee than French press, and allows for more natural coffee oils than drip brew. Start with a ratio of 1:15 or 1 gram of coffee per 15 grams of water, and then adjust per one's preference of strength. The overall brewing process should take between 3.5 and 4.5 minutes. Figure 13.22 displays a carafe of coffee using the pour-over method.

1. Place the Chemex filter in the decanter with a single fold away from the spout. Rinse and pre-heat the filter and decanter by pouring some hot water all the way around the filter to get a nice seal against the glass. Discard the pre-heat water in a sink.
2. Position the decanter over a scale and add 50 grams (1.8 ounces) of coffee in the filter. This amount of coffee will require 750 grams (about 27 ounces) of water. This is using a ratio of 1:15, but can be adjusted based on desired strength. Be sure to zero out the scale once the grounds have been placed into the filter. The grind of the beans should be medium to coarse, comparable to the consistency of kosher salt.
3. The initial pour is very important. Remove hot water (between 195 and 205°F) from the heat source and set timer for 4 minutes. Instead of pouring all the water onto the grounds, begin to slowly pour just enough water to saturate the beans (about 150 grams) and stir for about 5 seconds with a spoon or chopstick.

FIGURE 13.23
Manual pour-over method of brewing. Courtesy of Big Shoulders Coffee.

FIGURE 13.24
Manual pour-over method of brewing (2). Courtesy of Big Shoulders Coffee.

Wait for about 20–40 seconds for the ground coffee to "bloom." During this time, trapped carbon dioxide (CO_2) escapes from the grounds which allows for maximum flavor extraction. Figures 13.23 and 13.24 display a pour-over coffee being prepared as the hot water is slowly poured over the ground coffee.

4. Begin the second pour by slowly pouring the remaining contents of the water over the grounds until the volume reaches about a fingertip below the top of the rim or about 600 grams of water. When pouring, be sure to use a circular or back-and-forth motion to ensure an even soaking of the grounds.
5. Make sure very little to no water remains in the filter when pouring the coffee into a cup. The entire brewing process should have taken between 3.5 and 4.5 minutes.

Final Thoughts on Coffee

It takes a lot of practice for the beverage manager to develop and enhance their coffee tasting abilities, not much different than it is with wine and beer tasting. For coffee professionals, they typically engage in daily cuppings. These cuppings are used by professional tasters to perform sensory evaluation of samples of coffee beans from different origins, ones that have undergone different processing, different levels of roasting, and/or different levels of grind size. The beans are ground, water is poured over the grounds, and the liquid is tasted both hot and as it cools. Similar to assessing other beverages, the key evaluation characteristics are focused on aromas/flavors and structural characteristics. Figure 13.25 displays coffee prepared accordingly for a cupping assessment.

FIGURE 13.25
Coffee cupping. Courtesy of John Peter Laloganes.

Tea

Learning Objective 8
Identify some origins for specialty tea production

Tea is often touted as the second most consumed beverage in the world, just after water. Unfortunately, in too many restaurants and bars, tea is treated as an afterthought. With just a little effort, the beverage manager can assemble a respectable tea program with a thoughtful menu and gather success from tea's healthy profit margins.

Tea is a beverage best understood through breaking down its botanical name—*Camellia* (related to flower) and *Sinensis* (means from China). Figure 13.26 illustrates a Korean tea garden where the Camellia sinensis flourishes. Camellia sinensis is an evergreen bush, if left to grow wild, it will grow into a 60 feet tree or taller. However, it is more common to have the bush pruned to about 3–5 feet that allows for its tender young leaves to be hand cultivated. Tea plants thrive in the tropical and subtropical climates with warm temperatures, ample shade, and an abundance of rainfall. Higher-quality teas grow at higher elevations between 3,000 and 6,000 feet, which works to cool the climate and slow the growth of the leaves. Tea is grown all over the world; however, it has its origin in Southeast Asia, more notably in China where it was often used for medicinal purposes. Other areas include India, Tibet, Burma (Myanmar), Laos, Bangladesh, Thailand, and Vietnam. These countries have used tea throughout centuries in many applications in daily practice—initially, they used tea leaves as a food and eventually, these were infused into water to create the drink that we know it today. Later on in the 17th century, tea became a stronghold in the United Kingdom (UK). The creation of the "tea time" with the low and high teas became prominent throughout much of the 18th–20th centuries. The five classic tea production countries (China, India, Kenya, Sri Lanka, and Vietnam) have set the standard for specialty tea. Specialty tea is represented by quality and flavor of the tea leaves. While these origins are classic to the tea plant, it is also capable of being grown in less obvious places such as New Zealand and Hawaii.

FIGURE 13.26
Korean tea gardens. Courtesy of Suzette Hammond, Being Tea.

FIGURE 13.27
Herbal teas. Courtesy of Suzette Hammond, Being Tea.

Tea has been used and consumed in many ways throughout history and in different cultures around the globe. Many beverages are referred to as tea; however, the authentic drink is the one deriving from the plant, Camellia sinensis. Other beverages referred to as "tea," but not deriving from this plant are derived from a combination of either herbs and/or botanicals. Figure 13.27 identifies some herbal teas. These alternative tea beverages are quite common, such as chamomile, peppermint, and mint. In some uncommon cases, tea blenders may combine some of these botanicals and herbs with an actual portion of the Camellia sinensis plant.

Categories of Tea

Learning Objective 9
Identify and explain the five categories of tea

Tea is chemically complex with some estimated 150–500 different chemical compounds that can provide a range of simplicity to complexity in a finished cup of tea. There are five main categories of tea that are made from the Camellia sinensis plant which include: Green, white, oolong, black, and *puerh* (poo-EHR) tea. Like wine, different geographical areas and production methods can alter the styles and nuances of these categories of tea.

FIGURE 13.28
Oxidized tea leaves. Courtesy of Suzette Hammond, Being Tea.

Oxidation

This is a concept that is central to the production process and defining to the five main categories of tea. Oxidation is the enzymatic reaction that occurs in tea leaves when they have been rolled or bruised: Releasing its surface area allows it to react with oxygen, causing it to turn color and alter its aromatic compounds. Figure 13.28 showcases some oxidized tea leaves. This oxidation concept can be compared to the same process of a bruised banana or chopped apple turning brown. The greater degree of oxidation will create more prominent and complex aroma/flavor profiles with a darker color.

None-to-Lightly Oxidized	**Green tea** (unoxidized)	Fresh herbal, grassy, and vegetal aromas/flavors. Green teas are steamed or pan-fired immediately after harvest in order to halt their oxidation process.
	White tea (very lightly oxidized)	Delicate and floral. These teas are air-dried on bamboo racks, just allowed to wilt, similar to a fresh flower or herbs.
	Oolong tea (light to heavy oxidation) or **Wulong tea**	Offers diverse range of aromas/flavors from perfume like and honey to slightly nutty.
	Black tea (heavy oxidation)	Can range from citrusy-to-malty. Offer a fuller body with more intense and complex flavors.
Fully Oxidized	**Dark tea** (heavily oxidized and fermented) or Puerh tea	These teas are quite unique as they are allowed to age and ferment. Puerh offers earthy complex profiles.

FIGURE 13.29
Green tea. Courtesy of John Peter Laloganes.

Green Tea

Green tea is one of the least manipulated and minimally oxidized tea categories. The leaves are usually withered, but not rolled. The leaves are steamed, typical for Japanese types or pan-fried (fired) which is typical for Chinese types. The process of steaming or firing works to halt any form of oxidation. Figure 13.29 illustrates a brewed green tea, loved for its vibrant green color.

Brew green teas between 170 and 180°F for about two minutes. Green teas are highly sensitive to brewing temperatures and extraction times. Green teas can easily go from bright and vibrant green colors to army green overcooked broccoli appearance. Water temps and extraction times must be adhered to, to ensure the integrity of these delicate teas.

- Famous Green Teas: Gunpowder, Matcha, and Sencha are examples of steamed green tea, while Dragon Well is an example of pan-fired green tea.

White Tea

White tea is the most delicate and one of the least manipulated tea categories. Instead of being exposed to an artificial heat as other tea categories, white tea leaves are simply allowed to wither and dry in a carefully controlled environment. The leaves are minimally oxidized; therefore, it yields a very light and precocious taste profile. When harvesting white tea leaves, only the immature buds and tops of the leaves are obtained.

Like green teas, white teas should be brewed between 170 and 180°F for about 2–3 minutes. White teas are highly sensitive to brewing temperatures and extraction times. Water temps and extraction times must be adhered to, to ensure the integrity of these delicate teas.

- Famous White Teas: Silver Needle, White Peony, and Darjeeling White

Oolong Tea

Oolong teas are withered and then rolled or shaken. They are oxidized about half the time for typical black tea styles. These teas offer a slight nutty and toasty quality due to the light oxidation. Figure 13.30 displays a before-and-after (the before is located on the right-side) oolong tea infusion.

Brew at 190–205°F for 3–6 minutes. Oolong teas are often considered very forgiving when dealing with brewing temps, extraction times, as well as number of extractions (or flushings). Figure 13.31 shows a pot of oolong tea being brewed. They are hardier teas that can withstand higher water temps and multiple flushings without a loss of quality. And some people even prefer the differences between a first and second flushing versus a third and fourth flushing.

- Famous Oolong Teas: Coconut Oolong, Iron Goddess of Mercy, and Eastern Beauty

FIGURE 13.30

Oolong before and after infusion. Courtesy of Suzette Hammond, Being Tea.

FIGURE 13.31

Oolong tea pot. Courtesy of Suzette Hammond, Being Tea.

Black Tea

Black tea is the most processed of all the categories and types of teas. They are fully oxidized, as the fresh leaves are withered for several hours and then rolled. This manipulation encourages the chemical reaction of oxidation that results in the browning of the tea leaves. After these tea leaves have been sufficiently oxidized, the leaves are heated and then dried in wood fires.

Brew at 195–212°F for 3–5 minutes. Black teas are the most forgiving when dealing with brewing temps. They are hardier, more intense teas that can withstand higher water temperatures. Extraction time should be honored, as extensive times can lead to more bitterness in the taste. Sometimes, black teas are served with milk and sugar as a way to balance out their intensity. This method is common with certain strong teas in the marketplace such as Assam tea (from India).

- Famous Black Teas: Assam, Ceylon, Darjeeling, Keemum, Himalayan, and English Breakfast
- Lapsang Souchong – Originates from Fujian China is one of the most unusual and distinctive black tea. Lapsang Souchong has an intensely smoky aroma and flavor because of the pine wood fire used in the final drying process of the tea leaves.

Dark Tea or Puerh Tea

Dark tea is a general name for Puerh (sometimes spelled pu-erh) tea that is named after its place of origin in the Yunnan province of China. It is marketed in a similar way (geographical) that Champagne (sparkling wine from Champagne, France) is marketed. These tea leaves are picked and dampened and then fermented for about 60 days. Through this time, they develop a rich color and deep, malty, earthy aroma and flavor characteristics.

Brew dark teas in a similar manner that black teas are brewed. They should be brewed at 195–212°F for 3–5 minutes.

Alternative Teas

While these drinks are not technically tea, they are still important and great contributors of enjoyable beverages. Any sort of botanicals, herbs, flowers, or spiced concoctions have been known to offer numerous health benefits along with serving as possible medicinal remedies. Some specialty tea producers are using herbs and spices that may be sourced near distribution as a way to incorporate an element of locality.

Herbs and Plants

- Chamomile – Thought to promote calm and stress releasing tendencies. Also can sooth upset stomach.
- Rooibos – This derives from an African native bush which contains natural electrolytes and minerals such as iron, potassium, and calcium. It is anti-inflammatory and rich in antioxidants.

Spices and Rhizomes

- Ginger – Has natural anti-nausea properties and assists in reducing an upset stomach. Can also be an immune booster.

Flowers

- Hibiscus Flowers – Anti-inflammatory and rich in antioxidants.
- Jasmine Flowers – Believed to have aphrodisiac qualities.

Brewing Hot Tea

Learning Objective 10
Understand the variables of proper brewing of tea

When attempting to brew a great cup of tea, it is critical to always start with fresh, filtered water. For most municipalities, the faucet or tap water is not a good source; they often contain chlorine, excessive minerals, and many other components that contribute an off-flavor in the finished tea. Using a carbon filtering system—such as those found in a typical water pitcher with a filter—will improve the quality of the tea substantially. Using bottled spring water works excellent, but stay away from using distilled or mineral-heavy water.

To brew a perfect cup of tea, there are three variables that need to be considered. These variables can be slightly manipulated depending upon the category and type of tea. The three variables are:

1. **Type and Quantity of Tea** The typical proportion of tea to water is 1 tablespoon of tea (3–5 grams) per 8 ounces (but can be stretched to 10–12 ounces) of water. To be most accurate, tea should be weighed because quantities can vary dramatically due to the size and shape of leaves.

2. **Temperature of the Water** The temperature of the water is most critical with lighter, less oxidized categories of tea. Since they are more delicate, they run the risk of easily being burned if the water is too hot. For green and white tea the water temperature should range between 175 and 180°F. Oolong tea should be infused in a range of 190–200°F, while other categories of black and dark teas can withstand temperatures as high as boiling (212°F) without a loss of quality in the tea.
3. **Time of Infusion** For most delicate teas, the infusion time should be 2–3 minutes. When infusion goes well beyond that recommendation, the tea will gain more of a stewed, overcooked color and aroma profile. Oolong, black, dark, and herbal teas are more forgiving in terms of infusion time. The recommendation for these teas is 3–5 minutes, but again if it slightly over steeps, the tea will not have any significant loss of quality.

Building a Tea Program

Tea is unfortunately all too often one of the last or second to last consideration when building a beverage program in a restaurant. For some reason, it hasn't had as much love or appeal or attention as does the alcoholic beverages and coffee. With the changing of the marketplace, nonalcoholic beverages are becoming increasingly popular, coupled with their potential for high profit margins, it should be an additional consideration when creating a beverage menu. With slight effort, it is possible to charge more for the tea in each restaurant if one can heighten the quality, service delivery, and branding around the tea segment. One of the questions for managers to ask is, "Can my restaurant be known as a tea destination?" or at least "How can I put as much care in my tea program as I do in my wine or coffee program?" Suzette (Sooz) Hammond, consultant, trainer and founder of *Being Tea* makes an interesting point that, "it is ironic when a restaurant has a well curated wine list and then when someone orders tea, they bring out a bag of Lipton." Figure 13.32 illustrates a sample restaurant menu that emphasizes quality-oriented tea and coffee selections.

FIGURE 13.32
Tea and coffee menu. Courtesy of John Peter Laloganes.

Does It Make Financial Sense?

Sooz Hammond points out that given all the beverages commonly sold in restaurants, tea maintains the largest profit margins than any other. For example, Japanese Sencha (a *green tea* from Japan) may cost about \$30 for a bag with a quantity of 200 grams. The typical portion for an order of tea would be about 1 tablespoon (3–5 grams) which breaks down to 0.15 cents per gram. The portion cost of the tea will range between 0.45 and 0.75 cents. With a typical restaurant selling price of \$3 for a cup of tea, the gross profit margin (difference between product cost and selling price) is between \$2.25 and \$2.55 or converted to a percent basis, there is a 75–85 percent profit margin per portion of tea.

- *Example:* Japanese Sencha = \$30 for 200 grams
  - 200 grams/4g serving = 50 servings
  - \$30 (cost)/50 (servings) = \$0.60 per cup
  - Average pot of fine tea = \$3.00–\$5.00 ***Minimum \$2.40 profit***

Considerations in Building a Tea Program

There are many considerations for designing and arranging a tea menu, or at the very least, the section on the menu where tea is mentioned. Today's drinker has become more of a conscious consumer when selecting their beverage of choice. They are recognizing that origin and style are just as important in tea as they are in the world of wine. One example is to highlight the tea farms, or highlighting seasonal teas. Having a balanced offering of tea selections in a restaurant is important to meet the needs of the consumer. At minimum, a menu should offer at least 6–8 tea selections. These selections should be balanced in offering options from the different categories of tea, but also providing a variety of options within popular categories. The breakdown may look something like the list below:

1. Green tea – Steamed version
2. Green tea – Pan-fired version
3. Black tea – English (or Irish breakfast)
4. Black tea – Earl Grey
5. Herbal – Fruit
6. Herbal – Mint
7. Oolong
8. Seasonal option

As one can see, the 6–8 selections get filled up quick when considering the expectations that many consumers have in terms of the product offerings.

Storing and Preserving Tea

Tea goes stale quickly. It is recommended to use any opened package of tea within 45–60 days. If the tea is unopened and airtight, it can retain its freshness for about two years. For maximum freshness, keep all tea (opened or unopened) in a cool, dry environment that limits or eliminates light.

Check Your Knowledge

Directions: Use these questions to test your knowledge and understanding of the concepts presented in the chapter.

I. MULTIPLE CHOICE: Select the best possible answer from the options available.

1. A tea's level of oxidation can determine the
   a. category of tea
   b. aromas/flavors
   c. color
   d. all of the above
2. Brewing temperature range for most green teas is
   a. 180–195°F
   b. 212–220°F
   c. 175–180°F
   d. 100–120°F
3. For most teas, one tablespoon per ___ oz. of water is appropriate.
   a. 4
   b. 6
   c. 8
   d. 10

4. If the tea is steeped too long, what occurs?
   a. Over-extraction
   b. Under-extraction
   c. Stewed aromas and flavors
   d. Both a and c
   e. All of the above
5. The origins of coffee are believed to be from
   a. Brazil
   b. Columbia
   c. Ethiopia
   d. Hawaii
6. Using a coarse-to-medium grind is ideal for which coffee brewing method?
   a. Espresso
   b. Mocha pot
   c. French press
   d. Both a and b
7. Using a fine to superfine grind is ideal for which coffee brewing method?
   a. Espresso
   b. Drip
   c. French press
   d. Pour-over
8. Over-extraction of coffee may occur when
   a. too many unpleasant elements of bitterness and harshness are extracted
   b. the coffee may be ground too coarse
   c. the espresso shot may have run too short or the coffee may have been ground too fine
   d. not enough coffee has been used, resulting in a dilute, sour, and underdeveloped taste
   e. both a and c
9. Under-extraction of coffee may occur when
   a. the coffee may be ground too fine
   b. not enough coffee has been used, resulting in a dilute, sour, and underdeveloped taste
   c. the brewing process may be run too short or the coffee may have been ground to coarse
   d. too many unpleasant elements of bitterness and harshness are extracted
   e. both b and c
10. Lapsang Souchong is a type of which tea?
    a. Green tea
    b. White tea
    c. Oolong tea
    d. Black tea
11. Sencha is a type of which tea?
    a. Green tea
    b. White tea
    c. Oolong tea
    d. Black tea
12. Espresso can mean which of the following?
    a. A type of coffee roast
    b. A method of making coffee
    c. A type of grind
    d. Both a and b

13. During the roasting of coffee beans, the 1st crack yields
    a. an audible sizzling sound
    b. oily sheen begins to appear on the bean
    c. sugars are browning
    d. sugars are burning
14. Darker roast levels for coffee share which characteristic?
    a. Have more acid
    b. Have less acid
    c. Have a grassy smell
    d. Have no oiliness on the beans
15. Which method of brewing uses steeping as part of its production process?
    a. Drip brew
    b. Espresso
    c. Turkish coffee
    d. French press

II. DISCUSSION QUESTIONS

16. Explain the different methods of processing coffee beans once they have been harvested.
17. Identify and explain each of the five categories of tea.
18. Identify and explain the five variables in coffee brewing.
19. Identify and explain two specialty coffee drinks.
20. Identify two of the most popular manual methods of brewing coffee.

CHAPTER 14

Constructing the Beverage Concept

CHAPTER 14 LEARNING OBJECTIVES

After reading this chapter, the learner will be able to:

1. Describe the essential ingredients to the success of a beverage establishment
2. Identify the difference between a prospectus and a business plan
3. Compare and contrast the beverage offerings between an upscale and casual restaurant
4. Identify the related areas to understand when conducting a market research
5. Consider some benefits of using daily specials and seasonal products
6. Recognize some styles of service
7. Identify the broad forms of business ownership
8. Identify some ways that atmosphere can assist in defining the concept
9. Identify the purpose of each of the four essential financial statements

Our goals can only be reached through a vehicle of a plan in which we must fervently believe, and upon which we must vigorously act. There is no other route to success.

—Pablo Picasso

The Essential Ingredients to Success

Learning Objective 1
Describe the essential ingredients to the success of a beverage establishment

According to the National Restaurant Association, the restaurant industry has reached sales of $782.7 billion with over 1 million locations in the United States in 2016. Most of these more successful organizations have some form of strategic plan—a dynamic force that acts to shape the function of the business concept. This plan usually is documented in a prospectus or business plan, consisting of a vision and mission statement, short-term and long-term goals, and actions for achieving these goals. The strategic plan is documented in the initial phases of concept formation to serve as a blue print and foundation for the organization to develop a competitive advantage and remain profitable. The strategic plan should encompass strategies on how to accomplish the following successful business practices:

- *Produce Profit:* The manager's primary responsibility and accountability is to operate an organization in a manner that, at least in the long term, is profitable.
- *Generate Revenue and Control Costs:* Profit is achieved through producing revenue (money coming into the establishment) and controlling the expenses (money going out of the establishment) necessary to operate and sustain the business.
- *Encourage Repeat Business:* According to the National Restaurant Association (NRA), "60–80 percent of a hospitality organization's sales (or revenue) is based on repeat business." Repeat business occurs when a guest (or customer) has sought satisfaction of fulfilling their needs and the business has met or exceeded them and encouraging the probability of continued business in the future.

- *Satisfy Customers:* To obtain repeat business, the manager creates and communicates a vision that encourages employees to meet or exceed customer expectations in order to achieve a satisfied customer.

Prospectus Primer

Learning Objective 2
Identify the difference between a prospectus and a business plan

Regardless of forming a new or expanding an existing business, the prospectus is one of the most critical preliminary and foundational documents that outline the business concept. The business concept is expressed in the prospectus—a living document that identifies and outlines the proposed critical aspects relevant to the configuration of the overall business. The prospectus is an abridged version of a business plan that identifies the necessary information needed by investors and other stakeholders to make an educated decision on whether to invest their resources in a business venture. Pictured in Figure 14.1 is a busy city scene with restaurant.

FIGURE 14.1
Restaurant row. Courtesy of Erika Cespedes.

To effectively operate any business, the manager must stimulate revenue and control expenses—therefore, it becomes imperative for the manager to clearly understand the *concept* of the establishment. The concept is a combination of various factors that form the character and uniqueness of a business—which can vary its components to tailor the needs to a location and/or set of customers. The most significant and defining factors that characterize a concept include the following:

1. Creating the ***company description***
2. Determining the ***operational format***
3. Formulation of a ***vision*** and ***mission*** of the organization
4. Conducting ***market research*** for
   - recognizing a primary ***target market***
   - determining a suitable ***location*** to attract the target market
   - assessing the ***competition***
5. Developing the ***products and services*** based off the market research and vision of the concept
6. Designing an ***atmosphere*** that reflects the needs of the target market
7. Defining the level and type of ***service***
8. Developing a ***marketing strategy*** (expanded upon in Chapter 16)

Most successful business establishments are fluid, adaptable, and will evolve along with the changing needs of both internal and external factors of the business—the prospectus is a work in progress that clearly identifies the business intentions at a certain point in time. There is no single formula for developing a prospectus or business plan, but some elements tend to always be inclusive. The business prospectus should concisely and undoubtedly illustrate these factors beginning with a clearly conveyed vision and mission of the organization and the projected investment. Figure 14.2 illustrates Il Mulino restaurant in New York tradition, which offers classic Italian cuisine where it melds with Art Deco scenes on beautiful South Beach in Florida. Blending the sophistication of New York with elegance at the heart of the South of Fifth neighborhood, enjoy exquisite dishes paired with an extensive selection of wine.

FIGURE 14.2
Dining room at Il Mulino. Courtesy of Erika Cespedes.

The prospectus precisely defines the business, identifies goals, and serves as the firm's resume—it also acts as an on-going management tool. The prospectus may vary in length but the essential details must be covered to highlight the business concept to potential investors. When possible, statistics should be cited from referenced reputable sources that support the needs and justification for the existence of the business. Often, the prospectus will reference other similar successful businesses to identify potential areas of opportunity in the marketplace. The prospectus is crucial because it helps gauge the potential of the long-term sustainability for any business—they help potential investors become informed and educated in operating and investment decisions.

Business plans (as opposed to only a prospectus) become increasingly necessary as the investment may be perceived as more volatile and uncertain. While this study is slightly antiquated, an interesting and related research article was published in the August 2005 issue of the *Cornell Quarterly* (the journal of applied research serving hospitality practitioners and scholars), titled "Why Restaurants Fail," by H.G. Parsa, John T. Self, David Njite, and Tiffany King, which illustrates the risk associated with the food and beverage industry by indicating failure rates. Using actual restaurant statistics for independent restaurants operating between 1996 and 1999 in Columbus, Ohio, the researchers found

- the failure rate for independent restaurants was *26 percent in the first year of opening*
- the total failure rate for the three-year period from opening was 60 percent

Reasons for failure rates of businesses, particularly beverage establishments, are too numerous to mention; however, research consistently identifies ownership and management not having thorough preparation and business knowledge as a commonly cited factor. Many of these factors are intended to be addressed and contemplated in a well thought-out prospectus.

Company Description

Learning Objective 3
Compare and contrast the beverage offerings between an upscale and casual restaurant

The company description states the core nature of the product and/or services that will be provided. It will also identify a list of the marketplace needs trying to satisfy the consumers through the execution of an establishment's products or services—identification of ways to provide a competitive advantage. There will be a focus made on how the operation will be unique and in essence, "different and better" D&B. Some examples may include a superior ability to satisfy the customers' needs, unique product offerings, highly efficient methods of delivering the product or service, outstanding personnel, or a key location.

Operational Formats

The prospective ownership typically defines the operational format of the business in these preliminary stages of concept development. This very important decision of operational format will influence all other components of the concept—from the layout of the concept to the skill-set of employees hired, the food and beverage offerings, and so on. Not only do the details of ownership and responsibilities need to be documented, the very important details of defining the operation's days and hours of operation, number of seats, expected number of covers (guests or customers), check average for all meal periods, and annual sales projection must also take place. Most food and beverage establishments fall into one of these business model formats:

- *Beverage driven operations* offer a focus on a particular beverage category. These types of establishments can be small or large and independent or chain operated. The offerings include: brew pubs, beer-centric establishments, wine, whiskey, cider, sake, and cocktail bars. There are winery restaurants such as City Winery or Cooper's Hawk, and there are Tiki bars such as Three Dots and a Dash or Lost Lake. All these beverage-driven establishments have sprung up across America in both suburban and urban locations. Food is offered in these operations, but it is thoughtfully selected to enhance and not overshadow the beverage component.
- *Fine dining restaurants* offer full table service accompanied by an upscale atmosphere, high quality of beverage and food options, and exceptional service(s) with higher price points.

  The beverage focus in many fine dining restaurants will include an extensive wine selection of Old and New World options, with an emphasis on producers that have smaller production and limited or "exclusive" offerings. Upscale operations will likely offer some local or national craft beer and high quality import options. There will be a cocktail menu that offers a full selection of high-end spirits complete with a selection of classic cocktails, but also some original or "signature" creations that are unique to the operation.
- *Casual dining establishments* offer full table service that is more upscale than quick-casual restaurants, but also more affordable than fine dining restaurants. They appeal to a wide customer base and are usually family-friendly. Casual dining establishments focus on providing good value, while offering an extensive selection of food and beverage items.

  The beverage focus in many casual restaurants will often have more of a limited selection of wine, mostly focused on large-production, brand oriented, domestic wines or, in general, wines of the New World. Casual restaurants will commonly offer a large selection of domestic beers (many on draft), with an emphasis on large producers or faux craft beers with a smaller sampling of craft beer options. This product mix will change if the operation is an independent restaurant versus a chain casual restaurant. The independent operation will likely emphasize more local craft beers. There will be a spirits list that offers a full selection of products with varying prices points complete with a selection of unique cocktails that are fun, sweet, and fruit-driven.
- *Quick-casual establishments* are typically perceived to offer better beverage and food quality and improved service over quick-service restaurant (QSR) places. Their menus tend to be less extensive but also less expensive than casual dining restaurants.

  The beverage focus in most quick-casual restaurants are limited to nonalcoholic options such as soda, milkshakes, coffee (and espresso-based drinks), tea, juice, and possibly smoothies. A small selection of operators will offer wines in smaller formats such as splits (187 ml) and by the half-bottle (375 ml). A few operators offer a modest selection of beer.
- *Quick-service restaurants (QSR)* provide a convenience of location and speed of service at a low to moderate price point. These restaurants typically have simple décor, inexpensive food items, and fast counter-service.

  The beverage focus in most quick-service restaurants will emphasize non-alcoholic options. These operators will offer soda, milkshakes, coffee, tea, juice, and possibly smoothies.

Other decisions that are significant to the overall concept are whether an establishment has acquired a liquor license, dram shop insurance, deciding to provide take-out or delivery service, offering in-house banquets or off-site catering services, etc.

Vision and Mission

The vision and mission of an organization can provide guidance and direction—they consist of a set of values that help an organization align its actions with its purpose. The vision identifies who the establishment strives to be, and the mission describes how the establishment will get there. They help to shape the future as they articulate a dream into a reality allowing everyone the opportunity to align their efforts. The vision and mission work simultaneously to demonstrate the unique purpose of the organization and to capture the qualities that are most desirable. The ancient proverb of *Where there is no vision, the people perish* has such relevance—when the job of line-level employees can easily become diluted through day-to-day repetition. It is human nature to lose sight of the big conceptual view at times. Therefore, a sense of purpose helps to diminish this mentality.

Vision statements offer an aspirational, vivid, and idealized phrase or description of what an organization is striving to accomplish in the near and long-term future. The vision creates an empowering framework for individuals to conduct their behaviors. John F. Kennedy clearly defined a vision of, "We will put a man on the moon before the end of the decade." Certainly, this famous vision was instrumental for guiding the American people in the right direction to ultimately achieve and convert that dream into reality.

For Example: "We inspire and nurture the human spirit—one person, one cup, and one neighborhood at a time"—Starbucks

For Example: "Be the world's beer company. Through all our products, services and relationships, we will add to life's enjoyment"—Anheuser-Busch (AB-InBev)

Mission statements offer an intended sense of purpose and direction—they provide a declaration of an organization's core purpose and focus that normally remains unchanged over time. The mission statement is worded in a manner that provides employees with a larger sense of purpose—so they can see themselves as "building a highway" rather than "laying down asphalt." The mission outlines the necessary attributes needed to achieve the vision.

If properly crafted, vision statements are something to be pursued while mission statements are something to be accomplished. Creating a shared vision and mission among employees of the business is a key to engaging their consistent level of passion, devotion, and expertise to the result—the creation of satisfied customers and ultimately building repeat business.

Market Research

Learning Objective 4
Identify the related areas to understand when conducting a market research

Market research is imperative for any organization which desires to make solid, educated business decisions. This is the process of gathering and analyzing valuable consumer and economic data. Successful businesses will utilize market research to provide insight to assist in identifying the potential target markets, suitable location, and current or potential competition.

Customer

Customers are the lifeblood of any organization—money is the medium used in exchange for goods and services. Effective businesses generally focus on a customer—one that is most likely to consume the establishment's products.

A target market is simply the groups of customers that are the most desirable people for an establishment to direct its actions. When defining target markets, it is important to narrow them to a manageable scope and size. Many businesses make the mistake of trying to be everything to everybody. The philosophy, "If you try to please everyone, you will end up satisfying no one" is pertinent for many unsuccessful failing establishments. Instead, by concentrating efforts on a few key market segments, it is possible to reap the most from even minimal investments in attempting to satisfy the select customers.

FIGURE 14.3
Millenial. Courtesy of Erika Cespedes.

Identification of a specific target market involves analyzing and grouping customers into "like" characteristics and profiling them into segments. The business then selects a few key segments (primary and secondary) to concentrate their organizational and marketing efforts. Market research is exhaustively conducted to accurately determine the preferences of the chosen target market. By analyzing the *trading area* (the areas and locations where much of the customers are coming from), a business can most effectively meet the needs and expectations of the market. Once a target market's similarities and differences are identified, customer groups can be identified according to a combination of demographics and psychographic characteristics. Knowing these distinctive traits enables the leaders of organizations to make more intelligent decisions regarding the vision and mission of the concept. With a greater understanding of the target market characteristics, the manager can select the appropriate type and styles of beverages, food, price point, atmosphere, and level and type of service. Ultimately, the "brand" should appeal and resonate with the target market(s). Pictured in Figure 14.3 is a young twenty–thirty something woman who can be categorized as part of a target market.

There are two broad methods used to segment a market: geographic segmentation and customer segmentation. Each of the methods attempts to profile people according to some common characteristics.

- *Geographic Segmentation:* This form of target marketing focuses on serving the needs of customers in a geographical location. The target market may be a somewhat captive audience, perhaps with limited accessibility, and/or very little alternatives. Using Starbucks as an example, many of their store locations are strategically positioned so as to attract their target customers within ½ mile distance or less.

- *Customer Segmentation:* This form of target marketing focuses on identifying those people most likely to consume the products and/or services. This profiling is often based upon some form of demographic (socioeconomic status) and psychographic characteristics.

  Demographic characteristics consist of statistical distinctiveness of people in specific geographic areas. Individuals concentrated in each area may be grouped on the basis of age, gender, income level, marital status, traveling distance to and from work, type of household, employment, etc. This information can be even further divided according to zip codes. Groupings of individuals may even be given titles to recognize and differentiate them from other groups. For example, baby boomers are generally identified as those born between 1946 and 1964 and represent about 40 percent of the population.

DEMOGRAPHIC SNAP-SHOT BASED ON BIRTH YEARS

Generation	Born
The silent	1925–1945
Baby boomers	1946–1964
Generation X	1965–1980
Generation Y/millennial	1981–2000

Psychographic characteristics are based on uniqueness of people's lifestyles. They depict motivations of consumer behavior and include areas such as personality types, habits, leisure activities, ideologies, values, beliefs, and attitudes. Particular groups may be lumped together because of their beliefs. For example, "achievers" have many wants and needs and are dynamic in the marketplace. Their image is vital to them, and they favor established, prestige products and services that demonstrate achievement to their peers.

Location

Most beverage establishments attract customers from what is known as a *trading area*—the vicinity around an establishment from which the clear majority of customers (or revenue) will derive from. The trading area encompasses a radius surrounding the location that can range from a single building to a few blocks to several miles to across the country, depending on the type of establishment.

- *Demand Generators:* Demand generators are places or events that cause people to be near a particular establishment. Additionally, *population centers* are formed when demand generators attract large groups of people (whether for single events or consistently throughout a day). Population centers may contain groupings of potential customers and include train stations, bus stops, concert venues, shopping malls, strip malls, the downtowns of cities, and convention centers. The characteristics of trading areas and the type of demand generators will likely define the type of customer who will be inclined to visit a business venue. For example, a trading area and demand generator may attract customers who desire a relaxed, informal atmosphere and drinking experience, while another type of demand generator may encourage more sophisticated consumers seeking a more formal atmosphere.
- *Ease of Access:* Access or accessibility is the ease of entrance to the facility. Many successful beverage establishments are in an accessible and somewhat visible locations. The degree of accessibility to a beverage operation is a crucial consideration that may impact the quantity or volume of customers. If a particular establishment is considered a "destination type," consider how people get around in the area where the business will be located. If the location is in a suburban area, most people may get around by car, and then ample parking may need to be provided. On the other hand, if a location is in an urban area, public transportation hubs or foot traffic may need to be prominent. Both traffic barriers and availability of parking can impact the ease of access. Having on-street parking or the availability of valet parking may be imperative in city locations, versus having a large parking lot in suburban sites.
- *Visibility:* Visibility is the degree of exposure to the public. Generally, most food and beverage establishments strive to seek locations that provide ample exposure and are somewhat easy or at least accessible by potential customers. The location or building should have a certain "curb appeal" that appears pleasing from the outside. Having adequate exterior lighting and signage are both important to providing a sense of belonging. Figure 14.4 identifies ample signage and visibility in this iconic Chicago restaurant.

FIGURE 14.4
The Berghoff. Courtesy of Erika Cespedes.

- *Proximity to Competitors:* While it may seem counter-intuitive, operating a business close to competitors can be beneficial—especially true if the business relies heavily on foot traffic. Shopping malls are a good example of why proximity to the competitors is an important factor. The number of potential customers increases exponentially on a per-store basis around a concentration of similar businesses. For example, while one store might attract fifty customers, there is a considerable draw from surrounding businesses that can serve to market for future visits.
- *Zoning and Signs:* Zoning and allowance for signage for many enterprises can be a deal maker or breaker. Check with local zoning policies and city ordinances for parking, hours of operation, and sign requirements. Many communities set restrictions regarding these aspects to "preserve" a certain character of a neighborhood.

Competition

Competition is an effort of two or more organizations acting independently to obtain the business of the same or similar group of target customers. Regardless of the type of business, competition is inevitable and exists for almost all products and services. To learn about the competition in a selected marketplace, a *competitive analysis* is often conducted during the market analysis phase when developing a prospectus or business plan. This practice is useful to assess the strengths and weaknesses of current and potential direct competitors. This analysis provides both offensive and defensive strategic contexts through which to identify the potential opportunities and threats in the marketplace. Two levels of economic competition are often classified:

- *Direct Competition:* This type of competition offers the same or very similar products and services in the marketplace. Example: McDonald's is a competition to Burger King.
- *Indirect (or secondary) Competition:* This type of competition offers products and services that may serve as close substitutes for one another. Example: McDonald's is a competition to Subway.

When the competitive analysis is being conducted, each direct competitor's products and service should be identified according to market segment. Each competitor should be assessed and profiled in terms of their strengths and weaknesses which can ultimately provide areas of opportunities and threats in the marketplace.

Products and Services

Learning Objective 5
Consider some benefits of using daily specials and seasonal products

Products and services are the main sources of revenue for most businesses. The beverage and/or food items are the tangible products listed on a menu that tend to be the driving force or influence of the business—also they serve as the foundation on which the other components of the concept are based. Each product and service should be described in detail to emphasize the potential benefits for the selected target market. In recent years, the beverage menu has become just as integral to the

FIGURE 14.5
French restaurant. Courtesy of John Peter Laloganes.

success of an establishment as the food menu (and even more important in the case of a brew pub, cocktail lounge, wine bar, or wine store). The beverage and food menus should be designed to enhance one another and assisted to feed the concept of the establishment. Pictured in Figure 14.5 is Piqu'Boeuf Restaurant and Grill located in Beaune, France, which offers local beef with a highly compatible local wine list to enhance the dining experience.

The first task before designing menu items is to determine the type of cuisine that will be featured—it is this element that drives revenue. Simultaneously, the corresponding beverage focus should begin to solidify. If this is a beverage driven concept, then food will be selected in order to enhance the beverage component. In food establishments, beverages generally account for approximately one-fourth to one-third of revenue, yet contribute to a greater percentage of profit than food does. Beverages, regardless of whether they are wine, beer, or spirits, will play either a starring or a supporting role for the type of food cuisine. Ideally, beverage menus must provide choices that complement the food focus and cuisine of the establishment, as well as the price points of the varying clientele.

Special Beverages and Seasonal Items

When determining the types of products, consider incorporating some innovative approaches and variations of the core product(s). For example, a wine bar can increase both their product selection and greater appeal to their customer base by offering wine cocktails (Mimosas, Bellinis and Kir Royale) not just wine. A coffee/tea cafe can offer Sparkling Teas, Hibiscus Tea Sangria or a Green Matcha Tea Latte.

Another consideration is to include seasonal beverage offerings. There is no reason the concept of seasonality should only impact the culinary side of restaurant. Beverages can also work to create feelings of nostalgia, conjuring up images and joyful memories of the past. These items also create variety, general excitement and interest from the customers. It gives them another reason to keep coming back or come back more often. For example, Starbucks features Pumpkin Lattes in the month of October and Peppermint Lattes in December. Many full-service casual and upscale restaurants will offer signature winter cocktails using coffee and hot chocolate as a base while adding spirits such as rum, whiskey, and various liqueurs like butterscotch or Baileys Irish cream.

Service Styles

Learning Objective 6
Recognize some styles of service

Service transactions can be electronic, indirect, and/or direct (face-to-face), yet it is the service employees that represent a large part of the guest's experience. Face-to-face service by each employee and the delivery of personal interaction between the guest and the employee are critical. Customer service is the cornerstone of being able to build revenue, and without friendly interaction in the beverage establishment repeat business is often sacrificed. All employees should contribute to the atmosphere, energy and expectation of service in the restaurant. The employee behaviors (and their uniforms) should mirror the concept in all their interactions with the guests. The employee serves as an ambassador to the "brand." For example: If the restaurant is an intimate Italian place where couples come to spend a romantic evening, then staff would probably be attentive, but low-key. If the restaurant is a sport bar, then the experience is possibly more of a loud, interactive experience.

The level and type of services are the intangible items that are often decided simultaneously while the other factors of the concept are being constructed. If owners have decided to create a formal, fine dining operation, then the level and type of service should adhere and be appropriate to that vision. There are several broad types of service formats that can be modified with different levels of formality to fit an individual establishment. Consider the following basic styles of service:

- *Counter/Self Service:* This extremely informal type of service requires customers *to place and pick up* their own orders. Many quick casual restaurants, such as *Noodles & Company* and *Blaze Pizza*, incorporate this type of service format along with offering a modest selection of wine with their food options.
- *Bar Service:* Bar service requires customers to sit at a counter to place an order, which a server or bartender will then deliver to them. This type of service is somewhat informal and casual.
- *American/Table Service:* This is one of the most common types of service formats. It allows customers to be seated and communicate an order to a waitperson. The order is then prepared and delivered to the seated customer. This type of service can be formal, semi-formal, or casual.
- *French/Tableside Service:* French service involves partially preparing food in the kitchen, while final preparation and serving are completed tableside on a *guéridon* (gay-ree-DOHN), or mobile cart. This type of service is formal and is often combined with other service styles. Some high-end fine-dining restaurants have incorporated this type of service, with certain dishes such as Caesar salad or bananas foster prepared tableside for presentation purposes. Beverage service can also be conducted in this manner—the preparation of tableside margaritas and bottled wine service provides a heightened visual element to the experience.
- *Family Style Service:* Family style service involves delivering food to the dining table on platters and bowls. The customers will serve themselves and then pass the food around the table. This type of service is informal and sometimes is offered in combination with American/table style service. Restaurants such as *Maggiano's Little Italy* have experienced success with this format. The type of service can dictate a level of formality that will match a beverage menu and the various other factors that form the foundation of the concept.

Organizational Structure

Learning Objective 7
Identify the broad forms of business ownership

This section of the prospectus clearly identifies the company's organizational structure, details about ownership, and profiles the qualifications—roles and responsibilities of the management team. Organizational structure refers to the format of business ownership as well as how management aligns their departments, employees, products, and services.

Ownership Information

Ownership information begins with the identification of the legal "ownership" structure and format of the business along with the subsequent ownership information. Important ownership information that should be incorporated into the business plan includes names of owners, percentage of ownership, and extent of involvement with the company. The following are common formats of business ownership:

- *Sole Proprietorships* are a legal entity that are owned and run by one individual and in which there is no legal distinction between them and the business. The owner receives all profits (subject to taxation specific to the business) and has unlimited responsibility for all losses and debts.

- *General Partnerships* are a legal entity created by an agreement of two or more people. The owners share equally in both responsibility and liability for any legal actions and debts the company may experience.
- *Limited Partnerships* are a legal entity that is like a general partnership. There is at least one general partner and all other owners are limited partners—that is, in regard to their limited liability to the degree of their investment.
- *Limited Liability Companies* (LLC) are a legal entity that blends elements of partnership and corporate structures. This form of business ownership provides limited liability for its owners and avoids being subjected to double taxation.
- *Corporations* are a legal entity having its own privileges and liabilities distinct from those of its members. An important feature of a corporation is its limited liability aspects. If a corporation goes bankrupt, shareholders normally only stand to lose their investment. There are primarily two different forms of corporations: S corporations and C corporations.
  - *S Corporations* have the legal right to pass corporate income and losses through to their shareholders for federal tax purposes. S corporations do not pay any federal income taxes and (unlike a C corporation) they avoid double taxation on the corporate income.
  - *C Corporations* have no regard to any limit on the number of shareholders, foreign or domestic. However, they are subject to double taxation of the corporation's income and the separate taxation on their dividends.

Management Profiles

One of the strongest factors for success and growth in most companies is the capability of its owner/management team. They should provide resumes that include the following information:

- Name
- Position (include brief position description along with primary duties)
- Education and/or special training and prior employment
- Industry recognition and community involvement
- Compensation basis

Organizational Chart and Staffing

One of the most common and effective methods to arrange the companies structure begins through the creation of an organizational chart with a narrative description. There needs to be someone responsible and accountable for each function in the business.

- *Organizational Chart:* An organizational chart is a visual representation that illustrates the structure of an organization. It identifies the different departments (or sub-systems) and their respective managers and employees. The chart identifies the relationships among the different positions and their relative ranking in relation to one another. Organization charts are an effective way to communicate responsibilities, dependencies, relationships, and in the process, assist in alleviating conflict.
- *Span of Control:* The span of control principle advocates that there is an optimal number of employees one can effectively manage. The larger quantity of employees that one must manage, will correlate with the ability to effectively manage them well. Several factors influence one's span of control, such as the size of location, expertise of the employees, and similarity of task. Each variable has a direct impact on how one manages—and manages effectively.

- *Selection and Hiring Process:* This process begins with outlining desirable employee traits and performance capabilities necessary for each position in the establishment. Employee handbooks are created to delineate important company policies and expectations upon employment. It is also important that salary and benefits packages be defined at this stage.

Staffing Requirements

To generate repeat business and realize the maximum profitability, it's necessary for managers to organize and control staffing as well as their affiliated labor expense. Each food and beverage establishment will have varying needs when it comes to staffing—these needs will partially be determined by the type of cuisine, beverage menu, level and type of service, target market, location, price point, and hours of operation.

Any given food and beverage establishment will choose to staff many of the numerous front-of-the-house service positions from servers to server assistants and bartenders to barbacks. Service employees directly impact the guest's experience by providing food and/or drinks along with tending to their ancillary needs. The maître d' or host ensures smooth customer flow by greeting and seating guests upon arrival; the bartender or sommelier prepares and serves wine, beer, and/or spirits; the waiter/waitress provides food service along with additional beverage service as needed; and finally, the busser clears and resets the guest's table. Depending upon the type of establishment, managers may choose to staff these positions, or instead choose to streamline some and have workers perform multiple job functions associated with the various positions.

Job Descriptions

One of the first and most important methods for organizing beverage service staff (and employees in general) is to develop and implement the use of *job descriptions.* Job descriptions are a written outline that describes an employee's most significant duties, responsibilities, and the necessary qualifications for a given position.

Job descriptions are of value to both the employees and the employer. From the employees' standpoint, job descriptions can be used to help them learn their job duties and to remind them of the results they are expected to achieve. They also establish management's right to take corrective action when the duties covered by the job description are not performed as required. From the employer's standpoint, written job descriptions can serve as a basis for minimizing the misunderstandings that occur between managers and their workers concerning job requirements.

If an establishment has updated job descriptions in use, it will be easier for management to determine how many staff members are needed to perform the necessary tasks for a shift and indirectly impact revenue and expenses and ultimate profitability.

Job Analysis

On a regular basis, managers should analyze each job that is performed at their establishments. A job analysis is a process used to gather information about the current performance of workers within a given job and compare that to the intended performance as identified on the job description and performance standards. This is essentially conducted to determine any performance gaps, which can occur between what the employees are doing and what they are supposed to be doing. This job analysis process can help to develop, assess, and/or revise job descriptions—ensuring the desired performance is clearly stated. Actual performance is measured in comparison with required performance standards established by benchmarks for acceptable performance in that position. Depending on the results of the performance gap, a plan of action will need to be developed to address the disparity between actual and desired performance. The plan to address the disparity may involve:

- Providing additional staff training
- Correcting deficient employee performance
- Rewriting the expected performance standards
- Rewriting the job description
- Purchasing new or additional supplies and equipment

A great deal of information for this analysis can be gathered through observing employees who currently perform in the related positions. Once this information is gathered, it can be used to develop or refine job descriptions and determine staffing needs that can be used throughout the many phases of human resource management.

Performance Standards

Since most food and beverage establishments are labor intensive, creating and clearly communicating *performance standards* and expectations are the cornerstone of any successful business. Performance standards are statements written for each employee job function, which identify the desired behaviors or outcomes described in specific, objective verifiable terms. They are written in job specific areas of knowledge, skill, and/or attitude that establish results-oriented behaviors. Performance standards state what behaviors or results are expected for employee performance to be considered satisfactory—these standards are the criteria against which actual performance is judged against. These standards identify "how" and "how well" the specific job tasks should be performed. This information would be included in a written job description or certainly at least within the training program. For example, the standard garnish for a martini might be two green olives; therefore, in order to perform the task of garnishing this drink properly, a bartender would use two green olives, skewered on a toothpick. More details on developing performance standards will be discussed in chapter 15.

Atmosphere

Learning Objective 8
Identify some ways that atmosphere can assist in defining the concept

The atmosphere is a defining element in the creation of a food and beverage establishment. The environment communicates and attracts a customer based on the design, style, and impression of the atmosphere. As per the National Restaurant Association, American consumers spend roughly 49 percent of their money for food and beverages outside their own dwelling. Regardless of type of establishment—from fast food to fine dining—the customer is often looking for an experience that can't be replicated in their home.

Effects of Atmosphere

Most successful businesses attempt to create an atmosphere that an intended target market(s) can identify with and possibly form some connection—often considered a prerequisite to creating loyal customers. Customers often seek to mirror a style or feel based according to who they are and what they are seeking. They want to feel connected to the environment they are choosing to spend time; the atmosphere helps to define the personality and presence of a concept. The atmosphere is a combination of factors that evoke an overall feel—mood, attitude—or emotion throughout the design of a space. The kind of environment created can determine whether the concept will be formal or informal, festive or intimate, bright or dim. These aspects are often created through tangible aesthetic type elements such as décor, but intangible variables as well—such as the mood, service, and uniform of the employees assist to foster a particular type of atmosphere.

FIGURE 14.6
The Radler. Courtesy of Erika Cespedes.

Aspects of Decoration

Décor (short for decoration) stretches throughout a physical space and adds to the overall atmosphere. Several aspects can be used to create the atmosphere, such as the lighting, table and bar top accessories, music, drapes, flooring, linen, and so on. Pictured in Figure 14.6 is an attractive back bar at the *Radler*, a contemporary German restaurant in Chicago's Logan Square neighborhood. Some establishments have a noisy, energetic environment, whereas others offer a quiet dining environment designed to appeal to a different target market. Therefore, understanding the needs and motivations of the market segment is indispensable in creating the type of environment and concept appropriate for the target market. Below is a sample list of essential and distinguishable décor aspects that provide both atmosphere and functionality to a space.

- *Lighting:* Lighting creates an instant sense of mood and feeling. Lighting fixtures should be chosen to match the design of the concept. The type of lighting will help create a feeling of comfort and highlight the dining and drinking areas with an aura of conviviality and intimacy, or social revelry. Use of soft, low lighting such as candles or table lamps creates an elegance and coziness, while brighter lighting can be effective in creating a more casual, festive atmosphere.
- *Tabletop:* The table and bar tops are some of the most obvious, apparent factors that customers will experience. When choosing table and bar top supplies, select dinnerware and decorations that will complement the food and dining room décor. Make certain the following elements match the design concept. Factors such as flatware, glassware, and linens (or lack of) create a certain feel.
- *Music:* Music is the auditory element that can cause both conscious and subconscious reactions—it is a subjectively perceived phenomenon by the people who hear it. Certain styles and genres of music are chosen to induce some desirable response in customers—whether it is to relax or liven them up—to make them revisit nostalgia as a teenager or feel wealthy and sophisticated. Food and beverage establishments are supplied most often with ambient music to provide some means of dampening out individual conversations from others and the general clatter amongst the space. Particular types of establishment host live music—certainly a way to emphasize the distinction of a concept. Ultimately, the most important goal is to play music enjoyed by the selected target market(s) and assist to differentiate the concept.
- *Walls and Ceilings:* Walls and ceilings provide an excellent source of adding color and design into the space. Besides the obvious paint colors, walls can also be used to feature local artwork or any art that is appropriate to the concept. The walls are too often ignored as a merchandising aspect—they can be used for enhancing the atmosphere or promoting products or the seasonality of the time of year.
- *Flooring:* Flooring can vary with the number of options available—from wood to carpet and tile to stone. There are numerous advantages and disadvantages to consider that go beyond the scope of this text. Overall, they should be durable and strive toward enhancing the look and feel of the concept.

Financial Data

Learning Objective 9
Identify the purpose of each of the four essential financial statements

Requesting funding from one or more sources is almost an expectation with any new or expanding business—particularly a food and beverage establishment. The request for funding is clearly defined and highly detailed in the business plan yet it generally plays a smaller role in the prospectus. The financial data is commonly developed after the market has been analyzed and clear objectives have been created. It is at this stage that projected financial statements are drafted and financial resources will theoretically be allocated. Generally, it is useful to identify some different funding scenarios, such as a worst, best, and optimal case scenario that should correspond to any projected financial statements. One of the clear signs every investor wants to see in the financial statements is *at what point the business will become profitable*—simply put, when do they obtain their *return on investment* or ROI.

Regardless of a startup or expanding enterprise, all businesses will be required to supply prospective or *forecasted* financial data. Forecast or forecasting is a projection of the expected financial position and the results of operation. *Historical data* provides an excellent reference point for making forecasts; however, information is unavailable for any new business. In these situations, use of industry statistics, benchmarks or competitor information can be quite valuable. Most creditors request the following projected financial reports for at least two to three years prior to any approval of funds:

- *Income Statements* (or profit and loss): The profit and loss statement displays the *revenues* (produced from the sale of products and services) or cash inflows, related *expenses* (costs incurred to sell the products and services) or cash outflows, and ultimate profit or loss for specified time period (often monthly).
- *Balance Sheets* (or statement of financial position): The balance sheet is a summary of the financial balances (assets, liabilities, and the owner's equity or net worth) of the organization at a specific point in time. Balance sheets are often described as a "snapshot" of a company's financial condition.
- *Cash Flow Statements:* The cash flow statement is concerned with the flow of *cash in* and *cash out* of the business. The statement captures both the current operating results and the accompanying changes in the balance sheet. The cash flow statement is useful in determining the short-term viability of a company, particularly its ability to pay bills.
- *Capital Expenditure Budgets:* The capital expenditure budget identifies the amount of money a company needs to invest in renovation, major equipment, land, buildings, and long-term assets that are projected to generate future income.

Check Your Knowledge

Directions: Use these questions to test your knowledge and understanding of the concepts presented in the chapter.

I. MULTIPLE CHOICE: Select the best possible answer from the options available.

1. This practice is useful to assess the strengths and weaknesses of current and potential direct competitors.
   a. Trading area
   b. Direct competition
   c. Secondary competition
   d. Competitive analysis

2. A legal entity that is owned and run by one individual and in which there is no legal distinction between them and the business.
   a. Corporation
   b. Limited partnership
   c. Sole proprietorship
   d. S Corporation
3. This type of ownership is a legal entity having its own privileges and liabilities distinct from those of its shareholders or owners.
   a. Corporation
   b. Limited partnership
   c. Sole proprietorship
   d. Limited partnership
4. Places or events that cause people to be near a particular establishment.
   a. Demand generators
   b. Trading areas
   c. Populations centers
   d. Market segmentation
5. This precisely defines the business, identifies goals, and serves as the firm's resume—it also acts as an on-going management tool.
   a. Competitive analysis
   b. Prospectus
   c. Atmosphere
   d. Management profiles
6. This financial statement displays the revenues (produced from the sale of products and services), related expenses (costs incurred to sell the products and services), and ultimate profit or loss.
   a. Balance sheet
   b. Statement of cash flow
   c. Capital expenditure budget
   d. Income statements
7. A manager's and owner's primary responsibility and accountability is to operate an organization in a manner that, at least in the long term, produces
   a. revenue
   b. controlled costs
   c. sales
   d. profit
8. This type of service requires customers to sit at a counter to place an order, which a service employee will then deliver to them. It is often considered somewhat informal and casual.
   a. Table service
   b. Family style service
   c. French service
   d. Bar service
9. An operational format that provides a high perceived value for their guests, defined by the upscale atmosphere, high beverage and food quality, and exceptional service(s) with higher price points.
   a. Quick service
   b. Quick casual
   c. Casual
   d. Fine dining

10. The purpose of *establishing standards and procedures* is to be used as a basis to compare
    a. the actual performance
    b. the expert performance
    c. the ideal performance
    d. none of the above
11. Which aspect of décor will least likely influence the overall atmosphere?
    a. Lighting
    b. Table and bar top accessories
    c. Music
    d. Food
12. Span of control is the principle that
    a. identifies the hierarchy of jobs
    b. advocates there is an optimal number of employees one can effectively manage
    c. illustrates who reports to whom
    d. none of the above
13. An organizational chart
    a. is a visual representation that illustrates the structure of an organization
    b. advocates there is an optimal number of employees one can effectively manage
    c. illustrates ideal areas that contain the desired target market
    d. identifies the logical progression of the prospectus
14. A written outline of a given job that identifies the major duties, and specifications of the job that needs to be performed.
    a. Position Description
    b. Job Description
    c. Performance Standard
    d. Position
15. The *performance gap* is the difference between
    a. desired performance and the standards
    b. actual performance and desired performance
    c. actual performance and the expert performance
    d. none of the above
16. Psychographics are characteristics that consist of
    a. statistical information based on age, gender, income level, and marital status.
    b. focus on serving the needs of customers in a geographical location
    c. characteristics that are based on uniqueness of people's lifestyles
    d. both a and b

II. DISCUSSION QUESTIONS

17. Explain the ingredients to success for a food and beverage establishment.
18. Explain the benefits of what market research can provide.
19. Provide some aspects of décor and their related considerations.
20. How is financial data important as part of the prospectus or business plan? What are some areas to include when presenting the financial data?

CHAPTER 15

Managing for Profit

CHAPTER 15 LEARNING OBJECTIVES

After reading this chapter, the learner will be able to:

1. Explain the importance of understanding the flow of beverages
2. Identify factors to consider when choosing wines, beers, and spirits
3. Identify some standards in receiving practices
4. Explain some proper storage techniques for wines, beers, and spirits
5. Identify control concerns during the issuing process
6. Calculate proper costing out of wines, beers, and spirits
7. Describe some control issues during the service control point

Be a yardstick of quality. Some people aren't used to an environment where excellence is expected.

—Steve Jobs

The Profitability of Alcoholic Beverages

Learning Objective 1
Explain the importance of understanding the flow of beverages

The demographics of the beverage industry have shifted quite dramatically over previous years. Today's drinkers are more affluent and sophisticated—and increasingly health conscious. The combination of these trends is curbing overall alcohol consumption but driving the desire for premium and super premium products. Consumers are demanding better-quality wines and premium ales and lagers to single malt whiskeys and triple-distilled vodkas—products that enthusiasts are consuming in smaller quantities but are willing to pay more for quality. Pictured in Figures 15.1 and 15.2 is *Liquor Park*, a beer, wine, and spirits retail store that specializes in craft beverages with an emphasis on small production and offering selections that are difficult to obtain elsewhere.

Beverage Sales is a term found on financial statements that is used to indicate the sale of wines, beers, and spirits. These sales may account for only 25–30 percent of total sales, yet significantly impact the profitability of any food and beverage establishment. Beverages have always played a significant part of the dining experience; however, culturally they are becoming more of an accepted daily part of life. Beverage sales represent a major profit center for many establishments due to their highly obtainable markup—not to mention—they maintain a much lower labor cost associated with their production as compared to food. In many beverage establishments, it is not unusual to markup alcohol anywhere from 100 to 300 percent from the product cost—making the gross margin far greater for beverage than for food.

This chapter will discuss the pertinent control points of beverage management: purchasing, receiving, storing, issuing, production and service procedures as they pertain to alcoholic beverages. These areas are cornerstone to what the managers spend a great deal of their time performing. While these steps are like those used for food products,

FIGURE 15.1
Liquor park. Courtesy of Erika Cespedes.

FIGURE 15.2
Liquor Park's wine shelving. Courtesy of Erika Cespedes.

the control procedures can be quite different than for food, and, in some respects, much more challenging. Ultimately, beverage costs must be controlled if an operation is to reach maximum potential of gross profit associated with its sales.

The Flow of Beverages

Once alcohol is purchased, the beverages progress through different points—in contact with different people—until they are sold and ultimately consumed. This progression is the path that beverages travel throughout an establishment where it consists of several "points of concern" where the potential for limited control or loss of product is more volatile—these areas are identified as *control points*. These points are recognized as key areas for the beverage manager to exercise great control—otherwise these points may result in increased expenses and/or a loss of revenue. The points flow from: Purchasing → Receiving → Storing → Issuing → Production → Service. Figure 15.3 illustrates the flow of beverage and its corresponding control points.

FIGURE 15.3
Flow of beverages. Courtesy of John Peter Laloganes.

Listed above are the common control points within the flow of beverages. Due to concerns regarding a loss of beverage revenue and profit, it becomes increasingly more important for the beverage manager to implement some means of *control systems*. These control systems should contain "Standard Operating Procedures" (SOPs) to safeguard the products during these stages of the flow. The success of cost control will depend largely on how well the control systems are applied throughout the flow of beverages. When control systems are created, some considerations for effective results include cost effectiveness, ease of implementation and monitoring, consistency of results, and getting a return on investment. To ensure success in any control system, four standard steps (as identified in the subsequent section) should be established throughout the flow of beverages, explained in detail in the following section.

The Foundation to Control Systems

The beverage manager must ensure that effective and efficient controls are established along the flow of beverages. They are

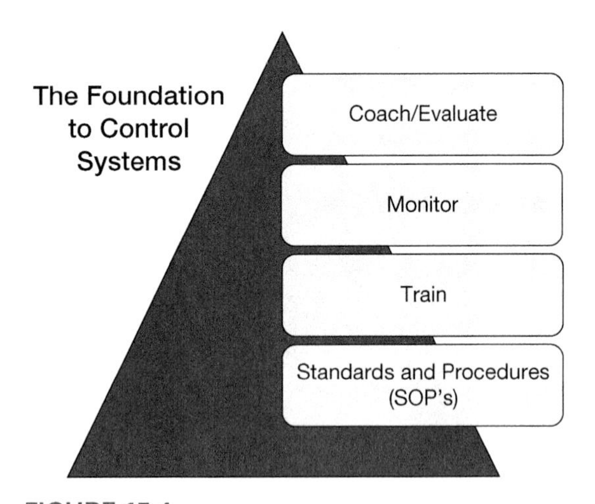

FIGURE 15.4
The foundation to control systems. Courtesy of John Peter Laloganes.

the cornerstones to any successful beverage establishment. While these control points may seem daunting, it is incredibly important to apply standards and procedures to maximize the establishment's ability to produce profit. Losing alcoholic beverages to employee pilferage and theft and product spoilage and breakage can have a significantly negative impact on the bottom line. Figure 15.4 depicts the steps needed for a beverage control system to provide quality assurance.

1. **Standard Operating Procedures (SOPs)** SOPs should be created for each point within the flow of beverages. SOPs act as a means of communicating the expectations between management and the line-level employees. They assist with maintaining the level of quality and encourage a consistent product and service. Quality is defined *as a degree or grade of excellence.* Establishing quality standards is used as a way for beverage managers to ensure that desired outcomes are predictable. Therefore, quality is often quantified in order to make it observable in nature, which makes it easier to know when it happens, and when it doesn't. Translating a desired behavioral outcome in terms of *time, count, weight, or volume* is helpful. For example: quantifying the greet time of a serving approaching a guest table in a dining room within a 2-minute time frame is a manner used to translate quality into a standard of time. SOPs are necessary to the foundation of any training program and are consequentially used as an evaluation tool by comparing the actual employee performance against them. Figure 15.5 illustrates the difference (or gap) of performance between the desired standard of performance versus the actual or current performance. This gap identifies the disparity between what is desired and what is reality, therefore illustrating the need for closing the gap through training or some other methodology.
2. **Train** Training is essential for all employees who work with alcoholic beverages. Effective training will be able to translate employee behavior into desired performance. Establishing clearly defined and measurable performance objectives is the foundation to a quality training program. *Performance standards* are statements that describe what the learner should be able to do upon completion of training—they act as the blue print or foundation to the entire design, development, and delivery of the training program. Acting as the blue print to the entire design, development, and delivery of the training program, the performance standards includes three components: (1) condition or environment, (2) task, and (3) measurable criterion. By incorporating each of these three components into a single statement, it becomes easier to ensure clarity and consistency. Below are several performance standards that contain each of the three components:
   - Upon the customer ordering a bottle of wine, the server must be able to execute the proper ten steps of wine service with 100 percent accuracy
   - Given a bottle of beer, the server must be able to pour the beer into a glass using the two-step pouring technique
   - When pouring a beer, the server/bartender must obtain a 1-inch head every time

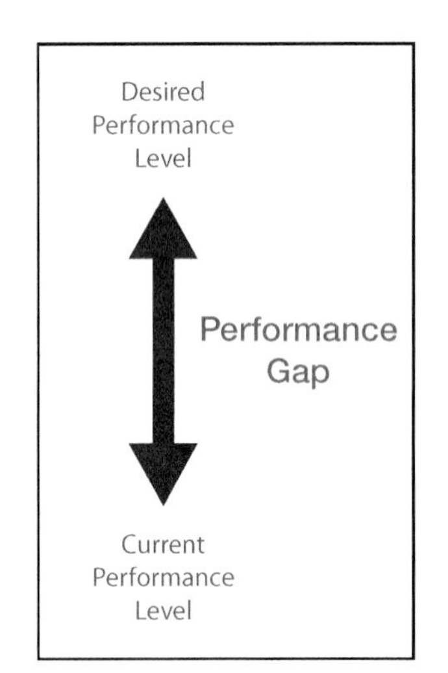

FIGURE 15.5
Performance gap. Courtesy of John Peter Laloganes.

3. **Monitor** Monitoring employee performance is the most effective manner to ensure they meet the SOPs. Monitoring allows management to identify any gaps between training (or in many cases a lack of training) and the

employees' performance. There are two methods of monitoring employee performance: direct and indirect. Direct monitoring involves management practicing *management by walking around* (or MBWA) to directly observe employee performance. Indirect monitoring involves management using indirect observation through the feedback of others, web-cam security, secret shoppers, and so on.

Secret Shopper A person employed by an organization to secretly pose as a customer to evaluate the quality of products, service, and overall experience. With shoppers acting as normal customers, they will complete a written assessment of their experience and submit it to management soon after. This provides managers a rare glimpse of the service from a guest's perspective. Secret shopper reports can be used as a robust form of feedback that allows management to address training needs and assist in "keeping employees honest" about following the regulations as set forth by the organization.

4. **Coach** Coaching is a management technique employed to correct or reinforce employee behavior. By taking appropriate actions to correct deviations from the SOPs, the employee is more likely to align their performance appropriately. Effective management acts quickly and consistently to adjust employee performance when SOPs are not being met.

Purchasing Control Point

Learning Objective 2
Identify factors to consider when choosing wines, beers, and spirits

Purchasing is a generic term used to indicate the process of getting the appropriate products desired by the establishment. Three other terms used in discussing or performing purchasing responsibilities are *selection, sourcing,* and *procurement*, all of which play an important role in the purchasing process.

The dizzying array of beverage options can be overwhelming for the consumer, even for the experienced beverage buyer. Thankfully, to some degree, not every beverage is appropriate to buy for all establishments. For example, certain Belgian beers might not sell as well at the local sports bar as they would at a full-service restaurant or upscale bar. Management first needs to determine what kind of alcoholic beverages are appropriate to offer. This decision will largely depend upon the type of operation, target customer, storage space, and budgetary constraints associated with the establishment.

Selection The selection process involves deciding the kinds and types of beverages that are appropriate for a beverage concept. This question will most likely have previously been decided during the planning of the prospectus. Selection entails choosing among alternatives such as Australian Chardonnay versus California Chardonnay. Most distributors provide beverage samples and tasting notes to assist the buyer in making the selection. Additionally, industry trade tastings are large events where distributors pour and promote their products for buyers to assist in making future selection decisions. These events are typically available year round in many of the larger cities throughout the United States. Pictured in Figure 15.6 is a small sample of the numerous (and often overwhelming) beverage options available in today's marketplace. The manager has many decisions to make when considering the breadth and depth of their beverage menu and its appropriateness given the concept of the establishment.

FIGURE 15.6
Array of beverages. Courtesy of Erika Cespedes.

Sourcing The sourcing of alcoholic beverages can be somewhat challenging; certainly, it is not as simple as ordering food products. In most states, the laws allow middlemen the sole rights to a territory, with no form of competition. Therefore, the buyer must search various suppliers to obtain the desired products. For example, suppose that ABC distributor carries a brand of Australian Chardonnay, and XYZ distributor carries a different one—yet neither distributor will carry the same Chardonnay.

Procurement Procurement is the process of ordering products and maintaining an orderly, systematic exchange between the buyer (the retailer) and the seller (the intermediary). Procurement defines (whether formally or informally) the procedures for obtaining the necessary products that were selected and sourced. For instance, deliveries may be made on Wednesday prior to 11:00 A.M., payment is by cash or check on delivery, and the manager must inspect and sign all invoices.

The Three-Tier Distribution System

The distribution system for alcoholic beverages is a bit complex, tangled with varying state and local regulations. Unlike the ease of buying food products, alcoholic beverages are governed by the 21st Amendment (which repealed Prohibition, established under the 18th Amendment), which gives each individual state the jurisdiction to regulate the sale and distribution of alcoholic beverages. After Prohibition, this system was created to prevent an imbalance of power between small and large producers and to provide layers of control throughout the distribution process. Throughout America, each state has some form and degree of governmental control. The Alcohol Beverage Commission (ABC) can exercise greater or lesser control, depending on the laws established within the individual state. The ABC (or some derivative) acts to control the licensing, purchasing, transportation, and sale of alcohol. Since the repeal of Prohibition, states have been allowed to classify themselves as either a control state or a license state, allowing each one to create its own sets of laws regarding the production, transportation, and sale of alcohol. Therefore, the distribution gives way to the three-tier distribution system composed of several levels, or tiers, which an alcoholic beverage is required to travel through to get to the end-user. Overall, since state laws vary, the United States in effect has fifty distribution systems, each with its own rules and laws. The three-tier distribution system consists of three levels: (1) primary sources (producers); (2) intermediaries (wholesalers and distributors); and (3) retailers (beverage establishments).

Primary → Intermediaries → Retailers

Primary Source

Primary sources consist of the producer or creator of the alcohol products. The primary sources can range in size from a small independent producer to a large, multinational commercial one. *Wineries* produce wine, *breweries* produce beer, and *distilleries* produce spirits.

Intermediary

Intermediaries (also known as middlemen) are often of one or two different types. One type is an importer (or wholesaler) who is legally licensed to transport alcohol into the United States. The second is a distributor, who services a particular state or grouping of states. Distributors market (on behalf of a producer) and transport alcohol from an importer or primary source for resale to a beverage establishment. It is not unusual for beverage establishments to utilize anywhere from half a dozen to a dozen beverage distributors at any given time to obtain appropriate beverage products needed for their establishment. In some states, small-scale producers can self-distribute directly to the retailer. However, the clear majority of beverage producers struggle with no-to-minimal representation in their desired markets. If a given producer has minimal production, then it may not be advantageous for a distributor to carry its products due to inconsistent availability to supply a given market(s).

Since the repeal of Prohibition, states have been allowed to classify themselves as either a control state or a license state.

- In control states, the state government sells some or all the alcoholic beverages through its network of stores, thus exercising complete control over prices and distribution. Control states often exercise greater control in the process of transporting alcohol as compared to licensed states. The chart below identifies a list of control states.

CONTROL STATES

Alabama	Montana	Utah
Idaho	New Hampshire	Vermont
Iowa	North Carolina	Virginia
Maine	Ohio	Washington
Michigan	Oregon	West Virginia
Mississippi	Pennsylvania	Wyoming

- In licensed states, the distributors/purveyors are licensed companies in the business of transporting and selling products and/or services from importers to beverage establishments. Licensed states allow authorized wholesalers and distributors to sell alcoholic beverages directly to the retailer. Primary sources (producers) give the distributors exclusive rights to market and sell their products within a specified geographic area. In control states, the distributors/purveyors are government-operated entities. California, Illinois, and Minnesota are among the licensed states.

Retailers

The retailer is the final major carrier and barrier of control prior to the products being purchased by the consumer. Retailers are the beverage establishments that sell wine, beer, and spirits directly to consumers. Retailers are classified as off-premise or on-premise—each one requiring a different form of licensing, insurance, and tax requirements to operate legally.

- *On-premise* is a term used to indicate a restaurant or bar. This is where the consumer will be visiting the establishment's premises to consume the alcoholic product(s).
- *Off-premise* is a term used to indicate a grocery store or liquor store, where the customer will purchase alcohol and then consume the product off the premise—separate from where the product was purchased.

The relevance of the 80-year-old three-tier system has been called into question. In November 2011, Washington state approved an initiative that essentially dismantled the three-tier distribution system and the state-operated retailing system in the state. The system in Washington now allows the retailers to bypass distributors to purchase directly from producers. It is even possible to negotiate volume discounts and warehouse their own inventories.

Furthermore, depending on the state, it is possible for personal individuals (as opposed to businesses) to personally order alcohol (with some restrictions) directly through primary sources. This process is known as *direct shipping*. This type of shipping is ideal for the individual consumer—yet it is prohibited for the licensed-beverage establishments. It has allowed for the smaller, lesser known producer to sell their products using unconventional means such as promoting through online forums as opposed to being forced to sell products through traditional retail or restaurant scenarios.

Buying Wine

When buying wine, the appropriateness of the brands and styles of wines are important considerations—these decisions should be based upon the vision of the establishment. A combination of factors such as target market and their degree of wine sophistication, price points, and cuisine, all should be considered. Instead, a typical practice consists of a buyer getting a "great deal" on a certain wine—this wine may not necessarily "fit into the vision" of the establishment. Not only does a well-thought-out wine list serve as an important sales tool, it also enhances the overall experience of an establishment.

When purchasing wine, managers should consider the breadth and depth of the beverage menu. The *breadth* of the wine list offers a wide range or scope of numerous wines from various grapes and locations, such as offering multiple options of a certain style of crisp and youthful white wines (as discussed in Chapter 5). Figure 15.7 displays a collection of crisp and youthful wine styles.

The *depth* is the degree of thoroughness offered on an establishment's beverage menu. An example of depth is offering several wines within a single category, such as a diverse wine list that contains Chardonnay or Sauvignon Blanc from several locations and offering multiple vintages. Figure 15.8 displays a collection of Sauvignon Blanc from famous growing locations. The type of establishment will significantly influence the degree of depth and breadth of any beverage menu. Fine dining establishments will want to offer greater depth and breadth of options as compared to casual and quick casual operations. Primarily, because the fine dining establishments have a customer base that demands varied options and therefore will be more able and willing to purchase and store more expensive, high price point wines.

When creating, or considering the revision of a wine list, the manager should at least consider the recent explosion of affordable, high-quality wine options. It is astonishing to find that so many beverage establishments, even in big cities, are not allowing the wine list to evolve with the increasingly educated,

FIGURE 15.7
Breadth of crisp and youthful white wines. Courtesy of Erika Cespedes.

FIGURE 15.8
Depth of Sauvignon Blanc. Courtesy of Erika Cespedes.

adventurous, and well-informed consumer. Many establishments (both chain and independently owned) are still clinging to the same-old mass-produced brand names and ordinary grape varietals. Volumes of research indicate that consumers are willing to experiment, yet many beverage establishments fail to modernize themselves and to comply with the needs in the marketplace. The astounding value of white wines from New Zealand and South Africa and the red wine options from Argentina and Chile are so likable not only because of their price but also for their fruit-forward personality. In addition, customers are becoming more likely to indulge in the surging demand for *sustainable* or *biodynamic* produced wines.

Wine is universally ordered by the case, which typically contains twelve bottles. Each bottle commonly holds 750 ml, equivalent to 25.4 oz. per bottle. Most distributors offer a case discount pricing as opposed to "breaking a case" and ordering bottles individually; however, the establishment needs to have the "proper" storage space if managers are going to buy in that quantity. Most suppliers will charge a nominal fee (0.40 cents to $1.00 per bottle), if a full case has not been ordered.

Alternative bottle sizes have soared in popularity over the last decade. Half-bottles (375 ml or 12.7 oz.) and splits (187 ml or 6.3 oz.) have become more available, and these allow the wine consumer to purchase a good-quality wine without committing to a more expensive traditional-sized bottle. Splits have become a great alternative for beverage establishments that may not offer a sparkling wine by the glass. Splits encourage the customer to buy a sparkling wine when the price of a full bottle may be cost prohibitive. Selling splits also lessens the cost to the restaurant if a sparkling-wine-by-the-glass program is not popular, otherwise forcing a manager to discard an open bottle that has lost its carbonation.

Buying Beer

Beer is the most popular alcoholic beverage consumed by Americans. Just about every establishment will want to have some type of beer on the drink menu, and, depending upon the establishment, it can account for almost 40 percent or more of total alcoholic beverage sales. Because of its increased popularity and the myriad number of brands—both domestic and import—deciding exactly which beers to offer is somewhat more complex than in days past. The vast assortment of beers that are accessible to drinkers today includes one for almost every palate (whether the consumers know this or not). With the rise in popularity of craft beer, the beverage manager should decide whether offering these beers would be profitable and appropriate for their type of establishment. In addition, gluten-free, low-alcohol, or nonalcoholic beers may also be considered, particularly with the public's increased awareness of health and of the hazards associated with drinking and driving.

Beer is primarily purchased in bottles and kegs (a full keg contains 1,984 ounce) with some small offerings available in can and 64 ounce growlers as sold in brewpubs. Each receptacle has its own advantages and disadvantages—which will most likely be based on customer preferences and the operation's available refrigerated and nonrefrigerated storage space. Some of the advantages to offering draft beer include the relatively small space required to store it and the ease and quickness with which it can be served; possibly the most important advantage for some is its incredible profit margin. One disadvantage of serving draft beer is that, once tapped, the beer has a short shelf life (about three weeks) for optimal quality. In addition, draft beer lines must be kept clean, and the product can be messy and easily wasted if the bartender is not trained in proper drawing, or pouring, techniques. Generally, draft beer has approximately 7 percent loss once it has been tapped. Figure 15.9 showcases some draft handles indicating the draft beer selection at a given restaurant. Figure 15.10 shows a menu featuring draft beers; most restaurants and bars prominently promote these types of beers as they have higher profit margins than bottled or canned beer.

FIGURE 15.9
Tap handles showing the draft beer selection. Courtesy of Erika Cespedes.

FIGURE 15.10
Draft selections. Courtesy of Erika Cespedes.

FIGURE 15.11
Pilsner Urquell. Courtesy of Erika Cespedes.

Bottled and canned beers have a longer shelf life, but they require greater storage space. Buying bottled and canned beer are very appropriate for establishments that do not forecast large sales of certain types of beers but would still like to offer them to their customers. As long as these beers are stored properly (away from heat and light), they maintain 100 percent yield. Pictured in Figure 15.11 is a collection of canned beer from Pilsner Urquell.

Buying Spirits

Purchasing decisions for spirits primarily depend upon the depth of options—does one have the space for the well over 300 types of vodka in the market place? Spirits are widely consumed, and they have an extremely long shelf life. Like beer and wine, the popularity of premium distilled spirits and the variety available to consumers has increased dramatically. Most spirits are sold in liter bottles containing 33.8 oz. of product; they may also be purchased in fifths containing 25.4 oz. Not very efficient in beverage establishments, spirits are also available in 1.75-liter or 59.2-oz. containers. The larger bottles tend to be more cost effective but become too difficult for bar employees to hold and make drinks in an efficient and "spill free" manner.

In deciding what brands of spirits to carry, the establishment must first determine which brands will determine the *well* and which ones will constitute the *call* (premium and super [or ultra] premium) options. Figure 15.12 showcases an elaborate back bar that houses all of the establishment's premium and ultra-premium spirit options. Well brands are ones the bartender pours when the customer does not specify a particular preference. For example, if a customer orders a vodka and tonic, the bartender would prepare the drink using the well vodka. These products are usually contained behind the bar (not on display) in what is known as a speed rail. The speed rail is a shelf that is positioned for easy access by the bartender for being efficient when making a large production of drinks.

The decision of which well brands to carry is an important one because of price and consumer expectations. At the bar, well brands are poured most frequently, so they are purchased more often and in large quantities. The availability and price of

FIGURE 15.12
The Radler's back bar. Courtesy of Erika Cespedes.

FIGURE 15.13
Cocktail selection. Courtesy of Erika Cespedes.

brands chosen will determine overall profit margins as well as customers perceived quality of the operation. Pictured in Figure 15.13 is a cocktail menu, which dictates the types of spirits the beverage manager will need to purchase. The establishment will certainly define the types of beverages offered and their corresponding selling prices.

Call brands are products the customer orders by a brand name, such as a Bombay Sapphire Martini, specifying "Bombay" as the brand of gin. Because most operations today continue to adhere to a two-tier, and, in some cases, a three-tier pricing structure, the choice of which call (or premium and super premium) brands to carry is an important one. Like the "well" brands, "call" brands significantly reflect on the overall quality of the establishment, and choosing which brands to offer is an important matter from the perspective of the customers.

Ordering Control Techniques

There are several ordering techniques that are widely used in the industry that can be advantageous for a beverage buyer to consider.

- *Optimal Ordering:* Inventory control is vital to the success of any business. Therefore, ordering effectively can assist in maintaining optimal levels of inventory. There are two extremes of inventory that may cause concern: The first is running out of product, referred to as *stock-outs*; and the second is having too much inventory, known as *surplus*. Either extreme jeopardizes the success of any organization. Maintaining an optimal level of inventory is a constant effort that requires the full attention of the beverage manager. When stock-outs (running out of an item) occur, it is important to notify all service staff promptly to alleviate a customer from ordering an out of stock product.

It is always a better service experience when a customer is informed of stock outs prior to ordering. Regardless, when stock outs occur too often, it reflects poorly on the buyer and may cause frustration on behalf of the staff—but more concerning is the continued disappointment of repeat customers. When surplus occur, they are generally caused by inadequate attention to ordering. Current on-hand inventory should always be known prior to placing an order for additional product. A constant surplus of items may lead to inadequate money-management and cash-flow problems, which may destroy the financial stability of a company.

- *Purchase Order (PO):* A PO is a form that lists the products and quantities ordered, and possibly the current purchase price, with selected intermediaries. POs should be utilized to create an audit trail that will track the product from the time it is ordered to the time it is received and stored. Some operations date stamp or specially mark a product as it is delivered before placing it in the storeroom inventory. An empty-for-full system can then be implemented that exchanges used, empty bottles from the bar for full, fresh product from the spirits storeroom.

  The PO form is often used to verbally place an order by telephone, faxing, or sending it electronically through the Internet. The PO creates a paper trail and communicates to other employees, both inside and outside the establishment, the products that have been ordered and will be delivered to the establishment.
- *Opportunity Buys:* An opportunity buy is purchasing a large quantity of a single product, or a large cumulative order. This kind of buying is rewarded by the seller with some form of discounting or another perk. Opportunity buys can appear like a "good thing" by saving money in the long term; however, it can also temporarily hurt the establishment's cash flow and cause inability to pay other bills. This will cause the operation to have more money tied up in their inventory, but the savings per bottle can sometimes make it worth the effort.

Ordering Methods

Calculating the appropriate amount of product necessary to order can be a challenging practice complete with many fluid and somewhat volatile factors. Some unknown elements include the peaks and valleys in business because of weather, time of year, neighborhood events, and so on. Looking at historical information for previous trends in consumption—whether by day of the week or time of the year—can provide some solid data for future ordering references. Having a sales history is an example of utilizing historical information to be used as a basis for more accurate ordering.

Par Stock Order Method

This ordering method is used for determining the quantities of day-to-day items. The par stock order method is based on a pre-established par stock, which is the amount of product needed in inventory to last until the next delivery date. Par stocks are created primarily based on the operation's *sales history* which tracks the total of customers' selection preferences over a period. Inventory levels and par stock should be carefully developed to avoid surplus and stock outages in order to ensure optimal quality of product and ensure good customer service. Generally, a good rule of thumb is to stock just enough product in the storeroom to prevent running out between deliveries. At the bar, sufficient product should be on hand to prevent running out between shifts or in a single day, depending upon the operation. The par stock ordering method is not static; instead it is fluid and needs to be constantly re-evaluated. During certain times of year, the par stock may need to be increased or decreased, according to the demands of business volume.

The process of par stock ordering may proceed like the following: the buyer visits the storage area(s) to count and document the current on-hand inventory of products. Next, the buyer subtracts the predetermined par stock level from the current on-hand amount. The differential between the two indicates the quantity necessary to be ordered.

Formula

Par Stock (amount needed until next delivery)
– Subtract the Inventory (what is on hand)
- -
= Amount to Order.

Example

24 bottles (ABC Sauvignon Blanc)
– 12 bottles (ABC Sauvignon Blanc)
- -
= 12 bottles or 1 case (ABC Sauvignon Blanc)

The Other Order Approach

This other approach is an alternative method used for determining quantities to order for special, one-time events such as banquets or catering with known guest counts in advance. As with most ordering methods, there will always be some presumption based on the customer's consumption—unless they are restricted through a system of requiring chits or coins to obtain their drink. Yield is also subjective—ideally, the bartender will obtain a 100 percent yield when preparing and transferring drinks into glassware. However, realistically allowing for a small percentage of spillage or over-pouring builds a safety factor into the process.

Formula

Step #1

$$\text{Quantity to Order in ounces} = \frac{\text{\# of portions} \times \text{Portion Size}}{\text{Yield \%}}$$

Step #2

$$\text{\# of bottles} = \frac{\text{Quantity to Order in Ounces (from answer in step \#1)}}{\text{25.4 (amount of ounces in a standard bottle)}}$$

Example

Step #1

$$\text{510.2 oz. to order} = \frac{\text{100 portions} \times \text{5-oz. portion (50 people with 2 drinks each)}}{98\%}$$

Step #2

$$\text{20.08 bottles, or 20 bottles} = \frac{\text{510.2 oz. (from answer in step \#1)}}{\text{25.4 (amount of ounces in a standard bottle)}}$$

The buyer may choose to purchase one full case plus eight additional bottles by breaking a case; or the buyer simply may order two full cases, yielding 24 bottles. Ordering the two cases allows the buyer to build in some inventory of extra bottles that can be used at other times if they are not consumed at the event they were intended for.

Primary Factors that Influence Ordering Frequency

It is quite common for food and beverage establishments to place weekly product orders—though the frequency may vary based on the numerous factors identified below.

1. **Storage Space** Storage space is usually limited for most organizations. Therefore, to maximize space, buyers may have to order smaller amounts of products more frequently. If an organization happens to have larger storage areas, some buyers will choose to take advantage of opportunity buys—where they purchase a larger number of products to obtain volume discounts.
2. **Funds Available** For control reasons, some organizations set price limits on the dollar limit of either single orders or total weekly orders based on projected sales volume. This may be to adhere to certain budget or cash-flow constraints.
3. **Delivery Schedules** Buyers are limited by the delivery schedule set by suppliers. Often, larger distributors deliver daily, except for Sundays, while smaller boutique type distributors deliver only one or two days a week.
4. **Minimum-Order Requirements** Distributors often set minimum-order requirements to discourage beverage establishments from ordering a single bottle or single case of product. Generally, requirements may be a $150.00 or a two-case minimum.
5. **Price Limits per Budget** Beverage organizations may set maximum price limits per bottle or case of wine. This is a means of control which may limit buyers from purchasing or being tempted to purchase a high-priced product that may not fit the vision or budget of the concept.
6. **Limited Supply of Products** Certain products have limited availability because they are either highly subject to seasonality, small-production items, or tightly allocated products for select sites in different markets.

Receiving Control Point

Learning Objective 3
Identify some standards in receiving practices

Receiving is the act of inspecting delivered products and either accepting or rejecting them based on meeting a predetermined set of standards as determined from the purchasing control point. The criteria for acceptable standards are centered on the core elements of quality, quantity, and price.

During the receiving process, beverages are delivered by the case or by the bottle. Therefore, it is not too difficult to determine whether a case or a bottle is partially empty or whether beer being delivered is past its expiration date for freshness. Managers should have procedures in place to monitor incoming inventory; for example, employees should verify whether the proper items are delivered and ensure that the items are undamaged. Managers should ensure that all orders are quickly inspected and referenced on both the invoice and the PO and then promptly transported to the appropriate storage areas.

Receiving Control Techniques

All delivered products should be accompanied with an *invoice*—a document that lists all products delivered, as well as the quantity, price, and, possibly, quality level of each product. Invoices alone cannot control received products; competent personnel and other elements in the receiving process are also necessary.

1. **Competent Personnel** The personnel to receive alcohol products should be trained in the SOPs of the flow of beverages.

2. **Appropriate Receiving Hours** The best receiving hours are ones that are staggered and not during peak service periods, to allow the order to be properly inspected and put securely away into storage areas.
3. **Invoice Receiving** Invoice receiving is the most effective method to implement for receiving control. This process references the physically delivered products against the invoice and the PO. It is not uncommon to have intentional or unintentional errors between the products ordered and what was delivered to the establishment.
4. **Daily Receiving Report** The daily receiving report is a summary of all deliveries for a single day. This report is another control technique that forms a paper trail to assist with tracking orders if there is a future discrepancy.

DAILY RECEIVING REPORT

Distributor	Date	Invoice #	Total $	Receiver
Southern wine	1/2	3244144	$346.98	Albert Schmid
Heritage	1/2	86577	$1,223.76	Albert Schmid
Wirtz beverage	1/2	32241	$2,111.90	Albert Schmid

The Receiving Process

1. **Once a delivery arrives, all products should be inspected for quality, quantity, and price.**
   a. **Quality** Verify that any product ordered matches the product that is delivered in terms of (if appropriate) producer, grape name, geographic region, quality classification, and vintage.
   b. **Quantity** Ensure the amount of the product as identified on the PO matches the correct amount, both on the invoice—but also the actual physical products.
   c. **Price** Validate that any price stated either on the PO or supplier contract agreement matches the amount stated on the invoice.
2. **Acceptance or Rejection of Delivery** Accepting or rejecting a product is not always a simplistic black-and-white decision. If products do not meet the standards, they can and should be rejected. When part of an order is rejected, it is vital that both the buyer and delivery driver initial the invoice. If a product is accepted that later turns out to be incorrect or not up to the SOPs, contact the salesperson immediately. Then, if merchandise is returned, ensure that the driver provides a credit memorandum when the product is removed from the premises.

Storage Control Points

Learning Objective 4
Explain some proper storage techniques for wines, beers, and spirits

Storage is the process of holding products under desirable conditions until utilization during the production and service control points. Storage management involves the active intention of maintaining (and in some cases, creating additional revenue from) the safe investment of the beverage items, whether the products being stored are only a few cases of beer or a wine cellar containing 35,000 bottles. Optimal storage conditions entail that beverage items are stored in a manner that protects them from *pilferage, theft, and undesirable environmental conditions* that may cause spoilage or waste.

The storage locations must be considered in respect to their amount of space and accessibility specific to the philosophy of the beverage establishment—all of which are vital to the health of any successful business.

Managing the Storeroom

Managing the storeroom involves being aware of the possibilities of losing products for different reasons. Managers should be able to grasp these areas of concern in order to then establish a plan on how to minimize loss of product and to ensure freshness and quality.

Pilferage and Theft

Pilferage and theft are consistent concerns throughout the industry. The act of stealing is fairly easy to execute, yet hard to detect, and extremely difficult to prevent on an ongoing basis. The temptations posed by constantly handling large sums of cash and dealing with a liquid inventory can often prove overwhelming for the employee and tiresome to control for the manager. Pilferage and theft are inevitable and probably never 100 percent avoidable. However, there are some practical recommendations that can reduce the establishment's vulnerability to these concerns.

Pilferage is often associated with inventory shrinkage by small-scale theft. However minor and insignificant, pilferage can be damaging, particularly in the long term. For example, if a bartender drinks a "free" glass of wine while working over the course of weeks to months, the costs affiliated with such behavior can be enormous.

Theft is a predetermined and relatively large-scale act of thievery. For example, employees may be charging the customers, but never ringing the sale into the cash register—then pocketing the cash. Sometimes, theft involves *collusion*, which may incorporate multiple individuals—perhaps a combination of various employees and even customers to conduct the act of theft. With the onset of technology, theft has become easier to carry out and more difficult to detect.

FIGURE 15.14
Wine shelving. Courtesy of Erika Cespedes.

Environmental Conditions

To maintain optimal conditions, wine and beer must be stored according to some basic guidelines. Figure 15.14 identifies proper storage of wine being placed on their sides to ensure the cork is moist, through being in contact with the wine in the bottle. In rough order of importance, there are five primary approaches to consider when beverages are being stored: (1) light, (2) vibrations, (3) temperature, (4) humidity, and (5) placement.

- *Light:* Ideally, wine and beer should be stored in a dark location or, at the least, in minimal direct and indirect light. Over a period of prolonged exposure to light (weeks to months), chemical changes may occur and alter the aroma, flavor, and taste of beer and wine.
- *Vibrations:* Beer and wine should be stored in a quiet location. Constant vibrations may cause chemical changes that alter the aroma, flavor, and taste of both products.
- *Temperature:* Ideally, bottled beer and wine should be stored at a consistent temperature of 55–65°F and draft beer at 38°F. Beverages are relatively stable even if temperatures vary gradually within a small range—it is sudden changes or prolonged warm temperatures that may damage them. Temperature becomes

more important for the lengthier periods that wine and beer will be stored. Wine evolves best at a consistent temperature between 55 and 62°F. The lower end of the temperature range slows down a wine's development, and the higher end speeds up development. If wine is stored for a short period and used within a couple of months, then a room temperature of 72°F would be adequate. A basement, closet, refrigerator, or, even better, a cooler will suffice, as long as it is absent from extreme temperature swings.

- *Humidity:* Humidity is an important consideration for wine and beer that are sealed with a cork closure. Particularly in medium- to long-term storage situations, humidity may become a problem. A relatively high level of humidity of 70–75 percent would be ideal. If wine and beer are stored in an environment with lower humidity, there is a risk of corks drying out and allowing oxygen to enter and spoil them. If humidity levels are much higher than 70–75 percent, there are the risks of a moldy cork and the bottle's label easily ripping or peeling away.
- *Placement:* All spirits and any wine sealed with screw-caps can be stored upright, and beer is stored in cases until further use. All wine and beer that is sealed using a cork should be stored on their sides—allowing the cork to remain in contact with the liquid and maintain a moist, swollen state forming a proper seal at all times during storage. Wine stored in cold temperatures (i.e., refrigerated) may survive unaffected while standing on its base for longer periods. For short-term storage (days to weeks), the cork won't dry out, but for medium- to long-term storage, placement of the wine on its sides becomes more of a concern.

Practicing MBWA

Management by Walking Around (MBWA) involves management being accessible, available, and visible throughout the establishment. Having the presence of an active management individual or team is important in keeping individuals honest and deterring bad motives. Setting up and implementing control systems throughout the flow of beverages is important and can be helpful in reducing pilferage and theft.

Practicing Inventory Management

Implementing methods of inventory management are vital to effective cost control. Inventory management can assist to ensure profitability, and they can be instituted per the level of control and time commitment available by management. Each method involves conducting physical inventories, or the counting and valuing of the beverage items in stock, on a regular basis. Commonly, three types of inventory management systems can be put in place: (1) perpetual, (2) periodic, and a (3) hybrid approach.

- *Perpetual Inventory Method:* Perpetual inventories are continuous, on-going records of the purchasing, "IN" to inventory, and what has been issued from storage, "OUT" of inventory, to the beverage production areas. Managers who employ a perpetual inventory system account for additions and deletions from inventory as they occur. The INs and OUTs can be recorded manually on a simple clipboard known as a *bin card* that is maintained in the storage area—or the record keeping can be done electronically. Either method allows the manager to know at any given moment the current quantity and value of inventory. Occasionally, perhaps monthly, the inventory is manually or physically counted to compare against the bin cards or electronic data to confirm accuracy and authenticity.

SAMPLE BIN CARD USED IN INVENTORY MANAGEMENT

Product	In	Date	Out	Date	Balance	Signature
Champagne Duval Leroy	24	1/3			24	Josh Kelly
Champagne Duval Leroy			2	1/4	22	Sandra Smith
Champagne Duval Leroy			2	1/6	20	George Schmoe

- *Periodic Inventory Method:* The periodic inventory method involves conducting a regular, or physical, inventory to communicate the quantity and value of inventory. By comparing inventory levels on a periodic basis (often monthly, but maybe weekly or even daily), the periodic method exercises less control, but consumes less time, than the perpetual method.

  Physical inventories refer to the actual quantity of each inventory item on hand, either in storage or in the bar production area. A physical inventory is conducted at regularly scheduled intervals, such as the last day of each month after the establishment has closed for business—it requires that all items on hand be accounted for and assigned a value. While this may seem like a tedious process, it is necessary for the periodic inventory method and for reconciling the perpetual inventory management method.

SAMPLE INVENTORY FORM USED IN INVENTORY MANAGEMENT

Product	Cellar	Wine Cooler	Total	Unit Price	Total Value
Kim Crawford, Sauvignon Blanc, Marlborough, New Zealand 2015	10	2	12	11.00	$132.00
Domaine Claude Riffault, Sancerre, Loire Valley 2015	4	2	6	18.50	$111.00
Ferrari-Carano, Fume Blanc, Sonoma Valley, California 2014	6	2	8	13.50	$108.00
Total White Wine	20	6	26		$351.00

- *Hybrid Inventory Method:* The hybrid approach takes into account the advantages of both previously mentioned systems. The hybrid method employs a perpetual up-to-date account of only the high-priced or most-sought-after inventory items that are most susceptible to theft and pilferage while applying the periodic method for all other items in stock. This allows the manager feedback regarding possible control issues pertaining to the more tempting higher priced inventory items while still allowing time to make necessary adjustments to procedures and systems before it is too late. The philosophy is to exercise greater control for the items that *need* control and apply less control on the items that do not need as much attention.

Understanding Cellar Management Philosophies

The type of establishment will largely determine the approach for storing or cellaring wine and some select vintage-dated beer. Depending on the philosophy, the operation may purchase primarily wines and beers that are ready to drink in the short term, or perhaps choosing the opposite extreme of investing heavily in long-term-aging wines and storing them for years through the cellaring process.

If the philosophy dictates buying a certain quantity of beverages for cellaring, or aging in house, then proper storage is essential to protect the financial investment, as it will allow the wine and beer to mature properly, progressing in quality and value.

- *Short-Term Aging* (weeks to months): These wines and beers can be consumed at any time during a period of less than a year. The focus in short-term aging is on buying wines with a drink-it or sell-it-now philosophy. Occasionally, managers may purchase a volume order for the sake of gaining a discount, which results in a larger-than-normal quantity of wine sitting in storage for a short time.
- *Intermediate Aging* (months to years): These wines and some select beers are consumed in a period of between one and three years. They have been moderately aged, and the wait will be rewarded by the components such as acid, tannin, alcohol, and fruit becoming subtler; at the same time, the wine becomes more complex and refined through the cellaring process.
- *Long-Term Aging* (years): These wines are consumed after a period of three years or more of cellaring. Long-term wines are from the best grapes and the best vintages that have the greatest aging potential, and they need long aging for their personalities to truly be expressed. Optimal environmental conditions are needed to allow the wine to undergo its chemical and physical changes slowly and undisturbed. If stored properly and opened at their peak, these wines will have appreciated in value. It is possible for wine stores or restaurants to purchase pre-aged wines from someone or somewhere else, but the practice comes with a cost.

Issuing Control Points

Learning Objective 5
Identify control concerns during the issuing process

Issuing is the process of transferring a product from the storeroom to the production area—in most cases, the bar and bar coolers. This control point should include some necessary checks-and-balances to control pilferage, theft, and loss of profit margins. The issuing control point involves two elements: (1) the physical movement, and (2) the record-keeping aspect. The issuing control point is imperative, as it ensures a safe passage from one point to another and acts as a means of inventory and financial accountability to the correct department within the operation. Larger establishments may have full-time storeroom employees officially responsible for issuing product from the storeroom to the bar areas—this often includes the use of formal requisitions that require the signature of management. Issuing creates an audit trail so that management can track where products are going and who is taking them. In smaller establishments, the issuing process may be as simple as asking the manager to unlock the cabinet, cooler, or storeroom in which the product is stored. The manager can take note of what is being removed from the cabinet and for what reason.

Issuing Control Techniques

Alcoholic beverages are a target for employee pilferage and theft because the product is easily concealed and is highly desirable. It is essential to implement procedures during the purchasing, storing, and issuing processes that will ensure product quality as well as eliminate all opportunities for theft. Managers can discourage theft by keeping alcohol storage areas locked at all times.

- *Requisition:* A requisition is a form that is used in high-volume establishments to establish greater control. An employee who needs a type and quantity of product from the secured and locked storage area completes this form. Establishments that require this form use it to create a paper trail when service stations need to be stocked for each shift or day of production. At the end of the day, management will often reconcile these forms against sales and the contents remaining in the bottles to ensure the actual product was needed and used.

- *Transfers:* Establishments may also require requisitions or other forms when beverage products need to be transferred between production areas. For example, suppose that the kitchen is making a sauce that requires a wine. The kitchen manager or other entitled employee will fill out an intra-unit transfer that documents the product. The product will be used in the kitchen, and the product cost will be moved and assessed to the kitchen. This process assists not only with accountability of product but also with financial accuracy.

Production Control Point

Learning Objective 6
Calculate proper costing out of wines, beers, and spirits

Production is the process of preparing products for sale to the customers. The objective of production is to ensure that all portions of any given beverage are identical to all other portions of the same item. This control point is important for both customer perception and cost control measures and in order, ultimately, to set an appropriate selling price that maximizes revenue.

Standards and Procedures in Production

For control and consistency purposes, it is necessary to develop a standardized portion size. Wine, beer, and spirits are often quantified per volume. For example, the portion size at restaurant ABC is a 16-oz. draft beer, meaning that each customer who orders a draft beer receives a sixteen-ounce portion. Rarely do bartenders measure the exact volume of beer or wine when pouring into the glass; instead, the type and size of glassware allows a close approximation of the desired standard portion size. Workers are trained to pour beer and wine to a certain fill level, or an invisible line on the glassware. With practice, it is possible to estimate pouring volume with surprising accuracy. On the other hand, spirits are commonly poured using jiggers, double-sided measuring devices. Once portion size standards have been determined, the customer is assured a drink of consistent quantity and quality each time it is ordered. Having a reliable standardized portion size leads to consistent costing for the establishment as well as a standard portion cost, or the dollar amount that a standard portion should cost each time it is served. Operating according to a standard portion cost is vital to cost control and achieving maximum revenue and ultimately projected profit.

As Purchased (AP) versus Edible Portion (EP)

As purchased, or AP, is an indicator of the gross quantity of an item purchased "as is" from the supplier. It is the quantity of product before being opened, or poured, or otherwise manipulated within the beverage establishment. Edible portion, or EP, is an indicator of the amount or cost of an item as it is served to a customer. This indicates that loss during processing most likely will occur or already has occurred. In most beverage establishments (when full bottles are sold), the AP is identical to the EP. There is no loss of product quantity or value when a customer purchases an entire bottle. However, there often is a small degree of loss associated with pouring wine by the glass. Portion sizes may not be always 100 percent accurate, as some portions are a bit over and others a bit under. It is likely that the yield of actual, sellable product quantity is 95–97 percent of the original 100 percent quantity.

The goal of costing-out beverages is to determine an appropriate selling price (SP). The manager pricing the alcohol assumes that bartenders and servers maintain the defined portion size, which should guarantee a portion cost that will result in determining an accurate SP. If the portion size for a glass of ABC Sauvignon Blanc is five ounces and the portion cost yields $2.00, then an accurate SP can be determined with reliable projected revenue that is easier to establish.

Cost Out "Wine By the Bottle" and "Wine By the Glass"

Step 1. Determining Bottle Cost To cost out a bottle of wine, the first step is to divide the cost per case by the number of bottles within the case. This will yield the cost per bottle, or bottle cost, as shown in the following step.

Formula:

$$\text{Determining Bottle Cost} = \frac{\text{Cost per Case}}{\text{\# of Bottles in Case Cost}}$$

Example:

$$\$28.80 = \frac{\$345.67}{12}$$

Step 2. Determining Cost per Ounce To cost out a bottle of wine by the glass, take the bottle cost as determined in the previous step and divide it by the number of ounces contained in the bottle. This yields the cost per ounce, as shown in the step below.

Formula:

$$\text{Determining Cost per Ounce} = \frac{\text{Bottle Cost}}{\text{\# of Ounces in Bottle}}$$

Example:

$$\$1.13 = \frac{\$28.80}{25.4}$$

Step 3. Determining Portion Cost Multiply the cost per ounce by the standard portion size. The typical portion size is five ounces, but it may vary by establishment. This yields the standardized portion cost, as shown in the step below.

Formula:

$$\text{Determining Portion Cost} = \text{Cost per Ounce} \times \text{Standard Portion Size}$$

Example:

$$\$5.65 = \$1.13 \times 5$$

Cost Out "Bottled Beer" and "Draft Beer"

Determining Bottled Beer Cost To cost out a bottle of beer, the first step is to divide the cost per case by the number of bottles within the case. There are typically,

24 bottles of beer to a case. This will yield the cost per bottle, or bottle cost as shown in the step below.

Formula:

$$\text{Determining Bottle Cost} = \frac{\text{Cost per Case}}{\text{\# of Bottles in Case Cost}}$$

Example:

$$\$1.50 = \frac{\$36.00}{24}$$

Determining Draft Beer Cost To cost out draft beer, which is typically purchased by the keg, the first step is to divide the cost per keg by the number of ounces within the keg. There are 15.5 gallons or equivalent to 1,984 ounces AP in a keg of beer. With the standard 7 percent loss on a keg of beer, there is 1,845 ounces of EP. This will yield the cost per EP ounce—the most accurate measure of cost per ounce as shown in the step below.

The second step is to determine the portion cost of a glass of beer. This portion cost will vary depending upon the size of draft beer. However, it is universal to offer a 12-oz. or 16-oz. draft beer.

Formula:

$$\text{Cost per EP ounce} = \frac{\text{Cost per Keg}}{\text{\# of EP ounces}}$$

Example:

$$\$0.05 = \frac{\$89.00}{1{,}845}$$

Cost Out "Cocktail"

Determining the Cost of a Cocktail In order to first cost out a cocktail, there needs to be a standardized recipe. Not necessarily an easy task when there are well over hundreds of cocktails and often a couple or more variations of a single one.

DRY VODKA OR GIN MARTINI $4.25

- 2 oz. of well vodka or gin
- Splash of dry vermouth (about 1/8 of an ounce)
- Olive or lemon twist (for garnish)

1. Pour the ingredients in a cocktail shaker filled with ice.
2. Shake and strain into a cocktail glass.
3. Garnish with olive or lemon twist.

The first step is to breakdown the bottle cost for each beverage—divide the cost per case by the number of bottles in the case. The second step is to breakdown the ounce cost for each beverage. The third step is to multiply the ounce cost by the portion size of each beverage. The fourth and final step is to add up each of the portion costs to obtain the total recipe cost.

Formula: Step #1

Cost per Bottle = $\frac{\text{Cost per Case}}{\text{\# of Bottles in Case}}$

Example: Vodka

\$20.00 = $\frac{\$240.00}{12}$

Example: Dry Vermouth

\$10.00 = $\frac{\$120.00}{12}$

Formula: Step #2

Cost per Ounce = $\frac{\text{Cost per Bottle}}{\text{\# of Ounces in the Bottle}}$

Example: Vodka

\$0.59 = $\frac{\$20.00}{33.8}$

Example: Dry Vermouth

\$0.30 = $\frac{\$10.00}{33.8}$

Formula: Step #3

Determining Portion Cost = Cost per Ounce × Standard Portion Size

Example: Vodka

\$1.59 = 2 oz. × \$0.59

Example: Dry Vermouth

\$.03 = 0.10 oz. × \$.30

Formula: Step #4

Total Recipe Cost = Portion Cost #1 + Portion Cost #2 + Nominal Fee

- Often a nominal fee is added for minor ingredients such as low cost garnishes or soda mixers.

Example: \$1.67 = \$1.59 (vodka) + \$.03(vermouth) + \$.05 (garnish)

Beverage Cost

Most establishments separate food and beverage sales in order to have greater insight of knowing where their sales and associated costs derive from. Costs for each category also are shown separately on most profit and loss statements (P&Ls).

In the food and beverage industry, "cost of sales" (or product costs) refers to the expense incurred to the establishment for purchasing raw product, to prepare it, and be able to resell the products to the customers. The cost of beverages includes not just the cost of the alcohol but also all other product costs such as juices, carbonated mixers, and fruit incurred in order to produce beverages. These costs are normally expressed both in dollar amounts and as a percentage of its selling price.

In the example below, food sales are \$220,000 and beverage sales are \$73,000. Costs for each category also are shown separately on most P&Ls. The costs incurred to produce the previously mentioned sales figures include \$75,000 for food and \$18,000 for beverages.

PROFIT AND LOSS STATEMENT

	$	% of Total Sales
Food sales	220,000	75
Beverage sales	73,000	25
Total sales	$293,000	100
Food costs	75,000	26
Beverage costs	18,000	6
Total costs of sales	$93,000	32
Gross profit	$200,000	68

Calculating Beverage "Cost-Percent" as a Percentage of Total Sales

The percentage of cost spent as a percentage of total sales can identify the breakdown of beverage and food sales. The formula for calculating beverage cost percentage as it relates to total sales:

Formula

$$\text{Beverage Cost \% to Total sales} = \frac{\text{Cost of Beverage Sold (Beverage Cost)}}{\text{Total sales}}$$

Example

$$\text{6\% to Total Sales} = \frac{\$18{,}000}{\$293{,}000}$$

Calculating Beverage "Cost-Percent" as a Percentage of Beverage Sales

The percentage of cost spent on beverages is one of the primary benchmarks by which an operation gauges its overall beverage performance. This formula shows a direct correlation between beverage sales and an associated beverage cost. The formula for calculating beverage cost percentage as it relates to beverage sales is as follows:

Formula

$$\text{Beverage Cost \%} = \frac{\text{Cost of Beverage Sold (Beverage Cost)}}{\text{Beverage sales}}$$

Example

$$25\% = \frac{\$18{,}000}{\$73{,}000}$$

Calculating Cost of Beverages Sold (or beverage cost)

Cost of beverages sold, or beverage cost, is calculated based on the value of the entire beverage inventory. Inventory is often conducted on a weekly or monthly basis for purposes of reconciling beverage cost and sales and to ensure control is taking place.

Beginning inventory (last period's ending inventory)		$_____
Plus this period's purchases	+	$_____
Equals goods available for sale	=	$_____
Less ending inventory (next period's beginning inventory)	–	$_____
Equals **Cost of Beverages Sold**	=	$_____

Once the cost of beverage sold has been calculated for a given period of time, divide this number by the beverage sales (ensuring the financial data is from the same period); the result will be the beverage cost percentage.

When performing these calculations, it is important to note that if the bar transfers any beverage products to the kitchen for cooking purposes, or to the bar for mixing purposes, those transfers need to be tracked and their value subtracted before totaling the cost of beverages sold.

Because of the relative ease with which a dishonest employee can manipulate inventory records, and, therefore, beverage cost percentage, most experts recommend that the duties of receiving, storing, issuing, and inventorying be separated. This is known as *separation of duties*—in other words, the individual who receives the product should not be the same individual who stores and issues the product. A different individual should be responsible for month-end inventories. While separating these duties is relatively easy for larger operations, it may be next to impossible for the smaller owner-operator who must rely on a limited staff. In cases, such as these, it is wise to assign these duties to the owner or manager of the operation.

Determining Selling Prices (SPs) to Ensure Profitability

Determining what prices to charge for beverage products is related to cost control and to an operation's overall intended profit. Charging too little for products can result in lowered profits; charging too much can result in lowered customer counts and/or less sales. Menu pricing for beverage sales is affected significantly by many factors, including local competition, customer demographics, product quality, and portion size. While a manager may not influence all of these factors, he or she can exercise control in determining an appropriate amount to charge customers for drinks.

Establishing an accurate SP is essential to producing a reliable estimate of revenue. The typical markup for alcohol from the wholesaler is about 35–40 percent and an additional 25–50 percent from the retailer. Many independent retailers use a higher markup than chain operations, because they tend to sell a lower volume.

There are several methods and approaches to pricing alcoholic beverages, but the concept of the establishment is the major determining factor. The concept defines who the customers are, what kind of alcohol they may desire, and the price they may be willing to pay. The appropriateness of the pricing should match that of the vision of the establishment. In general, managers use one of the two following concepts to determine what price to charge:

1. Product cost percentage
2. Contribution margin

Product Cost Percentage Method

The product cost percentage method of pricing is based on the idea that an item's cost should be a predetermined percentage of its selling price. For example, an operator wishes to achieve a 20-percent beverage cost-percent on a Martini; therefore, the remaining 80 percent of the sale of the martini is used for both covering other expenses and providing some element of profit. This remaining portion is otherwise known as contribution margin (or the CM). The product costs for the Martini total \$1.50. The Martini's SP can be determined by using the following formula:

Product cost | Desired cost percentage = Selling price

\$1.50 | 0.20 = \$7.50

If the Martini is sold at \$7.50, a 20-percent beverage cost (\$1.50) will be achieved and therefore, \$6.00 goes to cover other expenses and provide profit. The "common-sense" approach now takes over—the manager considers whether \$7.50 can be obtained in the market place given the concept and clientele. SPs should create a good price–value relationship in the mind of the consumer. Therefore, beverage pricing is usually not based on a mathematical equation alone. For that reason, the original \$7.50 SP may be adjusted slightly upward (perhaps \$7.95 is likely) or downward (may be perceived better at \$7.25 or \$6.95) to account for these reality factors that should be considered when determining SP for any item.

Following are three formulas that can be used to assist with determining cost, cost-percent, and SPs. The figures below identify the mathematical formulas used to achieve either answer of determining the cost of an item, a suggested selling price, or determining its cost-percent. For example,

- To determine the "beverage cost" of a given item, multiply the variables of "beverage selling price" and "beverage cost-percent."
- To determine the "beverage cost-percent" of a given item, divide the "beverage cost" by the "beverage selling price."
- To determine the "beverage selling price" of a given item, divide the beverage cost by the "beverage cost-percent."

The Contribution Margin Method

Another method of product pricing is to focus not on the item's cost percentage, but rather on its contribution margin—the difference between the item's product cost and its SP. Contribution margin is the gross profit or the amount that remains after product cost is subtracted from an item's SP. To obtain a product's contribution margin (CM), the formula is

Selling Price – Product Costs = Contribution Margin (CM)

When using this pricing approach for determining SPs, operators often establish "set" contribution margins for various beverage items or groups of items. The margin would be added to whatever the product cost has been determined to equal a suitable SP. For example, draft beer may be priced with a CM of \$3.00 each, cocktails with a CM of \$4.00, and bottled wines with various other CMs. Therefore, in this case, if draft beer costs \$2.00 per serving, its SP would be \$5.00.

Formula:

Product Cost + Contribution Margin = Selling Price

Example:

\$2.00 (product cost) + \$3.00 (pre-established margin) = \$5.00 (selling price)

Service Control at the Bar

Learning Objective 7
Describe some control issues during the service control point

To gain maximum revenue, it is critical for the establishment to ensure that a consistent customer experience takes place through these established SOPs. Having proper control systems will also help to ensure that management controls costs, and gains maximum revenue, leading to profitable operation. For example, control systems encourage and require bartenders to make drinks per standardized recipes. They also help to minimize giving away "free" drinks and ensure that all items are properly accounted for. Having standards and procedures are great safeguards to lessen the potential for employee pilferage and theft. Ensuring the bartenders have a ticket placed in front of each sitting guest at the bar, allows management to view the ticket guaranteeing that drinks have been accounted and rung through the register.

Pricing and Inventory Controls for Special Events

Beverage pricing for parties and receptions can seem daunting, but it need not be. Clients often have the choice of arranging a cash bar or a host bar. When a cash bar is requested, guests attending the function are expected to pay for their alcoholic beverages as they are consumed. For a host bar, the host is charged at the end of the function based upon the drinks consumed throughout the event.

- *Cash Bar Procedures:* The standard pricing procedures will suffice for a cash bar. It is important, however, to institute strict control procedures to prevent bartender theft. Many operations now use a ticket system rather than having cash exchange hands between bartenders and the customers attending the event, which requires that guests purchase drink tickets that can be exchanged at any of the satellite bars set up for the function. Some operations color code tickets. For example, they may use blue for beer, pink for wine, and green for mixed drinks. Other operators simply assign a set dollar value to each ticket. When the guest "buys" a beverage, beer might cost one ticket and a mixed drink might require two, depending on the cost of the drink.
- *Host Bar Procedures:* Many banquet clients prefer to pick up the entire beverage tab of their function. If this is the case, a host bar, or open bar, is generally arranged. There are many methods for setting prices and controlling inventory for such functions. Two of the most common methods are

1. Charging the host on a per person, per hour basis
2. Charging the host for the actual number of beverages consumed

If the per person, per hour basis is used, the beverage manager must estimate how much the average guest will consume during the event to establish a "per-person" charge. Clearly, various consumer groups will have different consumption habits differently when attending a hosted bar function; therefore, this pricing method is somewhat risky. Some managers, however, have had success by keeping historical consumption records that detail the average consumption of a wide variety of groups. These records can be used to establish pricing guidelines.

One of the most tried and true methods for controlling a host bar is to charge the client for the actual number of beverages consumed. This method requires that a beginning and ending inventory be taken at all satellite bars operating at the function. If there are any additions to inventory during the function, these must be recorded as well. A simple form can be devised for this process. Note that if assorted brands of each wine, spirit, and beer are to be offered, the form should be designed to reflect varying product costs.

SAMPLE HOST BAR INVENTORY CONTROL FORM

Beverage Type	Beginning Inventory	Plus (+) Additions from Inventory	Equal (=) Total Available to Serve	Minus (–) Any Ending Inventory	Equal (=) Total Usage for the Event	Unit Cost	Total Cost
Wines							
Beers							
Spirits							
TOTALS							

It is customary to allow the host to verify the beginning and ending inventory figures. In addition, some hosts will insist the manager provide empty bottles as proof of product consumed. If a product has been opened but not entirely consumed, some state liquor authorities allow the host to purchase the entire bottle and carry out the remaining contents in the bottle. Otherwise, the manager must employ a system of weighing or measuring the remaining contents to determine quantities consumed from partially used bottles.

Pilferage and Theft Issues at the Bar

The bar is the most profitable revenue producing area in most food and beverage establishments, but also one of the most vulnerable areas that is susceptible to pilferage and theft. Even the small-scale skimming, such as a free drink to the bartender's friends here-and-there ultimately add up to lost revenue and/or decreased profit margin due to increased costs.

- *The Short Ring:* Ringing incorrect items is one of the classic and easiest techniques for stealing behind the bar. The ring up occurs when a customer orders an alcoholic beverage from the bartender. The bartender will serve the drink, tell the customer its selling price, then ring up a soda on the register, and input the cash received for the drink into the register drawer. The difference between the drink and the soda may typically range anywhere from $3 to $7 cheaper. At the end of the night, the bartender will remove the extra money.
- *The Giveaway:* This technique involves the bartender giving away products (free drinks) to regular customers, co-workers, or friends/family who frequent the bar without recording the sale. The bartender's intention is to get repayment through a larger tip or just simply looking favorable. A variation of the "giveaway" is when a customer has several drinks throughout the course of their visit and the bartender conveniently forgets to update the guest check for the customer to obtain the drinks for free.
- *The Void:* This technique allows a bartender to ring up items on a guest check, present a check, and collect payment. Just prior to cashing the check out, the bartender will void the check and place the cash in the drawer to be removed at a later point.
- *The Dilution:* This allows the bartender to dilute or "add product" by pouring small amounts of water into the existing spirit bottles so there is not a discrepancy when management reconciles inventory. This allows the bartender to give drinks away or consume drinks while working without having beverage costs percentage be affected, or to identify any form of caution or concern to the manager.

- *The Phantom Bottles:* This technique involves the bartenders bringing in their own bottle of spirits to pour from and to pocket the cash sales. By doing this, once inventory or reconciling of beverage cost-percent will not show any caution or discrepancy.
- *The Phantom Drinks:* This technique is easy to carry out when there are large groups of people or an event that requires a host bar. The group of people order drinks sporadically throughout their visit, and at the end of their visit, they obtain a single tab/check with all the drinks. Illicit activity occurs when bartenders or management "add-on" drinks that were never ordered by customers to inflate the guest check.
- *The Substitute:* This method involves the bartender substituting a cheaper brand for a premium brand that usually sells for a much higher price, yet charging for the premium brand and pocketing the difference.
- *The Missing Bottle(s):* This technique may be carried out in collusion with bar backs and/or bus people. It involves carrying out full bottles of alcohol along with the empty bottles to the dumpster. Then removing the full bottles from the dumpster at the end of the evening.
- *The Short-Pour:* This method involves the bartender pouring less than the standardized amount of alcohol into a customer's glass. This short pour allows the bartender to over-pour later or for "the giveaway" for their desired customers to obtain bigger tips.

Control Techniques at the Bar

Control techniques are imperative because incidents of employee theft and misuse can be a frequent problem at the bar. Possibly the most vital system used to control bar costs is having some form of standardized portion control—the necessary factor to achieve desired profit margins. Whichever control system(s) are put in place, operators must enforce their use through training and constant monitoring and positive reinforcement.

- *Implement Standardized Recipes:* Developing and using standardized recipes/formulas when preparing beverages have significant impact on the overall consistency of products being served in terms of both quality and cost. It is recommended that mixed drinks conform to standardized recipes with standardized portions to achieve desired costs.
- *Utilize Portion Control Measures:* Since the sales price of a drink is based on its portion size, once the quantity begins to fluctuate, so will the drink's profit margin. Implementing an effective strategy to strictly control portioning is a crucial aspect to protecting the establishment's revenue and, ultimately, profit. Achieving proper portion control can be done in a variety of ways. Three of the most common are using jiggers, pour spouts, and liquor computer systems.

  As explained in previous chapters, a jigger is a double-sided measuring device used to accurately pour spirits; it typically measures in ounces or portions of an ounce. It is uncomplicated and inexpensive, and it is the long-standing choice of many establishments. Other operators use specially designed pour spouts that allow only a predetermined measure of liquor to flow from bottle to glass when the bartender prepares a mixed drink. These devices have become much more common in recent years as manufacturers have fine-tuned the measuring mechanisms located in the spouts.
- *Prohibit Bartenders from Reconciling Their Own Drawer:* At the end of a shift, bartenders should not be allowed to reconcile their cash drawer. If the bartenders are checking out their own cash drawer, after the cash is reconciled

with sales, the process allows them an ideal opportunity to simply pull out any cash overages that may have been obtained from not ringing in various items throughout the shift.

- *Create Tip Jar Procedures:* The bartender's tip jar should be placed well away from the operation's cash register or *point-of-sale system* (POS). It becomes too easy to divert funds from the register if the tip jar is placed in its vicinity. In addition, bartenders should be prohibited from making change out of their tip jar or taking currency from the tip jar and exchanging it for larger denominations out of the cash drawer. If a bartender is stealing from the cash drawer, it becomes too easy to retrieve the money from the register under the pretense of making change. For example, a bartender could take twenty $1.00 bills out of the tip jar, deposit the currency into the register, but instead of taking out a $20 bill in exchange—instead they could remove multiple bills beyond the money they put in the drawer.
- *Don't Allow Bartenders to Participate in the Physical Inventory Process:* The physical inventory should be reserved as a management responsibility. Bartenders who are stealing can use their participation in the physical inventory process to alter the recorded data so that it offsets theft. This can be accomplished by overstating the amount of alcohol on the inventory sheet—this will essentially have the same effect as if the theft never occurred.
- *Require Managerial Approval of Complimentary Items:* Bartenders should receive management approval before preparing any complimentary drink. This policy is intended to stop them from claiming, after the fact, that a drink was given away with management's consent, when in reality the drink was sold and the proceeds of the sale were pocketed.
- *Require Bartenders to Verify Cash Drawer Count:* Bartenders should be required to verify the amount of money used to comprise the bar register's opening bank. This practice will prevent the bartenders from claiming that their opening bank was either over or under the prescribed dollar amount to explain a cash shortage or overage in the register. Periodically, place an extra $10 or $20 bill in the bartender's bank and see if the person informs management of the cash overage. It is a good way to verify if the bartender is counting their bank prior to the shift and measure the person's degree of integrity.
- *Incorporate Secret Shoppers and Spotters:* Using a third-party verification can be quite enlightening. Secret shoppers are used to assess the quality of the service and products. Spotters are used to show up unannounced and unidentified, they proceed to observe the behaviors of bartenders and report any act of theft or suspicious behavior. Because most states' dramshop laws create immense liability for the operator, mystery shoppers also are trained to observe alcohol awareness and safety issues. Detailed written reports based on the observations of the spotters and mystery shoppers are submitted to management.

Check Your Knowledge

Directions: Use these questions to test your knowledge and understanding of the concepts presented in the chapter.

I. MULTIPLE CHOICE: Select the best possible answer from the options available.

1. In the distribution system for alcoholic beverages, which of the following are considered intermediaries?
   a. Distributors
   b. Growers
   c. Manufacturers
   d. Processors

2. In the distribution system for alcoholic beverages, which of the following transports alcohol into the United States?
   a. Distributors
   b. Wineries
   c. Retailers
   d. Importers
3. The amount of an item on hand that will carry an operation from one delivery date to the next is called
   a. par stock
   b. blanket order
   c. safety stock
   d. purchase order draft
4. The first step of the four-step control process is to
   a. monitor employee performance
   b. coach performance
   c. train employees
   d. create standards and procedures
5. Which is not an acceptable storage practice for wine?
   a. Lay it on its side
   b. Keep it in cool temperatures
   c. Keep it in fluctuating temperatures
   d. The storeroom should have low humidity
   e. All of the above
   f. Both c and d
6. If ABC Chardonnay costs $120 per case, and there are 12 bottles in a case, then the cost per bottle is
   a. $12
   b. $20
   c. $10
   d. $13
7. If ABC Chardonnay costs $120 per case, and there are 12 bottles per case, determine the cost per oz.
   a. 39 cents
   b. $254
   c. $2.54
   d. $1.95
8. If ABC Chardonnay costs $120 per case and there are 12 bottles per case, the cost per glass (assume 5-oz. portions) is
   a. $1.97
   b. $1.50
   c. 39 cents
   d. 30 cents

II. CALCULATIONS:

9. Refer to the information below to compute "Albert's Pub" cost of beverages sold:

Beginning inventory (Dec 1st 2012)	$26,000.00
Purchases	$34,256.00
Goods available for sale	$________
Ending inventory (Dec 31st 2012)	$22,849.00
Cost of beverages sold	$________

10. Albert's Bar has beverages sales totaling $178,129.00. Based on the answer to Question 14, calculate the bar's beverage cost percentage.

11. A restaurant operator desires a 24-percent beverage cost on a bottle of wine that costs the operator $12.00. What is the most appropriate selling price for the bottle of wine?
12. A restaurant operator desires a $10.00 contribution margin on a bottle of wine that costs the operator $8.00. What is the most appropriate selling price for the bottle of wine?

III. DISCUSSION QUESTIONS

13. Explain some ways that buying wine is different from buying beer and spirits.
14. List and discuss two beverage pricing methods for an open bar (host bar).
15. List and discuss at least three variables that will influence beverage selling prices.
16. Identify some of the factors that influence the depth and breadth of a wine list.
17. What factors must be considered when choosing purveyors of wines, beers, and spirits?
18. Describe the proper methods for storing the following:
    a. bottled wine
    b. bottled beer
    c. beer in kegs
19. Identify some control techniques that can be implemented during the storage, issuing, and production control points to ensure maximum revenue and profit margins are obtained.
20. Identify at least four theft and pilferage issues at the bar.

CHAPTER 16

Marketing the Beverage Establishment

CHAPTER 16 LEARNING OBJECTIVES

After reading this chapter, the learner will be able to:

1. Understand the basic concept of marketing as it relates to a beverage related establishment
2. Explain how "word of mouth" and social media can impact the food and beverage establishment
3. Provide some insight into creating a marketing strategy and how market penetration can influence the business
4. Provide insight into how creating an effective beverage program can positively influence the business
5. Explain how the beverage menu is more than just a list of beverage items
6. View the sample beverage menu and recognize its composition, breadth, and depth
7. Analyze the two beverage concepts illustrated and identify their appropriate marketing strategies

The aim of marketing is to know and understand the customer so well the product or service fits him and sells itself.

—Peter Drucker

The Essential Primer on Marketing

Learning Objective 1
Understand the basic concept of marketing as it relates to a beverage related establishment

Marketing is an organization-wide effort—a process that is directed toward communicating and promoting a message relevant to a beverage establishment, while attempting to reach an intended target market. Marketing is an integrated process through which companies build strong and lasting customer relationships. Successful establishments understand that marketing entails much more than advertising and sales promotion—the marketing plan is an integral part of an operation's overall business plan. Today's businesses are reorienting their marketing approach—most often incorporating digital social media and web-based formats as a means of targeting their intended customer. The rise of wireless technology and the pervasive use of the Internet have radically altered the way that contemporary businesses advertise and market their products. There has been a gradual shift over the last decade from the use of traditional media to one of digital. While some businesses still approach marketing in an "old-school" mentality, many more have reoriented their approach to capitalize on the popularity of the Internet and are embracing digital advertising to attract their most desirable target markets. Websites have become incredible marketing tools that merge advertising, promotions, news and information, and special events. In addition, these sites are often linked to a Facebook page, Snapchat, Instagram, YouTube, and Twitter account.

To successfully grow a business, it needs to attract and then strive to retain a large basc of satisfied customers. The consumer attempts to find perceived value in some form, whether it is based on money, quantity of food, quality of service, etc., for every return experience they have with any business. *Perceived value* is their overall benefits gained using the products and services offered by an operation. According to the National Restaurant Association (NRA) estimate, anywhere from 60–80 percent of an establishment's revenue is based on repeat business. The most effective and cheapest form of marketing and advertising is ensuring the captive customer is satisfied before they leave the "four walls" of the establishment. Therefore, ensuring the continued satisfaction of current clientele is paramount! Having satisfied customers can create positive word of mouth—one of the most significant sources of marketing and advertising that leads to long-term sustainability.

Advertising is a form of communication intended to persuade or remind a potential customer to frequent and/or purchase certain products and services. The advertisement will identify the name of a product or service and how consumer can benefit from consuming a brand. A well-lit sign is usually a necessity to provide visibility and advertise for most hospitality establishments.

Word of Mouth and the Influence of Social Media

Learning Objective 2
Explain how "word of mouth" and social media can impact the food and beverage establishment

Word-of-mouth advertising is the constantly evolving, ancient form of advertising and marketing that carries one of the strongest and most influential messages. This advertising involves a *testimonial*, where an individual praises the virtues of a business. Word of mouth remains one of the strongest and most influential forms of marketing and advertising. As an example, *Berghoff Restaurant* opened in 1898 and was used to showcase the Dortmunder-style beer created by the founder Herman Joseph Berghoff. The restaurant largely relied on word of mouth to remain in business for 107 years until the doors closed on February 28, 2006. In the current day, this type of advertising can be conveyed orally (the most traditional way) or written, which now frequently takes place through a digital environment. Pictured in Figure 16.1 is the exterior of the famous Berghoff restaurant.

FIGURE 16.1
The Berghoff. Courtesy of Erika Cespedes.

Written or digital forms of word-of-mouth advertising take place largely through the application of social media websites such as Facebook, Yelp, Instagram, Trip Advisor, Pinterest, Twitter, and YouTube. These applications use web-based and mobile technologies to turn communication into interactive dialogue amongst its users. Social media has the capability of reaching a small or large global audience with little to no cost that allows for a one- or two-way means of communication about a brand. Social media creates an immediacy, which can be capable of instantaneous responses if desired.

In the physical world, dissatisfied customers may complain to six of their friends—with the advent of social media, the disgruntled customer can now complain to 6,000 friends while still within the disconcerted food and beverage establishment.

One of the key components in successful social media marketing is building "social authority." This is developed when an individual or organization establishes themselves as an expert or connoisseur in their given field or area and can garner a following—and consequently strengthen a brand. Two notable examples are the following:

- *DMK Burger Bar*, located in Chicago, Illinois, strives for effective use of social media application. They advertise on their website, Facebook, and Twitter account—"The Chicago Burger Bar with exceptional Quality, Grass-Fed Beef, Fresh Baked Buns, Artisan Cheeses + Love." http://www.dmkburgerbar.com/. DMK manages their site to engage clientele with updated posts, pictures, and even staff information that assists in personalizing the customer experience that encourages repeat business.

- *Cellar Angels* is a web-based organization that leverages the power of the Internet to bring small wine producers and wine lovers together for a common good. Each week a different winery features a new wine with a brief discounted offer that is time sensitive—like a flash site. A portion of Cellar Angels' proceeds is donated to their charity partners, which their exclusive registered members get to select at check-out. Pictured in Figure 16.2 is Cellar Angels' logo.

CELLAR
ANGELS

FIGURE 16.2
Cellar Angels Logo. Courtesy of Cellar Angels.

According to co-proprietor Denise Cody, "Cellar Angels was created by a compassionate group of wine loving friends that are intent on changing the world. Our mission is simple: connect small and family-run wineries to a larger audience, expose wine lovers to incredible purchase opportunities offered exclusively to Cellar Angels members from partnering vineyards, and help a select group of charities. We love wine. We love introducing others to great wine and we love helping others. Cellar Angels provides an opportunity to accomplish all three." (© 2010 Cellar Angels, LLC. http://cellarangels.com/about-us)

Cellar Angels utilizes a combination of guest bloggers and their own video creations featuring winery owners and winemakers throughout their website. Cellar Angels markets and reinforces their message through social media outlets like YouTube, Facebook, Twitter, and their own blog, which are appropriate mediums to connect with their intended target market.

Marketing and Sales Strategy

Learning Objective 3
Provide some insight into creating a marketing strategy and how market penetration can influence the business

Marketing consists of precise, carefully measured and coordinated plan of action. One should think of marketing as a blueprint for the future. In general, marketing is divided into two categories: external marketing and internal marketing.

- *Internal marketing* refers to providing a user-friendly beverage menu, effective customer service, suggestive selling, and other in-house promotions.
- *External marketing* or advertising refers to activities undertaken to bring the customer into the establishment, such as radio or print advertising and use of billboards or coupons.

Marketing and sales strategies are an integral part of the prospectus and ultimately any business plan. Marketing programs, though widely varied, are all aimed at convincing people to try out or to continue using products and/or services. The first step in any marketing program is to identify the potential customer, or target market. Methods for defining the target market were largely discussed in *Chapter 14—Constructing the Beverage Concept.* Every successful beverage establishment seeks to identify and serve a specific target market(s), or group of people that support the business concept. Therefore, identifying a specific target market involves analyzing and grouping customers into "like" characteristics and profiling them into segments.

Many factors must be understood when determining the target market for a given business. Managers must determine the buying behavior, customer satisfaction level, attitudes, and lifestyles of their customers. The study of psychographics is relatively new for many marketers; however, it is critical in understanding these variables. Psychographics can enrich the basic demographic and geographic data by providing detailed information about their audience, like personality traits, sexual orientation, political leaning, religion, work environment, hobbies, television-watching habits, participation in sports and arts, vacation plans, and frequency of socializing. These could be thought of as "lifestyle" behaviors. Some examples include: heavy users of social media or customers more willing to make purchasing decisions based on the appearance of wine labels. A combination of psychographics along with geographic data and demographic characteristics can be used to segment and target specific markets. The more intimate the customer's wants and needs are known, the greater likelihood a business can work toward satisfying them. There is no single way to approach a marketing strategy but overall, they may address these different areas:

- *Market Penetration Strategy:* Creating strategies for market penetration are necessary if sales of a respective business appear to be "flat" or relatively consistent from same period in the previous year(s). Therefore, when revenue has not increased (possibly even dropped) is a sign that the business may be approaching stagnation. To increase market penetration, a business can employ any number of strategies to encourage potential customers to convert their use—therefore increasing market share of the business and taking away sales from competitors. Companies may choose to increase market penetration through greater promotional efforts. The business may launch an advertising campaign to generate greater brand awareness or implement a short-term promotion for a limited period.

  Using price as a means of penetration is the practice of dropping prices temporarily to stimulate market penetration. Although it's important to note that when using a price penetration approach, it's likely to attract customers in the short term, but can have damaging effects to the perception of quality associated with the brand over the long term.

  Bundling products is possibly a more effective approach than based solely on price penetration. Bundling can be used to help gain more lasting traction in previously untapped portions of the market. An effective example of bundling in the hospitality industry is the strategy of "Restaurant week." This is a marketing strategy used in several cities around the United States during the typical months of the year when slower sales (late January through early February) are inevitable. This involves a concerted effort of dozens or even hundreds of food and beverage businesses offering discounted prix fixe menus as an alternative to their existing bill of fare. These menus are attractively priced to showcase their concept and appeal to a broader audience. The restaurant week strategy has altered an otherwise slower period of business into a much more vibrant hospitality scene.

  An additional strategy for market penetration is applying a renewal approach where businesses redesign the menu, offer seasonal items which can be useful in appealing to new audiences, or strengthen the business with the existing consumer base. Either way can potentially create positive referrals for repeat business.

- *Communication Strategy:* Communication strategy focuses on attempting to send the appropriate messages through the correct medium (form of communication) to the selected target customers. The communication medium may be radio, print, or digital. An unconventional approach to this strategy can also be extended to include the physical appearance of both the service staff and the appearance of the food and beverage establishment. As a customer, they perceive an image of the establishment based on the message that is communicated through the type of physical elements they observe and experience. Consider the difference of perception between service staffs wearing a polo shirt versus a white button down shirt with a tie. Both are appropriate types of uniform, however, one will be perceived as less formal and the other as more formal. This can be one form of communicating a message, whether it is the intended one or not.

- *Sales Strategy:* Sales strategy focuses on how the actual products and services will be sold on a day-to-day basis. It should include the most influential primary elements: customer service and suggestive selling. Other selling and promotional ideas may include table tents or other signage. This will be discussed in a much broader and detailed context within the next sections of this chapter, titled "Internal Marketing" and "Creating an Effective Beverage Program."

Internal Marketing

Internal marketing is a proactive form of a sales strategy and can have a powerful influence on sales, and the overall long-term success of a given business. Sales can be boosted by conducting promotions with table tents, mailings, and guest databases. As with external marketing, beverage promotions are effective and enduring only with planning. The "one-shot specials" may only create a temporary impression unless management develops methods to keep the excitement and attraction of obtaining said products and services alive.

Many operators defer to their suppliers and distributors for merchandising ideas and assistance, and suppliers are only too glad to help. Today, it is not uncommon to see suppliers and operators working together to offer promotions geared around the exquisite pairing of a special menu with two or three types of the supplier's wines. Craft brewers are also anxious to get their products into the hands of local consumers. They do this by joining forces with beverage operators interested in co-promoting products and events.

Selling is considered part of the internal marketing process because it is possible to create value, provide additional forms of communication, and strengthen the customer connection. Selling both products and services includes both tangible and intangible elements. Selling is more than "just" servicing the customer with goods and services—additionally, it includes managing "how" they feel in the process of buying. The element of service impacts the believability or skepticism of marketing efforts. Effective customer service will help assist in creating and strengthening the marketing strategy—the goal is getting consumers to desire the benefits of the establishment's products and services and ultimately build long-term relationships.

Suggestive selling refers to the practice of service staff offering recommendations on complementary or enhancement items and ideas to the customer. It is the practice of providing the consumers with options of items that may not be known are available. For instance, if a customer orders a Martini, the server might ask "Would you like that made with Tanqueray or Beefeaters gin?" Or if a customer orders a draft beer, the response from the bartender might be, "would you like the 12-oz. or 16-oz. size glass?" Suggestive selling will build sales as the customer may trade up on their order, or possibly purchase something additional. However, for suggestive selling to be effective, employees must be knowledgeable about the operation's products and services, and they must be willing to engage the customer. In the end, to make this type of internal marketing successful, managers must commit to the training and development of their employees.

Creating an Effective Beverage Program

Learning Objective 4
Provide insight into how creating an effective beverage program can positively influence the business

An *effective* beverage program should be designed with both the salespeople and the end user in mind—it should mirror the concept of the food and beverage establishment. "Effectiveness" is defined as one that offers a potential for revenue-generating ideas that maximizes profitability, offers value, and encourages repeat business. Building an effective beverage program is another form of a sales strategy, but one that is completely internally based—made with the intention of satisfying the captive, existing customer while they are visiting and experiencing the business in "real-time." An effective program is interrelated to the food and other product offerings of the respective business. The two approaches most significant to influencing an effective beverage program include (1) merchandising techniques and (2) quality service. These techniques can work simultaneously to generate revenue and build repeat business and ultimately be used to increase an operation.

Merchandising Techniques

There are several merchandising techniques that can be used to differentiate the business from its competitors and distinguish an appeal to its customer base. Ultimately, in the short term, the goal is to sell more beverages and increase an operation's revenue, and in the long-term, encourage repeat business for sustained growth. Having a wine-by-the-glass program, various bottle sizes, wine flights, wine dinners, and a trained and motivated service staff can work simultaneously to generate revenue and build repeat business.

Wine by the Glass (BTG)

Wines by the glass (also called house wines) remain very popular among wine drinkers and very profitable for restaurants. The BTG program will often consist of two price-quality tiers of wines by the glass.

- Entry-level price tier for those seeking value and sometimes referred to the antiquated term, "house" wine.
- Premium tier consisting of better-quality and higher-priced options.

The initial level provides a reasonably priced alternative to paying for and committing to a full bottle of wine, which may be five times the cost and quantity of an individual glass. The average profit on wines sold by the glass is relatively high, and this can help to offset the potential spoilage that can occur with serving wine by the glass. A modest BTG program will include, at a minimum, two white wine and two red wine selections, but, more commonly, a selection of six to eight wines. A large selection provides greater variety and adaptability when pairing with foods on the menu.

The standard portion size for wines BTG is approximately 5 oz. (but can also be 4 oz. or even 6 oz. depending upon the desired standard). A typical bottle contains 25.4 oz. of wine, which means that each bottle contains roughly five glasses of wine. With a 4-oz. pour, there are roughly six servings per bottle, and if the restaurant offers a 6-oz. pour, then there are roughly four servings per bottle. Glassware can significantly influence value perception. If a standard, 5-oz. pour is served in an oversized 15-oz. glass, it may not appear like a value to the guest—that same customer may perceive greater value if the identical 5-oz. pour was served in an 8-oz. sized wine glass. Knowing these kinds of quirks can lead to proactively managing the beverage program rather than reacting to the market.

Various Bottle Sizes

The standard wine bottle size is 750 ml, containing 25.4 ounces. Providing alternative bottle sizes can provide variety for the wine drinker. Gaining popularity is the half-bottle (equivalent to 375 ml/12.8 oz.). This half-bottle allows for adaptability by the server to suggest multiple half-bottles for diners at the same table who are eating varied dishes where a single wine wouldn't suffice. It also allows for single diners to purchase a bottle without the commitment or price tag of a standard sized version.

The split (187 ml/6.4 oz.) also known as *Quarter-Bottle* or *Piccolo* (Italian for *small*), is a tiny bottle mainly used for single-servings of sparkling wines. It allows the single-bottle portion to be sold to the consumer without any loss to the restaurant. They are often used on airlines and in the mini bar of a hotel.

The magnum (1.5 liter) sized bottle contains the equivalent of two 750 ml standard wine bottles and is considered the most popular of the large formats. The name originates from the Latin word *Magnus*, simply meaning great. Like in all large-format wine bottles, wine ages more slowly, generally retaining fresher aromas for a longer period as less oxygen enters the bottle through the cork, relative to the volume of wine. The magnum bottle size is certainly useful for larger parties, assuming they all desire to drink the same wine, but also ideal for older vintages of a given wine. Since the size of a wine bottle affects the wine's aging process and thus the cellaring potential, the magnum is superior for aged wines.

Flights

Wine, beer, and spirit flights are a sampling of three to four smaller portions of drinks that are selected with a theme or connection to each other. This provides variety, but it can also be a learning experience through making comparisons with the other beverages arranged in the flight. For example, a sampling of four Chardonnays from around the world, perhaps from Sonoma, California; Clare Valley, Australia; Cote d' Beaune, France; and Chablis, France may be offered in the wine flight. Other flight options might be: Exotic Whites, Wines of Eastern Europe, and Spicy Reds. Some creative approaches could be arranged based on the styles of Jazz musicians or the individual personality of the actors who have played the character James Bond.

Beverage Dinners

Beverage dinners are a technique used to showcase a winery, brewery, or distillery by offering a multicourse menu, each course being paired with a wine or beer from the producer. Any beverage dinner, regardless of promoting wine, beer, or spirits paired with a food operation's cuisine can act to promote and market the business establishment.

BYOB or "Bring Your Own Bottle"

In some establishments, customers can bring their own bottle(s) of alcohol into the establishment. Some customers do this because they have a special wine that the restaurant does not carry. But BYOB may also be offered because the restaurant's wine list may not have adequate options. Another reason for BYOB is if an operation lacks a liquor license that allows it to legally sell alcohol. Most restaurants charge the customer a nominal service fee, known as a *corkage fee*, to compensate for the lack of revenue resulting from allowing this type of service.

Quality Service

For quality service to occur, it must be well managed. Managers play a vital and integral role in the delivery of customer service. They establish and model the climate and service standards necessary to encourage service personnel to follow. Despite this critical managerial role, it is the service staff who ultimately delivers the standards and procedures as previously determined by management. At the point of each service encounter with the customer, it's the service provider who is in control of the guest's experience. Making an investment in wine, beer, and spirits coupled with staff training can give an establishment's beverage program a strong competitive edge and develop loyalty through satisfied clientele. Servers of beverage maintain the role of acting as a compass—providing guidance and navigating the customer to locate and select beverage options that would best enhance their individual buying experience.

To achieve and maintain a well-trained service staff, regularly scheduled training sessions should be conducted that not only address service expectations and responsibilities but also product knowledge. The significance of effective training can't be overstated enough. An effectively trained service staff can interpret customer needs, influence decisions, and create lasting impressions that can encourage positive word-of-mouth and foster repeat business. Improperly trained employees risk being viewed as sloppy or unprofessional, while the well-trained service staff with refined skills and depth of knowledge can convert a good beverage program into an excellent one!

The service staff carries significant influence when it comes to a consumer's beverage selections. According to a slightly dated, yet relevant 2010 report from Technomic (a leading food industry consulting and research company), "nearly one quarter (23 percent) of consumers in a recent online survey say they would consider ordering a beverage they had not tried before if the server recommended it." For heavier consumption users, it was acknowledged that 30 percent of consumers would select a server's recommendation. When servers assume a user-friendly selling approach, they can dramatically reduce the intimidation between the beverage menu and the consumers. Service should be approachable with a perspective of, there is no bad beverage; instead, everyone has the right to like what they like based on their culture, experiences, and likes/dislikes, and preference of certain aroma/flavors and structural components. Attitude and pretentiousness have minimal application in the service business and a well-mannered server should follow suit. It's baffling to these authors that in this day in age, the industry still suffers from "the lost art of service." What has happened to the "hospitality" in the Hospitality Industry?

Some customers may lack confidence in their own beverage knowledge, and in the absence of any guidance from the server, they may not order beverages at all or may not be completely satisfied with their own choices. It is imperative that service staff can read the guest and gauge their level of beverage sophistication and make appropriate suggestions to help satisfy their needs. When a server attempts to guide too much, the knowledgeable customer might be disappointed or insulted—and yet if servers provide limited guidance, the uninformed consumer might be too intimidated to order at all. Here is a simplistic three-step approach that can be applied attempting to sell beverages:

- Approach the customer and bring up the subject of beverages. "Have you had any time to look at the beverage menu?"
- Discover what the guests' beverage preferences are (if any) by asking questions. "What are you interested in or what do you normally drink?"
- Recommend a beverage that meets the guests' preference. "We have several wine selections that offer the crisp and youthful characteristics that you mentioned, how about our Mohua Sauvignon Blanc from New Zealand or our Alois Lageder Pinot Grigio from Northeast Italy?"

It is helpful for service staff to maintain several useful scripts consisting of open-ended questions or key phrases in addition to the application of simple user-friendly terminology that encourages engaging dialogue with the customer. Service staff should attempt to simplify the beverage-buying experience by making it less intimidating and therefore more easily understand customer preferences and ultimately satisfy their needs. It is always critical to listen carefully and *read the guest*, gauging the interest and knowledge level of the customer. Effective servers can adapt their service approach as per the customer's level of beverage sophistication and desired degree of dialogue. Sometimes, the guest would rather decide independently while other customers seek engagement of the knowledgeable server. The server's manner will greatly influence the perception of the beverage program and resulting sales and guest satisfaction. To start with, assume the customer has minimal beverage knowledge when first approaching them. It is recommended to maintain the following basic questions:

1. Are you looking for a beverage to start or one to carry throughout the entire meal?
2. What do you normally like to drink? Are you looking to stay in a comfort zone or be a little adventurous?
3. If wine:
   3a. Would you like red, rosé, or white? Proceed to identifying some of the wine stylings. For example: Oh! you would like a white wine. Do you prefer "crisp and youthful whites," "silky and smooth whites," or more "rich and voluptuous whites"?
4. If beer:
   4a. Would you like something lighter, medium, or more full bodied? Proceed to identifying some beer styles. For example: Do you prefer "Saisons and Belgium Ales," "Pale Ales and IPAs," or "Darker Ales"?
5. If spirits:
   5a. Do you have a cocktail in mind? Or would you like a classic cocktail or one of our own unique signature cocktails?
6. Is there a price range in mind? This question should only be asked when the customer has made some reference to price, or for finding "value" options off the menu. Service staff should always be respectful with this scenario as to not bring about unwanted attention or embarrassment to the guest and/or their dining/drinking companions.
7. Were you looking for a wine to specifically enhance the dishes that you will be ordering or one to enjoy regardless of the food?

Pictured in Figure 16.3 is a young woman who displays satisfaction with her wine selection—the goal of the service staff.

FIGURE 16.3
Enjoying wine. Courtesy of Erika Cespedes.

The Beverage Menu

Learning Objective 5
Explain how the beverage menu is more than just a list of beverage items

The beverage menu is the driving force used to produce revenue, maximize profit, and generate repeat business. Just like a food menu, it is the controlling document that acts as a marketing tool to inform the customers about the beverages options. The menu emphasizes the drinks that are an integral part of the experience and therefore should complement the cuisine while harmonizing and reinforcing the identity of the overall restaurant/bar concept. The ultimate goals of the beverage menu are to:

- Advertise and promote wines (and other beverages) that enhance the food and overall dining and drinking experience
- Generate revenue and repeat business

The format and presentation of the menu should complement not only the food menu, but also the décor and atmosphere of the establishment.

Consistent consumer studies identify that user-friendly beverage menus are recommended; most guests do not want to spend significant time navigating a menu that resembles a stamp or coin collection. Keeping this in mind, each beverage establishment has different needs and preferences; there is no one best type of list because each concept is unique. Therefore, beverage menus are often varied in presentation, content, and even purpose. All too often, managers confuse the beverage menu as a form of a self-expression and lose sight of the needs of their customer. Gerald Asher, the respected English wine personality, makes an excellent point by mentioning that, "A [restaurant] wine list is praised and given awards for reasons that have little to do with its real purpose, as if it existed only to be admired passively, like a stamp collection. A wine list is good only when it functions well in tandem with a menu." Instead, assembling a thoughtful and appealing list with limited choices can provide less intimidation for the consumer and service staff. The goal of any menu is to sell the items listed on them.

The effective beverage menu begins with a strong selection of beverage options that make sense for the type of business, corresponding food menu, intended target market, and location. It is critical to remember that having great beverage options doesn't ensure success—the selection must be guided by a knowledgeable staff. The menu acts a map, or guide for the service staff to promote and sell. The "quality" of a good wine list used to be measured in weight and size of the list, but there is a growing number of restaurants and wine bars that are presenting a more thoughtful, smaller beverage menu that still offers an outstanding selection, just less breadth than in the past. Smaller lists can offer more depth in a specialized area or can simply be more conducive given the smaller size of storage space available. These beverage lists can also require less financial value tied up in inventory and encourage ease of navigation of the list for both servers and customers. Additionally, these smaller menus may be more easily revised on a daily, weekly, or monthly basis to keep offerings new and exciting for the customer. Since the millennial generation are savvier about wine and beer than previous generations at the same age, having a smaller list that changes more frequently allows them to explore and encourage repeat business.

One of the first steps in building an effective beverage menu is in developing the product mix or composition (types and styles of wine, beer, spirits, and other beverages) that complements the cuisine or is suitable for the overall concept. As a general guide, seafood restaurants will emphasize white wines, and steakhouses will emphasize reds wines. But of course, restaurants and cuisines in modern day are not always so simple. In the case of a beverage driven concept, there is a beverage that stands out as the driver, this naturally dictates the composition of the menu. Even if a single beverage tends to be the driver, the beverage manager still must decide on the breadth and depth of the menu. Below are some variables to consider when planning a menu.

- Breadth is a having a wide range of alternatives or scope of wines (or other beverages) being offered. It means building a menu that has wines represented from several different wine-producing countries around the world.
- Depth is complete in detail and dimension; thoroughness. This means offering a large selection of wines from several producers and vintages from a given

wine-producing country. Depth may also encompass the idea of offering several different expressions of a grape variety or style of wine. How many selections are appropriate?

The Grey Plume Beverage Menu: An Illustrative Example

The Grey Plume restaurant seeks to inspire and elevate the way Omaha thinks about food through culinary excellence, the promotion of local foods and growers, and a commitment to community.

—Chef/Owner, Clayton Chapman

Learning Objective 6
View the sample beverage menu and recognize its composition, breadth, and depth

The Grey Plume restaurant, located in Omaha, Nebraska, opened its doors in December of 2010. Figure 16.4 is the front of the Grey Plume. The chef/owner has since been nominated for a James Beard award and the restaurant was awarded the "Greenest Restaurant" in 2010 by the *Green Restaurant Association.* The chef/owner Clayton Chapman and managing partner Michael Howe have integrated their vision and mission throughout every element of the restaurant's construction and design, including furnishing and building materials; water and energy; recycling; pollution and chemical reduction; and disposables. In addition, the beverage program was created with the cuisine and the "green" concept in mind.

The Grey Plume restaurant is an illustration of how a restaurant concept can parallel its beverage menu. The wine list has since been nationally recognized by *Bon Appétit* magazine. David Lynch, a James Beard Award–winning Sommelier and author of *Vino Italiano: The Regional Wines of Italy*, recognized the Grey Plume wine list as one of his favorite top five condensed wine lists. The beverage program was initially constructed by Beverage Director, John Peter Laloganes.

The beverage menu was created to consist of roughly 50/50 of varietal-based labeling (common in the New World) versus geographical labeling (as common in the Old World). The average consumer is comfortable with understanding wines that employ varietal labeling, like buying milk, juice, or any other product that clearly identifies its contents on the front of the packaging. Geographical labeling tends to be a bit of an intimidation factor for many consumers. Geographical labeling most often does NOT indicate varietal identification; therefore, it becomes more of an educated hand-sold product. Most wine labels (and therefore wine lists) speak of grape varietals and geographical location but often this is confusing and, ultimately, customers care more about a wine's aromas/flavor components and/or structural components, which cumulatively form a certain "wine style."

FIGURE 16.4
Front of Grey Plume. Courtesy of Grey Plume.

Therefore, to reduce intimidation of the Grey Plume customer and gain greater clarification and communication (and educate in the process), wines are classified per structural components or broad "style" categories, rather than purely by their grape variety or geographical origin. This approach provides wine drinkers with a sensible template to allow them to easily peruse a modest-sized wine list and identify a selection—this style of menu also acts to become a helpful training, communication, and selling tool for the service staff.

The list begins with sparkling wine, progressing to white, rosé/red wine, and then beer, and finally dessert wine options. The wine styling approach divides the white wines into three structural categories and rosé/red wines into three structural categories as well. Within each broad style category, varietal-based wines and geographical-based wines will be clearly recognized using bold and italicized font style to make it easier for the guests to scan the type of wine they may be searching for. The wine list includes a modest selection of bottle options with numerous wines by the glass. Each wine by the glass consists of a 5.5-oz. pour with a price range of $8.25–$13.95 per selection. All wines by the glass were selected as having the ability to provide a good margin, having a few "green" options, showcasing different style categories, and providing a selling price that ultimately isn't met with shock or dismay by the potential buyer.

Customers pay a bit more for a wine by the glass as compared to buying the same quantity of wine through purchasing a bottle, therefore encouraging a slight price break to encourage bottle purchases. Each bottle contains 25.4 oz. of wine; therefore, a bottle contains roughly four portions (25.4 oz./5.5 oz. = 4.6 portions, or roughly 4+ glasses per bottle).

Identified below is an example of presenting wine and other beverage options in an unpretentious approach that encourages repeat buying in future visits. The menu was not only intended to create a user-friendly approach for the customer experience but was created to coordinate and parallel the food menu and overall feel of the restaurant concept.

BUBBLES

revitalizing ... lively ... festive

Anselmi, **Prosecco,** *Extra Brut, Friuli, Italy ... 30/8.25*

Llopart "Leopardi" **Cava Rosé,** *Penedès, Spain ... 45*

Argyle, **Sparkling Brut,** *2003, Oregon ... 64*

Graham Beck, **Cap Classique,** *South Africa ... 38/10.95*

Ruinart **Blanc de Blanc** *Brut, Champagne (½ bottle) ... 150*

Bollinger, **Champagne,** *Brut, Special Cuveé ... 79*

Brutell, **Franciacorta,** *Lombardia, Italy ... 62*

Duval-Leroy, **Champagne,** *Brut (½ bottle) ... 42*

Gruet, **Blanc de Noir,** *New Mexico (½ bottle) ... 26*

CRISP AND YOUTHFUL WHITES

zesty … clean … vibrant

Saint Clair Family Estate, **Sauvignon Blanc,** *Marlborough New Zealand, 2014 … 43*

Weingut Fred Loimer, "Lois" **Grüner-Veltliner,** *Kamptal, Austria, 2012 … 38*

Domaine Des Buissonnes, **Sancerre,** *Loire Valley, France, 2012 … 54*

Domaine Gerard Tremblay, **Chablis,** *"1er Cru Fourchaume" Burgundy, France, 2011 … 78*

Pewsey Vale, **Riesling,** *Eden Valley, Australia, 2015 … 38/10.5*

Venica & Venica, **Pinot Grigio,** *"Collio" Friuli-Venezia-Giulia, Italy, 2014 … 45*

SILKY AND SMOOTH WHITES

refreshing … bright … velvety

Foris Vineyards, **Pinot Blanc,** *Rogue Valley, Oregon, 2014 … 35/9.95*

Sokol Blosser, **Pinot Gris,** *Oregon, 2015 … 49*

Weingut Leitz, **Riesling,** *Spatlese "Rüdesheim," Rheingau, Germany, 2011 … 47*

Bodegas La Cana, **Albariño,** *Rias Biaxes, Spain, 2014 … 38/10.5*

RICH AND VOLUPTIOUS WHITES

lavish … elegant … voluptuous

Laetitia, **Viognier** *and* **Grenache Blanc,** *"Nadia" Santa Barbara Highlands Vineyards, 2014 … 36*

Alma Rosa, **Chardonnay,** *Santa Barbara County, California, 2012 … 44/11.95*

Bodegas Muga, **White Rioja,** *Spain, 2008 … 46*

Bouchard Aîné & Fils, **Meursault,** *Burgundy, France, 2011 … 120*

L'Ecole, **Semillion** *and* **Sauvignon Blanc,** *Columbia Valley, Washington, 2013 … 36*

FRUITY ROSÉ AND VIBRANT REDS

youthful … lively … charming

Alois Lageder, **Lagrein,** *Rosé, Trentino-Alto-Adige, Italy, 2012 … 36/8.95*

Bergström, **Pinot Noir,** *"Cumberland Reserve," Willamette Valley, Oregon, 2012 (½ bottle) … 63*

Paolo Scavino, **Dolcetto** *d' Alba, Piemonte, Italy, 2014 … 44*

Domain Dichon, **Moulin-a-Vent,** *Beaujolais, France, 2013 … 49*

Albert Bichot, **Mercurey,** *Burgundy, France, 2011 … 48/13.95*

Hayman & Hill, **Pinot Noir,** *Santa Lucia Highlands, California, 2013 … 43/10.95*

MELLOW AND COMPLEX REDS

rich ... smooth ... velvety

La Posta, **Malbec,** *"Pizzella Family Vineyard," Mendoza, Argentina, 2013 ... 38/10.95*

Château Gonin, **Bordeaux** *Superiore, France, 2013 ... 33*

Chandon, **Pinot Meunier,** *Napa Valley, California, 2014 ... 69*

Domaine Lamarche, **Vosne Romanee,** *Burgundy, France, 2011 ... 149*

Ransom, **Pinot Noir,** *Willamette Valley, Oregon, 2013 ... 59*

Charles Joguet, **Chinon,** *Loire Valley, France, 2013 ... 52*

Substance, **Syrah,** *Columbia Valley, Washington State, 2014 ... 38/10.95*

Mazzi, **Valpolicella Classico Superiore,** *Veneto, Italy, 2012 ... 39*

Bodegas Muga, **Rioja,** *Reserva "Unfiltered," Spain, 2008 ... 59*

BOLD AND INTENSE REDS

complex ... concentrated ... evolved

Whitehall Lane, **Merlot,** *Napa Valley, California, 2010 (½ bottle) ... 39*

Qupé, **Syrah,** *"Bien Nacido Vineyard," Central Coast, California, 2012 ... 48*

Ladera, **Cabernet Sauvignon,** *Napa Valley, California, 2011 ... 63*

Ridge Vineyards, **Zinfandel,** *"East Bench," Sonoma Valley, California, 2012 ... 42/12.95*

Millbrandt Vineyards, **Cabernet Sauvignon,** *Columbia Valley, Washington State, 2012 ... 34/8.95*

BEER

frothy ... lively ... satisfying

Nebraska Brewing, **Pale Ale** *Omaha, Nebraska (Draft) ... 5*

Duvel, **Belgian Ale** *Belgium ... 7*

Saison Dupont **Farmhouse Ale,** *Belgium (750 ml) ... 15*

Omegang **Witte Ale,** *Cooperstown, New York (750 ml) ... 13*

Hoegaarden **Wit** *Belgium ... 5*

Mc Chouffe **India Pale Ale** *Belgium ... 8*

Orval **Trappist Ale** *Belgium ... 9*

Goose Island, Matilda, **Belgian Style** *Chicago, Illinois ... 6*

Anchor, Old Foghorn **Barley Wine** *San Francisco, California ... 6*

Left Hand Brewing **Milk Stout** *Longmont, Colorado ... 5*

SWEETS

Seductive ... Rich ... Satisfying

Yalumba **Late Harvest Viognier,** *Australia, 2005*

Kopke **Ruby Port** *Portugal*

Kopke **20 Year Aged Tawny** *Portugal*

Vinhos Barbeito—The Rare Wine Company "Historic Series—New York" **Malmsey Madeira**

Vinchio-Vaglio Serra, **Brachetto,** *Piedmonte, Italy, 2009*

A Tale of Two Beverage Establishments

Learning Objective 7
Analyze the two beverage concepts illustrated and identify their appropriate marketing strategies

The methods used for marketing and selling beverage products are largely coordinated with the type of establishment. Consider two different beverage concepts: *The Library* and *The Office*. These establishments are in the same town. The Library is located next to a large university, whereas The Office is in a downtown business district. Identified in the following are the demographics and psychographics of the two establishments' target markets:

Name of Concept	The Library	The Office
Age	Under 25	Over 25
Marital status	Single	Mostly single, but some married
Income	Limited income	Fair to high disposable levels of income
Employment	College students or recent graduates	Professionals
Motivation/intent	"Looking to party"	"Looking to unwind"

The Library's target market will obviously be quite different from that of The Office. The Library should expect to draw most of its customers from the university, whereas The Office draws its customers from the downtown business area.

Referring to the customer demographic information, consider the following questions:

1. What types of drinks would *The Library* sell?
2. What types of drinks would *The Office* sell?

The Library would probably sell very little wine, but a good deal of mass-produced draft beer such as *Budweiser*, *Miller*, and *Coors*. They would also most likely sell mixed drinks with suggestive names such as *Sex on the Beach* or *Slow Comfortable Screw*. The Library food menu might include traditional American pub grub such as hot wings, potato skins, or personal pan pizzas.

The Office, on the other hand, would sell more wine and craft beers, such as *Sam Adams*, *Anchor Brewing*, *Dog Fish Head*, *Flying Dog, and Sierra Nevada*. In addition, The Office would probably sell classic mixed drinks such as *Martinis, Manhattans*, or *Whiskey Sours*. The Office food menu might include American bistro cuisine such as light pasta dishes, soup, and sandwich combinations, or grilled pizzas.

The Library and The Office also might choose different strategies to attract their respective customers during different holidays. The Library and The Office may emphasize different days to attract their customers, and if they did celebrate the same holiday, they would probably celebrate in different fashions. For example, each one of these establishments most likely would celebrate or acknowledge Halloween, but they would do it very differently. The Library might have a live band and a "wild" costume party. The Office might have something less outrageous, but just as enticing to its customers.

Other questions that may prove interesting: What kind of entertainment can be found at each establishment? What kind of dress code might be expected? What might the décor consist of? What type of service? What kind of uniforms might the employees be wearing?

Check Your Knowledge

Directions: Use these questions to test your knowledge and understanding of the concepts presented in the chapter.

I. MULTIPLE CHOICE: Select the best possible answer from the options available.

1. According to the National Restaurant Association (NRA) estimate, anywhere from _________ of an establishment's revenue is based on repeat business.
   a. 20–40 percent
   b. 40–60 percent
   c. 60–80 percent
   d. 80–100 percent
2. Social authority is developed when an individual or organization establishes themselves as a/an _____________ in their given field or area and can garner a following and consequently strengthen a brand.
   a. novice
   b. connoisseur
   c. authority
   d. all of the above
3. Marketing is an activity or process that is directed toward:
   a. communicating and promoting a message and building a long-term relationship with a target market
   b. intending to persuade a potential customer to frequent and purchase food and beverage products
   c. promoting a new product
   d. none of the above
4. The *communication strategy* attempts to
   a. focus on sending the appropriate messages through the correct medium to the intended target market
   b. focus on how a business will enter a market place and become differentiated amongst its competition
   c. focus on how the actual products and services will be sold on a day-to-day basis
   d. focus on building the business both through internal and external marketing and advertising
5. Marketing is divided into two categories:
   a. internal marketing and external marketing
   b. customer service and suggestive selling
   c. advertising and promotions
   d. verbal and non-verbal

6. Marketing to a *specific* "target market" involves analyzing and grouping customers and is best described as
   a. categorizing them into "like" characteristics and profiling them into segments
   b. categorizing them by age and gender
   c. categorizing them by hobbies and interests
   d. categorizing them by income and religion
7. External marketing is a proactive form of a sale strategy that refers to activities undertaken to influence sales, and the overall long-term success of a given business such as
   a. radio or print advertising
   b. use of billboards or coupons
   c. providing effective customer service
   d. suggestive selling
   e. both a and b
   f. both c and d
8. Suggestive selling refers to the practice of service staff
   a. offering recommendations on complementary items
   b. providing ideas of enhancement products
   c. providing awareness of options the customer may not know are offered
   d. all of the above

II. TRUE/FALSE Circle the best answer.

9. True/False Advertising is intended to emphasize the significance and impact of the customer or long-term relationship to the business.
10. True/False Marketing is a form of communication intended to persuade a potential customer to frequent and purchase food and beverage products.
11. True/False The beverage menu not only lists the items available to sell, but it also acts as a marketing tool.
12. True/False Beverage menus should be created in a manner that features the sommeliers' favorite wines without regard to the needs of the customer.
13. True/False Internal marketing utilizes flyers and posters to advertise events within the establishment.
14. True/False Demographic data is associated with consumers' hobbies, interests, and overall lifestyle choices.

III. DISCUSSION QUESTIONS

15. How is "word of mouth" the new and yet ancient form of advertising and marketing?
16. Referencing the section "A Tale of Two Beverage Establishments" and the concept of "holidays" in this chapter, create a marketing plan for The Library and for The Office. Compare and contrast what the two establishments might do for the holidays.
17. Provide some examples of internal versus external marketing?
18. Identify some important aspects to consider when creating a marketing strategy for a new food and beverage establishment?
19. Describe the difference between composition, breadth, and depth of a beverage menu.
20. Why is it important to identify an intended target market(s) when preparing a marketing strategy?

IV. ACTIVITY

Based on the chapter discussion of the both "The Library" and "The Office" beverage establishments, complete the chart below utilizing the information discussed and learned throughout this textbook.

Month	Name of Holiday	The Library's Selling Strategies	The Office's Selling Strategies
January	New Year's Day		
	The Super Bowl		
February	Valentine's Day		
	Mardi Gras (Fat Tuesday)		
	American Wine Appreciation Month		
March	St. Patrick's Day		
April	April Fool's Day		
	Easter		
	Passover		
	National Secretary Day		
May	May Day		
	Kentucky Derby or The Triple Crown		
	Memorial Day		
	Mother's Day		
	Cinco de Mayo		
June	Father's Day		
	Flag Day		
July	Independence Day		
August	Football tailgating parties		
September	Labor Day		
October	Oktoberfest		
	World Series		
	Halloween		
November	Election Day		
	Thanksgiving		
	Beaujolais Nouveau		
December	Christmas Season		
	New Year's Eve		

APPENDIX A

The Science of Fermentation

. . . to conduct [drinking] in the most rational and agreeable manner is one of the great arts of living.

—James Boswell, 1775

The natural process of fermentation precedes human history. The earliest evidence of fermented beverage production dates back by about eight thousand years, but it's highly likely the consumption of fermented food and beverage items was happening much earlier. Fruits such as apples or pears, or flowers with honey could have easily been exposed to airborne yeast and began a fermentation process. The simple sugars present in these food items could be easily converted to alcohol. Ever since ancient times, humans have been attempting to understand and control the fermentation process. However, it was not until Louis Pasteur (1822–1895), a French chemist and biologist, made significant contributions to chemistry, medicine, and indirectly the food and beverage industry, and has subsequently greatly benefited civilization.

In 1849, Pasteur began studying fermentation, a chemical process that breaks down organic materials. For example, fermentation occurs when yeast breaks down and converts a sugar source into ethyl alcohol and carbon dioxide gas. Fermentation occurs in both beer and wine, as well as in the base of a distilled spirit. Fermentation is essential in many aspects of the food and beverage industry—the production of bread, cheese, and yogurt relies on the chemical conversion of fermentation.

The sugars that are present in wine grapes are stored in the pulp along with water, acids, and flavor compounds. Once the grapes are pressed, the sugars of glucose and fructose are the main fermentable sugars made desirable for the yeast. Each type of sugar exists in approximately equal proportions in wine, although fructose is roughly twice as sweet as glucose. Glucose is also fermented at a faster rate, which means that a wine fermented to "dryness" will have less residual glucose than fructose. During alcoholic fermentation, yeast feeds on the sugar found in grape juice and converts it to ethanol and carbon dioxide. While not all sugars are completely digested by the yeast, there is such a small quantity remaining in a beverage that it is usually not perceptible when being tasted.

The simple sugars—glucose and fructose—are both made from three types of molecules: carbon (C), hydrogen (H), and oxygen (O). The way these molecules are arranged determines the form of the sugar, much like ingredients in a recipe. Scientists write their "recipes" in a string of letters and numbers. The numbers serve two purposes: to show how many of the molecules are present. Take for instance the chemical formula for water: H_2O. When a letter is prefaced by a small number, such as the "2" in H_2O, it represents the presence of two hydrogen molecules. In other cases, when there is no number following a letter, such as the "O" in H_2O, it represents that there is only one part molecule. In water, therefore, there are two hydrogen molecules and one molecule of oxygen.

In another example—glucose is scientifically written as $C_6H_{12}O_6$. This formula representing glucose is composed of "6" carbon atoms, "12" hydrogen atoms, and "6" oxygen atoms linked together to form a chemical agent. When yeast is added to the simple sugar (glucose, fructose, maltose, or sucrose), it consumes the sugar. When the living yeast is finished consuming the simple sugar, it expels the digested sugar in a different form. After the sugar is digested, the individual molecules "unravel" and reconfigure themselves into a new set. This change allows the yeast to expel two different forms of waste: alcohol (C_2H_5OH) and carbon dioxide (CO_2).

To summarize, when yeast is added to a glucose molecule, it eats the sugar and changes the configuration of the molecules. The yeast's waste is reorganized into molecules of carbon dioxide and alcohol. In scientific shorthand, scientists write this process down so that it looks like a mathematical equation:

	Fermentation
Scientific formula:	Yeast + $C_6H_{12}0_6$ = C_2H_5OH + $2CO_2$ + Energy
Laymen formula:	Yeast + Sugar = Alcohol, Carbon Dioxide, and Energy

Notice how the chemical reaction does not alter how many molecules of carbon, hydrogen, or oxygen are involved; the process cannot add or subtract new molecules, but they are arranged differently on each side of the equation. When this exchange takes place, energy is expelled, as shown in the equation.

The amount of alcohol from fermentation depends on the amount of sugar available for the yeast to consume. If there is limited sugar (or food for the yeast to consume), there won't be much alcohol expelled by the yeast. If the yeast is mixed with a liquid that has a high sugar content, however, the yeast will continue to eat the sugar and produce alcohol until the alcohol level reaches roughly 15 percent of the total volume. The yeast recognizes a somewhat "inhospitable environment" and stops functioning. At this point, because the alcohol is relatively high, it creates a toxic environment for the yeast; the alcohol reaches an incapacitating level, the yeast is overwhelmed, and alcohol production ceases.

APPENDIX B

Alcohol Safety and Liability

Pleasure which must be enjoyed at the expense of another's pain, can never be enjoyed by a worthy mind. Pleasure's couch is virtues grave.

—Augustine J. Duganne

Since the beginning of civilization, society has been trying to protect itself from abuses brought on by overindulgences of all kinds, including alcoholic beverages. Back in ancient Babylon, a city-state of early Mesopotamia and modern day Iraq, laws such as "Hammurabi's code" were designed to restrict possible violence and damage potentially caused by alcohol abuse. In moderate doses, ethanol (the type of alcohol found in wine, beer, and spirits) has beneficial effects—but in large amounts, it is toxic and can be fatal. It's true that alcohol relaxes inhibitions and allows for a bit of social lubrication, but it also impairs judgment, slows reaction time, and diminishes motor coordination. With the appropriate training, attentive service staff can observe and identify the behavioral signs of consumers who overindulge in the consumption of alcohol. The beverage manager who offers the sale of alcohol faces many challenges regarding this activity.

The beverage manager is continually balancing the needs of the establishment—to produce revenue—with the need to limit the establishment from unnecessary risk. Serving alcohol goes hand in hand with being responsible, making intelligent and appropriate decisions, and showing reasonable care for the safety and welfare of the customer and the public at large. The beverage manager's responsibility is to protect the reputation of the establishment and to limit personal and organizational liability by reducing the number of injuries and deaths associated with inappropriately serving alcohol beverages. Given the highly volatile nature of selling these types of products, whenever situations are in question, management should always error toward safety.

Alcohol Legislation and External Forces

Under the 21st Amendment to the U.S. Constitution (ratified on December 5, 1933), each state has their independent right to define the scope and control the sale of alcoholic beverages. Each state can choose to classify themselves according to a *licensed state* or a *control state*. Licensed states allow for the distribution of alcohol through licensed third-party companies, while control states distribute part or all of the alcoholic beverages through state-operated liquor stores. Either designation requires an operator who conducts the sale and/or consumption of alcohol on premise of any establishment to obtain a liquor license awarded via the state and/or local governing municipality. A liquor license, once granted, may be revoked or suspended by the state if the licensee violates prescribed laws. In general, there are two primary classifications of establishments that serve alcoholic beverages. The first includes full-service restaurants, in which beer, wine, and spirits are served primarily as an accompaniment to food and equates to less than 50 percent total revenue of the establishment. The second includes establishments such as bars, taverns, and nightclubs in which beverages are sold as the primary offering and equate to 50 percent or more of the total revenue of the establishment.

The alcohol beverage control boards operating in each city and county generally have specific powers and responsibilities—although the final control over alcohol sales usually rests with the state liquor authority or the Alcohol Beverage Commission (ABC). Matters regulated by some or all states include: licensing, illegal sales, hours of operation, dram shop liability, and alcohol service training.

MADD (originally called "Mothers Against Drunk Driving") is an organization that began in 1980, from a handful of mothers with a mission to stop drunk driving. MADD has evolved into one of the most widely influential nonprofit organizations in America as they have strived to help save thousands of lives. This organization has helped steer legislation as well as largely shape how society views drunk driving. As a result, over the last thirty years, a large amount of traffic safety and victims' rights legislation has been passed. According to MADD, "annual alcohol-related traffic fatalities have dropped from an estimated 30,000 in 1980 to fewer than 17,000 today." This is a commendable effort based on persistence and pure passion for the desire to ensure the safety of America's roadways.

Many state and local law enforcement officers support increased enforcement during high-risk holidays including Labor Day and New Year's Eve. In order to help remove drunk drivers off the road and promote public safety, law enforcement takes an assertive approach with sobriety checkpoints. Law enforcement officials set up specific check points on the roadway to evaluate drivers for signs of alcohol or drug impairment. A primary goal of sobriety checkpoints is to deter people from committing *driving under the influence* (DUI). According to MADD, this technique reduces fatalities by 20 percent on such holidays.

This appendix should NOT be thought of as any form of legal advice; instead, it applies a general overview and approximate guidelines regarding the patchwork of laws that exist in the United States. It's always expected that each manager researches and comprehends the laws associated with their particular establishment within its particular legal jurisdiction.

Alcohol Regulations and Liability

The serving, sale, and transportation of alcohol in the United States is more complex than other goods and services. Alcohol beverages do not enjoy the same treatment under the free trade provisions of the U.S. Constitution. Instead, alcoholic beverages are governed by the 21st Amendment (the act that repealed Prohibition, established under the 18th Amendment), which yields each individual state the jurisdiction to interpret and regulate the sale and distribution of alcoholic beverages. Overall, since state laws vary, the United States has in effect fifty states with their own distinct patchwork of laws.

In addition to the inconsistent alcohol-related laws across the country, it's also common for different counties inside of single states to have variations of laws. Counties can opt to be *dry*—signifying the ban of selling alcohol beverages or they can be *wet*—making the sale legal. If a county chooses to remain "dry," however, this does not mean that a resident of that county cannot drive to purchase alcohol in another nearby county. Some states have what are called *moist* counties—these counties may allow the sale of beer, but not the sale of wines or spirits. And sometimes these sales may only be limited to restaurants.

Driving Under the Influence

Laws regarding the sale and consumption of alcohol are passed at the state level. In all fifty states, a person is considered intoxicated if their *blood alcohol content* (*BAC*) is at .08 or above. Even though each state agrees on BAC levels, the laws regarding alcohol consumption in the United States are not uniform. Each state has different caveats.

For example, some states have what's called an "anti-plea bargaining" statute. According to this law, someone who is caught driving with a BAC above a certain level cannot plea-bargain, to reduce the offense to a non-alcohol-related offense. Again, the blood alcohol level for this law to take effect varies from state to state. The terms *driving under the influence* (DUI) or *driving while intoxicated* (DWI) are synonymous terms that represent an illegal act of operating, or in some jurisdictions merely being in physical control of, a motor vehicle while being under the influence of alcohol and/or other drugs.

Child Endangerment

Related to drinking and driving laws, an additional law related to "child endangerment" creates a separate offense that enhances an existing *driving under the influence* (DUI) penalty. This law may apply to any offender under the influence of alcohol with a minor present in the car. The specifics of these laws vary from state to state, but thirty-nine states have such laws.

Selling and Serving Restrictions

The selling and serving of alcohol beverages are a very lucrative activity for many restaurants, bars, and retail stores. However, the caveat is the risk of liability involved with such activity. It becomes vital that appropriate alcohol service to be executed with safety and concern versus the short-term temptation of profit. Two main areas of alcohol liability are the minimum legal drinking age and drivers who operate motor vehicles while under the influence of alcohol. Astute management creates pre-established standards and procedures for responsible alcohol service and ensures expectations are conveyed through training the front-of-house staff. Furthermore, the service staff needs to be empowered with the appropriate tools and management support in order to identify and ensure customers are of the appropriate age and sound judgment when entering and leaving the establishment when alcohol is involved. In most states, the following types of sales are considered illegal:

- The sale of alcohol to minors (under age twenty-one)
- The sale of alcohol to visibly intoxicated persons
- The sale of alcohol to habitual drunkards

Sales to anyone on this prohibited list may result in the suspension or revocation of a liquor license. Illegal sales can also lead to civil liability for resulting injuries and, particularly in the case of serving a minor, criminal liability for which penalties could include jail and a fine. The prudent foodservice operator will take care to avoid illegal sales.

Selling to Minors

In order to purchase, sell, serve, and consume alcohol, one has to be of certain age. It is unlawful to sell, serve, deliver, or give alcoholic beverages to any person(s) under twenty-one years of age or to an already intoxicated person. Restaurants and bars must carefully check identification cards, such as driver's licenses, in order to verify that every patron is of legal drinking age. Commonly, if individuals don't look the appropriate age, (or even at least thirty or forty years old) the service staff is often (rightfully so) instructed to check identification for verification purposes. Sometimes customers may feel offended by being asked for identification; however, it is much better to err on the side of caution and risk an unhappy customer rather than violate the law and risk a fine of suspension/revocation of your alcohol license.

Most states allow any three primary forms of acceptable identification (assuming they appear legitimate) that are issued by a federal, state, county, or municipal government agency as an acceptable means but are not limited to: a driver's license, passport, and/or military ID. The penalties for serving or selling to minors for first offense could be classified as a "Class A misdemeanor" punishable by $500–$2,500 fine and/or less than one-year imprisonment. Second and subsequent offenses could be punishable by $2,000–$2,500 fine and/or less than one-year imprisonment. If death occurs as result of violation, the infraction could be considered a felony punishable by up to $25,000 fine and/or one–three years imprisonment. A person under the age of twenty-one is prohibited from possessing alcoholic beverages on the street, highway, or any setting open to the public. To prevent sales and/or service of alcoholic beverages to individuals under the age of twenty-one, the licensee, its agent, or an employee has the right to refuse to sell or serve alcoholic beverages to anyone unable to produce adequate documentable proof of identity and age from one of the aforementioned forms of identification. Liquor licenses may be revoked and criminal sanctions imposed on licensees and individuals for violating provisions of the law.

Selling to an Intoxicated Person

It is unlawful to sell, give, or deliver alcohol to an intoxicated person. Violation of this provision is a Class A misdemeanor. The penalty for this infraction is often a minimum $500 fine (maximum $2,500) and/or jail sentence of up to one year. An establishment can also be fined or have its liquor license suspended or revoked as a result of not enforcing the appropriate legalities. Violation of this provision is a Class B misdemeanor. In some cases, such as a banquet event or open tab at a wedding, there is an open bar. The concept of an open bar is often misunderstood from a customer's perspective as they often believe it means an unlimited consumption of alcohol. Instead, it simply means that the guest is not responsible for paying for the alcohol they consume.

Happy Hour Laws

Happy hours have been banned in approximately half of all states across the country. These laws limit restaurants and bars from offering reduced price or multiple drink sales during a designated time period. More specifically, many state "happy hour" laws prohibit the licensed establishments from serving more than one drink at a time (except during product sampling), offer reduced price drinks for certain time, serve unlimited drinks for a fixed price (except at private functions), or increase the amount of alcohol or size of drink without proportionately increasing the price. These restrictions came about over the last thirty years as the legislation began implementing laws in order to reduce binge drinking. Without these laws in place, the practice of binge drinking during short time periods could create a greater risk of drinking and driving.

Hours of Operation

The hours during which a food and/or beverage establishment can serve alcoholic drinks are strictly regulated by the state, the city, and/or the county in which the establishment is located. The days on which alcoholic beverages can be sold—both in foodservice establishments and in retail outlets are also commonly regulated. Some locations prohibit Sunday sales, while other areas allow by-the-drink sales only after 12:00 p.m. Some areas allow only the sale of beer on Sundays, while other areas allow the sale of all alcoholic beverages. In many jurisdictions, alcohol cannot be sold until after 6:00 p.m. on local and national political election days. Clearly, the prudent operator must rely on state and local authorities when determining on what days and at what times it is legal to sell alcoholic beverages.

Dram Shop and Common Law Liability

As previously mentioned, every state has strict laws forbidding intoxicated people to drive motorized vehicles. Many states have now developed third-party liability legislation (or commonly called *dram shop laws*) that holds the beverage manager responsible, under certain conditions, for the actions of his or her intoxicated patrons. Most of these state laws are lengthy and complex. The "dram," or drink shop laws have become more rigorously enforced over the past twenty years. Food and beverage establishments can be held partially or fully responsible for the effects and damage of anyone who was harmed from the overconsumption of an intoxicated individual. There are no cut-and-dry situations in the eyes of the law. However, if an establishment has shown negligence in serving alcohol, it and the individuals who served the drinks can be subject to severe legal and civil penalties. These can range from fines of hundreds and thousands of dollars and jail time to lawsuits for thousands to millions of dollars. In many states, there are financial limits to the damages a jury may award an injured party, though in some states there are no limits.

The objectives of dram shop laws are to discourage owner/operators from selling alcohol illegally and to afford some kind of compensation to those victims whose injuries are a result of an unlawful sale of alcohol. The potential liability is very significant. Some illegal sales have resulted in verdicts that have financially ruined the bar or the restaurant that wrongfully served the alcohol. Because of these laws, operators are becoming increasingly concerned with alcohol awareness and abuse.

A person who is injured by the acts of an intoxicated individual may also have the right to bring a lawsuit based on the *common law* theory of negligence. In some states, such lawsuits may be filed against the operation that made the illegal sale, independent of any claim under a state's dram shop laws. Under the common law theory of negligence, operators must reasonably foresee that a sale to an obviously intoxicated customer could create a risk of harm to others. The foodservice operation must provide *reasonable care* to prevent such occurrences. Reasonable care is the degree of care that under normal circumstances would ordinarily be exercised by or might reasonably be expected of a normal prudent person. Because the general public is demanding responsible alcohol service, those who serve alcohol are being held to higher standards of care. Since dram shop legislation and common law liability vary from state to state, the prudent beverage manager should seek the advice of qualified counsel.

The Composition and Serving Size of a Drink

Upon consumption, alcohol quickly absorbs into the bloodstream; alternatively, it wears off and diminishes from the body very slowly. It takes just ten minutes for an individual to absorb 50 percent of any alcohol consumed—or an hour for the entire consumption to enter the bloodstream. Alcohol's path through the body can affect brain function in just a few minutes. Small amounts are absorbed into the mouth and excreted in breath, sweat, and urine, but 95 percent of the alcohol is metabolized by the liver. Blood alcohol content, or BAC, is a common means of measuring how much alcohol someone has consumed. A BAC of .10 is equivalent to one drop of alcohol in 1,000 drops of blood. If a person's BAC rises to .30, there is a high risk of coma, and a BAC of .40 can be fatal.

Factors That Affect a Person's BAC

The liver can metabolize alcohol at a consistent rate of about one standardized drink (one-third to one-half ounce of pure alcohol) per hour. Any additional quantity consumed in that time frame causes a build-up, with intoxicating effects. The length of time and quantity of alcohol consumed are significant variables to consider—How fast did one drink? What are the intervals between drinks? It all adds up, and is best to pace oneself. To put this in perspective, if five pint glasses of a 5-percent beer are consumed within a short period of time, it would take an individual's "Blood Alcohol Content" (BAC) approximately fifteen hours to return back to normal. Alcohol is absorbed into the bloodstream at different rates based upon different individual factors.

- **Individual Size** Individuals who weigh less or maintain a higher percentage of body fat will be more affected by alcohol. Men typically have less percentage of body fat than women, thus they tend to have a higher alcohol tolerance. Given all the potential variables, an individual's *body size* tends to be one of the more influential factors that affect the blood alcohol content.

- **Type of Food** The type of food consumed prior to or during the drinking of alcohol can assist in slowing the absorption and the effects of alcohol. Fatty and high-protein food items (such as French fries, cheese, burgers) provide a fuller stomach and take longer to digest therefore slowing the effects of intoxication.
- **The Presence of Carbonation** The bubbles or CO_2 assist alcohol content in any given drink to speed up its absorption process upon being consumed. Therefore, carbonation can reduce the time between consumption and intoxication and leads to the negative effects quickly.
- **Individual Health** An individual's mental and physical status such as mood, illness, depression, stress, and fatigue, can enhance the effects of alcohol. For any medication, the instructions or doctor's advice should be consulted prior to drinking. Some medications when mixed with alcohol consumption can be deadly.

Determining the Composition of an Individual Drink

The beverage manager should understand how to decipher the composition of an individual drink. Regardless of drink size, the composition consists of the "actual" or "pure" alcohol content. The concentration of alcohol is different according to the type of beverage. Keep in mind, it has become widespread practice that "a drink" has more than one standard drink, such as, a Long Island Iced Tea, with more than 3.25 oz. of varying spirits can actually equate to two standard drinks. The composition of a "standard drink" regardless of portion size consists of the following formula:

(A) The # of oz. multiplied by the (B) % of alcohol =
(C) the pure alcohol content of a drink (or its composition)

Identified below are the three common alcohol beverages and their potential strength based on volume and alcohol percent. This example illustrates that regardless of a drink's portion size, calculating the composition of pure alcohol of a given drink leads a more accurate understanding when comparing the strength of different drinks.

- **Beer** (A) 12 oz. beer × (B) 4% alcohol (abv) = (C) .48 oz. of pure alcohol (approximately ½ oz.)
- **Wine** (A) 5 oz. wine × (B) 13% alcohol (abv) = (C) .65 oz. of pure alcohol (approximately ½ oz.)
- **Spirits/Cocktails** (A) 1¼ oz. spirit × (B) 43% alcohol (86 proof) = (C) .54 oz. of pure alcohol (approximately ½ oz.). Spirits are rated according to this equation: PROOF/2 = Percentage of alcohol; for example, 80 proof/2 = 40% alcohol.

General Guidelines for Estimating Alcohol Limitations

The beverage manager should have some form of understanding what a typical person can consume within a given time period. A typical person can be broadly categorized according to size (given a rough stereotype of how society views small-, medium-, or large-sized people). In the first hour of consumption, generally an individual can consume the following amounts without extreme adverse reaction to the effects of intoxication:

- 1–2 drinks for a *small*-sized person
- 2–3 drinks for a *medium*-sized person
- 3–4 drinks for a *large*-sized person

After the first hour, the rate of subsequent hours of consumption becomes different. Since alcohol is already present in a person's system, more than one drink (consisting of a maximum of ½ oz. of pure alcohol) per hour regardless of body size can boost an individual to a yellow or red stage of drinking. These color stages are used to provide clues to the beverage service. Yellow is a representation of caution while red is indicative of stopping the service of alcohol.

An individual's body size and other factors (as previously identified) determine how much the level of alcohol in one's blood (or BAC) rises with each standard drink consumed. Even though BAC increases at different rates for varying individuals, each person's liver metabolizes alcohol at the same rate. Each hour, BAC of .016 can be metabolized. Examine the figures below—men and women and both individuals at varying weights have different BAC and are not affected at the same level by one drink. This chart can be used to ESTIMATE an individual's blood alcohol content (BAC) as it increases for each standard drink consumed.

Men

Weight	100	120	140	160	180	200	220	240	260
Increase in BAC per standard drink	.037	.031	.026	.023	.020	.018	.017	.015	.013

Women

In most weight ranges, women's BAC rises more per drink than men's

Weight	100	120	140	160	180	200	220	240	260
Increase in BAC per standard drink	.045	.037	.032	.028	.025	.022	.20	.018	.016

Training and Industry Alcohol Safety Certifications

Proper training of responsible alcohol service for the beverage manager and service staff is essential and should include specific instruction for recognizing the signs of customer intoxication. Many municipalities encourage voluntary (if not required in some jurisdictions) participation and the acquisition of a certificate in an alcohol safety training program as a condition for employment in a food and beverage establishment. In addition, some states require anyone who serves or sells alcohol to pay a nominal fee to renew a license and/or certification each year. These alcohol safety programs are an educational and training tool for sellers/servers of alcoholic beverages to serve responsibly and know the limits of the law. They act as a preventive measure to discourage over-consumption and keep intoxicated drivers off the roads. It is the responsibility of all beverage-related establishments to be aware of state and local liquor laws, rules, and regulations. Commonly, the training focuses on the core elements of

- training and educating sellers/servers to engage in responsible alcohol service
- identifying signs of intoxication and utilize various intervention techniques
- preventing DUIs and alcohol-related fatalities
- stopping underage sales and underage drinking

Educating owners, managers, and staff on dram shop laws and local ordinances regarding alcohol service is paramount to a solid, legality free organization.

While only a few states *require* server education, those who have obtained this type of certification have gained information about alcohol and its effect on people, the common signs of intoxication, and how to help patrons avoid becoming intoxicated. In some cases, server certification in alcohol awareness may assist in a *reasonable care* defense should the establishment be sued under common law theory of negligence. Frequent refresher courses are an important component of alcohol service training. Employee meetings provide a good opportunity to reinforce the message that alcohol must be served responsibly. Some states, Maryland or Illinois, for example, have laws that require all establishments that serve alcohol or engage in happy hour laws to be certified in an alcohol awareness training program. Increasingly, cities and states are formulating policies that promote the intolerance of individuals operating motorized vehicle while intoxicated.

Ultimately, all owners, managers, and servers must be informed and remain up to date on local ordinances for alcohol server training requirements. Some reputable industry training programs include:

- ServSafe Alcohol sponsored by the National Restaurant Association, offers on-line and traditional classroom training options for serving alcohol responsibly
- TIPS (Training for Intervention Procedures for Servers), offered by Health Communications Incorporated
- C.A.R.E. (Controlling Alcohol Risks Effectively), sponsored by the American Hotel and Lodging Association

Reading the Guest

In all states, selling alcohol to people who are already intoxicated is illegal. Though determining one's level of intoxication is not easy as the effects of alcohol will differ greatly from person to person. To qualify as *illegal*, the person's appearance or actions must indicate he or she is intoxicated. Although intoxication is sometimes difficult to detect, this difficulty may not be a reasonable defense of an illegal sale. Instead, one of the best defenses is having the bartender being able to count drinks and being able to assess the customer's behavior and stages of intoxication. These indicators include slurred speech, bloodshot and watery eyes, flushed face, and poor coordination, all of which can be evidenced by difficulty in performing such acts as making change or handling money, lighting a cigarette, or walking without staggering or stumbling. Behavioral evidence of intoxication may also include being overly friendly, boisterous, loud, argumentative, crude, and/or annoying to other customers.

Reading the guest involves recognizing behavior signs caused by the effects of alcohol. The process is expected to be carried out by service staff and goes along with responsible serving of alcohol. Identifying stages of behavior allows servers and bartenders to make a determination of the point at which the customer has had enough. The signs are categorized according to the traffic lights associated at an intersection. Green lights mean "go," yellow lights mean "slow down," and red lights means "stop."

- In the *GREEN*, the guest is relaxed, comfortable, and talkative. *Note*: Servers could offer alcoholic beverages, food, other beverages, and upsell drinks. But as with all levels of reading the guest, it's important to count the drinks a guest consumes as they visit the establishment.
- In the *YELLOW*, the guest is talkative or laughing louder than normal, arguing, antagonizing, or careless with money. *Note*: Servers should not avoid the guest, but offer water and high-protein food and, possibly, delay beverage service. Ensure that the guest does not reach the red level.
- In the *RED*, the guest is making irrational statements, stumbling or falling down, or unable to sit up straight. *Note*: If a customer is determined to be

in the "red," servers/bartenders should stop serving alcoholic beverages—to continue selling them when a customer is in the red is illegal. As a beverage server/seller, and certainly management of any food and/or beverage establishments should comprehend that *drinking is a privilege, not a right, and that privilege can be taken away (by the management) at any point.* Management's responsibility is to prevent a customer from ever reaching this level. Certainly, if a customer does happen to reach the red level, it's the management's obligation to prevent them from driving away.

Intervention Techniques

An intervention is the act of deliberately intervening into a situation or dispute where a guest has consumed excessive alcohol in order to prevent undesirable consequences. An intervention is never an easy situation, but there are some "tried and true" approaches that can make the process of "cutting off a guest" less hostile. Generally speaking, intoxicated customers don't like being cut off, but as representatives of the beverage establishment, the manager and servers/bartenders have an obligation to promote customer safety and lessen potential for liability. Below are some possible approaches to consider when having to deny a guest alcohol service.

- **Wait Until the Guest Orders** It's best to allow a customer to consume their beverage and then refuse service before serving another drink, never after the drink has been delivered. Also, never take a drink away from a customer, as it most likely shouldn't have been served in the first place, any attempt to remove the drink will escalate the risk of conflict.
- **Alert a Backup** Always inform at least one fellow employee when an intervention is going to take place. The co-worker can assist by contacting police if any behavior by the intoxicated consumer becomes inappropriate such as being overly aggressive or abusive in any manner.
- **Isolate the Guest if Possible** When having to deny an individual the service of alcohol, it may be helpful to isolate the guest if possible. This assists the individual in averting possible embarrassment and may prevent a heightened conflict. Management and staff should always be assertive to avoid any miscommunication by speaking firmly and calmly and certainly tactfully, when informing the guest that service is being stopped.
- **Do Not Be Judgmental** When conducting an intervention, it's helpful to lessen conflict by avoiding such phrases as "You are drunk!" This only heightens the potential for conflict. A more effective comment may be "Unfortunately, we won't be able to serve any more alcoholic beverages this evening."
- **Contact the Police** Contact the authorities immediately if at any point the intoxicated customer uses strong verbal abuse, uses any form of physical abuse, or begins to drive away. All staff should be comfortable in this option as a preventive measure in order to lessen any further disaster.
- **Don't Make Contact** It's extremely important to never touch or attempt to physically restrain an intoxicated guest. The natural reaction of many intoxicated individuals is to become aggressive and attack. Also, to some individuals, the contact may be perceived as sexual or hostile.

Given a lawsuit, intoxication of an individual is usually proved in one of two ways. The first one is used primarily when charged with driving while intoxicated, when there is an ability to illustrate an elevated blood alcohol content (BAC)—the second is to utilize a collection of anecdotal evidence. The BAC can *officially* be determined through analyzing the person's blood, breath, urine, or saliva. For example: In general, a 170-pound man with a fairly empty stomach would likely reach a BAC of .08 percent after drinking four servings of alcohol within an hour. A 137-pound female would reach the same level after drinking three servings of alcohol within an hour. This estimation can be proven given an analysis as described above.

The second method of asserting intoxication is through providing anecdotal evidence from a witness(s) who observed the behavioral indicators of intoxication. A witness can be anyone who observed the patron in an intoxicated state, such as a bartender, server, other customers, or the police.

The table below illustrates how BAC corresponds to the effects of intoxication for most people. Ultimately, each person can react in varying and multiple ways and may possess different capacities for metabolizing alcohol, based on genetics. Men and women metabolize or process alcohol at different rates. Women absorb and metabolize alcohol differently from men. They have higher BACs after consuming the same amount of alcohol as men and are more susceptible to the effects of intoxication quicker.

Blood Alcohol Content Stage

Green Stage	**Effects**
Normal behavior to mild intoxication	• Mood prior to drinking may be mildly intensified • Anxiety or inhibitions may be reduced, somewhat relaxed, and maybe lightheaded • Euphoria (feeling of well-being), increased sociability, and liveliness • Behavior and emotions may be exaggerated, making you louder, more intense, or faster or bolder than usual • Progressive decrease in attention, coordination, and judgment
Yellow Stage	**Effects**
Mild to moderate intoxication	• Progressive decrease in memory and comprehension • Difficulty paying attention and applying appropriate judgment • Progressive visual, verbal (slurred speech) impairment • Face pale or flushed • Emotional instability • Decreased reaction time, poor coordination, and slower reflexes
Red Stage	**Effects**
Severe to extreme intoxication (risk of coma or death, medical assistance required)	• Confusion, disorientation, and incoherence • Significant decrease in motor coordination, sensory impairment, and perception • Numbness, insensitivity to pain • Nausea and vomiting • Apathy, drowsiness, possible emotional outbursts • Temporary blindness, blackouts, loss of consciousness • Hypothermia (reduced body temperature) • Loss of reflexes, bladder, and bowel control • Risk of inhaling vomit • Respiratory depression (slowed breathing) • Coma or death due to respiratory arrest

APPENDIX C

The Tasting Process

Tasting wine, as opposed to just drinking it adds an extra dimension to the basic routines of eating and drinking; it turns obligation into pleasure.

—John Laloganes, *The Essentials of Wine with Food Pairing Techniques*

The Physiology of Alcohol Beverage Preferences

An individual's palate is conditioned by life experiences, with personal taste preferences being established as early as childhood. As people age, their tastes often evolve and expand in coordination with their additional food and cultural experiences. Collectively, these experiences will define their preferences of drink. In addition, social and cultural norms play a significant role in shaping individual and societal attitudes regarding the affiliation and appropriateness of alcohol.

Drinking is essentially a social act, subject to a variety of rules and norms regarding who may drink what and when—where and with whom. From culture to culture, there may be high degrees of social differentiation and acceptance within the different categories of drink. For some groups, wine may be the exclusive choice—while spirits or cocktails may be considered unwarranted. In some social groups, beer is the preferred "working-man" beverage—while wine can be viewed as elitist and be banned, just barring ridicule. Beyond the obvious fondness and pleasure for taste, moderate alcohol consumption can be used to promote social conviviality. But it is the conviviality, not the alcohol, which is of central importance. Most societies have specific, designated environments for communal drinking. The primary function of drinking establishments, in almost all cultures, appears to be the facilitation of social interaction and the fostering of dialogue. These are societies in which alcohol is traditionally an accepted, routine, and morally neutral element of everyday life.

Drinking versus Tasting

"Drinking" is a term that is erroneously and inappropriately used in place of "tasting." Drinking is something people do as a diversion or pastime—tasting involves a completely different set of rules and professional jargon. The United States in many respects has become an ambivalent drinking culture, characterized by the conflict between or among coexisting value structures of individual groups amplified by workplace and social norms. It's unfortunate that any reference to the word "drink" often connotes excessive use of alcoholic consumption and the evils of overindulgence. In the appropriate context, drink is merely a beverage or liquid that acts to refresh and socially lubricate interaction. Drinks can provide gestures of goodwill and celebration—in moderate doses, alcohol promotes relaxation and creative contemplation. At times, alcohol may be symbolic of an expression of gratitude or mark the resolution of a dispute. "Drinking" refers to the casual consumption of a beverage in a setting or situation with or without intent to "get drunk." Just as there are negative effects with excessive smoking, consuming a disproportionate amount of fast food, or being highly caffeinated from coffee, none are considered as evil as the overindulgence of drinking. Many extremists and self-proclaimed righteous individuals and groups look to constrain or chastise any drinking that is "immoral" or "unsuitable" according to their context of its application. These individuals or groups work toward placing considerable constraints by shaping the behavior of others and defining alcohol as an evil and stifle the judgment of "quantity" by an appropriate-of-age adult from consuming a legally allowable substance.

The "tasting process" incorporates a sensory examination and evaluation of a drink (wine, beer, spirits, coffee, and tea). The practice of tasting is as ancient as its production; a more formalized methodology has slowly become established from the 14th century onward. Modern, professional tasters such as sommeliers and beverage buyers use a set of formal progressive steps coupled with the application of specific terminology used to make an analytical assessment regarding the range of perceived visual, aromatic, and structural characteristics of a beverage. Figure C.1 shows a sample tasting sheet that can be used to assist in analyzing and deducing the essential characteristics of a given wine.

Producer: ______________________

Varietal(s): ______________ Vintage: ______________

Location: ______________ Alcohol % ______________

VISUAL COMPONENTS

White Wine Color Intensity:
- Watery ➧ Pale ➧ Medium

White Wine Color Hue:
- Greenish ➧ Straw ➧ Golden-Yellow ➧ Amber ➧ Brown

Red Wine Color Intensity:
- Medium ➧ Deep ➧ Opaque

Red Wine Color Hue:
- Purple ➧ Ruby-Red ➧ Red ➧ Brick-Red ➧ Brown

AROMATIC COMPONENTS

Healthy: Yes – No **Aroma Intensity** Muted ➧ Lightly Aromatic ➧ Fairly Aromatic ➧ Highly Aromatic

White Wine Aroma/Flavor Components

FRUIT
- Tree Fruit — Apricot, peach, pear, apple
- Citrus Fruit — Lemon, lime, grapefruit, orange, tangerine
- Tropical Fruit — Melon, banana, lychee, coconut, pineapple, passion fruit, fruit salad, mango, golden raisin

COFFEE/BAKE SHOP
- Nuts — Toasted hazelnut, walnuts, almond, nutmeg
- Bread — Yeast, toast, biscuit, dough
- Sauces — Caramel, toffee, vanilla, butterscotch, honey, cream, butter, custard
- Spices — Cinnamon, cloves, orange peel, anise, ginger, pepper (b or w), cardamom

MINERAL/CHEMICAL

Chalk, flint, petrol, ammonia, steel, wet stone

GARDEN
- Barnyard — Grass, hay, straw
- Herbal — Tomato vine, dill, fresh chives, tea, pine
- Vegetable — Cucumber, bell pepper, asparagus, tomato, olive
- Floral — Rose, peonies, orange blossom, honey-suckle, violets

Rosé And Red Wine Aroma/Flavor Components

FRUIT
- Red Fruit — Raspberry, red cherry, plum, red currant, cranberry, strawberry
- Black Fruit — Blueberry, blackberry, black cherry, blackcurrant
- Baked/Dried — Prune, raisin, jam, fig, plum, baked/dried red fruits, baked/dried black fruits

COFFEE/BAKE SHOP
- Nuts — Toasted hazelnut, walnuts, almond, nutmeg
- Bread — Yeast, toast, biscuit, dough
- Sauces — Caramel, toffee, vanilla, butterscotch, honey, cream, butter, custard, chocolate
- Spices — Cinnamon, cloves, orange peel, anise, ginger, anise, bubblegum, pepper (b or w), cardamom

TOBACCO SHOP

Pine, cigar, cigarettes, leather, cedar, tar

GARDEN
- Earth — Forest, mud, dirt, chalk, manure, dust
- Herbal — Eucalyptus, mint, lanolin, tea
- Vegetable — Green pepper, green olive, black olive, mushroom flower
- Floral — Rose, violet, geranium, lavender, lilac, orange blossom, dried

STRUCTURAL COMPONENTS
- **Carbonation** NA ➧ Flat ➧ Spritzy ➧ Lively
- **Dryness** Dry ➧ Off Dry ➧ Sweet
- **Acidity Level** Low ➧ Medium ➧ High
- **Tannin Level** NA ➧ Low ➧ Medium ➧ High
- **Body Level** Light ➧ Medium ➧ Full
- **Alcohol Level** Mild 11%– ➧ Warmth 11%–13.5% ➧ Spicy 13.5%+

CONCLUSION
- **Quality:** Poor ➧ Good ➧ Outstanding
- **Readiness:** Drink Now (within the year) ➧ Could Age ➧ Definitely Needs Aging ➧ Tired
- **Comments:** ______________________

FIGURE C.1

Wine tasting sheet. Courtesy of John Peter Laloganes.

Developing Tasting Skills

One of the most common concerns that individuals express during tastings is, "I don't smell anything." This is a common and expected response for novice tasters to experience. Often, people go through life eating and drinking purely for satiety reasons and uncommonly for purposes of exploration and understanding of the underlying personality attributes of a given food or drink. The following are some suggested techniques that can assist in evolving one's understanding and pleasure of tasting the different alcohol beverages:

Practice Tasting as an Alternative to Only Drinking Making a conscious effort every time there is an opportunity to smell and taste something (whether it be food, candy, juice, water, etc.) allows additional possibility to shape one's senses. The ability to isolate and identify specific sensory characteristics can be challenging, but will develop with a conscious and guided effort through experience. "Practice" is the most important advice for any taster who chooses to make an informed and educated choice regarding the products they are experiencing.

Maintain a Tasting Journal Maintaining a tasting journal is one of the most effective methods employed by professional tasters used to develop their palate. Writing down impressions and associations of the different beverages assists in establishing an aroma memory and understanding the corresponding typicity associated with each drink. Tasters will not always identify exact associations with which to identify a wine's personality; however, there are some common guidelines of visual, aroma, flavor, and structural components that are universal and can assist in providing consistent parameters. In describing the drink through the tasting process, it's important to express them in an objective, not subjective, manner. For example, "floral" is a more specific descriptive term and more useful to record than "It tastes good," which is a subjective opinion. It's not a bad practice to have opinions; however, those are generally reserved until after an initial objective assessment of the product. In addition, this is a significant distinction between tasting and drinking.

The Tasting Ritual

"Drinking" beverages is easy, as it involves an unguided action often conducted without thought, used to quench thirst. "Tasting," on the other hand, can be more of a challenge: it involves a concentrated sensory approach, a trained and refined technique used to determine the personality characteristics of a drink through the senses of sight, smell, and taste. Tasting is an attempt to capture an intimate and evocative sensory experience to understand the unique personality of a given product and then, most often, communicate it in uncomplicated terms to others. Tasting a beverage, as opposed to just drinking it, adds an extra dimension to the basic routines of eating and drinking; it turns obligation into comprehension and pleasure. The tasting process is about recognizing the unrecognizable, to maximize the enjoyment and understanding of a given drink, and to become more intimate and better acquainted with what one is about to consume. To be simplistic, yet incredibly user-friendly for communication purposes, is to dissect a beverage down to its more important variables that both consumers and novice enthusiasts care most about—how the beverage smells and how it feels in the mouth. These two categories are defined by (1) *aroma/flavor components*, the smells associated with a beverage, and (2) *structural components*, the mouthfeel sensations associated with a particular beverage. Figure C.2 depicts a tasting diagram that indicates key areas used during the tasting process.

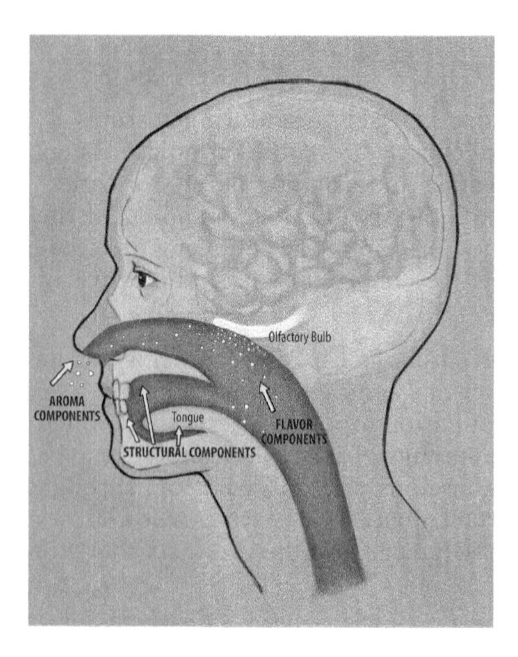

FIGURE C.2
Tasting diagram. Courtesy of Thomas Moore.

Aromas and flavors are detected through the nasal cavity while structural components are physical sensations detected on the tongue, gums, and back of the throat.

For more intermediate and advanced comprehension of beverages, the tasting process involves three essential steps of assessment that include *looking, smelling,* and *tasting*. For simplification purposes, think of the seven S's, which include (1) see, (2) swirl, (3) smell, (4) sip, (5) savor, and (6) spit or (7) swallow.

- See—Determining the visual components of the drink
- Swirl and Smell—Determining the aromatic components of the drink
- Sip and Savor (or simply taste)—Discerning the flavor and structural components associated with the drink.
- Spit or Swallow—Establishing an objective profile of the drink

See—Determining the Visual Components of the Drink

The appearance of any drink can create some initial expectations. The visual aspects provide the first impression, which can offer valuable clues to its character—just as the appearance of a person may suggest certain personality and value traits. Figures C.3 and C.4 identify the taster assessing the clarity, color hue, and color intensity of the wine by placing it against a neutral white background.

Wine: Clarity, Color Hue, and Color Intensity

The clarity, hue, and color suggest different clues about a wine—each requiring different approaches for locating them. A wine's clarity (or lack thereof) is a measure of

FIGURE C.3
Inspecting the wine. Courtesy of Erika Cespedes.

FIGURE C.4
Inspecting the wine (2). Courtesy of Erika Cespedes.

its degree of clarification during the production process. Historically, wine was intended to be clear and pristine, though more modern-day winemakers are practicing minimal clarification techniques in regard to their red wines. Therefore, it has become more expected for high-quality red wines to showcase subtle amounts of sediment when being viewed in a glass. Though it is largely anecdotal, the philosophy of a "hands-off" approach toward clarification is believed to be superior for wine therefore not stripping away some of its essential characteristics.

The color intensity (or level of saturation) is determined by the amount of color pigment (anthocyanin) derived from the fermentation of a red wine with its grape skins. Therefore, smaller grape berries contain a higher ratio of skin to juice, yielding a deeper, more concentrated color pigment—in contrast, a larger grape has a lower proportion of skin in relation to juice, yielding a wine with a lower color pigment. The color intensity of a white wine is largely a measure of exposure to oxygen and/or degree of residual sugar. The application of barrel aging (allowing small amounts of oxygen) and greater residual sugar left in a white wine will increase its color intensity from a watery or pale to golden yellow or amber.

The color hue (or color scale) associated with a wine is largely indicative of its evolution through unintentional or intentional effects of oxygen. This holds true for both white and red wine. In a red wine, as maturity increases, the color hue evolves (from purple, to ruby-red, to red, and then brick-red) and the color intensity declines. The loss of color intensity is due to the natural process of color pigment (along with tannin particles) falling out of the wine and forming sediment. In reference to a white wine, there is very limited color pigment; therefore, as maturity occurs, there is no loss of color intensity, yet the color hue begins to evolve (from a greenish-straw, to straw-yellow, to golden-yellow, and then to amber). Ultimately, a brown color is the final hue that can be apparent in both a red and a white wine. This color is most often associated with a wine that has moved beyond its usable or desirable span of life. Brown color is an indication (with few exceptions) of a tired wine, although there are a couple of exceptions in regard to certain fortified wines.

The Process of Sight

Pour a small amount of drink, about 1–2 ounces, in the glass. The proper quantity is necessary for best assessment of color and, subsequently, it helps to detect aroma. When conducting a flight of drinks, ensure that all drinks have the same quantity of liquid in each glass to accurately assess and compare the color intensity among them. Hold the stem of the glass between the thumb on one side and the index finger and middle finger on the other side. Tilt the wine glass and hold it over an opaque white background (such as a placemat or tablecloth) in a well-lit area. This is the best way to observe and assess any drink. Note the following characteristics: clarity is having a freedom from particles; color depth (or intensity) is measuring the level of color concentration; and color hue (or shade of color) is a measure of the age of the wine. It is more accurate to judge hue by viewing the wine as the glass is tilted. View the wine from its deepest color for intensity, which is in the core or center of the glass, to its very thinnest and lightest color for hue, located at the rim or edge of the glass, which is the first place to show signs of either youth or age.

White wines can range in color hue from a greenish tint to pale straw yellow—both are a possible sign of cool climate, youth, bone dryness, and stainless steel aging—to golden yellow, probably indicating a warmer climate, well-aged wine, exposure to oak barrels, or, possibly, some level of residual sugar. Amber is a sign of a highly evolved white wine, and brown shows signs of possible death or spoilage.

- *White Wine Clarity:* Clear ➧ Light Sediment ➧ Heavy Sediment
- *White Wine Color Intensity:* Watery ➧ Pale ➧ Medium
- *White Wine Color Hue:* Greenish ➧ Straw Yellow ➧ Golden Yellow ➧ Amber ➧ Brown

Red wine may range in color hue from purple to ruby red, both of which may indicate a youthful wine, to red, which means a wine is at its peak or height, then brick red, which indicates an evolved, more mellow wine, showing signs of age, and then finally to brown, which shows signs of death or spoilage.

- *Red Wine Clarity:* Clear ➧ Light Sediment ➧ Heavy Sediment
- *Red Wine Color Intensity:* Medium ➧ Deep ➧ Opaque
- *Red Wine Color Hue:* Purple ➧ Ruby Red ➧ Red ➧ Brick Red ➧ Brown

Beer: Head, Clarity, Color Hue, and Color Intensity

Properly poured beer, whether it's from a bottle or draft, should contain the formation of an adequate head, the foam at the top of the drinking vessel that is ideally 1 inch of height, though some beer enthusiasts prefer as much as 1¼ head. Head is comparable to the *crema* associated with a well-extracted espresso shot, a necessary and quality holistic component to the enjoyment of the drink. The head acts as a net—allowing the slow release of the beer's essential aromas—and provides a delicate, creamy mouthfeel to counterbalance the levels of carbonation once the beer is tasted. A beer's clarity is often a clue to its clarification techniques and/or whether it has been bottle-conditioned. Clarity is often viewed as a measure of a beer's typicity. For example, it would be expected for a Hefeweizen beer to be cloudy, since these are wheat beers that are "unfiltered" and "bottle-conditioned" leaving trace amounts of yeast in the bottle.

The color hue of beer derives mostly from the roasting level and degree of the malts. Generally, with more roasted malts, the beer will appear darker and more intense. The color of beer is measured per the Standard Reference Method (SRM) to specify beer color. A higher SRM indicates a darker beer color hue with corresponding color intensity. SRM is the standard measure used by American brewers for a ranking of degrees of color as related to beer and the grains used to brew it. The SRM typically runs from 1 to around 40+, with darker colors associated with higher numbers. For instance, pilsner malt measures 1–3, while black malts, commonly used in stouts, rates around 40+. This method involves the use of spectrophotometer to assign a number of degrees SRM to the allowance of light intensity.

- *Head Prominence:* Sparse ➧ Average ➧ Large ➧ Creamy
- *Head Color:* White ➧ Off-white ➧ Tan ➧ Brown
- *Beer Color Clarity:* Clear ➧ Hazy ➧ Cloudy ➧ Muddy
- *Beer Color Intensity:* Medium ➧ Deep ➧ Opaque
- *Beer Color Hue:* Straw Yellow ➧ Golden Yellow ➧ Amber ➧ Copper ➧ Brown ➧ Black

Spirits and Liqueurs: Clarity, Color Hue, and Color Intensity

The color of a spirit is determined through the aging process which takes place in wood barrels. The combination of wood (or lack of) and the slow passage of oxygen alter the otherwise clear spirit into a range from an amber color to a brown one. Initially, spirits are clear colored and gain some level of amber to brown color based on length of aging, allowance of caramel coloring, and toast of the barrels.

- *Spirit Color Clarity:* Clear ➧ Hazy ➧ Cloudy ➧ Muddy
- *Spirit Color Intensity:* Watery ➧ Pale ➧ Medium ➧ Deep ➧ Opaque
- *Spirit Color Hue:* Clear Golden ➧ Amber ➧ Copper ➧ Tawny ➧ Mahogany ➧ Brown
- *Liqueur Color Hue:* White ➧ Off-White ➧ Pink ➧ Red ➧ Orange ➧ Yellow ➧ Green ➧ Blue ➧ Purple ➧ Brown ➧ Black

The patterns and viscosity that remain on the sides of the glass of any drink are referred to as legs or tears, which provide a visual indication of the drink's body. Slower moving and more viscous legs indicate a drink with higher alcohol content or a high volume of sugar, both of which are characteristic of a drink with a fuller body. Quickly moving legs may indicate a drink is lower in alcohol or lower in sugar, signifying a drink with a lighter body.

Swirl and Smell—Determining the Aromatic Components of the Drink

Smell is probably the most important and certainly the most evocative sense people possess. This is a key sensory tool when tasting because the nose, depending on individual training and experience, allows an individual to detect and distinguish an estimated 10,000 different aromas. Even those lacking the ability to discern specific odors can often be prompted to learn them by repeated exposure through added experience. The focus of initial training is the repetition for the basis of building both a memory base for subsequent recognition of drinks and a solid foundation of common odor associations.

When smelling a drink, we can also determine its degree of aromatic intensity. The level of intensity provides an impression whether the drink has a somewhat diluted, muted aroma or a highly intense, concentrated one.

The Process of Smell

To get a strong sense of the wine's odors, use of the proper glassware can help to concentrate the aroma molecules in the wine. Fill the glass only about 1–2 ounces to allow enough space for the wine's vapors, or volatile aromatic compounds, to be released when swirled. Next, swirl or agitate the drink (due to carbonation, it is not necessary to do this [or at least minimally] for beer or sparkling wine) to release and intensify the aroma of the wine. Then immediately raise the glass to the nose and take several small sniffs. As the drink clings to the inside of the glass, the alcohol evaporates and carries with it volatile aromas of the drink. The first impression of aroma is important, as the nose is at its freshest point and maintains the ability to identify subtle nuances and discernible differences. Concentrate on the smell to form an initial association and description of the drink's aroma personality. The longer a beverage is smelled, the greater the risk of fatigue, as the nose temporarily loses its keen sense of perceptible subtle aromas. Figures C.5, C.6, C.7, and C.8 identify the

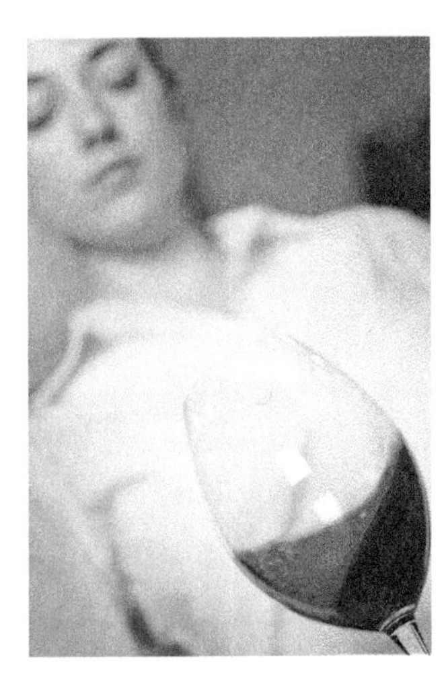

FIGURE C.5
Swirling the wine (1). Courtesy of Erika Cespedes.

FIGURE C.6
Swirling the wine (2). Courtesy of Erika Cespedes.

FIGURE C.7
Sniffing the wine. Courtesy of Erika Cespedes.

FIGURE C.8
Sniffing the wine (2). Courtesy of Erika Cespedes.

taster swirling the glass of wine in order to release the aroma compounds to detect smell characteristics of the wine.

- *Healthy:* Yes ➧ No
- *Aromatic Intensity:* Muted ➧ Lightly Aromatic ➧ Fairly Aromatic ➧ Highly Aromatic

The most common and detectable aromas associated with wine are broken down into four common aroma/flavor categories for white as well as rosé and red wines and into five common aroma/flavor categories for beer:

WHITE WINE AROMA COMPONENTS

FRUIT

- Tree Fruit — Apricot, peach, pear, apple
- Citrus Fruit — Lemon, lime, grapefruit, orange, tangerine
- Tropical Fruit — Melon, banana, lychee, coconut, pineapple, passion fruit, fruit salad, mango, golden raisin

COFFEE/BAKE SHOP

- Nuts — Toasted hazelnut, walnuts, almond, nutmeg
- Bread — Yeast, toast, biscuit, dough
- Sauces — Caramel, toffee, vanilla, butterscotch, honey, cream, butter, custard
- Spices — Cinnamon, cloves, orange peel, anise, ginger, pepper (b or w), cardamom

MINERAL/CHEMICAL

- Chalk, flint, petrol, ammonia, rubber, steel, wet stone

GARDEN

- Barnyard — Grass, hay, straw
- Herbal — Tomato vine, dill, fresh chives, tea, pine
- Vegetable — Cucumber, bell pepper, asparagus, tomato, olive
- Floral — Rose, peonies, orange blossom, honeysuckle, violets

ROSÉ and RED WINE AROMA COMPONENTS

FRUIT

- Fresh Red Fruit — Raspberry, red cherry, plum, redcurrant, cranberry, strawberry
- Fresh Black Fruit — Blueberry, blackberry, black cherry, blackcurrant
- Baked/Dried Fruit — Prune, raisin, jam, fig, plum, baked/dried red fruits, baked/dried black fruits

COFFEE/BAKE SHOP

- Nuts — Toasted hazelnut, walnuts, almond, nutmeg
- Bread — Yeast, toast, biscuit, dough
- Sauces — Caramel, toffee, vanilla, butterscotch, honey, cream, butter, custard, chocolate
- Spices — Cinnamon, cloves, orange peel, anise, ginger, licorice/anise, bubblegum, pepper (b or w), cardamom

TOBACCO SHOP

- Pine, cigar, cigarettes, leather, cedar, tar

GARDEN

- Earth — Forest, mud, dirt, chalk, manure, dust
- Herbal — Eucalyptus, mint, lanolin, tea
- Vegetable — Green pepper, green olive, black olive, mushroom
- Floral — Rose, violet, geranium, lavender, lilac, orange blossom, dried flowers

BEER, SPIRIT, and LIQUEUR AROMA COMPONENTS

FRUIT

- Tree Fruit — Apricot, peach, pear, apple
- Citrus Fruit — Lemon, lime, grapefruit, orange, tangerine

• Tropical Fruit	Melon, banana, lychee, coconut, pineapple, passion fruit, fruit salad, mango, golden raisin
• Fresh Red Fruit	Raspberry, red cherry, plum, redcurrant, cranberry, strawberry
• Fresh Black Fruit	Blueberry, blackberry, black cherry, blackcurrant
• Baked/Dried Fruit	Prune, raisin, jam, fig, plum, baked/dried red fruits, baked/dried black fruits

COFFEE/BAKE SHOP

• Nuts	Toasted hazelnut, walnuts, almond, nutmeg
• Bread	Yeast, toast, biscuit, dough
• Sauces	Caramel, toffee, vanilla, butterscotch, honey, cream, butter, custard, chocolate
• Spices	Cinnamon, cloves, orange peel, licorice/anise, ginger, bubblegum, pepper (b or w), cardamom

TOBACCO SHOP

- Pine, cigar, cigarettes, leather, cedar, tar

MINERAL/CHEMICAL

- Chalk, flint, petrol, ammonia, rubber, steel, wet stone

GARDEN

• Earth	Forest, mud, dirt, chalk, manure, dust
• Barnyard	Grass, hay, straw
• Herbal	Tomato vine, dill, fresh chives, tea, pine
• Vegetable	Cucumber, bell pepper, asparagus, tomato, olive
• Floral	Rose, peonies, orange blossom, honeysuckle, violets, dried flowers

Fatigue and Adaptation

Smell is the most easily stimulated sense, but it is also the most fragile. The nose will fatigue after a short period of smelling (primary olfaction) something and will become temporarily unable to detect additional aromas. Simultaneously, adaptation sets in, which is the self-adjustment to a constant level of stimulus in an environment. For example, most of us, from time to time, have applied a noticeable amount of cologne or perfume to our bodies. After a short time, however, the odors may no longer be noticeable to us. This experience of fatigue and adaptation is no different when one is sniffing wine and will certainly influence the tasting process. One solution is to use quick sniffs, make an assessment, and move on, rather than sniffing a single wine for an extended period.

Sip and Savor (or Simply Taste)—Discerning the Flavor and Structural Components of the Drink

There are two broad aspects of sensing a drink once it reaches the palate. First, the *aromas* that were originally detected through the nasal passage now can be sensed as *flavors* through the inside segment of the nose called the *retronasal passage*. Second, the drink yields certain mouthfeel, textural, or simply called—structural components. These structural components are sensed by nerve receptors called "buds," and there are about 9,000 of them on the average tongue. Once a beverage is present in the mouth, it is imperative to concentrate and be able to dissect the six possible structural components. Many novice tasters make immediate assumptions based on an obvious structural component that stands out while forgetting about the other, occasional subtle ones. For example, it is possible with a bold, tannic red wine that a taster recognizes the obvious intense and predominant drying sensation of tannin—causing a possible misdirection and inaccurate assessment that acidity is not present or that a drink's body is high. It is most important to understand that structural components are distinctively separate from one another, but also operate interdependently causing the ability to sense the accuracy of another component. Figures C.9 and C.10 identify the taster sipping on a small amount of wine to assess its structural components. Pictured in Figure C.11 is a tongue map, a diagram of some select structural zones. When tasting a drink, the palate can detect several structural components:

1. Carbonation	NA ➧ Flat ➧ Gentle Fizzy ➧ Lively ➧ Aggressive
2. Dryness/Sweetness	Dry ➧ Off Dry ➧ Sweet
3. Acidity	Low ➧ Medium ➧ High
4. Tannin Level	NA ➧ Low ➧ Medium ➧ High
5. Body	Light ➧ Medium ➧ Full
6. Alcohol Level	Undetectable to Mild ➧ Warmth ➧ Spicy

FIGURE C.9
Sipping the wine. Courtesy of Erika Cespedes.

Carbonation

Carbonation or effervescence refers to the bubbles present in sparkling wine and beer. The bubbles act to transport the aromas upward toward the nose and once in the mouth, the bubbles act to provide a textural sensation that livens and refreshes the palate. In combination with other structural components, the presence of bubbles provides more of a perceived lightness than would otherwise not be detected. The level of carbonation (or nitrogen, in "some" beers) varies from one beer style and sparkling wine style to another. For some drinks the carbonation may give the drink a thick and creamy mouthfeel, while for others it contributes a prickly, vibrant sensation.

Dryness/Sweetness

The degree of dryness or sweetness derives from a drink's quantity of residual sugar remaining after completion of fermentation or with the addition of sugar after distillation. While sugar can be present in varying amounts, it may not always be perceived because of the balancing effects of other structural components. In some beverages, such as a red wine, the additional presence of tannin in an already dry wine will work toward accentuating the drying feeling in the mouth. Or given a sweeter beverage in conjunction with a large quantity of sourness (from acidity) and/or spiciness (from alcohol) acts

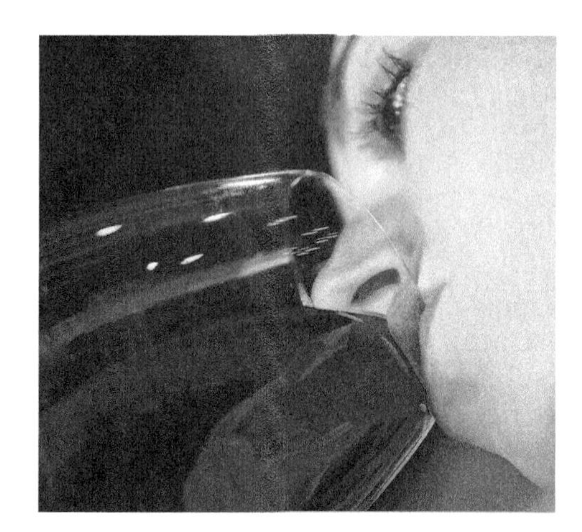

FIGURE C.10
Tasting the wine. Courtesy of Erika Cespedes.

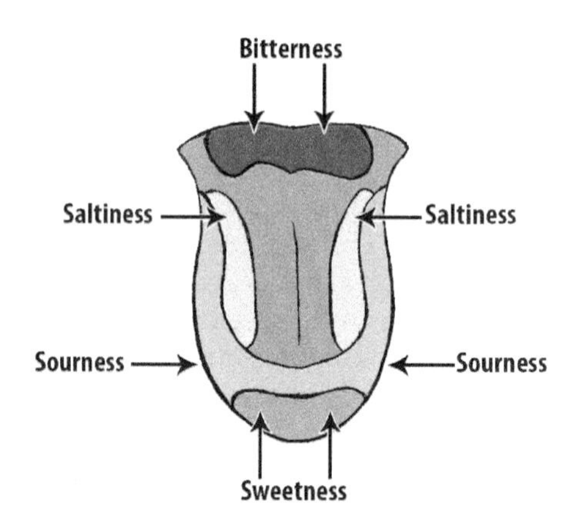

FIGURE C.11
Tongue Map. Courtesy of Thomas Moore.

collectively to counterbalance some of the cloying elements that sweetness would otherwise leave behind.

Sourness

Sourness derives from a drink's acidity levels and causes salivation. The degree and type of acidity can cause a mild to intense zestiness, tartness, or liveliness that works to keep the palate fresh. Acidity is sensed throughout the palate, but it is concentrated on the sides and underneath the tongue and remains a vital and fundamental structural component for most drinks. Beverages with ample acidity may convey less sweetness onto the palate. For example, it provides a counterbalance to sweetness of malt in beer, or residual sugar in wine, or the sweetness from the presence of simple syrup or a liqueur in a cocktail.

Bitterness

Bitterness derives from a drink's tannin that causes a drying sensation. Tannin is sensed throughout the palate, but it is concentrated toward the back of the tongue and around the gum line of the mouth. Tannins vary in potency; the degree of tannin can cause a lesser or greater drying sensation and when coupled with an already dry beverage, it perpetuates this sensation. For simplicity purposes, it can be compared to the sensation one might feel when feeling a fabric—silky (having a certain smoothness) versus felt (having a certain chewy or grittiness). Tannin derives from red wine grapes (due to their extended presence of skin and juice contact) in regard to wine, hops in reference to beer, and barrel aging in regard to both spirits and wine. In beer, the bitterness or acidity is measured according to the "International Bitterness Units" (IBU) scale.

Body

The body of any drink can be compared to a certain viscosity, or richness of mouthfeel. The weight of a drink can be felt as light-, medium-, or full-bodied, which is merely the impression of the viscosity of the drink in the mouth. Body can be comparable to the weight or mouthfeel associated with skim milk, 2 percent milk, or whole milk.

- Light body = skim milk
- Medium body = 2 percent milk
- Full body = whole milk

In beer, the measurement of the specific gravity (SG) has traditionally been used to estimate the strength of beer by measuring its density. This density is a direct correlation with a beer's mouthfeel or body. The higher a given beer's gravity, the fuller is its mouthfeel. This approach relies on the fact that dissolved sugars and alcohol each affect the density of beer differently. The concentration of sugars is directly proportional to the gravity; the original gravity gives a brewer an idea of the potential alcoholic strength of the final product. After fermentation, the difference between the final and original gravities indicates the amount of sugar converted into alcohol, allowing the concentration of alcoholic strength to be calculated.

Spiciness

Alcohol is perceived as a spiciness or heat sensation. The higher the alcohol, the higher is the sensation of spiciness. The spice is experienced toward the back of the mouth and down the throat. In conjunction with other structural elements such as carbonation and/or sweetness, they can act to lessen the perceived sensation of heat from the presence of alcohol. Alcohol plays a large role in the style, body, structure, and taste of a drink. Alcohol can act somewhat similarly to the way acid is sensed, by assisting with cleansing the palate and having the ability to cut through rich types of foods and sauces. For wine, the alcohol content is measured by an alcohol percent. For beer, the strength is measured by the term *abv* or *alcohol by volume*. It can be quantified either indirectly by measurement of specific gravity or more directly by determining the overall percentage of alcohol in the beer. For spirits and liqueurs, the alcohol content is measured according to the *proof*, which is equivalent to twice the alcohol percent. Since spirits yield higher alcohol content in comparison to beer and wine, the alcohol is commonly detected as an obvious warmth to a highly spicy sensation in the back of the throat.

The Process of Tasting

Place approximately one-half ounce of drink into the mouth, suck in some oxygen to open the drink's aromas and flavors, and gently swish the drink around the mouth. To experience the structural components, ensure the drink remains in the mouth long enough—for a minimum of five seconds to activate the sensors on the palate. The purpose of sucking oxygen is to release the volatile aromas and flavors of the drink—they are sent to the back of the nasal cavity, and then up into an interior nasal passage located in the back of the mouth called the retronasal passage. The smell and flavor are now intensified. This could be referred to as smelling the wine on the inside, which is associated with the term "flavor."

Occasionally, it's necessary to interpret a consumer's preference for structural components, because the thresholds of perceptible levels can vary significantly among individuals. It's important to remember that a drink's structural components are distinguishable from one another—yet the consumer may only perceive the drink in its entirety without the ability to identify or covey its constituents.

Spit or Swallow—Establishing the Objective Profile of the Drink

After spitting or swallowing, notice the flavors and sensations that remain as the finish or persistence. Figures C.12 and C.13 identify the taster spitting out the wine after having tasted it. Many associate a longer finish with a better-quality drink,

FIGURE C.12
Chewing the wine. Courtesy of Erika Cespedes.

FIGURE C.13
Spitting out the wine. Courtesy of Erika Cespedes.

however, that opinion is somewhat misguided. Some drinks are not meant to have a longer finish, as they were created to be simple and perhaps one dimensional. The better measure is to understand the typicity of a certain drink. For example, if a wine (such as Cabernet Sauvignon) typically has a long, lingering finish and, during a tasting, it has been observed that a glass of that wine has a short finish, then the quality of that wine should be called into question.

- Finish Quick ➧ Intermediate ➧ Lingering
- Maturation Unaged ➧ Short Aging ➧ Matured
- Readiness Drink now (within the year) ➧ could age (a couple of years) ➧ definitely needs aging ➧ tired
- Price $____________________
- Quality Poor ➧ acceptable ➧ good ➧ outstanding
- Identification: ____
- Comments: ____

____

____

The Serving Temperature's Influence

The aromas/flavors and structural components can obviously vary from drink to drink. Yet, it is not so obvious that an "identical" drink can vary considerably based on altering its temperature. Serving temperatures are one of the significant agents that can truly alter the entire profile of a drink. Serving temperatures should be in an optimal range that will allow a drink to best express itself. Temperatures will influence the perceptions of aromas/flavors and structural components by either emphasizing or de-emphasizing their sensations. For example, a white wine served very cold gives an impression that it is less aromatic and more acidic than the same wine at a warmer temperature. Conversely, a white wine served too warm will accentuate its alcohol content and de-emphasize its acidity, perceiving a flabbier wine than it is in reality. A red wine served too warm yields more perception of alcohol than the same wine served cooler.

The temperature of a beer has an influence on a drinker's experience; warmer temperatures reveal the range of flavors in a beer, but cooler temperatures are more refreshing. If a beer is too cold, the aromas and flavors can be muted yet the carbonation, acidity, and freshness can be accentuated. Many lightly colored and/or lower abv beers are served well chilled (35–45°F), while many darker colored and higher abv beers should be served at warmer (55–65°F) temperatures.

WINE TASTING SHEET

Producer: ______________________________

Varietal(s): ______________________ Vintage: ______________________

Location: ______________________ Alcohol % ______________________

VISUAL COMPONENTS

White Wine Color Intensity:
- Watery ▸ Pale ▸ Medium

White Wine Color Hue:
- Greenish ▸ Straw ▸ Golden-Yellow ▸ Amber ▸ Brown

Red Wine Color Intensity:
- Medium ▸ Deep ▸ Opaque

Red Wine Color Hue:
- Purple ▸ Ruby-Red ▸ Red ▸ Brick-Red ▸ Brown

AROMATIC COMPONENTS

Healthy: Yes – No **Aroma Intensity** Muted ▸ Lightly Aromatic ▸ Fairly Aromatic ▸ Highly Aromatic

White Wine Aroma/Flavor Components

FRUIT
- Tree Fruit: Apricot, peach, pear, apple
- Citrus Fruit: Lemon, lime, grapefruit, orange, tangerine
- Tropical Fruit: Melon, banana, lychee, coconut, pineapple, passion fruit, fruit salad, mango, golden raisin

COFFEE/BAKE SHOP
- Nuts: Toasted hazelnut, walnuts, almond, nutmeg
- Bread: Yeast, toast, biscuit, dough
- Sauces: Caramel, toffee, vanilla, butterscotch, honey, cream, butter, custard
- Spices: Cinnamon, cloves, orange peel, anise, ginger, pepper (b or w), cardamom

MINERAL/CHEMICAL

Chalk, flint, petrol, ammonia, steel, wet stone

GARDEN
- Barnyard: Grass, hay, straw
- Herbal: Tomato vine, dill, fresh chives, tea, pine
- Vegetable: Cucumber, bell pepper, asparagus, tomato, olive
- Floral: Rose, peonies, orange blossom, honeysuckle, violets

Rosé And Red Wine Aroma/Flavor Components

FRUIT
- Red Fruit: Raspberry, red cherry, plum, red currant, cranberry, strawberry
- Black Fruit: Blueberry, blackberry, blackcherry, blackcurrant
- Baked/Dried: Prune, raisin, jam, fig, plum, baked/dried red fruits, baked/dried black fruits

COFFEE/BAKE SHOP
- Nuts: Toasted hazelnut, walnuts, almond, nutmeg
- Bread: Yeast, toast, biscuit, dough
- Sauces: Caramel, toffee, vanilla, butterscotch, honey, cream, butter, custard, chocolate
- Spices: Cinnamon, cloves, orange peel, anise, ginger, anise, bubblegum, pepper (b or w), cardamom

TOBACCO SHOP

Pine, cigar, cigarettes, leather, cedar, tar

GARDEN
- Earth: Forest, mud, dirt, chalk, manure, dust
- Herbal: Eucalyptus, mint, lanolin, tea
- Vegetable: Green pepper, green olive, black olive, mushroom flower
- Floral: Rose, violet, geranium, lavender, lilac, orange blossom, dried

STRUCTURAL COMPONENTS

- **Carbonation** NA ▸ Flat ▸ Spritzy ▸ Lively
- **Dryness** Dry ▸ Off Dry ▸ Sweet
- **Acidity Level** Low ▸ Medium ▸ High
- **Tannin Level** NA ▸ Low ▸ Medium ▸ High
- **Body Level** Light ▸ Medium ▸ Full
- **Alcohol Level** Mild 11%– ▸ Warmth 11–13.5% ▸ Spicy 13.5%+

CONCLUSION

- **Quality:** Poor ▸ Good ▸ Outstanding
- **Readiness:** Drink Now (within the year) ▸ Could Age ▸ Definitely Needs Aging ▸ Tired
- **Comments:** ______________________________

____

BEER TASTING SHEET

Producer/Name: ______________________________

Lager or Ale: ____________________ Style: ____________________

Location/Origin: ____________________ ABV % ____________________

VISUAL COMPONENTS

Color Scale:
- Straw ➧ Golden ➧ Amber ➧ Copper ➧ Brown ➧ Black

Clarity:
- Brilliant ➧ Dull ➧ Cloudy

Head:
- Absent ➧ Present ➧ ➧ ➧

Head Size:
- Flat ➧ ½ in. or less ➧ 1 in. head ➧ 1 in. or more

AROMATIC COMPONENTS

Healthy: Yes – No **Aroma Intensity** Muted ➧ Lightly Aromatic ➧ Fairly Aromatic ➧ Highly Aromatic

AROMA/FLAVOR COMPONENTS

FRUIT/VEGETABLES

- Tree Fruit — Apricot, peach, pear, apple
- Citrus Fruit — Lemon, lime, grapefruit, orange, tangerine
- Tropical Fruit — Melon, banana, lychee, coconut, pineapple, passion fruit, fruit salad, mango, golden raisin
- Baked/Dried — Prune, raisin, jam, fig, plum, baked/dried red fruits, baked/dried black fruits
- Vegetable — Cucumber, bell pepper, asparagus, tomato, mushroom, olive

GARDEN
- Earth — Forest, mud, dirt, chalk, manure, dust
- Herbal — Herbs, eucalyptus, mint, lanolin, tea
- Barnyard — Grass, hay, straw
- Mineral/ Chemical — Chalk, flint, petrol, steel, wet stone

COFFEE/BAKE SHOP

- Nuts — Toasted hazelnut, walnuts, almond, nutmeg
- Bread — Yeast, toast, biscuit, dough
- Sauces — Caramel, toffee, vanilla, butterscotch, honey, cream, butter, custard, chocolate
- Spices — Cinnamon, cloves, orange peel, anise, ginger, anise, bubblegum, pepper (b or w), cardamom

TOBACCO SHOP

Pine, cigar, cigarettes, leather, cedar, tar

STRUCTURAL COMPONENTS
- **Carbonation Level** NA ➧ Flat ➧ Faint ➧ Spritz ➧ Lively
- **Dryness Level** Dry ➧ Off Dry ➧ Sweet
- **Acidity (sour) Level** NA ➧ Low ➧ Medium ➧ High
- **Tannin (drying/ bittering) Level** NA ➧ Low ➧ Medium ➧ High
- **Body Level** Watery ➧ Light ➧ Medium ➧ Full ➧ Viscous
- **Alcohol Level** Mild 5% or less ➧ Warmth 5–8% ➧ Spicy 8%+

CONCLUSION
- **Finish/Persistence:** Absent ➧ Short ➧ Medium ➧ Lengthy
- **Readiness:** Drink Now ➧ Could Age ➧ Tired
- **Comments:** ______________________________

____

APPENDIX D

Drink and Food Pairing

Think "simple" as my old master used to say—meaning reduce the whole of its parts into the simplest terms, getting back to first principles.

—Frank Lloyd Wright

The process of serving food and drink together has been such a natural combination since the beginning of civilization. Today's consumer is now enjoying their drink with a meal not only because of tradition, but to provide basic and heightened pleasures to the dining experience. Pairing a drink (whether extravagant or humble) with a food can elevate a meal and the dining experience from mundane to special occasion. Some people pair essentially for the benefits of refreshment and conviviality while others approach drink and food pairing partly as an art and science. Pairing a drink that is compatible with a food can transform a good food into something great. Regardless of one's motivation for pairing, *tradition* and *pleasure* are two specific reasons that individuals commonly choose to serve drinks and foods together.

Tradition

The tradition of pairing wine, beer, and spirits with food is quite possibly as old as the drinks themselves. In Old-World Europe, wine is treated as a food; another compliment and component to every day's meal. "Classical" food and drink pairings often derive from long-term trading relationships or historical political alliances. For example, the years of Portugal being ruled by the English sheds justification for the beloved connection of Portugal's Port wine pairing with the English Stilton cheese. Many cultures traditionally pair drinks (predominately beer and wine) with food because they are a natural extension of the meal. This pairing is not thought of as something unusual or overly complicated to conduct, instead, drinks are thought of as a condiment and simply, "what grows and evolves together—goes together" with not much more thought going into this process. For example, in Burgundy (or Bourgogne), France, classic dishes such as Escargot or Boeuf Bourgogne have originated and evolved as an extension that integrates incredibly well with their red Burgundy—which, to the unsuspecting wine consumer, this wine is created from the Pinot Noir grape from Burgundy, France.

Pleasure

Pairing drinks with foods is an individual's attempt at seeking personal enjoyment. Such an obvious reason—not much different than why one seeks jelly with their peanut butter or milk with their cookies. Personal choice and preferences or "drink what you like, with what you like to eat" is as common sense as looking both ways before crossing a street. With this pairing approach, there doesn't involve a lot of thought—nor should there have to be—for most consumers. That is why most pairing decisions are relegated to the trained service staff, who strives to heighten the guest's dining experience. Regardless of one's professional opinion, whether one prefers a drink that enhances the meal or not, drinks can serve simply as a nice refreshment and accompaniment to a certain food type or meal. This is an "it's all good" philosophy.

Hedonism is an extension and enhancement to the "pleasure" justification of pairing food and drink. This approach is the ultimate pursuit of maximizing pleasure, particularly involving the senses. It involves an attempt at trying to achieve a "Nirvana-like" union between food and drink. Some would concur that it has an underlying motivation of obtaining sensual pleasure that allows one to indulge and find ultimate sensory fulfillment. The food and drink hedonist is one who seeks the most luxurious and seamless integrations of pairings or searching for the "perfect pairing." This approach takes drink and food pairing to the ultimate—a form of art and science. Knowing and communicating these types of pairings are critical for the beverage manager. The consumer wants guidance based on the intimate knowledge of the beverage and food menu of the beverage manager or his or her staff. The old-school axiom of "white wine with white meat and red wine with red meat" has always been one of the greatest oversimplifications in the world of gastronomy. If this is the approach taken by the beverage manager, it would be safe to say they are missing an opportunity to truly evolve with the modern expectation of today's customer. While the approach is a useful crutch for the uninitiated, there exists a more "contemporary" and useful approach (discussed in the following pages) that has been developed and refined by sommelier and author of the *Essentials of Wine with Food Pairing Techniques*, John Laloganes. Often, individuals discuss some ultimate hedonistic pleasures that are very classic to the Old-World wine-producing countries, such as "fresh chèvre from Chavignol France paired with a crisp, *Sancerre Blanc*" or "Seared foie gras paired with a *Sauternes*" or "roasted lamb paired with a red *Bordeaux*." However, through application of the *Analytical Drink and Food Pairing Approach*, it is possible to pair a beverage with any dish no matter how basic or simple.

An Effective Pairing Is...?

The challenges with "effective" pairings begin with the interpretation of how this concept is defined. From a professional perspective, successful pairings are based on understanding (1) the drink, (2) food ingredients (and the dishes they produce), (3) culinary techniques (the application of cooking methods, seasonings, and sauces), and (4) the application of the Analytical Drink and Food Pairing Approach. The comprehension of these variables is necessary to the success of assembling an effective drink and food pairing. The intricacies associated with these variables are why "effective" pairings are often obtained and left to the guidance of trained professionals. Every dish is dynamic and can comprise dozens of food ingredients and infinite combinations, which contribute to somewhat difficult subjective pairings. With the application and practice of drink and food pairing principles, it is possible to greatly improve the overall consistent consumer satisfaction. An effective pairing is one in which the interaction of drink and food doesn't diminish the pleasure of either partner, but instead enhances each other to become a more fulfilling whole. At the very least, drink and food should be able to co-mingle with one another. The most successful approach is to mirror a drink with a food's increasing intensity, body, and substance throughout a meal. The drink should parallel the flow of the meal; therefore, it makes sense that lighter drinks are paired with simple, lean foods prior to more robust food items being paired with heavier, bolder drinks.

When attempting to pair a wine, beer, or spirit with a food, it may be helpful to think of them as a condiment or just another ingredient to accompany food. Wine can act much the same way as relish enhancing the enjoyment of a hot dog, peanut butter providing a contrast to jelly, and cream providing richness in coffee. All these combinations are intended to enhance the main food item that is being consumed. Some beverages by their nature simply provide greater compatibility with food, though it is possible to ensure an easy transition to pairing with the application of a few ground rules. For example, beer and sparkling wine remain incredibly approachable and, quite possibly, are the most adaptable drinks to pair with food of all types. Their ample carbonation combined with moderate to low levels of alcohol and ample acidity work collaboratively to counterbalance many of the assertive ingredients found in many of today's popular cuisines such as Thai, Mexican, Chinese, and barbecue.

The Integration of Drink and Food

Drink and food compatibility begins with an understanding of a drink's core structural components to provide an effective pairing framework. The integration of pairing drink and food involves applying the "Analytical Approach"—a three-step process for the beverage professional or food and beverage enthusiast. This approach involves:

1. *Mirror the body and weight (or overall intensity)* of both the drink and the food to ensure neither one overwhelms the other.
2. *Harmonize the interactions of structural components* by comparing or contrasting them between the drink and food.
3. *Connect bridge ingredients* in the food with aromas and flavors in the drink.

Principle #1 **Mirror the body and weight (or overall intensity) of both the drink and the food to ensure neither one overwhelms the other.** This principle is *the most significant step* to forming the foundation of any successful drink and food pairing. It is focused on creating an equal balance or "mirroring effect" of the body of both items so that neither will likely overwhelm the other. The "like" characteristics allow the drink and food to remain compatible, and they work to keep the meal grounded. For example, a light- to medium-bodied white wine such as "Sauvignon Blanc" will be overwhelmed by a heavy dish such as a grilled porterhouse steak with melted blue cheese. In contrast, a medium- to full-bodied red wine, "Cabernet Sauvignon," may overshadow a delicate dish of poached scallops. Pictured in Figure D.1 is a matrix that identifies effective and not effective drink and food pairings.

A drink's body is one of the most important components that describes its impression of weight, fullness, or overall mouthfeel on the palate. It is usually the result of a combination of glycerin (deriving largely through maceration/fermentation and/or cold soak process), the degree of extract, alcohol content, and/or amount of residual sugar. Drinks can often be described as light-bodied, medium-bodied, or full-bodied.

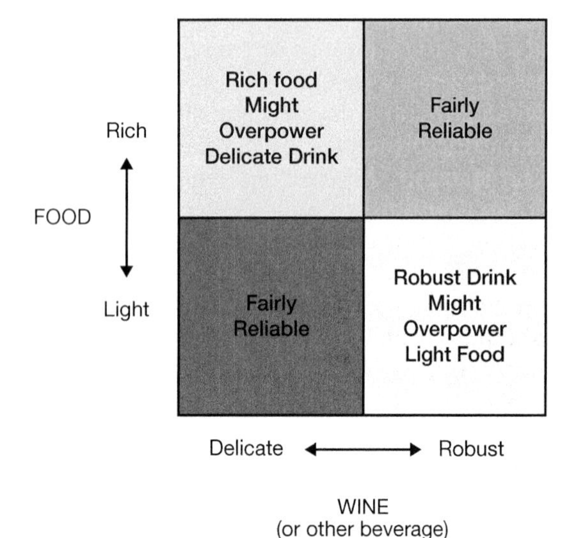

FIGURE D.1
Body mirroring grid. Courtesy of John Peter Laloganes.

The first part of applying principle #1 involves breaking down the plate of food and determining the *primary food type(s)* to match with a drink. The food types have a certain body and weight (or overall intensity) that need to be assessed to determine a wine's compatibility. Typically, the primary food types tend to be protein based, but there are several exceptions, such as salads, soups, pasta, and vegetarian dishes. Intensity can be described as a certain richness or concentration as sensed on the palate.

The second part of applying principle #1 is to consider any factors that may alter the body and weight of the primary food type. This impact on cooking methods, type and quantity of sauce, and other significant ingredients or accompaniments will considerably alter the appropriateness of a beverage pairing. The intensity of a food can be increased with the application of more robust cooking methods and/or intense seasonings. For example, scallops become more powerful going from poached to grilled or broiled. These additional considerations can serve to intensify and heighten the degree of concentration or power of the food. And as food intensifies and becomes rich and robust, the beverages should naturally follow suit. The more familiar tasters are with base products, cooking techniques, sauces, and other significant ingredients and accompaniments, the more effective they can be at modifying food's weight, flavor, and texture levels to pair beverages more effectively.

Principle #2 **Harmonize the interactions of structural components by comparing or contrasting them between the drink and food.** Harmony can be achieved—depending on the particular component—through either comparing or contrasting them with ones found in the food. Some pairings will be more dependent and effective when compared to "like" characteristics in food—other components rely more heavily on "contrasting" components. These specific characteristics of food and drink interact with each other in predictable ways. Taking advantage of these interactions ensures that the food and drink will strive to counterbalance one another.

- *Comparing components* can be used to counterbalance certain interactions between drinks and food items.
- *Contrasting components* can be used to offset or diminish the interactions between drinks and food items.

The six structural components include *body* (as discussed in the previous principle), *dryness/sweetness*, *acidity*, *tannin*, *alcohol content*, and *carbonation*. Depending upon the type of drink, all or at least several of these components can create mouthfeel and sensations on the palate and cause interactions with a food item. The *second principle* of the analytical pairing approach begins with assessing any significant overt *structural components* of the drink and harmonizing them with the food.

Dry/Off-Dry/Sweet Components

The dryness/sweetness levels of a drink can be detectable at subtle or obvious levels. When drinks are considered dry, they represent no (or very slight) perceptible levels of residual sugar. Drinks determined to be "dry" work best when compared to savory (non-sweet) food items. When drinks are determined to contain perceptible sweetness, they represent slight to obvious levels of residual sugar. Drinks with "off-dry" to "sweet" levels of residual sugar work incredibly well at striving to offset *contrasting* foods that contain considerable saltiness, spiciness, smokiness, and acidity (think of adding sugar to lemonade). Sweetened drinks can also be excellent accompaniments to sweet desserts. When *comparing* sweet drinks with sweet foods (often dessert), the drink must be as sweet, or sweeter than a food item. If the food is sweeter than the drink, there is a tendency for the drink to taste dull and flat.

Acidic Component

Acid is fundamental to both beer and a wine's structure, encouraging a crisp, fresh, and lively sensation. Acid is perceived as sourness or tartness on the palate and that causes salivation. Acidity is prevalent in light- to medium-bodied white wines as well as lager style and a handful of ales such as lambic and saison beers. This component can be either contrasted or compared depending upon the overt elements associated with the food items.

- *Contrasting* acidity is useful when given a food that contains sauces based in oil or light cream; acidic wines can work to offset their richness.
- *Comparing* acidity is a useful approach in high acid food items that can be counterbalanced by the acidity present in a drink. For example, crisp high-acid wines such as light- to medium-bodied white wines or crisp lager style Pilsner beers parallel the acidity found in fresh tomatoes or fresh goat cheese.

As described, either approach can work to compliment and find harmony between acidity in the drink and the agents found within a given dish of food.

Tannic Components

Tannin is fundamental structural component found in red wines and in highly hopped beers. It adds a firm texture to a drink and contributes to an increased level of bitterness causing a drying sensation. *Contrasting* tannin is the best approach by working to counterbalance or temper it through the use of pairing fatty, high-protein meat dishes. Higher levels of tannin are off-set or diminished with more uncoagulated (less cooking doneness) fatty, high-protein meats. Therefore, wines like Cabernet Sauvignon and beers such as Barley wine can integrate effectively with juicy steaks prepared rare to medium-rare degree of doneness.

Alcohol Components

Alcohol is present in varying levels in all three categories of drink—wines, beers, and spirits. The sensation of warmth and spiciness becomes more apparent as the alcohol content of a given drink increases. *Contrasting* spicy, high-alcohol drinks work well at being offset by fatty, rich foods or sauces. The richness works to calm the heat sensed in alcohol content.

Carbonation Components

Carbonation or bubbles that are present in sparkling wine and beer serve to refresh and cleanse the palate. *Contrasting* a food's richness and fattiness from butter and cream sauces, smokiness, slight spiciness, and saltiness can benefit from sparkling beverages.

Summary of Harmony

This chart offers a summary of how the beverage manager can choose to compare and/or contrast each structural component.

FINDING HARMONY – COMPARE AND/OR CONTRAST STRUCTURAL COMPONENTS

Structural Component	Compare	Contrast
Dry	Savory food	X
Off-Dry	Semi-sweet food	Spicy, fatty, smoky, salty or slightly sweet food
Sweet	Sweet food	X
Acid	Acidic ingredients	Vegetable/Dairy fat
Alcohol	X	Vegetable/Dairy/Animal fat
Tannin	X	Animal fat/High protein
Carbonation	X	Vegetable/Dairy fat

Principle #3 **Connect bridge ingredients in the food with aromas and flavors in the drink.** Applying this principle can work to further achieve and strengthen the compatibility of a pairing. In review of aromas and flavors:

- *Aroma*—The scent or smell of a drink inhaled via the nose.
- *Flavor*—Term used to describe the process of smelling the wine on the inside of one's mouth as the wine aromas are forced up the retronasal passage.

This principle involves finding a *bridge* ingredient(s) that food and drink have in common. Bridges can add an interesting dimension or validate the pairing experience as they assist to connect the base ingredient (food type), cooking method, or sauce of a dish to a particular beverage for a more effective pairing. Begin by assessing the primary aromas and flavors that are present in both the drink and/or food. For example, if the primary flavor of herbs can be evident in a food item, it is possible to pair a wine that has some of those same herbal qualities.

- **Food** Lean fish, such as halibut, placed in parchment paper with some fresh dill, lemon juice, and aromatic vegetables cooked en-papillote.
- **Wine** Sauvignon Blanc sometimes has a recognizable grassy and herbal aroma and flavor associated with these seasonings.

This pairing will work quite effectively if the prerequisite principle #1 of mirroring the weight or body is achieved—this principle of bridging aroma and flavor elements merely solidifies and strengthens any given pairing.

For example, the aromas and flavors associated with malty styled beers (caramel, chocolate, toast, and toffee) can bridge well with grilled, roasted, and smoked proteins food items. The aromas and flavors associated with hop styled beers (herbs, vegetables, and grass) can bridge well with pasta and grain-based dishes as well as lighter proteins such as seafood and poultry.

Other Considerations When Pairing

In addition to applying the analytical approach to drink and food pairing, the professional or beverage enthusiast may also consider some other pairing variables. Paying attention to the *season, occasion, and mood* (SOM) makes appropriate sense when pairing food and drink. *Seasonality* applies or pertains to an accompanying drink paired with a food to the "appropriateness" of a time of year. Some association of cooler versus warmer weather tends to make sense for most consumers—it is a timely concept that is applied year-in and year-out regarding all types of products and services throughout the world. As the weather dips in the fall and winter seasons—naturally, people crave warmer, richer foods to maintain some form of equilibrium from the cooler outside temperatures. As the weather rises in the spring and summer seasons, people strive to remain cool and refreshed—naturally, people attempt toward cooling off and gravitate toward crisp refreshing beverages. *Occasion* applies to pairing certain drinks and foods that are suitable, expected, or appropriate from some occurrence, event, and/or social gathering. *Mood* applies to matching drinks with food suitable to a quality or feeling given a moment—whim—or a distinctive emotional quality or prevailing emotional tone or general attitude.

Connecting with Wine Styles

Crisp and Youthful Whites

Crisp and youthful white wines are often described as zesty, clean, and vibrant. These wines are light-to-medium in body with medium plus to high levels of acidity. These wines are the least manipulated and are often likely to showcase not only the grape's primary characteristics but also the essential aspects of its location.

Some examples of wines that typically fall under this wine style category include: Pinot Grigio/Gris, Sauvignon Blanc, Grüner Veltliner, Muscat, Unoaked Chardonnay (Chablis), and Albariño.

FIGURE D.2
Mixed green salad with raspberries and almonds paired with Pinot Grigio. Courtesy of Erika Cespedes.

Pairing Strategies

1. The lean body and ample acidity harmonizes with appetizers, salads, soups, and dishes with acidic ingredients. Figure D.2 shows mixed green salad with raspberries and almonds paired with a Pinot Grigio.
2. These light-to-medium body wines pair with similar levels of protein such as lean poultry and raw or lean seafood and sushi.
3. The high malic and tartaric acids in the wines act to amplify a food's structure and flavors.
4. Vegetarian friendly foods such as pasta, grains with or without any combination of seafood, or poultry with oil or cream-based sauces. Figure D.3 shows spring risotto with asparagus paired with Sauvignon Blanc.
5. Fresh/soft cheese and pasta filata cheese: Fresh chèvre, mozzarella.

Silky and Smooth Whites

Silky and smooth whites are often described as refreshing, bright, and velvety. These wines contain a range of body, alcohol, and sweetness. Most often they share a noticeable level of phenolic ripeness in the aromatics and subtle weight (medium body) due to the slight residual sugar that may remain after fermentation. These wines always contain ample levels of tart, acidity that work to provide a balance to the highly-ripened characteristics.

Some examples of wines that typically fall under this wine style category include: Riesling, Gewürztraminer, Pinot Blanc/Bianco, and Chenin Blanc.

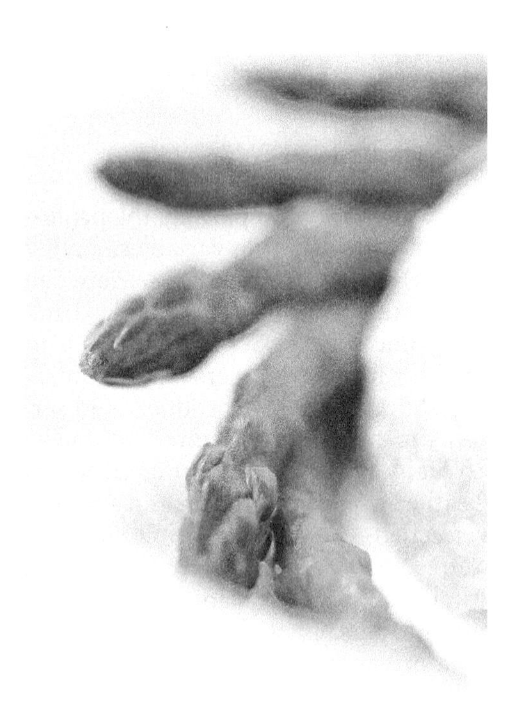

FIGURE D.3
Spring risotto with asparagus paired with Sauvignon Blanc. Courtesy of Erika Cespedes.

Pairing Strategies

1. Pairs well with salty, spicy, fatty, smoky, and sweet food items.
2. Contains varying levels of residual sugar that allows harmony in cuisines with spice and/or salty components such as: Chinese, Thai, and Japanese food.
3. Often intense concentration of aromas, flavors, and mouthfeel increase the density of the wine, which allows it to work with fatty *poultry* (duck, goose), *pork* (pork loin, barbecue ribs), and game birds (quail).
4. If adequate RS, they can couple with several dessert options such as: *Fruit based* (fruit tart, cobbler, pies, crisps, and fruit mousse cake), *chocolate based* (white chocolate and varying types of nuts), and *pastry based* (blueberry muffins, cranberry orange muffins, and pineapple upside down cake).

5. With high RS, these wines will have considerable sweetness and density (such as German Auslese) and works with *Blue-Vein cheese* (Bleu, Gorgonzola, and Stilton).

Rich and Voluptuous Whites

Rich and voluptuous white wines are often described as being lavish, elegant, and voluptuous. These wines have a heavier body (often medium plus to full body) providing enough weight to mirror the body of more substantial meals. Beyond seeing extended hang-time on the vine, these white wines have commonly undergone many vinification techniques—contributing to their opulent characteristics. Commonly these wines experience newer oak barrel aging, extended lees contact, and/or a conversion of the wine's malic acid to softer lactic acid.

Some examples of wines that commonly fall under this wine style category include: Chardonnay, Sémillon, Viognier, Marsanne, and Roussanne. Figure D.4 shows roasted corn and potato, bacon chowder paired with white Burgundy.

FIGURE D.4
Roasted corn and potato, bacon chowder paired with White Burgundy. Courtesy of Erika Cespedes.

Pairing Strategies

1. The medium-to-full body pairs well with richer meats such as veal and pork and fatty poultry such as pheasant, hen, duck, and goose.
2. Works well with richer fish and shellfish such as: crustacean. Also fatty and meaty finfish.
3. Can work with fatty poultry (chicken and turkey) with skin-on, seafood, or pasta that utilizes more robust cooking methods, and/or the incorporation of butter or cream-based sauces.
4. Rich and hearty pasta dishes with oil or cream based sauces. Figure D.5 shows fettucine alfredo with grilled chicken paired with White Burgundy.
5. Cream-based soups and chowders as well as dairy-based salad dressings.
6. Pairs well with richer, washed or *rind ripened and washed rind cheeses* and *fresh/soft cheese* and *semi-hard cheeses*.

FIGURE D.5
Fettucine alfredo with grilled chicken paired with Chardonnay. Courtesy of Erika Cespedes.

Fruity Rosé and Vibrant Reds

Fruity rosé and vibrant red wines are commonly described as being youthful, vibrant, and charming. These wines are often produced in a cool-to-warm growing climate allowing for ample levels of acidity to remain in the wine. The grapes may have a higher proportion of juice-to-skin ratio, yielding a lighter body with minimal levels of tannin. Additionally, they will likely have less color pigment in the skins, so they may appear and taste light. If oak aging is used, it's often with a light touch (if used at all) and serves as more of an undertone and enhancement of the wine's primary characteristics.

Wines that typically fall within this wine style category include: Rosé wines (from all over the world!), Gamay, Dolcetto, Barbera, and Pinot Noir.

FIGURE D.6
Grilled salmon and crispy Vidalia onion shavings with asparagus paired with Russian River Valley Pinot Noir.
Courtesy of Erika Cespedes.

Pairing Strategies

1. Although Rosés and some light reds are not complex (exceptions of course), they do have a place for matching with some more difficult food items.
2. They can work well with spicy Asian, Indian, Latin American, or Cajun foods.
3. Excellent with fatty fish with robust cooking methods, such as shark, tuna, and salmon, along with seafood stews such as Bouillabaisse and Cioppino. Figure D.6 shows grilled salmon and crispy Vidalia onion shavings with asparagus paired with Russian River Valley Pinot Noir.
4. Can pair with some vegetarian dishes that have heightened flavors, such as mushroom risotto enriched with a touch of cream or butter.
5. Vibrant reds are some of the most adaptable wines and can pair well with roasted and braised preparations of chicken, turkey, pheasant, and duck.
   - They work with meats such as poultry with skin, stewed or braised chicken such as Coq au Vin, duck, and game birds such as quail, turkey, veal, and pork.
6. Work excellent with coagulated beef such as Boeuf Bourguignon (beef stew), roast beef sandwiches, beef stroganoff or any beef that is cooked medium well.
7. Can pair with *rind ripened cheeses* (washed rind) and some *cheddar* and mountain/alpine types of cheeses

Mellow and Complex Reds

Mellow and complex red wines are often described as being rich, smooth, and velvety. This style category produces wines with a mellow mouthfeel—yet they yield complex layers of aromas and flavors. These wines offer medium to deep color intensity—moderate acidity, light to medium tannin, and a light plus to medium body. They often produce wines that have been oak-aged for a period of months to years that work to contribute ample dimensions and complexity to enhance the overall wine.

Some wines that commonly fall within this wine style category include: Merlot, Tempranillo, Malbec, Sangiovese, Cabernet Franc, Carménère, Pinotage, and Grenache/Garnacha.

Pairing Strategies

1. Wines can be paired with skin-on grilled and roasted poultry, duck, and game birds.
2. Roasted or grilled meats such as pork tenderloin, venison, veal, and lamb. Grilled steaks with less texture, fat, and flavor such as beef tenderloin and skirt steak. Figure D.7 shows grilled skirt steak with roasted corn, onions, and red bell peppers paired with Washington State Merlot.
3. Fruit-forward versions can pair well with pizza and burgers. Also, Latin and Tex-Mex such as chicken or steak fajitas and beef quesadillas. Figure D.8 shows grilled burger with hickory smoked bacon, aged cheddar cheese paired with Argentine Malbec.

FIGURE D.7
Grilled skirt steak with roasted corn, onions, and red bell peppers paired with Washington State Merlot. Courtesy of Erika Cespedes.

FIGURE D.8
Grilled burger with hickory smoked bacon, aged cheddar cheese paired with Argentine Malbec. Courtesy of Erika Cespedes.

4. Some New-World versions can offer an intense concentration of fruit (common in Washington and Napa Valley) and pair well with *chocolate-based* desserts.

Bold and Intense Reds

Bold and intense red wines are often described as being complex, concentrated, and evolved. This style category often produces big, often spicy, usually oak-aged—and matured for months to years in both barrel and bottle. These red wines have aromas and mouth filling flavors of tobacco shop, jam-like, and dried fruit and structural components of moderate acidity, medium plus to high tannins and body. These reds are often associated with warm to hot climates causing elevated sugar levels which results the fermentation of higher alcohol to levels upward of 13.8 percent toward 15.5 percent.

Some common wines that fall within this style category include: Cabernet Sauvignon, Syrah/Shiraz, Zinfandel, Touriga Nacional, and Nebbiolo.

Pairing Strategies

1. The firm tannins that are often present in these wines can be tempered with robust cooking method and uncoagulated meat protein and fat that are found in beef, particularly steaks that require quick dry heat cooking methods. Figure D.9 shows grilled porterhouse with blue cheese, mashed potatoes, fried onion strings, and red wine reduction paired with Napa Valley Cabernet Sauvignon.
2. These wines can partner well with other meats such as lamb chops or veal shank. Figure D.10 shows grilled lamb chop, red wine demi sauce, and roasted garlic rosemary potatoes paired with Central Coast Syrah.

FIGURE D.9
Grilled porterhouse with blue cheese, mashed potatoes, fried onion strings, and red wine reduction paired with Napa Valley Cabernet Sauvignon. Courtesy of Erika Cespedes.

FIGURE D.10
Grilled lamb chop, red wine demi sauce, and roasted garlic rosemary potatoes paired with Central Coast Syrah. Courtesy of Erika Cespedes.

3. Many of these wines can pair with a drier, darker chocolate that has a high cocoa content.
4. As these wines age, they become less intense as the tannins soften and flavors come together, requiring less fatty meat to be partnered successfully.
5. Intense and concentrated sauces and/or accompaniments can work to further connect the wine's intense body with the main protein item on the plate.
6. These wines can pair well with aged *cheddars* and *blue-vein* type cheese.

Putting It Together

It is critical for the beverage manager to understand the principles of wine (and other beverage) and food pairing. Daily, the manager must suitably pair drinks that can enhance a guest's meal and overall experience. Additionally, the manager will need to train servers in a simplistic, yet effective manner to suggest pairings dozens of times throughout the course of their work shift. The beverage manager may also be involved in working with the chef or kitchen manager in planning special events such as beverage dinners. For example, there may be an event that features a special 5–6 course prix fixe pairing menu of the wines from a winery featured with a different dish created by the restaurant.

The Analytical Drink and Food Pairing Approach offers a scheme that requires one to analyze both the dish and the drink to find suitable connections. One should not overpower the other. Through the application of critical thinking skills, it's possible to create a reasoning or justification once a wine has been chosen to pair with a dish. This is an advanced skill—one that happens best when there is an understanding of both beer styles, spirits, grape varietals and their typicity, but also a base understanding of food ingredients and cooking methods. Once this prerequisite exists, it becomes easier to synthesize and evaluate various styles of wine with various structural and aroma/flavor components of food to determine an effective matching.

Below is a sample 5-course prix fixe wine and food pairing menu. It offers a suggested wine with each course along with a reasoning as to why the wine can pair successfully with each course. The reasoning that is provided is not something that would ever be given to a guest; however, this reasoning is provided as an example of what would go on in the manager's mind as they go through the process of creating effective pairings.

SAMPLE WINE AND FOOD PAIRING 5-COURSE PRIX FIXE MENU

THAI CUISINE

Appetizer Course
Fresh Spring Rolls – Spring roll stuffed with eggs, cucumber, bean sprouts, and tofu served with peanut sauce and green onion.

Ravento's i Blanc | Cava | Spain | 2014 | 12%
This fresh, light-bodied sparkling wine mirrors the lightness of fresh, raw spring rolls. The ample acidity and bubbles in sparkling wine will work both to contrast the creaminess of the peanut sauce and cleanse the palate in the process.

Salad Course
Som Tum (young papaya salad) – Shredded young papaya, green beans, tomatoes, peanut, and dried shrimps mixed with fish sauce, palm sugar, lime juice, and chili.

Kim Crawford | Sauvignon Blanc | Marlborough, NZ | 2015| 13%
This crisp, zesty, light-bodied Sauvignon Blanc mirrors the lean, fresh papaya salad. The high acidity in the wine will be compared with the zesty acidity found in the fish sauce vinaigrette. The citrus and garden aromas of the wine will bridge the flavor from lime juice and tomatoes.

Soup Course
Tom Kha Kai – Chicken breast in coconut milk soup with lemon grass, galangal, kaffir lime leaves, fish sauce and lime juice.

Chateau Ste Michelle "Eroica" | Riesling | Columbia Valley, WA | 2015 | 11.5%
This medium body white wine mirrors the body of milky coconut soup. The slight residual sugar from the wine compliments the slight sweetness found in the coconut milk.

Entrée Course
Pad Thai – Stir fried thin rice noodles with shrimps, fried tofu, bean sprouts, eggs, green onion, ground peanut, palm sugar, fermented soy beans, fish sauce, lime juice, and chili flakes.

Argyle | Pinot Noir | Willamette Valley, Oregon | 2014 | 13.9%
The light-bodied red wine mirrors the body of the Pad Thai's concentrated sauce. The high acid in the wine will create harmony by contrasting the richness of the sauce and cleansing the palate. The slight earthy aromas found in the wine will bridge those found in the Pad Thai.

Dessert Course
Fried banana – Fried banana and coconut-filled spring roll topped with honey and sesame seeds.

Weingut Blees Ferber | Riesling | Mosel, Germany | 2015 | 11%
The full-bodied sweet Riesling mirrors the rich body of caramelized banana and coconut cream-filled deep-fried spring roll. The sweetness in the wine compliments the sweetness from this dessert. The tropical fruits and honey aroma in the wine bridge well with honey and caramelized banana from the dessert.

GLOSSARY

A

Acid/Acidity Acidity is a structural component found in wine, beer, and spirits. It's perceived as a tartness, sourness, or zesty sensation that causes salivation on the palate. Acidity is extremely important in determining the structure (or backbone) of a drink contributing to a multi-dimensional sensation. Drinks that are low in acidity are often described as tasting flat or flabby with a one-dimensional and simple presence.

Adaptation The temporary loss in one's ability to perceive and recognize distinctive aromas and flavors.

Adjuncts Fermentable material used as a substitute for traditional barley grains such as rice and/or corn to make beer lighter bodied and/or less expensive.

Aeration The deliberate choice of incorporating oxygen into a wine, allowing it to "breathe" in order to soften the tannins and allow aromas and flavors to integrate with one another. Red wines benefit most from aeration, which is accomplished by decanting or by swirling the wine in a glass.

Aggressive A beverage that is boldly assertive in terms of aroma/flavor and/or structural components.

Aging The process of storing wine, beer, or spirits in either reductive (stainless steel, concrete, etc.) or oxidative (oak, chestnut, etc.) vessels in order to preserve or contribute additional personality and allow the drink's constituents to integrate. The process of aging can take weeks to years depending upon the vision of the producer.

Alcohol Ethyl alcohol or ethanol (C_2H_5OH) is an intoxicating by-product derived from the fermentation process of yeast consuming a sugar source. The degree of alcohol affects the body, weight, or overall mouthfeel and personality of a beverage. Alcohol content is expressed as a percentage of volume for wine and beer or by proof (twice the amount of alcohol percent) for spirits.

Alcohol and Tobacco Tax and Trade Bureau (TTB) Previously the "Bureau of Alcohol, Tobacco, and Firearms" (BATF), it is the government body that oversees alcohol production and taxation in the United States.

Alcohol Beverage Commission (ABC) The name often given to the state or local government agency/office responsible for licensing of alcohol and related establishments.

Ale One of the two broad categories of beer made with a top-fermenting yeast.

Alsace (al-SASS) A small French wine region bordering eastern France and western Germany that produces mostly white wines from grapes that are of German origin but the wine is made in the "French style." The most prolific grapes include Riesling, Gewürztraminer, Pinot Blanc, and Pinot Gris grapes.

Altitude Otherwise known as elevation; refers to the vertical height of vineyards generally referencing above sea level. The higher altitude causes a decrease in pressure and therefore the air to expand creating cooler air.

American Oak American oak is an alternative to the expensive and more subdued French oak. Often marked by discernable vanilla aromas/flavors and is used primarily for aging Chardonnay and bold, intense red wines such as Cabernet Sauvignon, Merlot, and Zinfandel.

American Viticultural Area (AVA) A distinctive grape-growing geographical area within the United States. AVAs are officially designated by the Alcohol and Tobacco Tax and Trade Bureau (TTB). An AVA (such as Napa Valley or Sonoma County) guarantees that at least a minimum of 85 percent (with some exceptions) of the grapes came from the location as identified on the label.

Ampelographer (amp-pehl-ah-gruh-fer) An individual who practices ampelography; the study of the identification of grapevine botany.

Amphorae (ahm-FOR-uh) An ancient two-handled wine vessel used to transport wine. It was originally used during the Greek and Roman periods.

Analytical Drink and Pairing Approach This methodical three-step approach to pairing beverages and food involves (1) mirroring the body and weight (or overall intensity) of the drink and the food to ensure that neither one overwhelms the other, (2) harmonize the interactions of structural components by comparing or contrasting them between the drink and food, (3) connect/bridge ingredients in the food with aromas and flavors in the drink.

Antioxidants Antioxidants have recently been linked to reducing the risk of heart disease and certain types of cancer. They contain compounds that are believed to inhibit the formation of cancer cells and reduce the buildup of fat cells in the arteries.

Appellation A French term that identifies a grape's designated geographical growing area. The term has legal definition in France regarding what is grown, how it's grown, and how wine is made. In order to use an appellation on a wine label, the regulations vary from 75–100 percent of the grapes used to make the wine must be grown in the place as stated on the label. However, the term has been expanded and loosely applied across the wine industry to simply mean a place where grapes are grown.

Appellation d'Origine Contrôlée (AOC or AC) (ah-pehl lah-SYAHN daw-ree-JEEN kawn-traw-LAY) French concept for "controlled appellation of origin" and refers to wine, cheese, butter, etc. The appellation d'origine contrôlée is the foremost category that ensures the quality of wine (and other products) meet quality criteria in several growing and production steps. The designation is awarded and controlled by the French governmental agency Institut National des Appellation d' Origine (INAO) and guarantees that products to which it pertains have been held to a set of rigorous production standards.

Armagnac (ahr-mahn-yack) The oldest brandy; made in the area of Armagnac area just southeast of Bordeaux France.

Arnaud de Villeneuve Sometime in the 13th and 14th century, he taught distillation as part of alchemy at Avignon and Montpellier. He became recognized as the "Father of Distillation," even though the technology of distillation existed more than two millennia prior to his birth.

Aroma The scent or smell of a drink inhaled via the external part of the nose or nasal cavity.

Aromatic Intensity The degree of aroma concentration that can range from muted to highly aromatic.

Aspect Used to describe the direction in which a slope faces. For example, in the Northern Hemisphere, cooler regions benefit from south and south-east facing slopes that maximize heat and sunlight throughout the day.

Atmospheres (or atms) A term used to describe a unit of pressure equal to 14.69 pounds of force per square inch; often used in the production of sparkling wine, where it describes bottle pressure which can range anywhere from 5 to 7 atms.

Auslese (OWS-lay-zuh) German for "select picking"; refers to the selective hand harvesting of extremely ripe bunches of grapes, often with a touch of noble rot (called Edelfaule in German).

B

Bacchanalia A Roman celebration of Bacchus, the Roman god of wine.

Bacchus The Roman God of wine.

Backbone Often used to describe a drink with definable structural components—often specifically in reference to a drink's acidity and/or tannin levels.

Barrel Aging The length of time an alcoholic beverage spends in a barrel before being bottled. Barrel aging allows the beverage to be exposed to the slow passage of oxygen during which a small amount of evaporation occurs. These effects dramatically influence the personality of a drink through imparting of aromas/flavors (vanilla, spice, and tobacco) and darkening the color shade while softening many of its structural components.

Bar Spoon The bar spoon contains a long spiral handle that is ideal for reaching the bottom of tall glassware. This type of spoon is essential for the stirring and layering drink making techniques.

Bartender (also barkeeper, tapster, buddy, pal) is an experienced person who "tends the bar" by primarily making and serving alcohol beverage (wine, beer, spirits, and cocktails) from behind a counter as a principal responsibility of their job.

Beer The fermentation of starches mainly derived from grains—most commonly malted barley (although wheat, rye, and corn are used as well). Beer is enhanced with hops (and occasionally fruits) which add flavor, acidity, bitterness, and preserving qualities. Beer can be broadly categorized into Ales and Lagers.

Beerenauslese (BA) (BEHR-ehn-OWS-lay-zuh) The German term for select berries that have been handpicked. BA is a rich, sweet dessert wine made of overripe, shriveled berries that are almost always affected by noble rot.

Big Six The big six grapes are arguably the most noble, adaptable, and famous examples of international varietals produced around the world. The big six consist of three white wine grapes: Sauvignon Blanc, Riesling, and Chardonnay, and three red wine grapes: Pinot Noir, Merlot, and Cabernet Sauvignon.

Biodynamics Philosophical viewpoint asserting that the land is a living system and vineyards are an ecological self-sustaining whole.

Bitter A dry, puckery sensation that may be caused by tannin which is largely present in red wine from grape skins or from hops in beer. Slight bitterness may be a desirable trait used to provide a balance to the other structural components that may be present.

Blanc de Blanc (BLAHN duh BLAHN) Translates to "white from white," or a white wine made from white grapes. Most often used to describe sparkling wines made solely from Chardonnay or other white wine varietals.

Blanc de Noir (BLAHN duh NWAH) Translates to "white from red," or a white wine made from red grapes. Most often used to describe sparkling wines made from Pinot Noir and Pinot Meunier or other red wine varietals.

Blend The term can be used to indicate a blend of either different grape varietals into a single wine (such as a "red Bordeaux" is a blend of primarily Cabernet Sauvignon and Merlot) or used to indicate a blend of wines from multiple vineyards and/or years and may therefore be identified as a non-vintage wine such as with sparkling wine and fortified wine.

Blender The blender is an essential machine used to blend drinks and crush ice for making frozen drinks such as frozen margaritas or strawberry daiquiris.

Blood Alcohol Content (BAC) The most common system for measuring and reporting "blood alcohol content" or "BAC" uses the weight of alcohol (milligrams) and the volume of blood (deciliter). This yields a blood alcohol concentration that can be expressed as a percentage (e.g., 0.10 percent alcohol by volume).

Body A structural component that describes a drink's impression of weight, fullness, or overall mouthfeel on the palate. It is usually the result of a combination of glycerin (deriving largely through maceration/fermentation and/or cold soak process), the degree of extract, alcohol content, and/or amount of residual sugar. Drinks can often be described as light-bodied, medium-bodied, or full-bodied.

Bordeaux (bohr-DOH) The Bordeaux region of France produces blended red wine (primarily in varying quantities of Cabernet Sauvignon, Merlot, Cabernet Franc, and others) and white wine and dessert wine (both from a blend of various quantities of Sauvignon Blanc and Sémillon grape varietals).

Botrytis cinerea (boh-TRI-tis sihn-EAR-ee-uh) Also called noble rot, a beneficial mold that may grow on wine grapes, causing them to dehydrate and shrivel, resulting in the remaining juice becoming highly concentrated. This desired condition yields the honeyed richness of many classic dessert wines such as Sauternes, Trockenbeerenauslese, and Tokaji.

Bottle-Conditioning Also referred to as bottle-fermented; associated with sparkling wine or beer that undergoes a secondary fermentation and maturation within its bottle.

Bottom-Fermenting One of two broad categories of yeast used in fermentation for beer production. This yeast ferments at the bottom of the vessel and defines itself as the "Lager" category of beers.

Bourbon America's most famous whiskey, mostly associated with the state of Kentucky (but produced throughout the U.S), produced from at least 51 percent corn, aged for a minimum of two years in new, charred oak barrels.

Brandy A distilled beverage from a fermented mixture of grapes—in essence brandy is distilled wine.

Breathing Allowing a wine or spirit to come into contact with some desirable oxygen for a short period of time. Breathing allows the components of the drink to integrate.

Brewery A facility where beer is produced.

Brix An American system used to measure the sugar content of grapes upon harvest or the quantity of residual sugar left in the wine upon completion of fermentation. The brix multiplied by 0.55 equals the potential alcohol by volume content of the wine being produced.

Brut (BROOT) A term used to indicate a "dry" style of sparkling wine.

Burgundy (BER-gun-dee) The Burgundy region of France produces both red (primarily Pinot Noir with smaller amounts of Gamay in southern Burgundy) and white wines (predominately Chardonnay).

C

Call Brand A distilled spirit that is ordered and identified through the use of a brand name. For example, Bombay Sapphire Gin.

Calvados (kehl-vah-dose) The world's most famous and prestigious apple brandy produced in the Normandy region in northern France.

Canadian Whisky A whiskey produced in Canada made from a blend of grains in a manner that no single grain can exceed 49 percent of the total.

Canopy The foliage (leaves) that is produced on the grapevine.

Canopy Management The practice of adjusting or positioning a grape vine's leaves, shoots, and fruit as it grows, in order to gain such beneficial advantages as increased exposure to sunlight and movement of air.

Cap The thick layer of skin, stems, and seeds that collects at the top of the tank during the fermentation of red wine.

Carbonation A structural component that can be sensed in beer and sparkling wine with a "tingly" sensation from the bubbles or CO_2. The levels of carbonation can range from flat through aggressive.

Cash Bar A term associated with banquets and other catered functions at which attendees pay for their own drinks.

Cava Spain's most prestigious sparkling wine produced from the traditional French méthode champenoise. The majority of cava is produced in the Catalonia–Barcelona area of Spain.

Chalice (chaehl-uhs) A goblet or footed glass used to hold beer.

Champagne (sham-PAYN) Champagne is both a region and a type of wine. To be specific, Champagne is a famous sparkling wine that derives from the Champagne region of France and is made according to stringent AOC laws.

Character Used to describe a beverage with specific qualities related to its style or variety.

Charmat Method (shar-MAH) Also known as tank or bulk process; named for Frenchman Eugène Charmat, the developer of the method. The Charmat process is an inexpensive way to create a "fruit-forward" sparkling wine that limits complexity and preserves the wine's youth. The wine undergoes secondary fermentation in a stainless steel pressurized tank.

Chewy A descriptive term used to characterize a very tannic red wine.

Clarification The process of both removing undesirable particles in wine or beer and making it more stable by eliminating the chance for refermentation.

Clarity A wine-tasting term used to indicate a drink's freedom from particles.

Climate Refers to the general weather conditions prevailing in an area over a long period.

Clone The reproduction or replication of a grapevine usually produced through cuttings or grafting from some desirable parent vine.

Cloudy A term used to indicate a beer having been unfiltered prior to bottling.

Cocktail A generic name used to indicate a mixed drink consisting of predominately a distilled spirit as the base.

Cocktail Shaker The Boston shaker consists of two containers; usually at least one is stainless steel, and the other is glass (often a pint glass) that allows one to overlap the other. The standard shaker is a stainless steel tin with a removable strainer at the top.

Cocktail "Hawthorne" Strainer Strainers are a circular metal tool with a handle and metal spring on top. They are specially designed to block unwanted ice when pouring a drink into a glass after it has been shaken or stirred.

Cognac (kohn-yak) One of the world's most famous examples of brandy. It's produced in the town of Cognac and the areas surrounding it in western France, north of the Bordeaux region.

Cold Stabilization One of the common clarification techniques in which a wine's temperature is lowered to 32°F causing the natural tartrate crystals (commonly associated with white wines) and other insoluble solids to precipitate.

Collins Glass This type of glassware is tall and slender that is used for drinks such as the Tom Collins. Sometimes this glass is referred to as the highball glass and looks very similar to the rocks glass, but it is larger in size and therefore, utilized for larger mixed drinks such as the screwdriver.

Color Hue Otherwise known as color shade, identifying the range of color of a beverage.

Color Intensity Otherwise known as color depth, identifying the degree of color pigment present in a beverage.

Common Law A system of unwritten law not evidenced by statute but by traditions and the opinions and judgments of courts of law. It is generally agreed that the Law of England, as it existed at the time of the North American colonial settlements, is the basis of common law in the United States today, with the exception of Louisiana, which found its influence in the Napoleonic Code.

Components A drink's components can be broadly classified into aroma/flavor components (how a drink smells) and structural components (how a drink tastes).

Consejo Regulador (cohn-SAY-ho ray-goo-lah-DOOR) A Spanish governing, administrative body present in each wine region.

Contraetiquetas (con-trah-ett-ee-kAY-tahs) The back label on a bottle of Spanish wine that signifies the stamp of approval by the *consejo regulador*.

Contribution Margin The difference between a menu item's cost and its selling price. The profit, or margin that contributes to covering fixed costs and providing for a profit.

Controlling Alcohol Risks Effectively (CARE®) A professional certification course that provides alcohol awareness and certification for employees who serve and sell alcoholic beverages. The course is offered by the Educational Institute of the American Hotel and Lodging Association.

Control State One of two possible state classification in which the sale of alcoholic beverage products is directly controlled by the state authorities.

Cooper A professional barrel maker.

Cork The oldest and overall effective closure used to seal a bottle. Authentic cork is derived from the bark of the oak tree.

Corked A term used to indicate that a wine has been tainted with TCA and therefore not healthy. It will cause undesirable aromas and flavors reminiscent of "wet dog."

Corkscrew A tool used to remove a cork from a wine (and some beer) bottles. It consists of a metal spiral (the worm), lever (used for attaching on to the neck of the bottle), and a small hinged knife that is housed in the handle end (for removing the foil wrapping around the neck of many wine bottles).

Coupe (coop) A type of glass used for sparkling wine that is distinguishable by its small, short stem with a wide, shallow bowl.

Craft Beer The term is not legally defined; however, it's generally thought to be made with an annual production of less than 6 million barrels and always a definite devotion to the integrity of their product.

D

Decanter A glass vessel into which wine is decanted.

Decanting A technique used to remove sediment in an old red wine or to allow oxygen to soften the structural components and allow the wine to integrate as in young red wines. Decanting involves slowly pouring wine from the bottle into another container (typically a decanter) in order to separate the liquid from the sediment. The procedure may also be used to aerate the wine in order to soften the tannin and allow the wine to open up and the aromas and flavors to integrate.

Dégorgement (day-gorge-MAWN) French term for disgorging the removal of collected yeast that has settled in the neck of the bottle of sparkling wine during the méthode champenoise production process.

Denominazione d'Origine Controllata (DOC) (deh-NOH-mee-nah-TSYAW-neh dee oh-REE-jeh-neh con-traw-LAH-tah) The second highest ranking of the Italian wine classification system.

Denominazione d'Origine Controllata e Garantita (DOCG) (deh-NOHmee-nah-SYAW-neh dee oh-REE-jee-neh con-traw-LAH-tah eh gah-rahn-TEE-tah) The highest ranking of the Italian wine classification system.

Diluted Referencing a drink's lack of aroma, flavor and/or limiting mouthfeel appearing as if it has been watered down.

Dionysus (die-uh-ny-suhs) The Greek God of wine.

Distillation The process of heating a fermented mixture to separate and remove its water content by causing the alcohol to vaporize and then "re" condense with a higher alcohol strength and greater purity, upon which it may be referred to as a spirit.

Distillery A facility where spirits or distilled beverages are produced.

Dom Pierre Pérignon (d. 1715) The Benedictine monk performed great volumes of research and contributions about sparkling wine. He maintained detailed vineyard records that allowed for the technical expertise of blending that led to the significance of consistency and complexity in the finished bottle of Champagne.

Dosage (doh-ZAHJ) Denotes the addition of a small amount of sugar to adjust the dryness/sweetness levels of a sparkling wine.

Draft Otherwise spelled "draught" is the process of dispensing beer from a tank, cask, or keg, by hand pump, pressure from an air pump or, injected carbon dioxide inserted into the beer container prior to sealing.

Dram Simply, a "drink."

Dram shop Laws State laws that create a statutory cause of action against businesses and, in some cases, its employees, shifting the liability for acts committed by an individual under the influence of alcohol from that individual to the server or the establishment that supplied the intoxicating beverage.

Driving Under the Influence (DUI) or Driving While Intoxicated (DWI) Synonymous terms that represent an illegal act of operating (or in some jurisdictions merely being in physical control of) a motor vehicle while being under the influence of alcohol and or other drugs.

Dry A structural component referencing a drink with no perceptible level of residual sugar or sweetness.

Dry hopping Involves the process of additional "dry hops" to already fermenting or aging beer to increase its hop aroma and flavor as well as structural (bitter) characteristics.

E

Eau-de-Vie (OOH-duh-vee) Literally, "water of life"; an unaged brandy.

Eiswein (ICE-vyn) A German term for "ice-wine," which is a dessert wine made from grapes that are harvested and pressed while frozen therefore extracting water content leaving highly concentrated, sweetened juice for fermentation.

Enology (ee-NAHL-uh-jee) Also spelled oenology. The art, science, and practice of winemaking.

Estate Bottled A term used by such producers that make wine from their own vineyards (or where they have significant control with long-term contracted growers) and that are adjacent to the winery estate. The wines must also be produced and bottled at the winery.

Esters Natural chemical compounds produced from the fermentation process that contributes to many of the fruity aromas and flavors of an alcoholic beverage, particularly in beer and wine.

Extract Referring to the process of aggressively removing color, tannin, and flavor from grape skins during the fermentation process of rosé and red wines in order to contribute greater color, aroma, flavor, and body.

F

Fermentation The process by which yeast metabolizes sugar—producing ethyl alcohol, carbon dioxide, heat, and other by-products.

Finish Also called the "aftertaste" or "persistence" refers to the aroma/flavor and structural components remaining on the palate after the drink has been tasted. The aftertaste can range in extremes from short to lingering and is most appropriately measured according to the "typicity" of a given wine, beer, or spirit.

Flaming A mixed drink that is set on fire prior to serving it to a customer. Flaming provides an element of flair bartending.

Flat Also called "flabby," it is a term used to describe a beer or wine that is low or lacking in vibrancy often when they lack or have lost their acidity and/or carbonation. A flat beverage is very simple and one dimensional.

Flavonoids A group of chemical compounds found in grape seeds, stems, and skins that contribute color, aromas, flavors, and antioxidant benefits.

Flavor A term used to describe the process of smelling a beverage on the inside of one's mouth as the drink's aromas are forced up the retronasal passage.

Floating A drink making technique that layers spirits and/or liqueurs over one another based upon their level of gravity thereby creating layers.

Flor Otherwise known as the "flower"; A white yeast crust that forms on the surface of fino category of Sherry fortified wine during the aging process.

Flute (FLOOT) An ideally tall, slender-stemmed glass used for tasting and drinking sparkling wine.

Fortified Wine One of the three categories of wine in which table wine is the base, with the addition of added alcohol (in the form of a distilled spirit—often an unaged brandy) at some point during the fermentation process. Fortified wine typically contains between 15 and 22 percent alcohol.

Fortified Wine Glassware This smaller-sized glassware is designed to coordinate with a smaller portion size of fortified wine. The typical size of fortified wine glassware consists of an approximate 4 oz. capacity to allow for 2 oz. portion of wine.

Free Pouring A method of pouring spirits and liqueurs in which no actual measuring device is used. Instead, a silent count is used such as a "three or four count" pour equals 1 fluid ounce.

French Paradox In the 1980s, medical studies found a paradox in that French people who have a fatter diet also have a low incidence of heart disease. The study concluded that people who consume moderate amounts of red wine are less likely than nondrinkers to suffer from cardiovascular disease.

French Revolution A period (1789–1799) of extreme social and political upheaval in French history. France underwent an epic transformation from a monarchy to a democratic republic operated government.

G

Geographically Based Labeling Applies mostly to European wine labels, this concept simply refers to wines that are produced from strictly regulated areas.

Gin A distilled spirit made from the fermented mash of neutral grains with the additions of botanical agents (such as juniper berries, seeds, roots, and bark) incorporated during the distillation process.

Glucose One of the fermentable sugars found in the yeast's food source when making wine (from grapes) and beer (from malt).

Goblet This bowl-shaped, stemmed beer glass (often referred to as a chalice) contains a large surface area ideal for maintaining a healthy thick head on the top of the beer. These glasses are commonly used for French and Belgium beers.

Grand Cru (grahn-croo) This French quality term literally means, "great site" that refers to top-tier vineyards and their wines. This term is used to denote the highest classification of vineyards in Alsace, Burgundy, and Champagne.

Grappa (GRAHP-pah) This Italian spirit is distilled from the remains or pomace (PUHM-ess) of winemaking such as the grape's skins, seeds, and stems. Also known as marc in France.

Gravity A term often used to reference the body or weight of beer prior to and after fermentation.

H

Hammurabi's Code An ancient set of laws associated with "Hammurabi," the sixth king of the Amorite dynasty of Old Babylon. The code identifies many laws that probably evolved over a long period of time that provides clues to the attitudes and daily lives of the ancient Babylonians.

Hang-Time A concept that delays the grape harvest in order to increase ripeness and consequently also increasing higher sugar content and ultimately a "fruit-forward" wine with higher amounts of alcohol content.

Highball A cocktail containing spirit and some kind of carbonated mixer.

Hops One of the significant ingredients used in beer production. Hops are dried, cone-shaped flowers that are found on the catkin vine; related to the cannabis family that add bitterness to beer and provide antiseptic qualities to limit microbial growth.

Host Bar A term associated with a banquet or other catered event at which the client, or host, pays for all beverages consumed. The term is synonymous with open bar.

Hot A descriptive term used to indicate that a wine or beer contains an obvious perceptible level of alcohol content that causes a spicy or burning sensation in the back of the throat.

I

Indigenous Grape Varietals Grapes that are thought to be connected primarily with a specific location or homeland. Example: Barbera, Dolcetto, and Nebbiolo from Italy's Piedmont region.

Internal Marketing Promotional selling from within the establishment such as suggestive selling or using free samples to promote beverage items.

International Bitterness Units (IBU) A system used to indicate the quantity and intensity of hop bitterness in a finished beer.

International Grape Variety Grapes that are often referred to as a "classic variety" or "noble variety" which has both a long-established reputation and adaptability for producing high-quality wine throughout the world. Example: Chardonnay and Cabernet Sauvignon.

Intervention The act of deliberately intervening into a situation or dispute where a guest has consumed excessive alcohol in order to prevent undesirable consequences.

Intoxicated The state or condition of being drunk or inebriated.

Irish Whiskey A distilled whiskey from Ireland, mostly made from malted barley.

Irrigation The artificial application of water to land in order to assist in the production of its associated crops.

Issuing The process of a bar acquiring its alcoholic beverage items from the storeroom or other storage areas necessary for production and service.

J

Jigger A double-sided measuring device used for measuring specific quantities of spirits during drink making.

Job Descriptions These are written statements that describe an employee's job duties, qualifications, and most important outcomes needed from a given position.

Judgment of Paris The famous wine-tasting event, "1976 Judgment of Paris," that shocked the world and became the significant defining point for the American (and for the most part, the entire New World) wine industry. The competition was judged by nine French judges that involved blind tasting and scoring the quality of ten French and California Cabernet Sauvignon wines and ten California and French Chardonnay wines. The American wines *Warren Winiarski's* Cabernet Sauvignon from Stag's Leap Wine Cellars and *Mike Grigich's* Chardonnay from Chateau Montelena won both categories over their prestigious French counterparts.

Jug Wine Mass produced, non-descript, low-quality wine that is contained in jugs or boxes for dispensing. Many jug wine producers steal the names of famous Old-World wine-producing regions or countries in order to manipulate the unsuspecting consumer.

K

Keg A term used in the beverage industry to denote a half barrel of beer, equivalent to 15.5 U.S. gallons or 1,984 ounces. A half keg or, 7.75 U.S. gallons, is referred to as a pony-keg.

L

Lactic Acid An acid produced in high levels after a wine has undergone a production technique called malolactic fermentation. This acid has a dramatic influence on the style of the wine by contributing additional aromas and flavors (bakeshop), fuller body, and softening the tart acidic characteristics.

Lager One of the two broad categories of beer that uses bottom fermented yeast. Many American mass produced beers fall into this category.

Lagering Derived from the German word for "storage" that refers to a beer's maturation for several weeks or months at cold temperatures (near 32°F) in order to settle residual yeast, impart carbonation, and make for clean aromas and flavors.

Languedoc-Roussillon (lahng-DAWK roos-see-YAWN) and Provence (praw-VAHNS) These regions are located in southern France, just North of the Mediterranean Sea. The majority of production is red wine from Syrah, Mourvèdre, Grenache, and numerous other varietals in smaller quantities. In addition, these regions produce some of France's most famous versions of fortified wine known as *Vin Doux Naturel (VDN)* (van doo nah-tew-REHL).

Late Harvest Refers to wines made from grapes picked later than the normal harvest time and therefore with a higher sugar content (24 percent or above). Most late-harvest wines contain some perceptible to obvious levels of residual sugar, making them appropriate for/or with dessert.

Lees The decomposing or dead yeast cells.

Licensed State A state which allows licensed vendors and retailers to sell alcoholic beverages.

Liqueur de Tirage (tee-RAHZH) Method whereby, in Champagne production, the blended base wine is given a dose of sugar and yeast in order to induce a secondary fermentation within the original bottle.

Liqueurs (lee-kyoor) Also known as "cordials," it is a type of spirit used to describe an obvious amount of perceptible sugar density and flavoring has been added. Most liqueurs range between 34 and 60 proof, or between 16 and 30 percent alcohol by volume.

Loire Valley (LWAHR) The Loire, another famous wine region of France, is known primarily for their extraordinary white wines (primarily from Chenin Blanc and Sauvignon Blanc grapes), but also produces red wines (from Cabernet Franc and Pinot Noir), dessert wines, and sparkling wines.

Louis Pasteur (d. 1895) In the mid-19th century, he noted the connection between yeast and the process of the fermentation in which the yeast act as catalyst through a series of reactions that convert sugar into alcohol.

M

Maceration The contact time between the grape skins (and sometimes stems) and the grape must prior to and during the fermentation process in order to extract greater amounts of color, tannin, aroma, and flavor.

MADD Originally called "Mothers Against Drunk Driving," it is an organization that began in 1980, from a handful of mothers with a mission to stop drunk driving. MADD has evolved into one of the most widely influential non-profit organizations in America as they have strived to help save thousands of lives.

Malolactic Fermentation (MLF) A winemaking technique that induces a biochemical reaction to convert a wine's malic acid (fruit acids) into softer lactic acid (dairy acids). This process is used to impart additional aromas and flavors and fuller body and softens acidity.

Malt The germinated and roasted grain used to make beer and distilled spirits.

Martini Glass Otherwise known as the "cocktail" glass, it is a stemmed, triangular-shaped glass used to serve a martini.

Meritage (mehr-ih-tij) A particular kind of proprietary wine that was legally created in the 1980s. The name "Meritage" is a combination of two words, "merit" and "heritage," to symbolize and therefore replicate the quality and history associated with the origination of these wines made in a Bordeaux style.

Méthode Champenoise (may-TOAD cham-pen-WAHZ) Has been renamed méthode traditionelle. The traditional method for making Champagne and other high-quality sparkling wine that induces a secondary fermentation and traps the carbon dioxide within its original bottle.

Mezcal A distilled spirit similar to tequila, but made from any agave plant native to Mexico. Mezcal can be produced in several locations throughout Mexico, but the vast majority derives from Oaxaca.

Microbrewery A brewery that maintains small production, less than 15,000 barrels of beer per year with 75 percent or more of its beer sold off-site.

Mixology The practice of incorporating art and science into making cocktails.

Mocktail A nonalcoholic beverage made to mimic a cocktail.

Muddler Similar to the appearance of a miniature bat-like device, the muddler is a thick stick made of wood or stainless steel. It's used to crush ice, mash fruit, and express the essential oils from herbs.

Must The unfermented juice of grapes prior to being turned into wine.

N

New World References the significant countries that have a relatively brief history and culture associated with grape growing and wine production. In the New World, grapevines arrived by way of European settlers through immigration, exploration, trade, and war. The significant New-World wine-producing countries include America, Australia, Argentina, Chile, South Africa, and New Zealand. These countries were at most settled within the last 500 years or so.

Nicholas Longworth (d. 1863) Considered by many to be the founding father of American wine and is noted for owning the first commercially successful winery in the United States, in Cincinnati, Ohio. Longworth experimented with hundreds of different grape varietals and several vine species in his attempts at making wine an egalitarian beverage.

Noble Rot Also called *Botrytis cinerea*, it is a beneficial mold that may grow on wine grapes in moist climates, causing them to dehydrate and shrivel resulting in the remaining juice becoming highly concentrated.

Non-Vintage (NV) A term used to describe a wine (often sparkling and fortified wines) blended from multiple harvests in order to allow the winemaker to create an individual "house" style that can be fairly consistent from bottle to bottle, year after year.

O

Oenotria (own-eet-tree-ah) The Ancient Greek term meaning *land of wine*.

Off-Dry A structural component that is used to indicate a slightly sweet drink in which residual sugar is slightly perceptible.

Old World References the long-established tradition of winemaking within the European countries of France, Italy, Germany, and Spain but can also include other countries located around the Mediterranean basin. These countries have a long history of growing grapes and making wine, and are largely responsible for the nurturing—and development—of the grapevine.

Organization Refers to how management structures its organization—largely its department, employees, products, services, and work procedures and methods.

Oxidation The chemical reaction whereby a wine is unintentionally exposed or overexposed to oxygen, causing the wine to become tainted through oxidation. Oxidation causes chemical changes and deterioration that alters the colors, aromas, and flavors of wines. Oxidized wines are also referred to as "maderized."

Oxidative Aging The process of storing beer, wines, or spirits in a vessel (commonly wood barrels) that allows the slow passage of oxygen over time, therefore enhancing a drink's aromas and flavors and altering its color and mouthfeel.

Oxidized A term used to describe a wine that has been exposed to oxygen for too long of a period of time during storage and/or the bottle has been opened too long.

P

Par Stock The amount of product needed to be on hand (in inventory) in order to support daily sales.

Passito (pah-SEE-toh) An Italian technique that involves laying grapes on racks or hanging them to partially dry for weeks to months that causes the evaporation of the grape's water content. In the process, the grape's aromas, flavors, acid and sugar contents are intensified.

Performance Standard Measurable job performance requirements clearly identified for each specific job task. They identify "how" and "how well" the specific job tasks should be performed.

Perpetual Inventory A method of monitoring inventory whereby additions to and deletions from the inventory are recorded as they occur. The method is normally reserved for high-cost food or beverage products.

Phenolic Ripeness Otherwise known as flavor ripeness, represented by a group of compounds that contribute color, aroma, flavor, and tannin to a grape. This kind of ripeness allows the tannins to become softer as the growing season progresses. Phenolic ripeness often trails sugar ripeness, but is important for allowing the maximum flavor of the grape to be obtained.

Phenolics (fen-ahl-iks) Natural chemical compounds found in a grape's skins and seeds and extracted from oak barrels. Phenolics are responsible for the tannins, color pigments, and aroma/flavor compounds found in wine.

Phylloxera vastatrix (fil-LOX-er-uh) An insect infamous in the 1860s that was responsible for decimating nearly two-thirds of the vineyards in Europe. Phylloxera injects its saliva as it attacks and eats the root system of the grapevine. As a result, the vine has the inability to ingest its nutrients, thus destroying the vine within a couple of years. Most of the world's vineyards are now grafted on American rootstock, which is more resistant to *Phylloxera*.

Physical Inventory An inventory control system in which an actual physical count and cost assessment of all inventory on hand is taken at the close of each financial period (often monthly).

Pilsner Glass This tall, slender beer glass is shaped somewhat like a funnel, with a larger top than bottom. This shape allows the beer to "show off" and accentuate its bubbles.

Pint Glass This beer glass has become one of the most utilized vessels mainly because it's durable and stackable, deeming it easy to store and easy to carry. The pint glass is characterized as nearly cylindrical, with a slight taper and wide mouth. There are two standard sizes, the 16-ounce (common in the United States) or the 20-ounce Imperial, which has a slight ridge toward the top.

Polymerize (PUH-lym-err-ize) A natural effect that occurs in aging red wines that causes its tannins and color compounds to form large molecules and allow for the eventual process of these particles falling out of the suspended wine solution and becoming sediment in the bottom of the barrel and/or bottle.

Prohibition One of the most infamous American periods from 1920 to 1933 when it was illegal to produce, transport, sell, and consume alcohol (with some exceptions). The period was marked by the 18th Amendment to the U.S. Constitution which went into effect on January 16, 1920, and was repealed on December 5, 1933, by the 21st Amendment.

18th Amendment The 18th Amendment to the United States Constitution, also known as the Volstead Act that made the production, transportation, and sale of alcoholic beverages illegal in the United States.

21st Amendment The amendment to the constitution that repealed Prohibition (the 18th Amendment).

Proof A scale used to measure the alcohol in distilled spirits. One degree of proof equals one-half percent of alcohol.

Proprietary-based Labeling Some select wine producers have been creating a certain style of wines that uses a branded name that sound prestigious or unique to the particular winery. Sometimes a proprietary name may refer to an entire estate or a particular wine being produced as an estate.

Provence (praw-VAHNS) French wine region in southern France.

Pruning A viticultural practice that removes excessive grapes and foliage from the vine for the purpose of affecting yield, which influences character development in the grapes.

Pulp The inside part of the grape that contains the juice, acid, sugar, and flavor. Approximately 75 percent of a grape by weight—pulp plays a major role in providing acid (which is present in the juice) and is pivotal in giving both red and white wine good structure.

Q

Qualitätswein mit Prädikat (QmP) (kvah-lee-TAYTS-vine meet PRAY-dee-kaht) Renamed as Prädikat. Term that translates to "quality wine with special attributes" and represents the highest quality wines in the German classification system. There are six subcategories within the QmP system, ranked in ascending order according to their sugar content upon harvest: Kabinett, Spätlese, Auslese, Beerenauslese, Eiswein, and Trockenbeerenauslese.

R

Racking A method of clarification that is considered ideal for limiting the loss of desirable components in a wine or beer. Racking involves periodically draining the sediment, or dead yeast cells, by transferring the liquid from one container to another, leaving sediment behind in the original container.

Reading the Guest This involves the server and/or seller of alcohol to recognize behavior signs caused by the effects of alcohol. The process is expected to be carried out by service staff and goes along with responsible serving of alcohol.

Reasonable Care The degree of care that under normal circumstances would ordinarily be exercised by or might reasonably be expected of a normal prudent person.

Red Wine Glassware These types of glasses are characterized by their large rounded bowl and wide surface area toward the rim of the glass. The larger surface area promotes the increased the rate of aeration as oxygen beneficially interacts with the wine's aroma/flavor and structural components.

Reinheitsgebot (rhine-HITES-gah-bote) Laws Also known as the "German Purity Law" of 1516. These regulations are still followed by many German beer producers. In essence, they stated that the only ingredients allowed in beer production are malted grain, hops, water, and yeast.

Remuage (reh-moo-ajh) French term for riddling—that is, the process of shaking sparkling wine bottles to encourage the lees (or yeast cells) to move toward the neck of the bottle.

Residual Sugar (RS) Any leftover (or unfermented) and perceptible sugar remaining in a wine or beer after the fermentation process.

Resveratrol (rez-VEHR-ah-trawl) One of the phenolic compounds found largely in grape skins that has beneficial effects on cholesterol levels and cancer preventative qualities.

Retronasal Passage The nasal passageway that connects the throat with the nose that enables one to detect the flavors of a beverage inside the mouth.

Rhône Valley (ROHN) The Rhône Valley is located toward mid-central to southern France. It produces mostly red wines (either single varietal or blended wines) from Syrah, Grenache, and Mourvèdre, with white wines produced from the Viognier, Marsanne, and Roussanne grape varietals.

Riddled/Riddling The process of placing a sparkling wine bottle inverted into a rack to be gently shaken over a period of several weeks in order to encourage the lees to collect at the neck of the bottle.

Robert Mondavi (d. 2008) One of the most influential American winemakers who brought worldwide recognition to California wine. From an early period, Mondavi assertively promoted the prominence of varietal labeling as opposed to generic labeling as was the norm in the 1950s. "Robert Mondavi Winery" was the first major winery built in Napa Valley in post-Prohibition.

Rocks Glass This type of glassware is used for serving alcohol either "neat" or for mixed drinks served over cubed ice (rocks). It is also known as an Old-Fashioned glass.

Rosé French term for pink. Rosé wines range in color from pink to salmon and are made from red wine grapes through limited skin contact in order to extract a slight amount of color. Sometimes, a small amount of red wine may be added instead.

Rum A distilled spirit derived from sugar cane or its by-products.

S

Saccharomyces cerevisiae (sack-a-roe-MY-sees sair-ah-VIS-ee-eye) Single-celled organism that lives and thrives on simple sugar and remains the most common species of yeast used in alcohol production.

Saignée (san-YAY) A method of producing wine that allows some of the color from red grape skins to bleed into the fermenting juice, creating a pinkish color.

Sake A fermented beverage made from rice and through the aid of koji mold.

Scotch A whiskey deriving from Scotland that is made from barley that has been smoked over peat. Scotch is often aged in old Bourbon or Sherry barrels.

Sediment The color pigments and tannins that form together and naturally separate out from a red wine as it ages. The wine is removed from the sediment through the decanting process.

ServSafe Alcohol Sponsored by the "National Restaurant Association," they offer online and traditional classroom training options for serving alcohol responsibly.

Shaking One of the drink-making techniques that involves shaking its ingredients in a shaker (often a metal tin) with ice before being strained into a glass.

Shot Glass This is a small glass used to measure and/or quickly consume a distilled spirit.

Slope Refers to the degree of steepness or incline of a hillside. A higher slope indicates a steeper incline and therefore better drainage and greater exposure to the sunlight.

Snifter This glass contains a large bowl and a short stem, which encourages the drinker to hold the bowl of the glass cradled in their hand. Commonly used for consuming brandy, whiskey and/or other aged spirits such as Añejo Tequila.

Soil A mixture of minerals, organic matter, and particles that are of different sizes and textures that acts to support the root structure of the grapevine. Soil influences the drainage levels and amount of absorption levels of minerals and nutrients.

Sommelier (saw-muh-LYAY) A French term, otherwise known as the "wine steward," who is in charge of managing the wine (and often beer and spirits) program which may include all or any of the following: selection, purchasing, storage, educating, and serving wine in a variety of venues such as a restaurant, bar, or retail wine store.

Spätlese (SHPAYT-lay-zuh) Literally, "late picked"; the German word for the second level of QmP wines.

Sparkling Wine One of the three categories of wine that is identified by its carbon dioxide or, simply, its bubbles.

Specific Gravity The measure of density of a liquid or solid compared to that of water (1.000 at 39°F). Often used to indicate a beer's weight or body.

Spirit Referencing any alcohol beverage made from an initial fermentation and subsequent distillation in order to extract water content. Spirits can be made from various base fermented beverages and then infused with any number of herbs, spices, or flavoring agents.

Stainless Steel Aging An aging method used primarily for white aromatic wines whose primary flavors and crisp acidity desire to be preserved. Stainless steel aging preserves the wine and prevents the passage of oxygen that would otherwise alter the wine's personality.

Stirring A cocktail making process that mixes a drink's ingredients together by stirring them with a bar spoon.

Structural Components The six elements of mouthfeel that include carbonation, dryness/sweetness, acidity, tannin, body, and alcohol content that are sensed on the palate when tasting a beverage. All or some of the components may be sensed depending upon the type of drink being tasted and consumed.

Sur Lie Aging Sometimes spelled "sur lee" refers to a wine being aged for an extended period of time on its lees (or yeast cells) in order to gain increased complexity in aromas, flavors and body.

Sweet A structural component that can be applied to wine, beer, liqueurs, or cocktails. The use of this term is used to indicate a noticeable and obvious perceptible level of sugar/sweetness.

T

Tannin A structural component found in a grape's skins, seeds, and stems as well as from a beer's hops—one of the major ingredients in beer. Tannin is a natural chemical compound that causes an astringent, mouth-puckering sensation that causes a significant drying sensation on the back of the tongue and around the gums of one's mouth and a drying sensation on the palate and also acts a preservative for extended aging of a beverage.

Tartrates Harmless crystals of potassium bitartrate that may form in wine casks or bottles (often on the cork) or seen in the wine glass when poured, from the tartaric acid naturally present in wine.

Terroir (tehr-WHAR) French term that loosely translates to "the connection of the land" and encompasses all the environmental factors that affect the grapevine's interactions of soil, climate, topography, and grape variety within a specific vineyard.

Table Wine Otherwise known as still wine. One of the three categories of wine (and most popular and widespread) that gets its name from the historical belief of consuming wine at the table with the meal. The alcoholic content of table wine generally is between 8 and 14 percent with colors ranging from white, rosé (pink), or red wines that can be dry, off-dry, or sweet.

TCA Or technically "2,4,6-trichloroanisole" (try-clore-AN-iss-all) is a wine fault otherwise known as being "corked." The wine contains a disagreeable smell detectable in very low concentrations by imparting a "wet cardboard" character to wine.

Tennessee Whiskey Whiskey made in Tennessee from a special process that uses a charcoal, maple filter. Only two distilleries currently produce Tennessee whiskey: Jack Daniels and George Dickel.

Tequila A distilled spirit deriving from the Tequila region of Mexico produced from a minimum 51 percent of the blue agave plant.

Tired A term used to describe a wine that has surpassed its optimal peak of consumption.

Top-Fermenting One of two broad categories of yeast used in fermentation for beer production. This yeast ferments at the top of the vessel and defines itself according to the "Ale" category of beers.

Topography This concept references a land's surface and shape, particularly of importance for grape growing; it is regarding a vineyard's *slope, aspect, and altitude*.
Trappist Order This branch of Cistercian Monks had taken their name from the La Grande Trappe Abbey in Normandy, France. Many of these early monastic orders distilled, brewed, and vinified products that were used by the early Church as both a medicine (for which it was not very effective) to sterilize wounds (which it does rather well) and a source of prosperity.
Twist A long (about 2 inches) thin piece of citrus (commonly lemon or lime) peel twisted and used as a garnish for a cocktail.
Typicity (tuh-piss-ih-tee) Refers to a drink illustrating traditional and expected character in terms of aromas/flavor and structural components that are typical of a particular drink's style.

V

Varietal References a specific type of grape variety.
Varietal-Based Labeling A concept applied to most "New World" and some "Old World" wine labels that legally implies the wine is made from a dominant grape variety.
Vieilles Vignes (vee ay veen-yuh) A French term for "old vines." In theory, old vines should produce better-quality fruit with smaller berries and thicker grape skins, yet they also produce less yield.
Vigneron (vihn-yehr-RAWN) A French term for someone who grows grapes and cultivates a vineyard for the ultimate purpose of winemaking.
Vin Doux Naturel (VDN) (van doo nah-tew-REHL) Sweet fortified wine primarily coming from the regions of southern France.
Vineyard A grape-growing area that can vary in size and in some larger grape-growing areas is identified as an appellation or a region.
Vinification The science and practice of making wine.
Vintage Term that refers both to the year wine grapes were harvested and such grapes were converted into wine.
Vintner Wine producer or winery proprietor.
Viticulture (vit-uh-cull-ture) The study and practice of cultivating grapes.
Vitis labrusca (lah-BROO-skah) Grapevine that is indigenous to North America.
Vitis vinifera (vin-if-EHR-ah) The classic indigenous European grapevine species most responsible for producing the world's best wines, including Pinot Noir, Chardonnay, Cabernets, etc.
Vodka A neutral distilled spirit that can legally be made from anything but is most often made from potatoes or grain.
Volstead Act Another name for the 18th Amendment to the U.S. constitution that was enacted in 1920.

W

Well Brand A bar's standard pour of distilled spirits when a specific brand name has not been specified for a mixed drink.
Whiskey A distilled spirit made from grain in the United States or Ireland.
Whisky A distilled spirit made from grain in Scotland or Canada.
White Wine Glassware This kind of glassware has a moderately sized bowl, with a tapered rim at the top of the glass to allow for enhanced aroma concentration of a white wine's delicate nuances.
Wine The fermented juice of grapes ... unless otherwise legally specified. Can be broadly categorized according to table, sparkling, and fortified wine.
Wine Bottle A vessel that has been used for centuries and appear in a variety of shapes and sizes, but have been standardized to generally contain 25.4 oz. (750 ml) of liquid.
Winery A facility where wine is produced.

Winkler Heat Index System A system of classifying grape-growing areas using the heat index to determine the optimal site selection for different grape varietals. Dr. Albert Winkler of the University of California-Davis developed this system.

Wood-Barrel Aging A centuries-old tradition that uses wood vessels to store and age most red wines and many full-bodied white wines. The industry standard is to use French or American oak. Oak from other places, such as Slovenian oak, is sometimes still used. In the past, different wine regions have used different kinds of wood, such as mahogany, chestnut, and pine.

Wort An unfermented beer, wort is the liquid extracted from a mash after the boil which extracted sugars, colors, and flavor from the grains and alpha acids from the hops.

Y

Yard Glass A very tall, thin glass that is about a yard in length. The traditional yard glass holds 42 ounces. These days, half yard glassware can be found as well.

Yeast An important microorganism that causes fermentation by converting sugar to alcohol. The predominant yeast, *Saccharomyces cerevisiae* (sack-row-MY-cees sair-ah-VIS-ee-eye), is the same microorganism that ferments wine, beer, and bread.

INDEX